Security Systems Technologies and Innovation

Security Systems Technologies and Innovation: A Comprehensive History of Advances in the Industry examines the past 50 years and the expansion of the security industry, security technologies, and the field as a whole. An overarching theme throughout is that advances in security technology run parallel to developments in computers and telecommunications. The coverage emphasizes that the foundational "nuts and bolts" intent and tactics that preceded today's electronics don't change, although the methodologies, challenges, and threats do morph and change over time.

This encyclopedic work begins in the 1970s with a foundational background of the monopoly-controlled industry at the time, which relied on mechanical devices and people. Understanding these precepts not only puts present technology in perspective but also equips readers with the tools to creatively approach unique security problems. This reflects the reality that the true role of any security decision-maker is to provide situational awareness and decision-making, since almost every security problem is nuanced and context-based. Out-of-the-box solutions – including technology – though popular, are rarely the answer. Security is about protecting people and property from highly motivated criminals – a constant chess game in which only one side follows the rules.

Rapid technological developments in the last 30 years, in particular, ushered in an unprecedented period of growth. Over this time, the industry developed into a goliath, evolving in step with the transformation of the American criminal justice and law enforcement communities, with almost complete technical revision and advancement every three to five years. Concurrently, countering terror threats began to supersede the industry's original focus on trespassers and property theft. This book chronicles this ever-changing corporate, technological, and threat-based landscape. This book proves fascinating reading, providing insight to those currently in the field and offering a glimpse of future security threats, and those current and emerging technologies that will shape the next 50 years of the industry.

Security Systems Technologies and Innovation

A Comprehensive History of Advances in the Industry

Charles Schnabolk
with special contributions from Stefanie Coakes
Additional Text Reviews by Todd Bierman and Mark Beaudry

CRC Press
Taylor & Francis Group
Boca Raton London New York

CRC Press is an imprint of the
Taylor & Francis Group, an **informa** business

Designed cover image: Getty Images

First edition published 2026
by CRC Press
2385 NW Executive Center Drive, Suite 320, Boca Raton FL 33431

and by CRC Press
4 Park Square, Milton Park, Abingdon, Oxon, OX14 4RN

CRC Press is an imprint of Taylor & Francis Group, LLC

ISBN: 9780367548988 (hbk)
ISBN: 9781041213116 (pbk)
ISBN: 9781003091103 (ebk)

DOI: 10.4324/9781003091103

Typeset in Times
by codeMantra

Contents

SECTION VI Video Recording and Storage

SECTION VII Wireless Communication

SECTION XII Fire Safety and Code Compliance

SECTION XIII Explosives in Major Conflicts

SECTION XIV *Deadly Weapons of Apparent Mass Appeal*

SECTION XV *Nonlethal Weaponry*

SECTION XVI Lie Detectors and Truth Deception

SECTION XVII Odd Protection Devices

Author Biography

Charles Schnabolk E.E., MBA, P.E., PSP, Lic. Elec. is a recognized industry pioneer with credits including security design for the original World Trade Center. As a successful consulting engineer, he also designed and implemented security systems for major hospitals, schools, high-rise commercial buildings, and large residential complexes. This book is a compendium of his unique knowledge on almost every aspect of modern security – from its 1970s conception to the modern, high-tech industry that it had developed into over the last half-century and more.

Preface

Many devices described in these 42 chapters are obsolete after rapid changes in digital technology – but the concepts aren't. Since 1984, change has only escalated in the security industry. That year, the first handbook on security controls was published by this author. This book can be considered an update covering a more than 40-year period that witnessed almost complete technical revision.

Security technology advances parallel those in computers and telecommunications. To make this book more encompassing and valuable, however, a great deal of emphasis is focused on historic devices and historically valid security concepts – nuts and bolts techniques and tactics that preceded today's electronics. This foundational background began in the 1970s when a monopoly-controlled industry was largely based on mechanical devices and people. Understanding these precepts not only puts present technology in perspective but also gives the reader the tools to creatively approach unique security problems.

And that's the real job, because almost every security problem is nuanced and situational. Out-of-the-box solutions, though popular, are rarely the right answer. Security is about protecting people and property from often highly motivated criminals – a constant chess game where only one side follows the rules.

Notable security milestones, specifically those the author was involved in, include:

- Design of the NY & NJ Port Authority's World Trade Center complex security system in 1970, the largest security design contract ever awarded at that time.
- Outside Engineering Consultant on the Port Authority's Counterterrorism Task Force in 1985, the nation's first such study of threats that eventually took form with the first foreign terrorist attack on American soil in 1993. After the truck bomb exploded in the South Tower garage, 200 Tower tenants sued the Port Authority because it ignored recommendations made in the author's 1985 report.
- Designed and/or installed customized security systems for over 250 schools, museums, banks, commercial and residential building, including the nation's largest private cooperative, in a multi-use high-rise building (Olympic Towers on 5th Avenue in New York City), and dozens of hospitals, universities, and retail stores.
- Developed hardware and secured safety code changes to allow use of a fail-safe electro-magnetic locking arrangement to secure fire exits against intruders – a multi-year project that required comprehensive lobbying of the code authority. This locking arrangement was first installed in 1984 on school's secondary exits in Newark, NJ.
- Invented the ARTGUARD protection system employed by both US museums and foreign institutions (Louvre, British Museum, etc.).

Authored the first technical book on physical security published in 1984, "Physical Security: Practices and Technologies" and over 150 magazines and professional journals articles.

Prologue
Premature Birth of Modern Security

Rapidly changing technology combined with an aging infrastructure requires continuous design of new security applications, especially when upgrading often poorly maintained legacy systems. These frequently become obsolete as more effective technology reaches the market. Today, Homeland Security directives coupled with code compliance make the mastery of physical security even more crucial, especially when the first generation of electronic surveillance, video motion detection, and biometric identification systems is already "legacy" technology. The cycle repeats every five years, if not more frequently these days.

Obtaining a replacement system is complicated because it requires understanding the many changes at the nexus of electronics and security. Selecting the right fit can be a source of confusion, more so after consulting poorly trained salespeople whose knowledge either lags behind the technology or is limited to unnecessary features that they can upsell you on. It is therefore incumbent on security professionals to keep up with changes through continuing education courses, trade shows, and seminars. The takeaway: neighborhood locksmiths and big box home centers are no longer appropriate venues for security solutions.

Adding to the obsolescence problem, incompatible equipment from a plethora of manufacturers is often sold by poorly trained installation companies possessing insufficient experience to install and maintain complex security systems. There are a few educational or licensing requirements for alarm installers. Even worse, the end user security director or building manager is often unequal to an aggressive alarm salesmen's pitch. This sourcebook will help to fill that information void.

The title "Director of Security" first appeared within the largest corporations and banks in the early 1980s, and later spread to government agencies, schools, museums, hospitals, hotels, and virtually every building complex. Before this century, the main responsibility of a security department was traffic flow at the main entrance and installation of black and white surveillance cameras so that $1200 IBM typewriters didn't "walk away." Those simpler problems no longer describe the duties of a security administrator in the digital age. Information Technology Departments are now at the center of many security purchases.

Prior to 1970, the lone security employee was usually a retired police officer. This lasted until the 1990s, because police officers were streetwise and willing to take a smaller salary to supplement pensions and retirement income. It was a male-dominated position requiring little technical skill; that is, until rapidly changing electronics infiltrated the industry. Security directors must now possess wide expertise and mitigate much more serious threats than petty thievery. The job description requires a thorough knowledge of risk management, vulnerability assessment, code compliance, emergency preparedness, human resources, crisis communication, privacy issues, and evacuation procedures. Even the most competent director must contend with unappreciative management, uncooperative or belligerent visitors, and untrained, unmotivated employees with "attitude" – and not the "can do" kind!

Fears that sophisticated software and advanced equipment would limit the job market to savants have proven baseless. Technology advances have expanded opportunities, changed the complexity of the workforce, and created new occupations – especially so within the security industry.

Computers, while great at pattern recognition and repetitive tasks, are poorly designed for creative problem-solving—though recent AI advances are proving that such capabilities are increasing, and at a rapid rate. Robots may chauffeur people around in driverless cars, but they cannot perform functions outside a pre-programmed domain. For now, design professionals and experienced security system integrators embody the boots-on-the-ground creativity needed for effective security design and implementation.

DOI: 10.4324/9781003091103-1

Section I

The Early Years of Private Security

While video surveillance technology was invented and developed in the 1940s for military use in WWI, the modern electronic era of physical security systems developed on a more broad and expansive scale in the 1970s, at and around the time when the World Trade Center was completed and occupied. Before then, security alarms and other intruder detection devices received little public attention. The local locksmith was the primary source of advice on physical protection and interior security mainly consisted of tripwire light beams and door locking hardware.

Until the last part of the 20th century, businesses and institutions didn't generally employ a "Security Director," and architects mostly ignored security considerations. The job usually fell on eventual building tenants. This bias against incorporating security is as much about sensibilities as ignorance. Architects and designers sometimes dislike even doorknobs – anything that breaks the symmetry of long, graceful corridors. As of the writing of this book, design professionals still barely tolerate the boxy intrusion of protection devices and rarely integrate security at the building design level. Building owners are often saddled with adding alarm and protection systems as an inconvenient afterthought when they lease the space.

1 PRE-HISTORY: ANTI-TRUST

In 1966, the United States used the Sherman Anti-Trust Act to break up a security services monopoly controlled by Grinnell Corporation. Grinnell controlled ADT, Holmes Protection, and AFA Protective Systems, Inc. The antitrust decision paved the way for explosive business growth in modern security and protection systems. As of 2024, the industry employs 1.1 million Americans in guard services alone.

Security has always been seen as an additional expense and not a revenue-producing item. This often makes it among the first budget line items to be cut when money is tight. In the 1980s, the insular industry was manned by a small group of self-styled security consultants – mostly retired FBI agents, police officers, military, and the occasional civilian with a criminal justice degree. These retired individuals considered security a second career to supplement their pension. In some cases, they were unmotivated or apathetic about their roles and frequently ignorant of technology and the technical aspects of the job. The job title rarely ranked as senior administration.

DOI: 10.4324/9781003091103-2

To raise their relatively low status, a group of practitioners formed the American Society for Industrial Security (ASIS) in the 1950s. By the early 1970s, ASIS had about 100 members. In 1973, the organization held its first trade show in a small hotel suite for 40 attendees and 12 exhibitors. By 2015, the organization's annual trade show had grown to 20,000 attendees and 2,100 exhibitors. ASIS, since renamed *ASIS International*, counts over 34,000 active security professionals among its members.

2 SECURITY CONSULTANCY

In 1970, the NY and NJ Port Authority's World Trade Center project in lower Manhattan became the first major building project to employ an independent security engineering and consulting firm (Figure P0). The two-engineer firm was tasked with designing a security system for what were then the world's largest building complex and tallest towers. Although it was the firm's first contract, they were favored over more-established engineering and management consultants due to their experience in the new and evolving security and protection field. At the time, it was the largest contract fee ever awarded (Figure S1.1).

FIGURE S1.1 An aerial view of the World Trade Center from March 2001.

Source: Wikipedia, CC Share-Alike Unported 3.0, User: Jeffmock.

At a Smithsonian Institute conference many years back, several hundred attendees were given sage advice by the Author: To maintain similar year-on-year budgets for staff personnel and equipment, security departments must counter several major thefts or other criminal activity during each year. Without this "crime wave," the facility's administration often cut the budget. The speaker received a standing ovation.

Practitioners of any occupation fall into two categories; *Amateur* or *Professional*. Like many other labels, the meaning of these two nouns has tended to overlap. Consequently, the term "Professional" has lost some of its previous esteem. A growing number of occupations and special skills engendered by the computer revolution employ the term "Professional" regardless of education or experience.

Traditionally, a professional had an associated license, diploma, or certificate. Over time, the term has been applied to any occupation requiring special training in a specific field and required demonstration of competence and adherence to standards set by either a self-regulating group or government agency.

In the financial field for instance, a Series 7 is required for stockbrokers, though several other designations also cover stock and bond trading. In the security profession, the most prominent credential is the Certified Protection Professional awarded by ASIS International. Initially, this credential was only held by Security Directors, but its use has been expanded to anyone employed within the various disciplines that comprise the security profession.

Often, people refer to a plumber, electrician, or carpenter as a "pro." This signifies a certain level of ability, of excellence. In the security field, it represents extensive trade knowledge, cleverness, adherence to standards, and an ability to protect people and property on time and within budget. That's the true measure of a professional. Professionals are expected to stay abreast of the latest industry developments, often by subscribing to trade journals and continuing education curricula (re-certification in many fields). Many are active in local chapters of trade organization.

The continued growth of professionalism in the security field requires bringing design professionals into the fold. While the knowledge contained herein may not add initials to your signature line, it will greatly increase the value of your work product. And there it is. The final mark of a professional. A willingness to go the extra mile because it's the right thing to do – to add to the safety and prosperity of your client even if no obvious credit is attached to that activity.

Over the past several decades, a growing number of colleges offer courses relating to security management and technology within criminal justice and law enforcement degree programs – not to mention the addition of Homeland Security programs formed in the aftermath of 9/11. There are more than 400 traditional campus and online college programs that award degrees in one or more disciplines within criminal justice, security management, and/or Homeland Security.

- Pro-*fes-sion-al-ism*, n. – the practice, or methods of a professional, as distinguished from an amateur.
- Pro-*fes-sion,* n. – a vocation that requires extensive education or specialized training.

1 Evolution of Communication – Devices in Security

Security devices evolve in cycles: a given device is implemented, criminals learn to defeat it, the system is upgraded, and then eventually defeated again. Communications between the secured location and a central alarm center lie at the heart of this chess game. Designers benefit from understanding the history of communications technology because it will inhibit the naïve reinvention of already failed security strategies – an all-too- common error.

From before the written word, humankind has experimented with modes of communication. It began with cave drawings, followed by smoke signals, letters, the Pony Express, telegraph, and telegrams. The modern electronic era added analog phones, smartphones, text messages, and the Internet. Each new communication method eventually overwrote the previous mode, despite the growing pains that accompanied its first introduction. Smoke signals were effective in the old American West, and, for all their limitations, they didn't cause car accidents like our modern text-messaging drivers – though defenders of texting correctly point to its greater reliability and less soot. At the turn of the 21st century, the public became saturated with new communication devices and software: cell phones, wireless video, Wi-Fi – all running almost unlimited iOS and Android apps. There doesn't appear to be an end in sight for new communication devices and methods. Current upgrades to digital technology, due for release within the next five years, will make every device now in use obsolete; in fact, obsolescence is built into current business models.

1 HISTORY OF THE SECURITY INDUSTRY

The first commercial passenger elevator was installed in 1857. After serious technical problems were corrected, this humble, efficient lift ushered in an era of taller buildings, changing the face of city skylines. Still, it was not until 1924 that Otis Elevator developed the first automatic elevator requiring no attendants for residential apartment buildings. It was another 30 years before automatic elevators were installed in skyscrapers. This scenario is typical of any new technology and could be applied to the history of security alarms, protection, and the detection industry, which has gone through cycles that produced as many mistakes as successes. Electrical and communication-related technology, like security, has always experienced a series of setbacks before finally proving its value, a pattern that can be traced back to 1858 when the Atlantic Cable was laid on the ocean floor.

Many engineers considered the Atlantic Cable as the beginning of the Industrial Revolution because, for the first time, instantaneous communication between America and Europe became possible. Although the cable was a major engineering achievement, it was a complete technical failure and remained in service for only a few days before being abandoned. Another cable laid in 1866 was a complete success and quickly labeled a technical marvel, one which could be compared to events like the moon landings of a century later. The second cable remained in operation for over a hundred years.

2 MORSE CODE

Morse Code has faded from public view as the military largely abandoned its use in favor of newer technologies. Still, it was an essential part of civilian, maritime, and military radio communications.

DOI: 10.4324/9781003091103-3

The Coast Guard stopped listening for Morse SOS signals at sea in the 1990s. Even today, however, 1,650,000 Ham Radio operators still communicate with dots and dashes at 30 words-a-minute. These hobbyists are licensed by the FCC after demonstrating a knowledge of radio theory and passing a proficiency test requiring 15 words of Morse Code per minute minimum. With the advent of wireless communication devices in the last three decades, the number of Ham Radios declined so rapidly that in 2007 the code requirement for the qualification test was eliminated. Morse Code still lives among the "radioheads" but will eventually pass from memory, just as pioneering technologies such as crystal sets, vacuum tubes, capacitors, and resistors have.

Transmitting messages over a solid telegraph cable replaced the need for carrying messages across the country by Pony Express riders, who captured the imagination of young and old alike as they started to ride across the western half of the continent in 1860. The service was created to provide faster communication between cities in the Midwest (Missouri) and the Far West (California) and consisted of relays of men riding horses and carrying saddlebags of mail across a 2,000-mile trail. The first westbound trip took 9 days and 23 hours, while the eastbound journey took 11 days and 12 hours. The relay of Pony Express riders could cover 250 miles in a 24-hour day. At its peak, the Pony Express had more than 180 stations with 80 riders, between 400 and 500 horses and covered about 2,000 miles. Even though the territory that the riders passed through was extremely hazardous, only one mail delivery was ever lost –a record better than any airline can hope to achieve with the movement of passenger baggage. Yet, that amazing service depicted in literature and the media for over a century and a half was in existence for only 18 months before being replaced by the expansion of telegraph to the West Coast. Cross-country railroads subsequently forced the company into bankruptcy in 1861. After the company went out of business, several security firms adopted the Pony Express logo for their own use until 2006, when the US Postal Service announced it had trademarked "Pony Express" along with "Air Mail." Those names and their associated logos are no longer in the public domain.

For historical accuracy, it should be noted that the saddles held about 20 pounds of mail along with 20 pounds of water, a Bible, a horn for alerting the relay station master to prepare the next horse, a revolver, and a choice of a rifle or another revolver, allowing a total of 165 pounds on the horse's back. Riders could not weigh over 125 pounds, changed horses about every 75–100 miles, and rode day and night. There was a famous Pony Express advertisement that read, "Wanted: Young, skinny, wiry fellows over eighteen. Must be expert riders, willing to risk death daily. Orphans preferred."

The riders received $25 per week for their dangerous assignment, which was more than enough to attract many qualified, adventurous applicants since the comparable wage for unskilled labor at the time was about $1 a week.

2.1 A New Millennium

The conception and premature birth of modern security technology far precedes the Department of Homeland Security and its dozens of sub-agencies. Ancient people also had security concerns. An unknown Neanderthal figured out (without federal guidelines or municipal fire codes) that rolling a rock in front of the cave entrance would provide protection of premises and possessions against intrusion by marauding cavemen or oversized, unfriendly prehistoric mammals. It was a clever application of available materials and humankind has continued, to this very day, to refine the ways and methods that offer better protection devices for front entrances than a bulky rock. Even the pyramids were equipped with booby traps and other ingenious gadgets to protect ancient kings from evil spirits, and the not-so-spiritual crimes of theft and vandalism.

Over succeeding centuries, criminals became more proficient at removing rocks from the main entrances of caves. By the mid-19th century, the uncontrollable rise in the crime rate finally forced communities to band together and form police departments in the states and to construct holding pens (jails).

A group of mechanically skilled individuals appearing after the Civil War claimed that by using a newly developed device they called a "locking mechanism" on an entrance door, the premises would be protected from intruders and thieves. They were the first and only group at that time to deal exclusively with security issues and took the name "locksmiths" to indicate their unusual talents. In the 18th century, Britain had become the world's center of finance and produced huge metal bank vaults. These vaults eventually found a ready market in the United States where robbers in the Old West would hold up unguarded banks and make the manager divulge the safe combination. This scenario became the subject of thousands of cowboy films made in the 1920s and 1930s. Vault manufacturing, however, was not considered part of the "lock industry" and didn't establish itself until the late 19th century following the bitted key and cylinder-type inventions. Another significant date in lock history is 1851 when lock picks were perfected and demonstrated at an industry exhibition. In 1865, the combination lock was unveiled, which prevented picking but had inherent flaws.

Independent businessmen started the storefront lock hardware industry. Even after 150 years, they continue to limit their skills to making house and car keys and have shown no inclination to enter the electronic protection and access control industry. Protecting only the front entrance of a building is no longer as important as it was when prehistoric giants roamed the earth. Locksmiths have become tradesmen in a dwindling trade and are not a factor in the booming security industry because they have shown a reluctance to work with any device that has "wires" attached to it or learn how to program a computer or install a card access control system. Just like the blacksmiths who preceded them (who became experts in repairing the horseshoe but took no interest in fixing carburetors), locksmiths have become almost as obsolete as typewriter repairmen, television repairmen along with the milk, ice, and diaper deliverymen.

In the early 19th century, inadequate lighting created problems in large factories. Locks were useless since this was prior to the introduction of the pin-tumbler lock, and it was suspected that thefts were committed by employees. There were no other mechanical devices available, forcing the owners of the building to rely on human interaction. In 1889, someone invented the very first successful mechanical security device that was only slightly related to a locking mechanism. It was used to control bartender theft in a Dayton, Ohio, restaurant. It was first called a security register and later became known as a cash register. It provided more security than the watchman or any of the locking hardware available at the time. The company sold all rights to the National Cash Register Company.

Security technology may be described as starting with the cash register, since it was one of the first innovative concepts built specifically to protect valuable assets. However, security has grown into a major industry where small entrepreneurs have provided new concepts and innovative state-of-the-art systems that have defined and advanced the security sector. Edwin Holmes is regarded as the founding father of modern electronic security after purchasing a patent in 1857 titled "Improvement in Electromagnetic Alarms" from the Rev. Augustus Russell Pope, who was awarded the patent in 1853. That patent described a spring device installed at each door, connected through a wire which terminated in the owner's sleeping room and rang a bell if a door or window was opened. It did not sell well at first, since no one believed such a device could transmit a signal over a wire better than pulling strings, which was in popular use at the time. It used an indicator to alert the owner that a breach had been made but did not indicate where it occurred. After all, the use of "electricity" was only ten years old at the time with fewer than five manufacturers of electrical wire in the nation. A few years later, the annunciator was added to indicate the location of the alarm. The same system was eventually applied to hotels that used the same technology to maintain a connection to the guest rooms.

Wires were run over rooftops from protected premises to a central station owned by Holmes. Other central stations entered the field but were no competition to the Holmes Company. As the burglar alarm business grew, other central station operators tried to enter the business but could not successfully compete with Holmes who had opened dozens of central stations in all neighborhoods of New York City (NYC) to monitor premises. It should be noted that once the signal was received, there was no way to contact the police, so all central stations hired guards to respond to the alarm. At one time, there were twice as many guards at central stations as there were police officers in the city. There were no telephones in existence and central stations had to run their own wires through the city from rooftop to rooftop. When telephones were gradually introduced, central station wires ran over telephone poles. There were so many wires hanging overhead that eventually some cities outlawed overhead wires, and the telephone company and burglary alarm central services had to move overhead wires to underground ducts that they rented from the city.

For almost 50 years, Holmes was the only alarm installation and monitoring firm allowed to use the city's underground tunnels and conduits, forcing the competition to rent wire connections from them – a monopoly that lasted well into the 1940s.

From that inauspicious beginning, and through many mergers and acquisitions, the telephone and alarm industries combined until only the larger firms survived into the 20th century when a small cooperative group of firms controlled almost the entire protection/central station monitoring and telephone industry.

3 LOCAL ALARM TO POLICE – A HISTORIC PERSPECTIVE

In 1953, Somerville, Massachusetts, inventor Pope patented the first automatic burglary sensor consisting of a small electromagnetic contact device attached to doors and windows. These magnets acted as an on/off switch. They were connected through a battery that vibrated a bell when a window was opened, creating the nation's first local alarm system. Immediately following this innovative application of a simple on-premises concept, the legendary Edwin Holmes obtained a patent for a device that provided a revolutionary and practical method to remotely monitor sensors installed on numerous premises from a single "central station" location, rather than having the signal confined to the premises. This established Holmes as a pioneer of modern electronic protection systems, allowing the monitoring service to expedite police response to crimes in progress (Figure 1.1). By the mid-1860s, many wealthy homeowners in Boston and NYC retained the Holmes firm's service on a leased basis, the first known arrangement in the country where the end user did not own the system but "rented" it on a monthly basis.

- The local alarm intruder detection systems relied completely on the cooperation of nearby citizens to call the police when they heard or saw the alarm activation signal. More often than not, signals caused by a break-in were ignored, but it was generally felt that the noise probably served as a deterrent and scared away the intruder. Eventually, intruders became aware of the growing indifference felt by most communities in the inner city who would not act as the liaison between the premises and the local police station. In fact, neighbors often complained to the police about the loud noise of the alarms and the police sometimes issued a summons to the store owner for disturbing the peace. Intruders learned that a siren might be an indication that the security system was not connected to a central station. Once the local alarm was disabled, there was nothing left on the premises to detect thieves' movement. In general, the major advantage of a local alarm in a residential environment

FIGURE 1.1 A photograph of Edwin Holmes.

was its ability to awaken sleeping inhabitants. As a burglary prevention device, its use was limited except in certain situations where there was a likelihood that the local populace still cared about their neighbors (usually in rural areas far from the police station where, in all likelihood, no one would be close enough to hear a local alarm).

The selection of an alarm system should be made with some understanding of its function, and, as limited as they are in general, some local alarms are more effective than others.

Bells were the most popular sounding devices in the days when local alarms were the only means of alerting the public that an intrusion had taken place. They were gradually replaced by lights and loud sirens, but for many years, these bells were seen on the sides of retail stores but were often incorrectly used. The bells were not housed within a protective enclosure that had a tamper switch, which activated when the housing was lifted. Without these added safeguards, the entire enclosure was ripped off the wall, the connecting wires cut, and the system rendered useless. Installing an elaborate protective enclosure that was difficult to remove, while leaving the connecting wires exposed, was one of the more common errors found in most of the local signaling devices. Cutting the exposed wires transformed the bell into a hanging ornament compromising the entire security system.

- **Lights** mounted on the exterior of the building – especially flashing, high-intensity red light units – are far more effective than the bell units since they attract the attention of passersby without the intruder being aware of the signaling source.
- **Sirens** are superior audible devices since they are more associated with an emergency situation than the conventional bell. The siren decibel noise level is much louder than the bell and can be made to "warble" (see Table 1.1). The actual sound level is not measured as a pure number, and some manufacturers contend that their device is measured with a db meter at ten feet, a distance that has somehow become a voluntary standard. A more practical method to compare sound levels is to use the amount of power used by the siren, rather than the decibel (db) output. A 40-watt siren (fully powered) is louder than a 30-watt siren, and so on. Since standards are still a little vague or nonexistent, the most realistic measurement is to place units near each other in the same environment and simply listen to the emitted sound. Some of the astronomical db levels claimed by manufacturers are the result of wishful thinking.

TABLE 1.1
Decibel level chart for reference

Physical Reaction	Decibel Level [db] at 10 feet	Example
Hurts eardrum	150 db or higher	Concord jet
	150 db	Fire alarms
Uncomfortable	110 db	Thunder
	110 db	Shouting in the ear
Loud	105 db	Audio device
	85 db	Heavy city traffic
	80 db	Alarm clock
Normal	60 db	Conversation
Quiet	40 db and less	Hospital
Very Quiet	30 db	Soft whisper

2 Leased Telephone Loop Circuits

1 PARTY LINES

Local security alarm systems were proven to be ineffective, especially with a customer base primarily consisting of jewelry stores, banks, furriers, and other retailers of precious items. The industry needed something more effective than a bell to scare intruders away at night, particularly in high-crime areas with upscale retail establishments. Holmes developed an automatic alarm signaling device based on a galvanometer, commonly used to detect changes in current levels in electric circuits. Not only did the galvanometer indicate when a sensor was tripped, but it could also detect efforts by more enterprising burglars who dared to circumvent alarm systems by tampering with the circuitry. Each location was assigned a galvanometer, and a cable was strung by the alarm company (usually Holmes Protective) from each business to the central stations scattered throughout the city. This was prior to the installation of telephone lines mounted on poles.

Some of the operational features devised by Holmes can be found in models that are still in use today.

- Small indicating panels with colored tabs were placed in the circuit to reveal which door or window was opened. Once the system was operational, users would watch the needle on the meter. If it moved, the operator sent word to the police informing them of the break-in location.
- To keep servants from accidentally tripping the sensors when they entered the premises, a timer shut the alarm off in the morning.
- After the advent of the electric light bulb in the 1880s, a switch was installed to illuminate the premises on sensor activation.
- Eventually, Holmes retained a force of uniformed patrolmen who were dispatched to the scene, often arriving before the police. This Holmes process is the model of modern-day central stations.

  The alarm companies did not need to use the telephone company for a two-way conversation. They only needed a binary on/off signal from the protected premises (empty at night) to a central station across town.

Leased telephone lines took the place of over-the-roof wires. The leased line could be run into any building with telephone service. A pair of wires was "leased" from the phone company at a set monthly fee with no additional charges since voice calls weren't required. In the early 1900s, the telephone was first being introduced to American homes and in rural areas; people more often than not shared the same set of telephone wires that ran to each home and terminated in the phone company's central office. The calls were routed through a series of phone stations until they arrived at the proper destination. It was known as a party line.

The phonograph in Figure 2.1 held a prerecorded message indicating the location of the alarm.

The leased phone line was provided at a fixed monthly rate based on the cost of the distance between the premises and the central station with no additional revenue from long-distance calls or lengthy conversation time. Phone companies would rather use their lines for regular phone service than lease the wires to an alarm company. The single dedicated pair of wires provided to the alarm system could have been more profitable if they had instead been used to service hundreds of phone

DOI: 10.4324/9781003091103-4

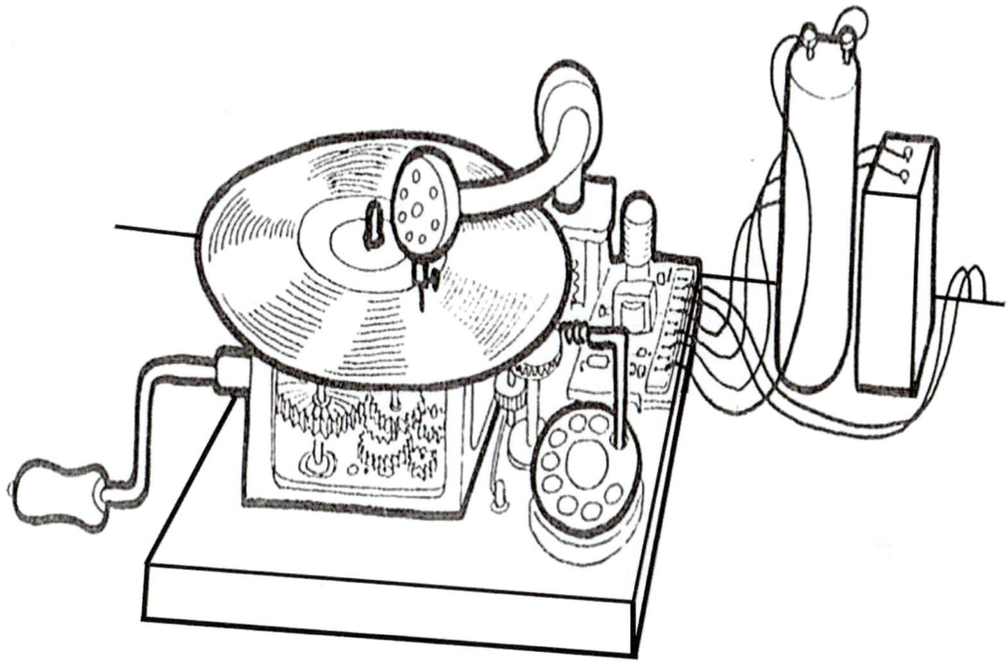

FIGURE 2.1 The phonograph was an ingenious device used by early alarm service companies.

conversations with per-call fees. Additionally, the alarm lines needed to be in perfect working order at all times, requiring close attention from the phone company.

2 MCCULLOUGH LOOP

The simple rotary-dial-party-line cord telephone is a relic of the pre-1970 era. There are still a few such systems in use, but there are no longer any manufacturers of the McCullough Loop system or any of its components. It was a modified version of the telephone party line that was used almost exclusively in many rural areas of the country. It was adopted by the big cities where a single pair of telephone wires ran in a "loop" to connect many sites to a central station. At the central station, each site was displayed on a huge panel where an operator would call the police or send a central station guard to the premises. It was used to protect homes, businesses, and even fire alarm pull stations on the streets. Instead of running a telephone to each individual site, the single loop accomplished the same thing at a fraction of the cost of a dedicated line to each site. When a burglar alarm was tripped at one of the sites, the McCullough Loop transmitter sent a series of coded signals (similar to a telegraph key) over the shared telephone line to a central station or to the telephone switchboard, and the site of the burglary was easily identified. It was, in essence, a modified telephone party line where each party in the loop could be identified.

Later versions used ballpoint pen refills that jumped on a strip of paper creating a series of readable blips. For example, a 1-1-1 notation identified one site and a 2-2-2 sequence designated another. This network of burglar alarm links was popular until the 1970s and brought with it the advent of digital communicators, which did not need a special pair of telephone wires to transmit a burglary in process.

- As a burglar alarm, the number of times the sequence was sent would provide additional information. For example, a three-time repeat signal meant the burglar alarm was tripped, and two repeats indicated the alarm was being activated as the owner left for the evening.

- When used in Fire Alarm Pull Boxes, the McCullough Loop signal was transmitted to a central station or directly to the fire department's in-house panel. Using volunteer firefighters, the incoming signal could trigger a large horn mounted over the firehouse. The horn would blast out a location code so that volunteers, who had memorized the codes, could go directly to the emergency location without checking into the firehouse. Of course, the loud blast would wake the entire neighborhood.

As a historical sidebar, it should be noted that in the early 1980s, the McCullough Loop was an important innovation in the security and fire alarm monitoring industry. In an informal survey of dozens of security directors, not one under the age of 50 recognized the terminology or its historic importance in the security field. This "loop" technology – essentially running a huge series circuit – is among the many important, if antiquated, concepts now all but forgotten since the advent of digital technology.

3 STREET FIRE ALARM BOXES

Part of the infrastructure of every large city was the red Fire Alarm Box (Figure 2.2). The box was a century-old traditional method of alerting the local dispatcher that a fire was in progress near the pull box location. In 2015, New York City (NYC) had 15,000 call boxes wired directly to the local station house (in compliance with the law that required one alarm box every four blocks). Some of the boxes with the familiar trademark ornamental torches molded on top are remnants of Old New York where street furniture was expected to be slightly ornate. In recent years, city officials have contended that these red boxes had become obsolete in the era of cell phone communication and cameras on street light poles. NYC mayors had tried to have all the boxes removed from the street corners because they cost over $9 million each year to maintain and were no longer needed.

FIGURE 2.2 A Fire Alarm Box.

Other factors for the removal of the boxes included a report from the Fire Department of New York that only 2.6% of all emergency calls come from the box, which connects directly to the fire dispatchers, as opposed to reports through 911 operators.

In 1996, NYC tried for the first time to remove the boxes, but a court order imposed an injunction after a federal judge ruled that no alternative had been proposed to allow disabled residents to call for help. In 2011, another effort was made to convince the court that even deaf people would be better served through the enhanced 911 communication network as it pinpointed the caller location even if they failed to speak. Furthermore, the new system used a special tapping code to distinguish between police and fire emergencies. Opponents of the request convinced the court that the code was too complex for emergency situations when people would have neither the time nor the presence of mind to think about a tapping system. The call boxes still remain, though they continue to be phased out, and some volunteers have painted the ornate boxes to restore them to their original appearance.

4 TAPE DIALER AND DIGITAL COMMUNICATOR

A revolution in communication technology occurred when the central station hookup began using an inexpensive digital communicator device in the early 1980s. It changed the need to obtain a leased telephone line and allowed additional central stations to enter the industry regardless of their location. Phone lines lost their importance once the revolutionary "digital dialers/communicators" were introduced and replaced almost all the leased lines previously used for security and fire alarm monitoring at remote central stations. This "dialer" even allowed the monitoring of premises on the other side of the North American continent since the device used the conventional telephone communication system to transmit the signal rather than a solid direct cable connection. With the advent of websites, email addresses, and satellite signals, the idea of leasing a direct-connect hard-wire telephone line for alarm communication purposes became a non-issue except for certain legacy fire alarm connections.

5 CENTRAL STATION MONITORING

The capabilities of central station monitoring services fall into many different categories. They may serve their own customers or the customers of alarm integrators throughout the nation. The number of clients serviced by a single firm ranges from 7,000 to over 900,000, and the larger firms employ over 800 people to monitor the following events and conditions:

- Burglary and Fire Alarm
- Carbon Monoxide Level
- Medical Emergency
- Temperature Control
- Video Verification
- Power Failure
- Two-way Conversation
- Flooding
- Appliance Control-

3 Security Alarm Cartel

1 ONE-COMPANY INDUSTRY (ILLEGAL MONOPOLY)

Obstacles to competition were created when each firm installed burglar and fire systems on their customers' premises under a "lease" arrangement and connected the alarms to their own central station. Note that, in those early days, alarm installers placed equipment and cabling on the client's premises while retaining ownership of the equipment, then rented it out to the client by the month on long-term contracts – much like some of today's residential solar panel agreements or cable service boxes. The installer/central station operator also charged a fee for the monitoring service, often providing a guard service to respond to the alarm. Unlike other services, when a dissatisfied customer decided to switch the monitoring service to a competitor, the original installer's lease agreement allowed them to remove not only the devices but also all the wiring and cabling. This would require any new firm to install new wire, significantly increasing costs.

This leasing practice became controversial and led to a landmark decision (Holmes Electric Protective v. Williams, 1922, NY Court of Appeals) after one installation firm in New York City's (NYC's) jewelry district claimed the practice was not only unfair but also a restraint-of-trade agreement and therefore illegal. They sued to force the original installer to leave the wiring in place. The court decided that, in what turned out to be a landmark decision, if the wire was removed, it had some value and should be reused and not discarded as scrap. This decision impacted the contractual arrangements, changing them from leasing the cable and wiring to selling the system, including cabling, outright to the customer. While the few independent alarm firms that tried to gain a foothold in the burgeoning market in the first half of the last century demonstrated both ingenuity and the American pioneering spirit, they did not survive for too long.

There existed an obscure NYC ordinance that in the early part of the century allowed only firms licensed as "transportation companies" to lease lines, and, not surprisingly, only one firm held that license. The company was Holmes, a name so popular at the time that it became a generic term for alarming the premises. The politics were such that any attempt by new firms to obtain this license to use the phone lines was rejected, and, if new alarm firms lacked the arteries for a central office operation, they could not provide the central station service required by the Fire Department of New York (FDNY) and insurance companies.

In order to build a central station to monitor burglar or fire alarms from businesses throughout the city, each station required a dedicated, direct telephone cable between the protected premises and the central station – which could be separated by many miles. Switching stations owned by the telephone company were scattered throughout the city and provided the communication network that allowed the central station to monitor clients' premises. Each premise was connected to the nearest telephone company switching station using a "drop" line. This drop line was connected to other switching stations, using the "backbone" line, until arriving at the central station many miles away, possibly in another borough. Having so many clients in every section of a city necessitated the development of a code that allowed groups of premises to share a single line rather than running a dedicated line to each premise. This allowed one central station to split the rental cost of the phone line among many customers.

> Breaking up the monopoly that controlled the evolving security industry was extremely difficult. No new company was able to compete without an expenditure of millions of dollars. Eventually, some firms were able to get the federal government to accomplish what no company was able to do for over 60 years. The government took the conglomerate to court and by applying anti-trust laws, broke up

DOI: 10.4324/9781003091103-5

> the monopoly. In 1965 the United States vs. Grinnell Corp trial resulted in a landmark decision by the U.S. Supreme Court in 1966 (United States v. Grinnell Corp:384 US 563). The court 's opinion came down against the idea (and reality) of 90% of the security industry remaining under the control of a tightly held cooperative venture, deciding it was a violation of anti-trust law. This forced Grinnell to either close their doors or decide to have the various entities that comprised it operate independently. Deciding upon the latter, the Grinnell behemoth's break-up marked the end of an era – and birthed the modern security industry.

During Supreme Court arguments made by the Grinnell defendants, one of the four firms testified against another, American District Telegraph (ADT), complaining about its national presence while his firm had only syndicate permission in just three cities. This was not unlike one mobster testifying against another's unfair labor practices. Within the auspices of the case, the court decided that defendants in the same action could not sue each other in the middle of a trial. That decision is precedent in anti-trust law.

While this highly profitable monopoly was in force in cities like NYC, Boston, and Philadelphia, security businesses in other parts of the nation were left to their own devices. They were not considered attractive to new ventures in those early years since small cities did not experience much criminal activity, except for a bank robbery every few years. Cars were scarce; the criminal element lacked an important tool for looting and a fast escape. The security industry was concentrated in larger cities and considered a sleeping giant until the 1966 anti-trust decision. As previously noted, this court decision was the impetus credited with creating the modern, highly competitive security alarm industry.

2 THE ANTI-TRUST DECREE OF 1968

Sherman's Antitrust Act was passed by Congress and was signed into law by President Harrison in 1890. Twelve years later, President Teddy Roosevelt used it against the Northern Securities Company, which dominated rail transportation in the Northwest. In 1891, President Taft broke up the Standard Oil empire.

When the government decreed the split-up of the four Grinnell entities, each individual firm had to reform under separate corporations. All four survived for over 40 years until one went out of business. Holmes Protective, one of the largest security monitoring services in the nation, suffered mismanagement after enduring acquisitions by several foreign firms. The 100-year-old company disappeared in 2008. Grinnell is still an independent company and Automatic Fire Alarm (AFA) remains a single entity owned primarily by the two sons of the man who purchased it after the anti-trust decree. ADT continues to expand and is once again part of a conglomerate.

ADT can trace its roots all the way back to 1867 when the stock ticker was introduced to record financial transactions. The inventor of the stock ticker also started a messenger service that sent instructions to and from the stock exchange floor, which was expanded to connect private homes to a central station. In 1874, 57 district telegraph delivery companies became ADT. As the telephone became more widely used, the ADT messenger business gradually declined. In 1901, ADT was incorporated into Western Union and separated its messenger service from the signaling business. In 1909, Western Union and ADT came under the control of American Telephone and Telegraph. ADT began to expand into new areas that included fire alarms and security alarms right up to the 1930s.

By the mid-1960s, ADT provided 80% of the central station alarm service in the United States, and in NYC, they controlled the entire central station market by price dumping to drive out the competition. They made up for the low prices by charging double in areas that offered no alternative service. In 1997, ADT was purchased by Tyco, a large conglomerate that made ADT an independent company in 2012. By 2015, ADT monitored the premises of 7 million customers and held a 25% market share of the residential market and 13% of the small business market in the United States.

AFA is a success story proving that the anti-trust decree wasn't necessarily fatal for the firms that were spun off from the conglomerate. All the newly formed companies not only survived after being set adrift but even expanded for over 40 years – until the cycle was repeated. Three of the four companies were eventually purchased by a conglomerate in the 1990s that operated much like the Grinnell monopoly that had lost the anti-trust suit in 1968 – a full cycle that took four decades to reunite the firm into a single company again. Of the four, only AFA remained independent.

Originally, when the US court anti-trust decision became final, stockholders in the conglomerate were given a certain number of stock shares in each of the four independent firms. The stock of AFA was purchased on the open market by the owner of a small telephone answering service with no background whatsoever in the security industry. In 1969, a new owner purchased enough stock in the AFA corporate entity to have majority control of the Board of Directors and was elected as the Chair and Chief Executive Officer of the now independent AFA company. The new owner expanded his existing answering service by installing additional switchboards and hiring more phone operators. The switchboards were modified to become a central station for fire alarm signals.

AFA has thrived for over four decades since being spun off, and in 1997, it changed its name from Automatic Fire Alarm to AFA Protective Systems. This public company has most of its stock under the control of one family with very little outstanding public stock. It has become the 12th largest central alarm monitoring service in the nation with 75,000 subscribers throughout the eastern seaboard.

3 OTHER FAILED LEGENDARY ACQUISITIONS

- **PINKERTON** tried to expand its 150-year-old guard service into the alarm installation business by acquiring several small alarm integrators at a cost that did not reflect the fact that once the owner of a family-owned installation firm took his money, the client retention rate disappeared along with the owner. The retention rate is the most important factor in the purchase price, and, once clients leave, so does the chance of survival for the new venture. Most of the acquisitions failed and Pinkerton was on the verge of bankruptcy when it was purchased in 2013 by the world's largest international guard service, Securitas AB of Sweden. Within two years, the famous Pinkerton name had lost much of its luster. Currently, it operates as a subsidiary of Securitas operating in two areas of focus: a risk and investigations division and another focusing on critical infrastructure.
- **MOSLER** was one of the most successful and oldest names in the security industry. Mosler manufactured safes since 1867 and installed security systems for both banks and private facilities before it was purchased by a conglomerate in the 1990s. The acquiring firm was operated by a new breed of Master of Business Administration (MBAs) who did not understand the business, nor did they have the experience to compete in the evolving security industry. Eventually, the legendary company went bankrupt in 2001, just as the industry exploded to record heights with companies operated by entrepreneurs with extensive experience and often without college degrees. The company and its century-old name have disappeared along with the likes of Holmes Protective.

Figure 3.1 details the more prominent consolidations that now control the physical security industry and are offered only as guidelines and is far from covering every major corporation that has redefined the industry.

4 MERGER AND CONGLOMERATES

An example from a relatively recent acquisition and merger campaign of the modern security industry is the expansion of Tyco Industries. It appeared on the marketplace in 1974 and, without any history in the security field, proceeded to purchase a long list of established security companies. It has grown into one of the largest security companies in the world. A brief chronological list of Tyco acquisitions since the early 1970s and before being acquired by Johnson Controls in 2017 is as follows:

1975	Simplex Technologies
1976	Grinnell Fire Protection Systems
1997	ADT (independent in 2012, then part of Apollo Global in 2016)
1999	Wells Fargo (the alarm company – not the banking empire)
2001	Sensormatic
2009	Brinks Home Security
2017	Purchased by Johnson Controls

5 NONCOMPETE CLAUSE

With the surge in acquisitions and mergers in the technology sector since the late 1980s, the use of noncompete clauses became an important part of most contracts, especially within the small cottage industry of security (Figure 3.1). The security industry had experienced phenomenal growth since the 1970s due to the nation becoming more security-conscious and the ease of starting a security firm. Until the large corporate conglomerates discovered the industry in the mid-1980s, small entrepreneurs began to sell their firms and within a few months started another business by retaining their loyal clients where friendships had been built that could not be easily duplicated by the conglomerates. The noncompete clause went into every acquisition and usually consisted of a two-year contract stipulating that the past owner stay on and assist during the transition period. After two years, that seller was legally permitted to start another firm in the security industry – and many did.

The major application of this clause was not limited to a few industrial mergers and has been part of employment agreements for decades. Companies use it to prevent employees from taking the company's trade secrets, business practices, or customer data to competing firms. This use of the noncompete clause has become more popular in the digital communication age where start-up firms in Silicon Valley develop new software packages that are easily stolen using USB drives. The noncompete clause in a contractual agreement appears legally binding but, in fact, is difficult to enforce. The courts do not generally enforce the agreement when no real protective interests are at stake or if the restrictions are unreasonable, such as time restraints lasting five years or longer.

In a few states such as California and North Dakota, this noncompete agreement is not legal. The only instance in which this clause is enforceable is when someone sells a business and agrees not to compete with the new owner. Even if the noncompete agreement is not legal, some employers still try to get their workforce to sign the document as an intimidation tactic to protect that for which they have no right to protect. The parent company counts on the employees not knowing or bothering to check on its legality. Noncompete clauses now appear in far-ranging fields beyond the worlds of technology, sales, and corporations with tightly held secrets. From event planners to chefs, investment fund managers to yoga instructors, employees are increasingly required to sign agreements that prohibit them from going to a rival company for a specific period of time.

TYCO BECOMES JOHNSON CONTROL
~~HOLMES PROTECTIVE~~ (gone)
SENSORMATIC
ROBOT
AMERICAN DYNAMICS
SOFTWARE HOUSE
~~HOLMES~~ [gone]
~~WELLS FARGO~~ (gone)
NATIONAL GUARDIAN
BRINKS HOME SECURITY
SIMPLEXGRINNELL FIRE SOLUTIONS
MOTOROLA
GRINNELL
~~ADT SOLD TO APOLLO 2012~~
NEXWATCH
NORTHERN COMPUTER
ULTRAC
ADEMCO
ADI - PITTWAY
JAVILON

BOSCH
PHILLIPS CAMERA
RADIONICS

MILLIDIUM
SYNAGISTICS
MARLOC

ASSA ABLOY
MEDICO
SECURITRON
HID
SARGENT
ADAM RITE
TIMELOX
CORBIN RUSSWIN
RIXSON
FOLGER ADAMS
INDAL
ARROW
YALE
VING CARD
SAFELOK
MULTI-T-LOCK

APOLLO GLOBAL
ADT
ASG SECURITY
PROTECTION 1

DIEBOLD NIXDORF
~~PINKERTON~~
~~BURNS SECURITY~~

INGERSOL RAND (IR)
LOCKNETICS
SCHLAGE

GENERAL ELECTRIC/BRADY ELECTRONICS
VON DUPRIN
IDENTICARD
LCN
MONARCH
STOPWARE
IVES
RECOGNITION SYSTEMS
XCEED ID EDWARDS FIRE ALARM

(sold security division in 2010)
IDENTICARD
PELCO to Schneider Electronics
IDENTICARD to Brady Electric

UNITED TECHNOLOGIES/JOHNSON CONTROLS
INFOGRAPHICS
CARD KEY

FALCONKALITEL
CISA SENTROL
LOCKNETICS
SCHLAGE CASI RUSCO
VON DUPRIN
INTERFLEX

STANLEY
NICHTIA
BALDWIN
TEKNOLOGIES
WANDERGUARD
HUGS & HALO
QUICKSET
SONITROL
BEST LOCK

MOTOROLA
INDALA
IONTRACK
MOOSE
LENEL
KIDDE
CHUBB

HONEYWELL
RSI VIDEO (VIDEOFIED)
XTRALIS
UNITED TECHNOLOGIES

SCHNEIDER ELECTRONICS
ANDOVER
INTEGRAL
PELCO

SINGLE PRODUCT INDEPENDENTS
HIRSCH ELECTRONIC VIDEX
VIDEO Q AMAG AWID
QUINTRON SALTO OSI
MORSE WATCHMEN MOBOTIC

FIGURE 3.1 Security industry consolidation (partial list of the more familiar firms).

4 Security Practices and Changing Technology

1 CHANGING PRIORITIES

Traditional security system providers have focused on a client base completely independent of related industries dealing with building management systems such as fire protection and environmental control. However, since the turn of this century, several conglomerates (Siemens, Johnson Controls, Honeywell, Tyco, etc.) have combined these diverse building management and protection industries into a one-source building management system. This change was based on the economic advantage of integrating the various disciplines that protect a building or complex, creating a synergistic approach. But that concept continues to be met with some reluctance from conventional security and fire protection industries that have operated on the business model of "going-it-alone." Many small entrepreneurial innovators in the 21st century have had trouble protecting their domain from corporate conglomerates that now compete by purchasing the smaller enterprises rather than expanding their basic product line. Security industry traditionalists have resisted intrusion but, since about 2003, another technology has evolved that may overwhelm the security, fire protection, and building management industries.

The physical security industry had rarely been credited with being at the forefront of advanced technology (it was, since the 1960s, considered a low-tech/copycat-type industry). Back in those pre-digital technology days, it was a common practice for people living in high-crime neighborhoods to protect their valuables by joining programs such as "Operation Identification." Through this program, a uniformed police officer came to the home and, using a stylus, carefully etched the owner's social security number onto their valuables (TV, clock radio, and any other tradable items). The "code" was intended to make it difficult for thieves to fence the appliances on the black market and easier for the police to find the true owner. In today's more sophisticated environment, it would be unthinkable to list a social security number on any portable device as that information is now more valuable to ID thieves than the item itself. To deter thieves, the police department encouraged homeowners who participated in operation identification to place a plaque on the front door with a message that read:

> Beware – We Participate in Operation Identification.

Not the kind of stuff that would have thieves quaking in their boots. Fortunately, the approach to security gradually changed, and the practice of etching items was abandoned. Rapid advances in security protection exploded after 9/11 when the security industry was placed in the public spotlight, forcing the industry to focus on developing new techniques to meet the demands of a paranoid public. The security industry responded to the increased demand by releasing untested security systems into the marketplace before their time and continuing to follow the practice of going through four sequential stages during the short cradle-to-grave life cycle that has defined the industry since 1970:

> ***(1) introduction – (2) slow decline - (3) rapid growth – (4) inevitable obsolescence***

Large corporations began establishing internal security departments with a Director-in-Charge who was usually a retired FBI agent (municipal and public agencies usually hired retired police officers). In 1980, only a few New York City (NYC) facilities had established a position that was designated as a Security Director. The salary for a Director was only slightly more than that of the average member of the guard force; since retired police officers could receive a pension after 20 years of service, they already had a substantial supplemental income lined up for retirement. Human

DOI: 10.4324/9781003091103-6

Resource Departments at hiring facilities would use that income to negotiate a lower salary for Security Director positions. It took almost a decade before the pension income was removed from the negotiation process.

The first college course in physical security was offered by Jersey City State College in 1978 (now Jersey City University). There were no textbooks available on the subject of physical security at that time, so the instructor used the Ademco Alarm catalog as the textbook. Of the few criminal justice programs offered by colleges and universities, most focused on areas relating to management and law enforcement. By the beginning of this century, the ASIS book catalog listed well over 150 technical books covering every area of physical and private security. There are now hundreds of colleges offering degrees in criminal justice, which usually include at least one class in physical security, security management, or the private security industry as a whole.

The 1990s witnessed a rapid growth in the number of medium-sized manufacturing firms entering the exploding security market, some grossing $50 to $150 million, with the more innovative companies going public or being purchased by Fortune 500 conglomerates. Along with private guard services, armored car services, and alarm installation firms, the security industry became a $120 billion market. One out of every 12 people in the nation's workforce was involved in some aspect of safety or security responsibility. In the last decade, many of the cottage industry firms began to disappear or were merged or purchased by larger firms. Gradually, throughout the 1980s, Japanese, South Korean, and Israeli firms entered the market. While the security industry is still growing, there are fewer small firms, and acquisitions and mergers have become a common business practice in the industry. With so many firms selling products under "private labeling" agreements with the manufacturer, it makes it difficult to determine the origin of many identical products that are sold under different names.

The communication and security technology industries have grown almost simultaneously as both fields exploded during the 1980s and 1990s and continue to reinvent themselves every five years. The public has accepted the fact that the five-year cycle of obsolescence is part of any developing technology – whether it be digital surveillance systems or smartphones. Slide rulers, suspenders, goulashes, typewriters, handwritten mail, and telegrams have been made utterly obsolete by new technology. Even Thomas Edison's creation of over 130 years ago, the traditional incandescent bulb, was added to the obsolete list in 2012 because after 2014 no incandescent bulb can be manufactured or imported into the United States.

2 TELEPHONE NAMES AND NUMBERS

In the not-too-distant past, and before the digital communication revolution, the wall-mounted rotary cord telephone along with the five-foot-tall Radio Corporation of America (RCA) radio and the mailbox were the most cherished means of linking Americans to the outside world. Radios were relegated to the car dashboard and probably the first significant technical advance in telephone communication was the once familiar dial tone that is no longer needed by 90% of the population who now rely on the ubiquitous cell phone, which has nearly replaced the landline phone.

The first commercial phone service was started around 1878, and within 11 years, the first coin-operated payphone was installed in Hartford, CT. According to the FCC, the number of payphones in the United States was at its highest of 2.2 million in 2000, dropping drastically at a steady rate of 10% every year since.

Less than 100,000 payphones now exist in the United States; the justification to abandon them, of course, is due to the fact that the number of wireless phone connections in the United States has surpassed the number of people.

For many years, American Telephone and Telegraph Company (AT&T) was the largest payphone provider, but they eventually left the payphone business to concentrate on the cellular communication market. Other communication companies have mostly sold their payphone businesses to small, independent regional companies. The new owners hope to slow the steady attrition rate of standalone payphone booths before they become extinct, much like corded phones and party line networks have. One factor that may prevent payphones from disappearing in one fell swoop is that they continue to generate profits, though not through their original intended purpose. Most of the remaining phone booths in major cities have, since 2003, been authorized to accept advertising, which is not only highly profitable but also due to their lower usage, don't need to be continually repaired.

Revenue from phone booth advertisements has outstripped earnings from actual calls with much of this newfound revenue going to the local municipal treasury. The familiar, strategically located phone booths have proven to be a bonanza to advertisers since they blanket the city at a very low cost. Critics feel that phone booths are eyesores that serve no purpose except as advertising displays. They argue that it would be more beneficial to plant trees in place of the phone booths. The ad industry has tried to convince city officials that, without the advertisement income, the phone booths would disappear entirely. The advertisements are keeping the payphone alive (barely). Some older booths have become collectors' items, attracting antique collectors across the nation.

Verizon was the last of the giant communication firms to leave the payphone marketplace following the departure of AT&T in 2008 and Sprint in 2006. Pacific Telemanagement Services ("PTS") purchased the remaining 50,000 Verizon payphones in 2011. These were added to PTS's existing network, mainly located at secure locations such as airports, Walmart's, and 7-Eleven convenience stores. These smaller companies found that there was a profit in the payphone business after the first 100 calls per month.

For travelers in days before the Internet or cell phones, the only way to make calls from a public location was a payphone. The public phone business became increasingly fraught with frustration starting around 1980 as public payphones became the symbol of a changing neighborhood. Traditional red painted booths gave way to graffiti-covered, vandalized, and often inoperable phone stations. If the headset were intact, it might not produce a working dial tone. Filthy mouthpieces and the nauseating stench of urine were common. Also, free-standing phone booths blocked pedestrian traffic and slowly assumed the look of a discarded museum piece.

In 2016, NYC announced the replacement of 6,400 remaining coin-operated payphones in favor of a high-tech replacement by LinkNYC. The new network of 9-foot-tall kiosks featured free Internet, no-cost domestic calls, and broadcasts 24-hour Wi-Fi within a 150-foot radius. The stations offered both keypad and touchscreen interfaces. The "Links" display public service announcements in emergencies and free USB device charging. The city kept only three of the old phone booths and auctioned off the rest. With 1 gb of bandwidth, the network is almost 100 times faster than average public Wi-Fi. The kiosks also let users make calls through third-party service providers like Vonage.

The project is a bellwether for the feasibility of citywide free Wi-Fi. The new stations include 13,356 pedestal payphones but only four of the old-style glass-enclosed sidewalk booths. Test locations were restricted to West End Avenue between 66th and 101st Streets. The Wi-Fi signal can be used by multiple device types, including laptops, smartphones, and tablets. Despite some initial resistance to the project, over 2,000 people logged onto the networks in one month for an average 38-minute session. Of course, Internet seekers will likely always have Starbucks available with the added benefit of free toilets.

3 OUTDATED COMMUNICATION LINKS

3.1 Cord Phones

Hard-wired cord phones began disappearing in the 1970s and are fast becoming relics of a bygone era. Soon, they'll only be remembered by those of us who watched the "Love Boat" and drank

Harvey Wall banger cocktails after a hard day operating the mimeograph machine under incandescent light bulbs. According to Chamberofcommerce.org:

> 7 in 10 adults are wireless-only phone users, which is about 183 million Americans nationwide. Northeast residents are loyal to landlines. More than 4 in 10 adults (41.2%) in the Northeast still live in a household with a landline phone, which is more than any other region in the country.

3.2 Mailboxes

Messenger delivery boys on bicycles streaking through traffic are going the way of the locksmith, blacksmith, and the green mailboxes on street corners. Email and tablets, along with chat rooms and Twitter, have replaced the need for collection boxes because very few people write letters anymore, resulting in the US Post Office's removal of the boxes at a rate of 40,000/year. The USPS has closed many of its small rural offices and, over the last two decades, has threatened to seek protection under the bankruptcy law. It is difficult to exaggerate how irrelevant the mail has become in the everyday life of many people under the age of 40. Older Americans used to check the mail daily, but after finding the mailbox filled with only junk mail, that ritual has, in many cases, become a semi-monthly visit.

The use of traditional mail has dropped significantly in the past two decades, as evidenced by Figure 4.1.

3.3 Western Union

The ever-evolving digital world has forced companies in the communication industry to be innovative in order to remain relevant. A case in point is the reinvention of Western Union, a fixture of American lore. It was started in 1851 by a Rochester, NY, businessman and became the first telegraph conglomerate, which linked the coasts by 1861. At one time, the company employed

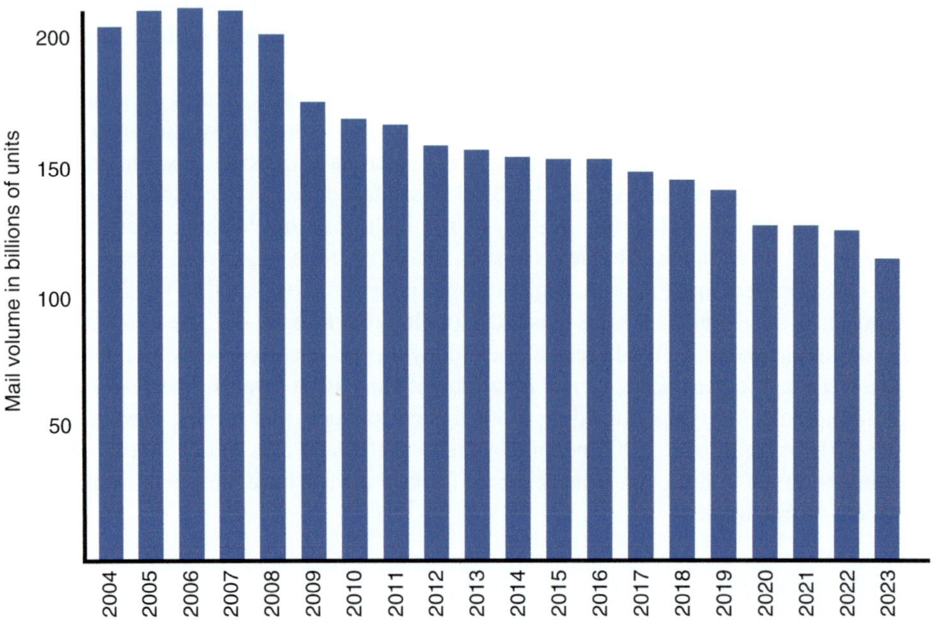

FIGURE 4.1 The US mail units delivered peaked in 2006 at roughly 213 billion units. It was just over 116 billion units as of 2023.

Source: Statista.com, 2024.

thousands of uniformed messengers who delivered telegrams (including the famous singing telegram) from hundreds of small offices to 200,000 Americans each year. They survived and even thrived for 80 years before succumbing to modern technology and email.

Western Union was created to break Samuel Morse's telegraph patent and grew into the first nationwide monopoly, once known as the "octopus of the wires." The company became so large and politically influential that, despite repeated attempts over the years to break its monopoly, the company actually grew larger and stronger. Most of the famous names in the history of American invention and finance were connected with Western Union in one way or another, including Thomas Edison, Cyrus Field, and Andrew Carnegie. The first president of Western Union left to form his own company, building the transcontinental telegraph in 1861 and then made a fortune by selling his company to Western Union. The company acquired all types of competitive companies but ignored the "bell and toy" made by a man named Alexander Graham Bell. It was the first of many future mistakes.

The company's daring attempt to string wires across North America and cables under the Atlantic Ocean captivated the imagination of a public that didn't understand Western Union's aggressive and unsavory business practices. Its anti-competitive tactic was to drive down rates and bankrupt the competition in the same way security alarm company American District Telegraph (ADT) did 50 years later. The company's brazenness and scope of operations were unheard of a hundred years ago. Western Union had many friends in Congress, the regulatory bureaucracy, and the press. The company sponsored and ran both the Republican and Democratic national conventions of 1872. Battles fought in business, politics, technology, and the courts all saw the company come out victorious. When the Postal Telegraph and Cable Company emerged as a rival, and couldn't be crushed like previous startups, it and Western Union merged in 1943 – whereupon the duopoly grew even larger. The new company's practices, in many ways, set the template for today's corporate raiders and the creation of large conglomerates.

The company was hit hard by both a recession and the increased popularity of the telephone. As phone giant AT&T ascended, Western Union declined. In its death throes in 1994, the company was acquired by a financial management firm at a bankruptcy auction. By then, its business had shrunk to money transfers. The business reinvented itself and relaunched as a public company in 2006. It terminated what remained of its telegraph service and became the world's largest money transfer company with a focus on servicing immigrant populations. Global migration had become the cornerstone of the reborn Western Union, and its in-house researchers outpace the Census Bureau in tracking migrant populations.

Western Union has 320,000 locations worldwide – five times as many as McDonald's, Starbucks, Burger King, and Walmart combined. In 2020, migrants transferred $508 billion, nearly three times the world's foreign aid budgets combined. The company affords $1 billion to marketing over a four-year period through fees that can run anywhere from 4 to 20%. Western Union is one of the very few companies that lost its way only to return and flourish in the Internet age by abandoning obsolete concepts and adapting to evolving markets.

The Western Union lesson should be a warning to the Google, Amazon, and Facebook of the world that now dominate world communication: glory and profit can be fleeting.

4 A SHORT LIFE CYCLE

Built-in obsolescence is frequently associated with industries that operate in any rapidly changing technology field, and that short cradle-to-grave life cycle defines the evolution of the security industry. This fact has made it difficult to select a system that will not be obsolete within a few years. When a security system reaches its seven-year milestone, it is considered a "legacy" system and choice of a replacement system is usually made at this time. Traditionally, most sophisticated electronic concepts are initially developed in the research labs of large firms undergoing many months of testing before their initial introduction into the marketplace (similar to the process in pharmaceutical firms when introducing a new drug).

That basic formula hasn't been followed by the average security industry manufacturer, which is still, to some extent, made up of hundreds of small firms in what is considered a cottage industry. These smaller firms appear to demonstrate an ability to produce advanced and unique alarm systems far quicker than the larger conglomerates but generally lack the funds to conduct a thorough research program. Small security firms can produce a system or a device in a matter of months, usually by testing the product after it is installed on the walls of the client's facility. This shortcut in the R&D cycle has produced systems that 50% of the time are amazingly effective, but 50% of the time have failed disastrously.

5 PROFITABLE FAILURES

Early on, the electronic-detection security industry was a money maker despite sales of mostly defective technology. Fortunately for the salespeople, the actual number of break-ins was so minuscule that the lack of proper system testing was rarely discovered. The average security system was sold on the assumption that a burglar would activate the alarm when breaching its surveillance zone. In practice, these systems might lie dormant for years without activation, so that few would ever know if it worked as promised or was completely useless.

A classic example: in the 1970s, a small importing firm sold automatic tape recorders designed to send a prerecorded message. When tripped, it would activate and dial a chosen phone number (usually the police) whereupon a 20-second voice message reported a robbery and location to the police. It was like an automatic call to the 911 operator a decade before that service existed. The system was inexpensive and simple to install. The only drawback was that it did not work! The system would remain dormant for years. When a rare emergency occurred, the tape mechanism, now frozen because the lubricant had solidified, failed to activate.

The firm discovered that only 5% of the 100,000 recorders sold were ever activated during the system's life. In those days, few conducted periodic testing of any electrical device except the front doorbell. The unit sold for $100 and the 5,000 units returned as being defective would receive a full return on the $100 purchase price with an apology from the distributor. Since 100,000 units produced $10,000,000 in sales, less the return cost of $500,000, the business model – if not the device itself – was extremely effective. In reality, there was no real security system; it was a placebo. Prompt, no-questions-asked defective device refunds quelled any curiosity. The firm remained in the security business, eventually going public, and continues to prosper. However, they no longer sell tape recorders.

The present marketplace caters to a savvier end user, and ethical firms populate the security system marketplace. Today, a 5% failure rate would likely destroy the reputation of most companies. Opportunities exist for smaller firms to have an impact in the security industry since larger publicly traded firms take so much time to develop new products. By the time the device is ready for the marketplace, the technology may have changed to such an extent that the device is obsolete. These large firms realized that in order to stay in the industry, they had to abandon in-house development efforts and began purchasing the newest alarm devices from the more innovative small manufacturers. Simply attaching their company logos to the products (private labeling) so that the public isn't aware of their origins has allowed the large firms to replace their devices periodically as newer ones are developed.

In recent years, the security industry has entered a consolidation phase and now those same large firms not only buy the product from the producers, but they also purchase the entire smaller firm outright. Some of these acquisitions have proven disappointing. The mere purchasing of a firm's client list and product line doesn't automatically include any of the entrepreneurial spirits the smaller firms were able to harness and build into that successful business. Once the small firm receives the money, many of the original owners disappear along with their creativity and business connections. If this trend of mergers and acquisitions continues, the security industry will become controlled by a few dozen firms. As recently as the mid-1980s, there were thousands of firms in the industry, and innovative ideas were encouraged. This same level of innovation usually doesn't occur when an industry is controlled by only a few conglomerates.

Small manufacturing entrepreneurs in the insular security industry centers have found it difficult to financially support the introduction of new products. There are no retail outlets such as Home Depot or Radio Shack to distribute complex security devices, and advertising costs are often more than a small firm can afford. Hiring an experienced sales force would be too expensive, and the financial commitment to exhibit at a tradeshow could be cost-prohibitive for a start-up company. The security industry, for the most part, relies on self-employed, independent salespeople known as "tech reps" pitching four or five uncompetitive security-related products in a protected region on commission. They provide sales assistance, don't do installations, and, while they may maintain a stock of product, do not service the systems post-installation. The only cost to the manufacturer is a commission on product sold. A tech rep's existence is precarious because success through regional profitability often justifies the parent firm's hiring of a full-time company employee to replace the rep.

Many businesses' financial officers, and even upper-level management, consider a security system as a non-revenue-producing line item. So, unless such systems and expenditures can demonstrably save money, the company will tend to limit security department resources.

Without noticeable criminal activity in or around a facility, security's status takes a low priority. Overhead expenditures like security systems cannot generally be increased indiscriminately in the way basic office equipment expenses might as it's more difficult to pass overhead costs onto the customer. That low priority status changed after 9/11, and security has become an essential part of any facility operation just as a heating, ventilation, and air conditioning (HVAC) system. The system selected must be subtle and not interfere with the normal functions of a facility and, most importantly, efforts should be made to ensure that the final system selected does not create a fortress-like environment which would alienate the very public the building was built to serve. Overkill is not a viable choice for any facility, even though some of the mandates from the Department of Homeland Security might seem to ignore this fact. There have been several situations where upgrading security is financially justified as a revenue-producing expenditure, especially when it preserves valuable inventory, decreases costly vandalism, or increases consumerism through a establishing a safer environment.

6 GROWTH INDUSTRY

Expansion of the security industry closely followed the trends in the communication and computer industries over the last three decades, becoming a multibillion-dollar market (Figure 4.2). The security business had slowly evolved from a cottage industry, made up of small manufacturers and installers, into a major industry with much of its impetus coming from world events and government regulations. Most of those original small manufacturers, which were active in the 1970s and 1980s with fewer than ten employees and surviving on gross annual revenues of less than $2 million, are all but disappearing. The problem that once defined this industry was that it grew too fast and often neglected basic research and testing procedures in the rush to outdo the competition. Expansion was a slow process at first since there were very few means of distribution, and there were very few security industry shows available until the mid-1980s. The limited number of trade magazines did

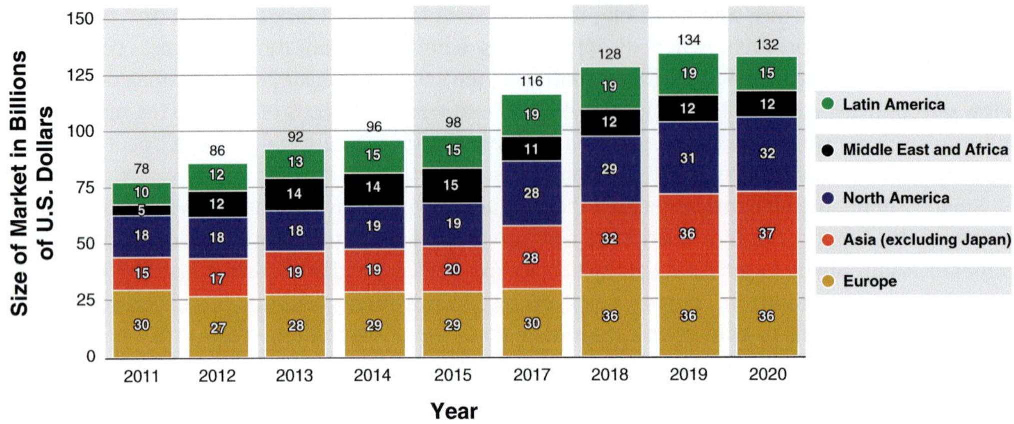

FIGURE 4.2 Size and growth of the security services industry globally. From 2011 to 2020.

Source: Data from Securitas AB, Statista.com.

not have the circulation to warrant the cost of an advertisement, and one catalog company carried 85% of the devices available. No single manufacturer controlled more than 2% of the marketplace even as recently as 1990.

The rise of the security industry began with an early group of determined entrepreneurs who found success in introducing new technology. Smaller firms were always more innovative and understood the marketplace better than the conglomerates. The larger firms would practically abandon their research efforts and purchase the systems from the small firms, repackaging and distributing the product under their logo (called private labeling) to the national marketplace. There are many financial justifications for this arrangement, with the main one being that the large firm could keep up with the newest technology and not waste money on research efforts by simply purchasing the newest and best system (always reselling it under their own logo) from any firm that had the most effective system. The modus operandi in the security industry is difficult to define since the technology is constantly changing. As the market matures, the focus has transitioned from hardware to selecting the best software to allow the connection of all security appliances on the IP network. The smaller installation and manufacturing firms are being forced to limit their products to a small niche market in order to fend off the larger firms from taking over the entire market. If they are not successful in developing new technology, these smaller firms will have to seriously consider marrying larger firms or merge with other smaller firms.

The type of firms that merged, consolidated, or were purchased outright has changed with the growing industry. In the 1980s, mergers mostly affected manufacturers of motion detectors and usually involved a purchase price of less than $20 million. That changed a decade later when the mergers involved CCTV, then card access control and biometric systems and, most recently, digital video recorders. Once they discovered the remarkable growth potential of this 50-year-old industry, companies outside of the industry began entering the market. IBM, looking to bolster its technology services, purchased Internet security firms in 2006 for over $3 billion, more money than the entire industry grossed in 1980. Other firms are spending over $2 billion to acquire similar Internet security firms. The smaller companies are going to have to become more innovative if they seek to remain in any of the many niche areas that make up this industry and fend off the larger firms controlling the marketplace.

Merging security and Information Technology (IT) functions, as well as the personalities involved in those two separate technical disciplines, has not been an easy task since the interaction between the security staff and computer technicians involves a huge cultural and educational gap. The two occupations have (1) diverse and often divergent objectives, (2) a difference in pay scale, and (3) a difference in knowledge as a computer programmer is generally better educated than a member of the uniformed security staff. While there is no proven method to encourage both parties to work in perfect harmony, there are many programs that have been successful in getting these diverse personalities to cooperate in upgrading the protection of the assets of the facility. Merging these two departments will remain a slow process and, in the opinion of many security experts, unless the security department cooperates with the IT department, the security staff will be swallowed up by the more technically oriented IT department and lose their independent identity.

7 VANISHING SMALL, INDEPENDENT ALARM INTEGRATORS

Prior to the late 1960s, the alarm industry was a small, low-profile industry controlled by one conglomerate powerful enough to stifle competition. After the Grinnell Corporation monopoly was split up under the 1968 Sherman Anti-Trust decision, the alarm industry exploded into a cottage business of small entrepreneurs that continued to grow through the 1970s and into the late 1980s. Small integrators and alarm manufacturers made an impact as the industry experienced a technology revolution. Imported Japanese black and white closed-circuit TV cameras allowed independent central station monitoring companies to flourish and, until 1995, the industry experienced a golden age.

If you could use a staple gun, it was easy to enter the security installation business particularly since the business was unencumbered by licensing requirements or standards and was not in compliance with local or national electrical codes. The industry grew so large that it attracted the interest of larger firms (both domestic and foreign). By the turn of the century, the atmosphere reversed and, once again, became a business of conglomerates. In a 40-year period, the security industry had gone full circle.

8 SMALL INTEGRATOR VS BIG CONGLOMERATE

When searching for the best integrator, the customer (end user) should not assume bigger firms are more likely to provide better results. Efficient service and maintenance are separate but equally important considerations and it is generally easier to contact someone in a small entrepreneurial firm than the local manager of a national conglomerate. No hard and fast rule can predict which size firm can better accomplish a particular task. Referrals by peers and discussions with a firm's principals often inform the choice. Simply relying on the promises of the salesmen can be disastrous since sales force turnover is generally high.

1. Those who are responsible for purchasing security systems, particularly those lacking a strong technical background, will often choose the nationally recognized provider. If the system doesn't perform to expectations, they can always point to Honeywell or General Electric's reputation. If a small firm were selected and the system failed, they might be fired. In general, big nationwide integrators prefer to do business with similar sized alarm installation companies. If the customer's sites are scattered across the country, a firm similar in size to Honeywell may be a better fit.

A complex of buildings that requires more sophisticated and flexible system design won't necessarily benefit from solutions offered by overzealous salesmen. Often, a security consultant with a background in the client's specific field or industry is a better choice. The consultant can design and select state-of-the-art components and write specifications without consideration of obsolete equipment lingering in a warehouse. Also, the right integrator will take the time to understand the facility's needs to avoid creating a fortress-like atmosphere that alienates the very population it was meant to serve. Far too many systems are installed without considering future expansion. Hundreds of access control systems are replaced because the software cannot accept additional readers or issue enough cards to a growing employee base.

There is no such thing as a perfect security system. That said, an industry where 50% of installations are inadequately designed and more than 50% are rarely serviced or maintained presents many unseen challenges to purchasing managers. The life cycle of any alarm system is six to eight years, regardless of who installs the system and changing integrators is always a strong possibility. The integrator (both the big and small firms) must provide user training until the client is fully comfortable using the system.

9 DO-IT-YOURSELF INSTALLATIONS

The demise of the independent security cottage industry was not only caused by re-emergent conglomerates. New self-destructive forces arose as technology advanced and easy-install wireless equipment created a new breed of home-security specialist: the homeowner. Security system installation, long the domain of professionals, was suddenly a job that could be done by uninformed workers charged with creating an electronic shield to be monitored from a distant command center. Surprisingly, low installation costs always included a monthly monitoring fee that became a perpetual profit center. Wireless technology and battery-operated sensors with a cellular connection completed the industry's do-it-yourself transformation. Self-monitoring plans allowed alarm reception by text or email, and everything could be remotely controlled from a computer with an Internet connection.

Webcams allow surveillance of the premises by placing cameras in or outdoors that send notifications when there is motion in the field of vision. There are several manufacturers of door locks which communicate wirelessly with a base station inside the home to allow the locking and unlocking of the door remotely over the Internet using a computer or a smartphone. By adding a camera, the person at the door is displayed on the computer screen. There are lock applications that have the ability to make temporary codes for other users (i.e., houseguests, meter readers, and contractors). When the person leaves, the codes can be deleted remotely to prevent reentry. These retail alarm systems provide more security for less than a thousand dollars and are similar to systems installed by integrators just 20 years ago for $20,000 (Table 4.1).

10 CURRENT LEADERS

Table 4.1 lists the Top 10 North American system integration service vendors listed by revenue.

TABLE 4.1
North American system integration service vendors Top 10 by revenue (in millions of US dollars)

Rank	Company and Headquarters Location	North American Systems Integration Revenue 2020
1	**Convergint Technologies** Schaumburg, Illinois	$1,920,220,000
2	**Securitas Electronic Security** Uniontown, Ohio	$1,053,000,000
3	**ADT** Boca Raton, Florida	$691,000,000
4	Pavion (formerly CTSI) Chantilly, Virginia	$438,230,000
5	**Allied Universal Technology Services** Santa Ana, California	$345,260,000
6	Paladin Technologies Vancouver, BC, Canada	$233,760,000
7	**Vector Security** Warrendale, Pennsylvania	$169,380,000
8	The Cook & Boardman Group, LLC Winston-Salem, North Carolina	$167,650,000
9	Kastle Systems International Falls Church, Virginia	$142,000,000
10	**Unlimited Technology Inc.** Chester Springs, Pennsylvania	$134,120,000

Source: Statista.com.

11 SELF-INSURANCE FOR THEFT

In the mid-1980s, a group of very large, independently owned appliance stores combined their purchasing power by forming an alliance in order to remain competitive against appliance chains such as Sears Roebuck and Walmart. This alliance comprised nine appliance chain store owners in different sections of the country totaling 350 retail stores combined. By purchasing their appliances under a single shipment from manufacturers, they could fill an entire freight boat of TV sets, washing machines, etc. in quantities that were often greater than the number of units purchased by big box stores. Once the appliances were delivered to the port city, the quantities ordered by each store were separated at the dock and shipped directly to each of the 350 outlets. It was so successful that they then purchased insurance for theft and inventory losses under a single, lower cost policy.

The risk management consultant devised an insurance plan that lowered their combined annual premium from $10 million to $4 million by creating a high deductible of $500,000 for each loss incident. The savings of $6 million were placed into a reserve fund, which would allow each of the nine owners to cover any losses under the $500,000 deductible amount. If a major disaster occurred (e.g., a fire), the store owner received the first $500,000 from the fund and the rest from the insurance company that wrote the high-deductible insurance policy. If the total money dispersed over a 12-month period was less than $4 million, the remaining funds would be dispersed in accordance with the number of stores involved in the alliance. If no losses occurred, the entire $4 million would be returned, which was equivalent to the basic premium.

Since the owners were fully protected from all losses, regardless of the reason or amount, they had no incentive to install an alarm system in their stores. The first issue with this arrangement was that, since it was common for retail stores to suffer undocumented inventory losses (employee theft, record-keeping errors, shoplifting, breakage, etc.), store owners covered these annual losses by claiming that a break-in had occurred and that the unaccounted inventory losses were stolen by thieves. Nearly the entire fund became a victim of this tactic. In order to save the program, they decided to hire a physical security consultant who would visit several of the sites and prepare a specification for a security system that each store had to install in order to remain in the insurance program. The minimum-security system for each store consisted of better locks, personnel background checks, sophisticated motion detectors, glass detectors, and better lighting. The average cost for the system was $8,000 per store or a total of a little less than $3 million. The alarm system eliminated the false claims made by the owners, and the fictional thefts disappeared. Losses went down to less than $250,000 and the kitty was able to return 80% of the premium to the owners after the first year.

12 INTERNET TECHNOLOGY AND SECURITY

With the influence of digital networks overwhelming all other communication functions, computer-driven security networks fell under the authority of the IT department. Cameras, card readers, and CCTV were all included. Due to the advent of the Web, the LAN and the WAN, coax cables and hard-wired connections between devices, and the central station have moved into history's dustbin.

The impact of 9/11 forever changed the security industry from an independent service to a business segment separated from IT by an invisible firewall. Technologies that underpin each department have become indistinguishable, and a growing number of public and private facilities have merged the two functions. Data storage is as critical a component as those used to protect the physical assets of the facility. With the advent of the Internet of Things, many common appliances and devices now have an IP address alongside the alarm systems and surveillance gear. Training security employees to understand and recognize threats to computer security became a problem as security personnel were not accustomed to being focused on computer technology. The security department was no longer limited to installing cameras and card readers and was implementing new login systems requiring employees to use physical cards to access the company network and purchasing software that prevents people from accessing the facility network. The security department has been an integral part of building operations for less than three decades and during that short time, it has experienced some of the most rapid changes in technology since the start of the industrial revolution.

Private firms, public institutions, and government agencies understand that sensitive data can be accessed through employees' PCs and laptops and there is probably as great a need in the digital age of communication to protect information as it is to protect physical assets. The major security threat is not the stealing of a physical piece of equipment, but rather the theft of an employee's password from a sticky note posted on the keyboard, or an employee simply walking off the premises with a laptop containing invaluable company secrets.

13 THE NEW ERA OF TECH STARTUPS

Seven of the top ten most visited websites are based in the San Francisco area, including Silicon Valley, an area that is considered the technology capital of the world. The start-up tech companies that started in Silicon Valley over the last three decades are giants in the digital age such as Google, Yahoo, and Facebook. Gradually, other cities and states have begun to attract prime digital companies away from the West Coast. According to the National Venture Capital Association, which ranked the top nine states for venture capital raised in 2022, after California at number one, the remaining in order are New York, Massachusetts, Illinois, Florida, Texas, Connecticut, Washington, the District of Columbia, and Georgia.

The ability to draw on an educated and trained workforce, while having reasonable rents but offering competitive salaries, makes choosing locations and certain cities a difficult but vital decision for start-up companies. Four cities have gained prominence in the high-tech field with Boston ranked second, after San Francisco, closely followed by NYC, Los Angeles, and Washington, D.C. Some of the hottest names in the high-tech marketplace avoid the top five cities due primarily to the high living expenses in each. For example, software maker Adobe is in Salt Lake City, chipmaker Intel is in Portland, Oregon, and Apple is in Austin, TX.

Despite the growing number of websites created since the start of the century, actual page views are becoming more concentrated. In 2000, the top ten websites accounted for a little over 30% of all daily page views in the United States, and by 2010, the top ten websites were viewed over 75% of the time. Standard operating systems like Microsoft Windows, Apple macOS, Google's Android, and Apple IOS (iPhone) are the biggest systems among the top ten.

Having so few companies with control of the market has significantly slowed the rate of newly formed enterprises in the nation. Major forces such as Apple, Microsoft, and Google own sweeping patents and employ fleets of lawyers to litigate against potential rivals and armies of lobbyists have helped to create formidable barriers to new startups. Patent law provides protection for 20 years to inventions that are new and useful, as decided by the Patent and Trademark Office.

The mere size of the giants in the communication field is enough to discourage young entrepreneurs from entering the marketplace. By making small improvements to existing systems and software warranting new patents, they can effectively make their intellectual property practically permanent. Those same companies flood the system with so many applications that a lone inventor will have to wait many years before a patent number is granted.

14 AMAZON SHIFTED JOBS AND MARKETS

By 2023, Amazon recorded an outstanding $574.8 billion in annual sales accounting for more than 40% of all online shopping in the United States.

See Table 4.2 for Amazon's sales trajectory over the past six years.

An online bookshop created in 1995, Amazon has since grown into one of the most powerful economic engines on earth, selling everything from airplane parts to dresses and has produced award-winning movies making the owner the second-richest person on earth with a net worth

TABLE 4.2
Estimated worldwide Amazon gross merchandise volume in Billions of US dollars, 2018–23

Year	First-party Sales (Amazon Retail)	Third-party Sales (Amazon Marketplace)	Total Combined
2023	$220B	$480B	$700B
2022	$210B	$430B	$640B
2021	$210B	$390B	$600B
2020	$190B	$300B	$490B
2019	$135B	$200B	$335B
2018	$117B	$160B	$277B

Source: Marketplace Pulse.

estimated to be nearly $200 billion. Twenty years after its initial public offering, Amazon.com is by far the largest retailer in the world, doubling the size of the second biggest retailer. It is also the fifth-most-valuable company in the United States.

Amazon employed roughly 1.6 million full-and part-time employees in 2024 and has captured 50 cents of every dollar spent online and is used for half the online shopping searches undertaken by the nation's consumers. With its intention of adding self-driving trucks, drone delivery, and automated grocery stores, the total number of lost jobs directly related to online sales controlled by the company could conceivably kill more American jobs than outsourcing jobs to China has.

Investors who purchased $100 of Amazon stock at the start, the value of their holding in 2024 would be nearly $9,500. When Google went public in 2004, a $100 investment today would be worth $6,539. Facebook had its IPO in 2012 and a $100 stock purchase would now be $683.

Section II

Motion Detection to Surveillance

The history of motion detection can be traced back well over 150 years to the period when Edwin Holmes received the first security system patent that relied on simple mechanical devices. Such motion protection systems were devices generally used to protect large open areas, door entrances, and roof penetrations and were limited to photoelectric light beams, foil, or glass break sensors on windows, magnetic switches on door frames, and a taut wire strung across the doors and windows and below the ceiling. It wasn't until the early 1970s that marked technological advances emerged. Modern-day volumetric motion-sensing devices – like passive infrared (PIR), ultrasonic, or microwave sensors – were first introduced to the security marketplace sequentially, with one technology completely replacing its predecessor.

SENSORS
Prior to 1970

1. **TAUT WIRE**
2. ***VIBRATION SENSORS***
3. ***LIGHT BEAMS (VISIBLE & INFRARED SPECTRUM)***

Mid-1970s to Mid-1980s

4. ***ULTRASONIC DETECTORS***
5. ***AUDIO/SOUND DISCRIMINATORS***

Early 1980s to about 2006

6. ***MICROWAVE DETECTORS***
7. ***DUAL DETECTORS (ULTRASONIC & MICROWAVE)***

DOI: 10.4324/9781003091103-7

After 1990

8. ***PASSIVE INFRARED DETECTORS (PIR)***
9. ***CCTV AND, SUBSEQUENTLY, INTELLIGENT NETWORKED VIDEO MOTION ANALYTICS***

Also covered are topics related to the rapid growth and application of CCTV surveillance motion detection techniques. Descriptions of video smoke detectors and special lenses developed over the last decade are included.

This section was written for the newly employed security practitioner as well as for the experienced professionals and students interested in the history and development of motion detectors. The material has been edited by many contributors who were involved in the introduction of many of the described detectors who appreciate the importance of understanding how these systems evolved and were replaced by advances in the modern security industry that began in 1970.

The observation and commentary are both objective and subjective since the contributors were witnesses to the short life cycle of each detection system from its introduction to the marketplace and its eventual demise due to the rapid advances in motion detection technology.

5 Volumetric Motion Technology – Short Life Cycle

1 A HISTORY OF DETECTION

The idea of detecting someone moving in a confined or designated area goes back well over 150 years to the period when Edwin Holmes received the first security system patent. Very few security professionals are familiar with any of those simple mechanical devices that preceded the introduction of the modern-day volumetric motion sensing devices like passive infrared (PIR), ultrasonic, or microwave sensors. The original area-or-motion protection systems were those crude mechanical devices generally used to protect large open areas, door entrances, and roof penetrations and were limited to photoelectric light beams, foil on the windows, magnetic switches on the door frame, and a taut wire strung across the doors and windows and below the ceiling.

Motion sensors fall into one of two operating technologies:

Active systems: Both the ultrasonic and microwave devices are called "active systems" since they use an emitter and a receiver to detect human motion in a confined area.

Passive systems: Motion detectors do not emit or transmit any impulses, but only receive some signal originated by humans. Examples would be the PIR detector, which is the most widely used motion detector, and the sound discriminator which was once very popular but is now considered technically obsolete except in rare circumstances (bank vaults, etc.). The analog or digital CCTV camera has software to detect any changes in the picture (intensity of light in the analog cameras and changes in pixels in the digital format). Audio detection could also be considered a passive system, but it has rarely been used in the 21st century.

Before the modern-day security industry began to apply new motion detection technology, most alarm devices were mechanical rather than electronic. Those early alarm systems were activated by a contact switch on the door frame when a beam of light was interrupted by a clumsy intruder, or when a window was broken or a taut wire was tripped. Very few technical advances occurred until a patent was issued for a volumetric (a room or area completely filled with the signal emitted by the sensor) motion detector in the early 1950s. That patent revolutionized the sensing of movement and was called an ultrasonic motion sensor. This motion sensor was based on the scientific phenomenon that a sound wave or echo can be measured whenever it is disturbed or altered by a moving object or figure within its short range. It soon became the standard of an evolving motion detection industry and is credited with being the electronic intrusion device that was responsible for rendering mechanical devices obsolete.

Motion detection technology was a revolutionary new technology that started in the mid-1970s with the introduction of audio detectors and ultrasonic sensors. However, individual systems had short life cycles with sensors becoming obsolete within five to ten years as newer technology was introduced.

2 INTERIOR INTRUSION SENSORS

The short life cycle has become a common occurrence in an industry where each new piece of equipment captures the imagination of the security community only to abruptly disappear as soon

DOI: 10.4324/9781003091103-8

as the next generation of systems is developed. This cycle has been more pronounced since the introduction of digital-based systems, upgraded software, and communication equipment.

2.1 Ultrasonic Sound Wave – Volumetric Detector

Sound waves are the physical phenomena that define the first commercial motion detector.

Sam Bagno, who was awarded a patent in 1952 for the ultrasonic alarm, invented a detector that changed the entire science of interior protection. This ultrasonic alarm was a transducer-type device that detected a change in the pitch of a traveling sound wave as it bounced off a moving body. This change in the pitch was a principle known as the Doppler Effect. This technology was also described as volumetric detection since it filled an entire room with sound.

The ultrasonic detector relied on two components: a unit to transmit a sound wave (a high-frequency speaker) and another to detect the frequency change of the returning sound. Both components (transducers) were housed in the same enclosure. If the room area was empty with no moving bodies, the frequency of the transmitted sound wave was exactly the same as the returning signal, with no motion detector activation. The first detectors were too sensitive, detecting anything that moved (intruders, curtains that moved with a breeze, etc.). Eventually, the problem was solved by adjusting the detector's sensitivity and by adding features that measured net movement in a room (curtains blowing out and returning to their original position created no net movement, while a burglar moving around the room did). Figure 5.1 illustrates various forms of sensors that have been used over the years for various interior settings.

Ultrasonic devices were usually mounted near the ceiling and would send out sound waves much like a radio speaker, but at frequencies higher than the frequency of sound. The ultrasonic unit was made to respond to changes in these waves caused by movement within a protected area the size of a classroom. Basically, a sound wave consists of air vibrations of varying pressure that move through space. Without air, sound cannot be transmitted from one point to another (only radio waves or light or microwave signals can be transmitted in a vacuum).

Ultrasonic detectors are considered an "active sensor" since they require a transmitter and a receiver device to cover the protected area. PIR detectors are "passive detectors" because they only receive signals – no transmission unit is involved.

The Doppler Effect is best described using a classic example of a person standing near the tracks watching a speeding locomotive. As the locomotive approaches, the person will hear the locomotive's whistle continue to rise in pitch or frequency. As the locomotive passes, the whistle will continue to wane to a lower pitch, eventually disappearing with the locomotive. In contrast, any passengers in the locomotive will hear a static pitch. The person standing near the track was experiencing a Doppler shift. As the locomotive approached, its forward movement compressed the sound waves producing a higher frequency or pitch. The faster an object or person moves toward a sound source, the greater and higher the pitch change (Figure 5.2). Reversing the direction decreases the perceived frequency. This pitch change is the basis of ultrasonic and sonic detection systems.

Pitch or tone is measured in waves per second. The higher the frequency, the higher the pitch. If only one wave travels through the air every second, it has a frequency of 1 Hertz (HZ). Audible sound can be defined as the range of frequencies between 30 Hz and approximately 18,000 Hz that can be heard by most humans. Any sound below or above this range of frequencies cannot be interpreted by the human ear. Sounds above this range of human hearing are called "ultrasound" or "ultrasonic." Hearing does not stop abruptly at 18,000 HZ, particularly if the sound is strong enough and the listener has good hearing. Frequencies up to 22,000 HZ can be heard by some people. Animals are very sensitive to extremely high frequencies, particularly dogs who respond to high-pitched whistles out of the human hearing range.

In the mid-to-late 1980s, ultrasonic detectors enjoyed a phenomenal growth, capturing 90% of the motion detection market. At security trade shows, there might be a dozen ultrasonic manufacturers

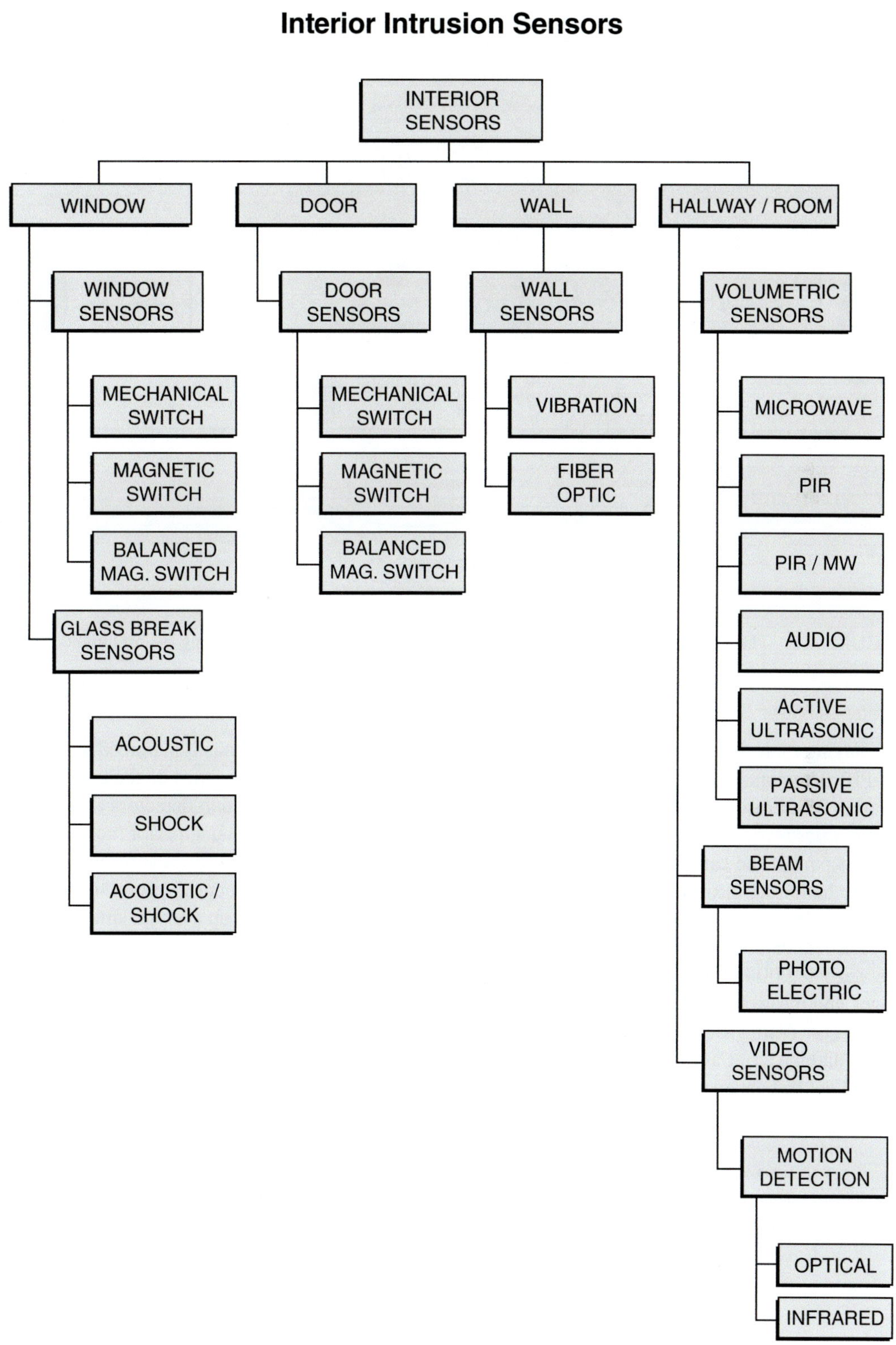

FIGURE 5.1 Interior sensor devices that have historically been used for various interior settings.

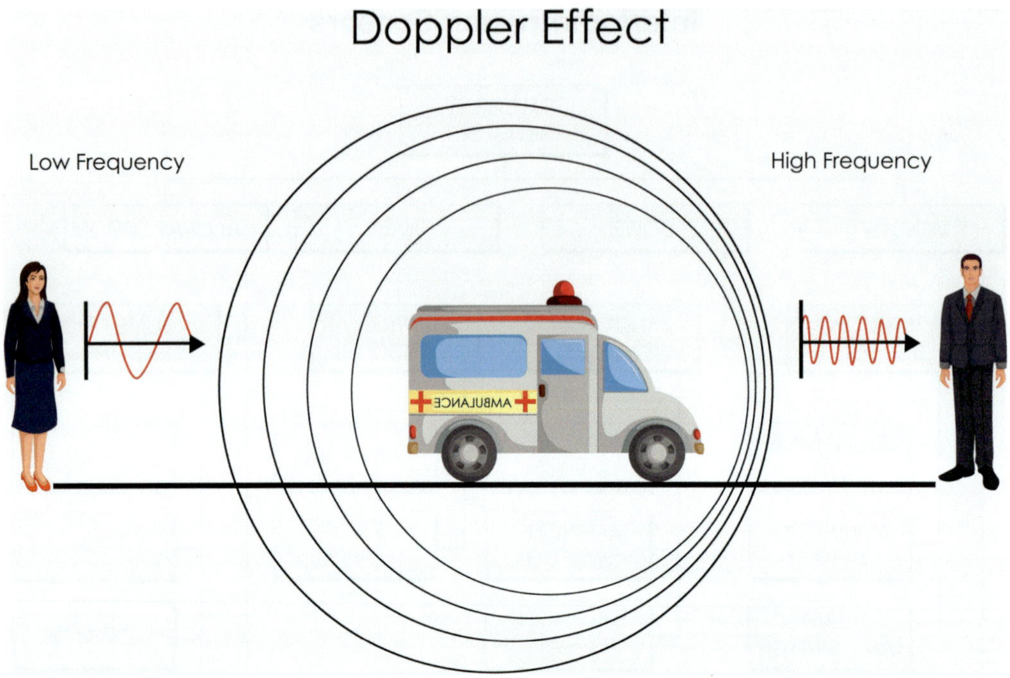

FIGURE 5.2 The Doppler Effect, and action of the sound waves, seen here using an ambulance as an example.

Source: Shutterstock.com.

displaying their wares but, by the early 1990s, those manufacturers were mostly replaced by makers of PIR detectors. These were far superior to ultrasonic devices and, by the turn of the century, the market for ultrasonic units all but disappeared. In fact, wall-mounted motion detectors (PIR, ultrasonic, microwave) were replaced by digital CCTV systems which offered far superior motion detection as part of the camera's basic circuitry. The description of ultrasonic devices is offered here as purely historical information and should no longer be of interest to any security professional. Still, it is a good example of how security technology changes every five years on average since 1980.

The standard ultrasonic sensor operated at frequencies slightly above human hearing; somewhere between 19,000 HZ and 21,000 HZ. There was no single "perfect" frequency since there are advantages and disadvantages to any given frequency. The final choice balances many technical considerations concerning sound wave propagation within an enclosed area. The industry tried to stay as close to the audible range as possible, but high enough so that ultrasound did not penetrate room barriers and would not be tripped off by common noises like hissing radiators or ringing telephones.

If someone was asleep in a room next to an apartment that had its stereo adjusted to maximum volume, the following would occur: the sleeper would be awakened by the sound of the drums and the low notes of the guitar but would have difficulty hearing the high "brassy" sounds of the horns. Those high notes would remain confined to the inside walls of the apartment where the sound source resided.

The most used frequency in the ultrasonic device was slightly above the hearing range of a burglar, so there would be no indication that an alarm system had been installed on the premises. However, the frequency was always lower than 50,000 Hz, since at the higher ranges the sound would easily be absorbed by the air (any absorption significantly reduces the distance a detector can cover). At the same time, these higher frequencies offered certain advantages since barriers (walls, ceilings, floors, and windows) cannot vibrate at higher frequencies. Therefore, more ultrasonic energy is forced back into the protected area at high frequency. Low frequencies within the audible range are not absorbed by the air but can lose some of the overall effectiveness by vibrating the walls and windows. Before the ultrasonic detector left the market, it usually operated at either 20,000 HZ or 26,000 Hz.

2.2 Sonic Detectors

Only one or two manufacturers offered sound-based motion detectors, and they were used almost exclusively in high-rise commercial buildings that housed businesses with costly merchandise including jewelry stores, art galleries, and appliance vendors mostly on upper floors as opposed to the street level. The devices were sold with the concept that the offices were so high up in the building that guard services and police would be slower to respond than at storefronts and single-story businesses.

The noise emitted from the sonic detector was very noticeable in the hallways and could be heard by any would-be thief. The concept never caught on as the average thief did not care if the motion was detected since the response time was long enough to permit the burglary.

The rebirth of ultrasonic principles occurred in 2006 when young students discovered a new tool that shifted the balance of power in the classroom. In school districts where cellular phone use was forbidden in the classroom, ultrasound became a perfect method for signaling the arrival of a text message without being detected by teachers or older individuals in the classroom. The technology was based on the fact that most adults gradually lose the ability to hear high-pitched sounds as they age. The medical term for the so-called aging ear is "presbycusis" and it served as the basis for the development of this simple device, "annoyancetech," by Kids Be Gone.

By transmitting a high-pitched, ear-splitting 17-KHz ringtone that's audible only to those under 25, the device can scatter a group of teens in four minutes, and shopkeepers used it to disperse young people loitering in front of their premises while leaving adults unaffected.

2.3 Microwave Radio Frequencies Detection

Radio frequency (RF) waves flood an area just like the ultrasonic detectors. However, instead of a weak sonic "noise" that doesn't penetrate a wall surface, the high-frequency microwave signals are powerful enough to travel for long distances offering both advantages and disadvantages over other types of motion detectors. Instead of transmitting high-frequency sound waves through a noise amplifier that remains within a small defined area (classroom size), the microwave signals, generated by a "Gunn diode," can transmit RF signals in all directions. It operates much like the radar systems used to detect the movement of aircraft flying miles above the terrain by bouncing radar signals off a plane's fuselage. In order to focus the microwave energy within a controllable direction, the transmitter and receiver are installed in a "cavity" which emits the high-frequency radio signal capable of penetrating walls and capturing movement and can protect an area as large as a football field. No other detector has ever been developed that is more powerful, but that caused its downfall after a brief period of popularity. These detectors are now primarily used by police officers in the form of radar guns and installed in building entrances to automatically open the doors when someone approaches.

The technological achievement most associated with World War II is, of course, the atomic bomb, but the development of radar involved a similarly intense scientific and engineering effort, which dramatically influenced the outcome of that war. Radar facilitated the Allied defeat of the entire Axis enterprise, aided Britain in their efforts against German bombing raids, and provided an advantage over enemy submarines. Unlike the famous Manhattan Project that ended the war in 1945, the lesser-known "Tizard Mission" was a British research effort that changed the course of World War II by developing a method to detect planes flying many hundreds of miles away and identify the direction in which they were headed. The first attempt was a massive "sound mirror" that picked up the noise of an approaching engine. Unfortunately, the device could not tell if the aircraft was a friend or foe. A more advanced technique was to measure the time that emitted and reflected sound waves took to travel to and return from an object in the sky. This method of radio detection and ranging came to be known as radar. All this work was carried out in 1940, just before America entered the war. Through a joint effort between America and Britain, they were able to convert electrical current into microwaves, and not only helped defeat the Nazis but eventually created the microwave oven and microwave motion detector, all using the same principle developed in 1940.

Powerful signals in the RF range went through walls separating groups of rooms and often penetrated the exterior walls and windows, detecting moving trucks outside the building perimeter. This extreme sensitivity practically destroyed the effectiveness of microwave detectors for the security marketplace. The first long-range microwave detector unit was introduced to the security industry in 1983, and a pilot program allowed it to be tested in a high school that had 400-foot-long hallways. It was the only motion detection device that was capable of covering this long hallway with a single unit, whereas ultrasonic units could not be used in open areas larger than 30¢, the size of an average classroom, since sound waves have limited range and are poorly suited for areas with air movement (e.g., a breeze or heat wave from a radiator). Microwaves' unique ability to penetrate walls and detect movement hundreds of feet away excited the industry but proved to be its undoing. The high school hallway had elevators, and the microwave units actually detected moving cables in the elevator shaft. Another problem at this test location was that the school, Brooklyn Technical High School, was highly competitive with a motivated, bright student body made up entirely of engineering students who found it a challenge to figure out the operating features of the unit. Within a few weeks, all gun diode transmitters contained within the unit were dismantled and dissected by the students.

A more serious setback that practically destroyed the microwave concept was a rumor that "microwaves" would render men sterile if they worked within the area covered by the RF sensor. It was an extension of the same fear that followed the introduction of the microwave oven when it entered the marketplace about a year before the sensor was mounted on the high school wall. While this rumor proved to be scientifically unfounded, it forced microwave oven manufacturers to modify their designs so that heating the food could only be done if the oven door was closed. The rumor, however, destroyed the microwave sensor's marketability, and it never gained a foothold in the security industry.

The microwave motion sensor did not replace the less powerful ultrasonic unit, but both devices shared a relatively short life cycle of fewer than eight years before PIR detectors appeared in the mid-1990s and took over the entire motion detection marketplace. Ultrasonic units have no practical application and are no longer made. Microwave detectors never became a factor in the security market but eventually found a huge market with their ability to capture movement and not be compromised by weather conditions. It has become the sensor of choice for millions of building entrances to automatically open doors, revolving or sliding, for anyone walking into the building.

- **LASER SENSORS** operate much like active motion detectors but are designed to measure distances by relying on a continuous beam of light that scans 190 degrees over a radius of about 100 feet. Though not a motion detector, it can detect an object's size, the flight time of the light beam to and from the object as it is reflected and is highly accurate.

2.4 PIR Radiation Motion Sensors

PIR motion detection systems operate using technology that is far superior to all the other active or passive detectors that were replaced by the mid-1990s. The development cycle of this detector came about through the same process that produced microwave technology since the PIR evolved from a wartime effort by the American government. It was developed by the US Department of Defense laboratories at Ft. Monmouth, NJ, in the late 1960s to detect enemy movement during the Vietnam conflict. It radically differed from all other detectors used in the security field since it did not require an energy-emitting source like those found with the active ultrasonic sensor or Gunn diode microwave transmitter. PIR are receivers only; human body heat is the transmitting source. When the first top-secret PIR units were used by the military, the sensor and cooling system required weighed over 500 pounds and was mounted directly above the open belly of low-flying military aircraft to detect enemy movement 2,000 feet below. It was able to record changes in infrared radiated by humans or any warm object such as running vehicles or campfires (Figure 5.3).

One of the many military advantages over other surveillance devices (cameras, radar, etc.) was that by capturing changes in thermal energy, the system was equally effective at night as during the day since light was not a factor. Conventional cameras captured images on their light-sensitive film while PIR sensors optically focused on the object they sensed change between the human body temperature and the temperature of the surrounding air. It did not capture a photo image, but instead recorded thermal images. The greater the temperature difference between the sensor and the individual's skin temperature, the greater the chance of detecting human presence (even in the absence of motion).

In highly sophisticated military applications, the sensor can be kept at cryogenic (sub-zero) temperatures as low as –300°F. This temperature is obtained by mounting a canister of liquid nitrogen above the small infrared sensor (the size of a thumb) and allowing it to drip onto the sensor every few seconds. It was also capable of forming recognizable images of the detected object or person when the thermal signals were converted to a series of photographic line patterns, a complicated process still used by the military but has no practical commercial application.

In the mid-1960s, the PIR sensor was still being tested in the military laboratory and was dispatched to Vietnam in 1968. During the testing phase, the US Department of Agriculture contacted the Department of Defense to borrow one of the test systems for the Forest Service Jump School

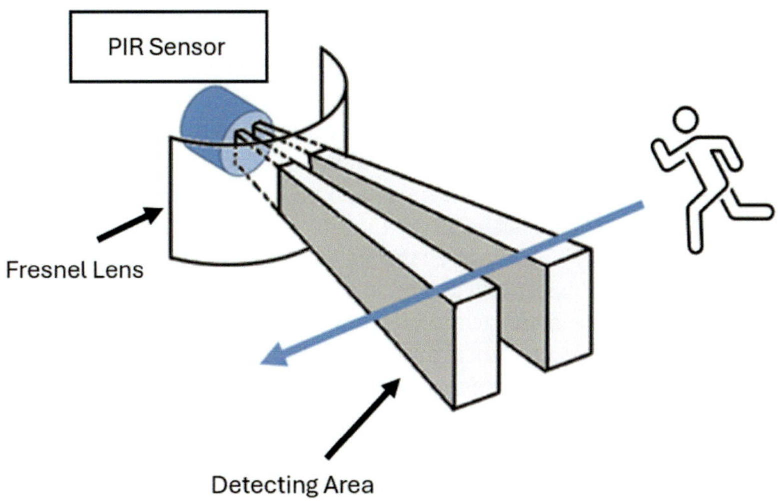

FIGURE 5.3 PIR Motion Sensors. The lens serves to divide the field into separate segments in order to detect motion across them.

in Missoula, MT. The experimental PIR system was installed on a modified small civilian aircraft and flew over mountains inaccessible from roads and susceptible to spontaneous fires that required fire-jumpers to parachute in to contain the fires before they grew out of control. The ideal situation was to spot the fire before it burst into flames – called incipient fires. Incipient means *before they appear.* Many, if not most fires in the Northwest start with lightning strikes that ignite trees, but an initial lighting strike usually causes sparks that slowly burn in tree trunks and the smoldering fire can't be seen by observers in a plane flying at 2,000 feet until these incipient fires become blazing infernos. The PIR sensor was able to detect an unseen hot spot in a tree trunk days before it became a destructive fire. The National Park Service is still using the same PIR technology at their jump schools.

Twenty-five years after PIR technology left the military laboratory, it was introduced to the security industry following a great deal of redesign to produce an infrared sensor capable of operating at room temperatures rather than the low cryogenic temperatures that heat-seeking missiles required. The cost was lowered from half a million dollars for the military version to less than $50 in the civilian market. The PIR motion sensor became the standard in the security industry once it proved to be more reliable and less susceptible to false signaling due to extraneous RF noise interference or false ultrasonic noise waves. The PIR unit can detect human skin temperatures and is effective up to 150 feet, depending on the optics and the difference between the body temperature of the person and the room temperature. In PIR motion detectors, the thermal energy source, skin temperature, is 88° and not 98.8° (body cavity temperature). If the room temperature rises to 88°F, then an individual would probably not be detected though that rarely occurs save in movies.

Converting a government-sponsored research project by refining the basic technology for the commercial marketplace has been a rare occurrence but has become more common in recent years. Previously, the practice in the security industry was to deliver the device to the end user without an intensive research and testing program and rarely was the environment where the system was to be used considered. But that lack of testing had historically been the source of many problems that have been associated with past security systems and innovations that had been untested or undertested prior to market.

Failure #1: A Fortune 500 company attempted to enter the security market in 1978 by taking advantage of its nationally recognized name along with its financial assets and state-of-the-art research laboratories in a West Virginia College town. It was one of the first PIR detectors ever made by a major manufacturer, which had the resources and marketing expertise the smaller motion detector manufacturers lacked. It had all the credentials to succeed except that the development lab was located in a rural area of West Virginia where its prototype devices passed all the important lab tests. While its PIR may have performed flawlessly in the test environment, it was never tested outside of a quiet rural countryside. The device's first exposure to the real world was when a chain of appliance stores in New York City purchased the first production models. While the operating frequency may have worked in the West Virginia wilderness, its operating frequency was identical to the RF used by New York City taxicabs. Every time a message was transmitted over the cab radio it triggered the detector. It was also discovered that although the device housing made for an appealing unit, it proved impractical because once the installer finished connecting the interior wiring, they would be severed when he attempted to close the cabinet. The cabinet won industrial design awards in England despite being a technical disaster for the installer. The company left the security marketplace forever and went back to making battleships.

Failure #2: Decades ago, the oldest, most famous, and largest contract guard service in the nation at the time, Pinkerton, had decided to expand into the electronic security business based on the fact that they already had the perfect client base (over 10,000 facilities for their guard service). They started a new division to develop a microwave motion detector. In attempting to avoid the technical problems faced by other new manufacturers, the unit was tested for over a year before being introduced to the end user. Once it was installed in a major office building in a major East

Coast city, it was discovered that it had a problem similar to that of the PIR detector previously mentioned. The operating RF of the system was not affected by taxi radios, but it operated at the frequencies used by airplane pilots preparing to land at the airport located in the heart of the city. Whenever a pilot contacted the airport tower with a request to land, the frequency would set the microwave detector into an alarm mode, rendering the device useless as a motion detector.

Both of these examples illustrate that the bigger the manufacturer, oftentimes the bigger the disaster. Placing laboratory-tested devices into real situations has always posed a challenge to companies, particularly those which do not understand that the security environment in a major city is not remotely related to a testing laboratory.

Post Script #1: Isolating RF frequency has challenged the security industry for decades. In 1995, a manufacturer of infant abduction systems discovered that waxing machines, computer terminals, and some medical devices operated at frequencies that set off the infant abduction security system. The system had to be removed from several hospitals and a complete redesign had to be funded to solve the problem.

Post Script #2: The above-described technical fiasco occurred in 1985 and similar problems continue to occur whenever newly introduced technology is sold to the public. In 2018, after almost 40 years of enormous advances in digital communication, major electric car manufacturers discovered that the motors that propel Teslas above 150 mph and allow Chevy Volts to go as far as 238 miles on a charge killed the AM radio reception. This unpredicted interference was due to the fact that the motors generate electromagnetic frequencies on the same wavelength as the AM radio signal. Instead of finding a solution to this problem, AM radio has essentially been eliminated by electric cars.

2.5 CCTV Video Motion Detector/Sensor

Detecting light changes on the TV monitor was the technology first used in the early 1980s when an unknown enterprising engineer embedded a photocell, taken from a 35 mm camera light meter, into a rubber suction cup and then stuck it directly onto the surface of a black and white or an early model color TV. Any change in the screen light intensity under the cup, a door opening, a person walking through the door, etc., would be recognized by the makeshift light meter as a change in the screen light intensity and trigger an alarm.

That simple application of changing light intensity to detect motion has advanced significantly since those early days. Digital cameras and software programs can now detect the slightest change in the pixel array on a monitor and then track an individual or vehicle as the object within the captured screen area. Programs can also track objects moving toward a building and ignore those moving away. The security industry has, in 30 years, come a long way from the light beam to IP cameras, digital recorders, and computer networks. The newest technology has rendered most conventional volumetric motion detectors obsolete.

When video motion detectors were first used with digital video recorders, they offered limited storage capacity and relied on high-speed, high-cost, and low-density disks, which could not store much more imagery than tape recorders. Once the technology became available to minimize the amount of space required to store quality digital video, the analog tape recorders began to be replaced. The obvious answer was to restrict the continuous recording that would burn up the disk space by recording on a time schedule or only record when the camera "saw" some unusual movement or change in scenery. Hence, the development of intelligent video. The next advancement was to follow an individual or object through the video scene by highlighting the individual with a colored box that remains on them as they move about the captured area. Object size, speed, and direction can be combined to ensure that a vehicle entering the exit lane will cause an alarm while a bicycle following the same pattern will not. Alerts can also be triggered by gathering crowds and falling, abandoned, or removed objects. Once the software is integrated into the system, the possibilities are almost limitless.

2.6 Audio Discriminator

Noise detectors, or discriminators, operate on an entirely different principle than conventional motion detection sensors. Sound detectors do not actually record or detect movement, rather these systems sense the noise caused by an intruder. The audio discriminator is credited with being the first truly motion-based sensing device, replacing earlier mechanical devices like light beams or wire-traps that were used to capture movement. A few years prior to the introduction of ultrasonic motion technology, an English firm entered the security market with a so-called motion detection system that wasn't a true motion detector. It was a sound detector that became popular for a few years until it was completely replaced when the ultrasonic system appeared on the market in the early 1980s.

Audio discriminators detected movement by simply capturing sounds (burglars, thieves, destructive teenagers, or anyone who made noise as they moved about the premises). It was particularly effective in sensing vandals attempting to vandalize school buildings. Microphones were installed throughout the building and are tuned to "hear" sound frequencies of footsteps or voices, or the throwing of furniture, and breaking equipment and glass windows. The sound was amplified and sent over a phone line to the central station or the police department which could listen in to determine its origin.

Threshold levels assigned to loud noises could be preset to a certain decibel level to trigger an alarm if the threshold frequency was exceeded, and the actual sound could be stored on tape recorders. Once the connection was made, the base station could listen to the sound to determine if it was a false alarm or an actual intrusion. It was the most popular detection system used in bank vaults and for school protection in many large school districts.

The better design systems incorporated sound discriminators that ignored sound frequencies that fell outside the range of breaking glass or the smashing of furniture. Discriminators were not that fine-tuned and would alarm if they sensed outside noise like people yelling in adjacent playgrounds or airplanes flying overhead.

The Newark, NJ, urban school district installed sound discriminators in all of its 100 school buildings in 1980–82, well before other motion detection devices were available. The total cost was less than $200,000 for the entire school district. The cost could have been ten times higher as each school building held two to three hundred classrooms. However, rather than installing a small microphone in each room and wiring it to the principal's office, the engineering firm that designed and installed the system reconfigured the existing public address speakers to function as microphones to detect noise in the classrooms or corridors. Since speakers and microphones operate on the same technical principle, there was no need to rewire the entire building.

Additional microphones were installed on the roofs to detect the sound of aircraft flying over the buildings prior to landing at nearby Newark Airport. The rooftop microphones detected engine sounds before they penetrated into the classrooms, at which point the system would be shut down for two minutes to allow the plane to pass over. Persistent false alarms from other extraneous noises were never solved, and although the sound discriminator was a successful solution to stop vandalism, it eventually went the way of tape recorders, vacuum tubes, black-and-white TVs, and film cameras. The sound detection system had a relatively short life cycle of about seven years before being replaced by the newly introduced ultrasonic, microwave, and finally PIR motion detectors.

Application #1: Sound detection came out of its unique niche in the banking industry. It was installed in quiet bank vaults in the 1950s and 1960s and is still found in some banks. It was the only practical detector system that would sense a potential attack even before anyone entered the vault. The system did not rely on detecting the actual movement of an intruder inside an enclosed area to be activated. This meant that the thief could be detected before setting foot in the vault because the noise detectors could sense the drilling through the vault door or the breaking up of the concrete under the vault, two events that occurred before anyone entered the vault. It was a highly effective detection system in this particular application.

Application #2: An entire industry was created around the sound discrimination system and franchising was used for the first time in the security industry to promote this simple technology. The parent company would sell exclusive locations and advertise the system throughout the nation. Hundreds of outlets were sold and for several years they were fairly successful, until more reliable monitoring techniques were developed. By placing microphones on the subscribers' premises, the monitoring service claimed that not only would they be able to respond to an alarm, but they could also listen in to the activity that caused the alarm activation. Supposedly, this would eliminate the false alarms that were plaguing the monitoring industry since sound detection could distinguish between a human voice and normal building environmental noises. Problems still arose when the operator could not be sure what caused the noise and would often treat the incident as an alarm, causing an intolerably high false alarm rate.

It should be noted that detection units and technology units for various illicit activities have always been popular, particularly in schools. In fact, there are currently vape detection technologies available with sensors placed in bathrooms by districts, with the ability to text administrators when such activity is detected!

2.7 Gunshot Acoustic Surveillance (ShotSpotter)

WIDE AREA. Determining the origin of gunfire was a concept conceived prior to World War I on an experimental basis but was not commercially available for use in urban high-crime areas by police departments until the early 1990s. The city of Palo Alto, California, was besieged with violent crime related to gun trafficking and in 1992 there were 42 homicides in that city, making it the per capita murder capital of the United States. The situation became so desperate that several noted seismologists were approached to establish a method of identifying gunfire. This resulted in the development of a system that determined the feasibility of identifying gunfire locations using acoustical or optical sensors. The prototype system was called "ShotSpotter," a commercial application from a manufacturer that had produced similar systems for the military.

Acoustic sensors have proven to be far superior to the **optical flash** detector and sensor technology that requires a line of sight to the discharge point. This older optical technology is not suitable for urban areas where such lines of sight are blocked by high-rise buildings. Optical systems have been successful in military environments where the immediacy of response is critical, and the areas are free of obstacles. Acoustic and optical sensors can be used in combination so that the data can be fused, enabling the gunshot location processing to have a more exact discharge time that can be used to calculate the distance of the discharge to the sensor with great precision. The main drawback of acoustic-only systems is that they typically generate alerts slower than optical sensors that react to the speed of light. Sound detectors rely on the propagation of sound waves moving at 768 miles/hour, which means that a sound reaching a sensor 1 mile from its origin will take almost five seconds. The acoustic sensors basically operate on two primary attributes of gunfire that enable the detection and exact location of the shot:

- A muzzle and optical blast that occurs when an explosive charge is ignited to propel a projectile from the chamber of the weapon. A typical muzzle blast generates an impulse sound wave with a noise level that ranges from 120 db to 160 db (the sound heard when standing next to a jet engine).
- A "snap" or "crack" that occurs as the projectile moves through the air at supersonic speeds.

ShotSpotter. Seconds matter, particularly when gunfire is involved, and a few seconds to accommodate pickup from distance sensors and discern the number of rounds fired (an indicator of incident severity) are both tolerable and an improvement over typical police dispatch responses. Without the gunfire tracker sensor, there was a lengthy delay of several minutes between the discharge of a weapon and the 911 phone call, and many times there was no 911 call. In cities without the gunfire detection system, a patrol officer would receive a gunshot call from the community, and it could take over 30 minutes of driving around within three or four blocks of the location to make sure there wasn't a victim in need of assistance.

How it works: When a gun goes off, three audio detectors will triangulate to pinpoint where the shots were fired within roughly 80 feet, sending key data to monitors located within the police department's jurisdiction or at a remote monitoring console. The sensors are so sensitive that it could conceivably detect conversations within the coverage area. Critics complain that the system could constitute an invasion of privacy, but for the most part, the complaints are ignored as an overreaction to another security surveillance system (See Figure 5.4).

Following experimental tests, the highly touted ShotSpotter system was installed in about a dozen cities in 2009 and by 2015 there were over 90 cities online. As of 2021, there were 127 cities and regions worldwide online. The system was promoted by the manufacturer as an effective crime-fighting tool, designed to identify the location of gunshots within a few feet. The initial systems, which cost $400,000 and an additional $40,000 per year to maintain and monitor, had been purchased and installed at police department headquarters. One police department reported that a test installation of a system loaned to them by the FBI was credited with only four arrests.

The slow acceptance of the system forced the manufacturer to offer a leasing arrangement along with 24/7 monitoring, which eliminated many of the budgetary issues.

The subscription plan has an annual fee of $40,000 to $60,000 per square mile that includes 24/7 monitoring by trained reviewers at the manufacturer's site in California. A typical installation requires placing 40 sensors along various walls within a two-square-mile area in high-crime neighborhoods. This setup would enable the system to pinpoint gunfire within 30 feet of where a shot originated. This acoustical detection system that triangulates sound picked up by sensors laced on building utility poles and other structures and has transformed how police officers do their jobs.

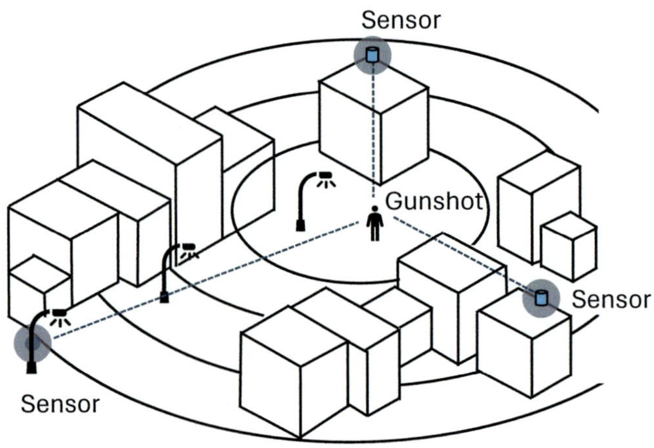

FIGURE 5.4 Sensors triangulate the gunfire. Analysts rule out false positives and notify police if it is determined to be actual gunfire.

Typical Scenario:

- 7:22.7 pm. – An alarm goes off in the control room of a suburban office building in California in 2012. A technician immediately focused on the computer screen, where the words "multiple gunshots" flashed across the monitor. The operator listened to a recording of the shots, tat-tat-tat-tat of five rounds from a small caliber weapon, then zoomed in on a satellite map to view where the gun had been fired. Milwaukee, 2,200 miles away.
- 7:23.5 pm. – The technician, satisfied that the sounds were gunfire, sends an alert to the Milwaukee Police Department.
- 7:5.1 pm. – A tactical team arrives at the Milwaukee location to find five.22 caliber shell casings and a bleeding 15-year-old boy who had been shot in the arm.
- Total elapsed time: 2 minutes, 55 seconds.

Many of the glitches that appeared when the first acoustic sensors were introduced occurred because the system worked too well, picking up noises such as firecrackers and backfiring cars and registering those sounds as gunfire. The sensors were also triggered by noise from helicopter rotors, the roar of downshifting trucks from nearby Interstate highways, doors slamming, birds chirping, cars on the highway, horns honking, and even arguments on the street. Listening to conversations between unsuspecting people on a street is considered by many in the criminal justice system to be an invasion of privacy, particularly when used as evidence in a criminal case. Defendants contend that the conversation falls under the state's wiretapping statures and can't be used against them in court.

Prosecutors respond to this argument by claiming that what is picked up by the acoustical gunshot recording was said in public without the expectation of privacy and can, therefore, be permitted as evidence in a criminal case. If the police are allowed to listen in on people talking on the street, then the issue becomes one of identifying the limit of this practice. It seems that the use of acoustical gunshot spotters may have helped police solve criminal activity, but it has also opened up a new can of worms. A *Boston Globe* article by Emily Spatz from May 2024 entitled "What is ShotSpotter? Controversial gunshot technology facing increased scrutiny" outlines some of the debates over the technology's use and implementation.

The Science and Technology (S&T) Directorate within the Department of Homeland Security (DHS) has worked closely with its industry partner Shooter Detection Systems (SDS), who have developed SDS Outdoor, as another gunshot detection product. S&T and first responders assessed the technology at Joint Base McGuire-Dix-Lakehurst in New Jersey in November 2022 (Figure 5.5).

3 OTHER SOUND EMITTING DEVICES

In recent years, sound detection systems have been developed that are based on the old principles of glass breaking, the ShotSpotter, drones, the sound of graffiti being sprayed, and more. Sound detection is normally accomplished by identifying a unique sound algorithm or pattern that breaks up the noise through analog-to-digital conversion technology. Once the sound is captured in digital form, the system is programmed to identify the particular sound it was designed to catch.

Glass Break: This device typically listens for a combination of low-frequency or infrasonic sounds emitted by the flexing of glass and higher, crisp sounds from the actual breaking of the glass.

Drone Detection: Helicopter-type drones emit unique sound patterns that can be detected. A device has been developed (droneshield.org) that will listen for drones, scan its database for common drone acoustic signatures, and ignore common sounds such as lawn mowers and leaf blowers. It is being used to detect guns being dropped into prison courtyards, illegal aerial video recordings in public and private venues, and for drones straying in airport space.

Graffiti Detection: The United States spends more than $25 billion annually on graffiti abatement. Potter Electric Signal, a company long known for its sound detection technology, produced a

FIGURE 5.5 Mounting the SDS Outdoor gunshot detection technology during S&T's field assessment.

Source: DHS S&T.

system called the Merlin Graffiti Detector. When placed high on exterior walls, the Merlin Graffiti Detector is able to detect aerosol can activity up to 25 feet away in a 120-foot detection zone.

Scream Detection: Scream Detection systems are designed to detect high-pitched vocalized sounds in the absence of other speech. There are such systems on the market that will automatically place a call to the security desk if a scream or shout is detected (jeron.com). These systems were designed mainly for large parking garages where panic buttons and call boxes were generally used to alert authorities to potentially dangerous situations. This device offers a faster response and eliminates the need for potential victims to attempt to find a call box in dark and/or poorly lit areas.

6 Home Entertainment B&W TV to CCTV Surveillance

1 THE TELEVISION BEGINNING

Although flat screen TVs are considered an amazing upgrade to those who remember the bulky, two-foot-thick heavy consoles of yesteryear, it was not as revolutionary as the ubiquitous remote-controlled device. Before the remote control, switching between channels meant that viewers had to get up from their chair and, as unbelievable as it sounds, walk across the room to manually turn a knob on the TV set. The remote control allowed TV viewers to change channels from a distance. In 1950, the Zenith television company released the first TV remote control called Lazy Bones that had one great drawback. The remote control had a long cable connecting it to the TV which many users tripped over. The first wireless TV remote was initially considered a luxury with only five or six channels to go through. With the introduction of hundreds of cable channels, the remote control became somewhat of a necessity, a fact that anyone who loses a remote can testify to.

Encouraged by the enthusiastic reception the remote-control concept received, in 1955, Zenith delivered the Flash-Matic. Each Flash-Matic set came with a gun-shaped contraption with a red trigger that emitted a visible ray of light which was pointed at a compatible TV set's channel dial. There was no longer a need for a connecting cable. Zenith promised, *Without budging your easy chair, you can turn the new TV set on or off or change channels.* Until this point, the public had long considered the toilet flush mechanism the most civilized invention ever devised. The remote control was treated as the next most important invention.

The Flash-Matic units required a photocell embedded in each corner of the screen; the viewer activated the cell by shooting it with the remote beam. One cell changed the channel up, another changed the channel down, a third muted the sound, and the fourth turned the set on and off. The sensational remote control helped sell 30,000 TV sets and the employee, who conceived and built the remote-control device, received a thousand-dollar bonus for the invention. However, once installed, several difficulties were discovered. Because the system was light-activated, sunlight hitting the TV screen could cause the channels to change spontaneously. Viewers also had trouble remembering which corner of the screen controlled which function. Within two years, the Flash-Matic was supplanted by a more efficient, more enduring, and far better-selling one.

The next upgrade in remote devices was the improved Space Command that used sound rather than light to change channels. The high-frequency, inaudible sound waves offered more reliability, though they were susceptible to the sound of jangling keys or rattling coins and were brought to the marketplace in 1956. From 1956 to the early 1980s, more than 9 million TV sets controlled by Space Command technology were sold. The light- and sound-generated devices were all replaced by the introduction of infrared (IR) remote controls which have become the standard of the industry.

The remote control changed the way Americans live just as the flush toilet did over a hundred and fifty years ago. No one who has grown up since the 1980s can believe there was once a society that actually got off the chair to change the TV channel. But along with this amazing life-altering device, came a source of numerous marital skirmishes and interfamily feuds ... the ongoing battle for control of the remote control that became an undeniable irritant in many households.

DOI: 10.4324/9781003091103-9

2 CLOSED-CIRCUIT TV – THE INITIAL BLACK & WHITE MONITORS

The introduction of closed-circuit television cameras (CCTV) can be credited with establishing the modern and thriving private security industry. Early black and white surveillance cameras may have produced grainy, out-of-focus images with poor resolution, but it represented a landmark in the evolution of the physical security business. Regardless of those early defects, the CCTV sector of the growing industry was universally accepted as the cure-all for most criminal activities since the culprits could be spotted and identified without the presence of human witnesses or a security guard. The surveillance cameras became an important component of a highly technical industry and did much to put to rest the century-old image of security as represented by the small storefront locksmith shop, which had been the model until about the middle of the 1970s.

By the mid-1970s, a few Japanese and American manufacturers had focused on the early use of closed-circuit security surveillance cameras for the evolving industry. The CCTV security cameras cost less than the consumer home units since they did not need the modulator circuit required to display commercial TV programs nor did they require an antenna or RF transmission for the off-site image monitoring. With CCTV, all pictures from each camera were connected to a TV monitor (one for each camera) through a dedicated direct connect cable by a coax cable (old) or a Cat 5 Internet Protocol (IP) addressable cable (new). The initial units, nine-inch black-and-white cameras and monitors, were relatively inexpensive, not very reliable, and had the same poor resolution as home units at that time. Between 1975 and 1985, there were just as many dummy cameras being sold as real cameras. At that time, the real value was in using a camera, dummy or otherwise, was to use it to deter crime rather than to record events.

Banking institutions were the first to embrace this new technology, installing black-and-white cameras all over their lobbies. Not only did the cameras identify the robbers, but they also became an effective deterrent when the would-be bank robber became nervous after spotting a camera. This deterrent factor encouraged many banks to install dummy cameras (empty housing that created the same effect as the real thing). Eventually, the robbers reacted to this new surveillance weapon by simply spraying paint over the lens or ignoring it entirely since they knew the picture would be so grainy that a positive identification was questionable. From this auspicious beginning, the CCTV system changed the way the public treated security.

Within a few years, the CCTV surveillance industry had grown exponentially and by the turn of the 21st century, CCTV cameras were found in almost every public and commercial building. With the development of higher resolution images and color analog cameras, the CCTV market became a major industry with millions of closed-circuit cameras in use, tracking the average citizen each time they ventured out of their home. This rapid expansion was not limited to the United States since CCTV cameras were even more popular in other countries, e.g., England's famous "ring of steel" in London, New York City (NYC), and the ubiquitous use of cameras throughout China are just a few examples where citizens in urban areas are filmed on both private and public cameras sometimes as much as 300 times a day. Certain transportation networks, like the NYC subway system, are monitoring over 15,000 cameras in their underground facilities scattered over 472 subway stations, through dozens of tunnels and on elevated tracks. The number of street-level cameras covering the exterior of buildings in the United States has quadrupled every five years and some cities, like London, have hundreds of thousands scattered in both public and private locations throughout that city.

1. Life in the 21st century is about the digital revolution, from talking cars to multitasking smartphones to high-definition digital 3D television sets. The surveillance camera industry has shifted very quickly toward IP network cameras due to their improved image quality, amazing flexibility that is far superior to analog cameras, and the ease and lower cost of installation and convenience of accessing and storing data. Intelligent IP cameras can be far more effective than DVR or other analog legacy systems, given their advanced features

such as built-in motion detection and alarm management. The cameras can be designed to decide when to send video, at what frame rate and resolution, and when to alert a specific operator for monitoring and/or response. Some cameras provide license plate recognition of both parked and moving vehicles at a speed of hundreds per minute and compare them to a vehicle database.

2. Many facilities purchase numerous IP cameras to satisfy an unsubstantiated belief that visual coverage of every area within or outside the building will solve all security issues. This boom in surveillance monitoring has gone largely unexamined and ignored by federal, state, and local laws and regulations and has given rise to issues of privacy and appropriateness (e.g., should videotaping people seeking addiction treatment be allowed? Who owns the images, who has access to them, and how long should they be archived?).

Cameras are often unsuitable for particular environments and/or unique situations. Surveillance systems must often be subtle and unobtrusive in order to prevent the creation of a fortress-like environment that would alienate the public. There should be a subtle awareness of security system presence without the public being made to feel as if they were intruders. Some systems are ill-conceived, installed in poor locations, and/or indiscriminately added with little forethought and without consideration of possible future expansion.

Large installations of IP surveillance systems are, in recent years, much further along in the cumbersome process of attempting to efficiently replace analog equipment with newer digital and network-compatible systems. Switching from analog to digital surveillance, and incorporating an IP-based network, has been a mammoth undertaking and the extent to which facilities have proceeded with this transition depends on the size and age of the existing security infrastructure and the owners' level of commitment to modernization. Compounding matters further was the realization that technology changes approximately every four to five years. It took four years to replace black-and-white cameras with color cameras and five years to completely install card access system (before discovering that the system could not add any additional readers to the system). Having said this, the digital revolution has definitely arrived and companies who have developed and sell the technology had and have a vested interest in making the latest products and software affordable and easy to implement and use.

3 INTELLIGENT VIDEO

IP Technology has revolutionized the surveillance industry. Each device in a building network connected to a local area network or the Internet is given an IP number or address which identifies the unique device among all other devices connected to the network. Each device shares the Internet path with the other network devices and can be viewed on a computer monitor located anywhere in the world. Other factors that make digital signal transmission superior to older analog technology include, but are not limited to, the following:

- The digital camera meets the needs of the facility with features such as pan/tilt/zoom, vandal-proof, weather-resistant and fixed dome cameras. Comparison of performance and durability is important consideration often overlooked when using cost factors as the only criterion. The ease of system expansion and unique ability to interface different components of a system top the list of advantages.
- Consideration of storage capacity includes frame rate, total time the video needs to be stored, and, most importantly, the capacity to allow for future expansion.
- Integrating existing analog cameras into the new network configuration is of utmost importance. There are many devices that will digitize the analog signal and transmit it over the network as if it were an IP addressable camera. However, this conversion is rather

costly and there are instances in which the conversion of analog video streams into digital formats results in poor picture quality.

- Wireless connections should be evaluated. Although they may be very cost-effective, they may not be suitable for use in hospital environments where signal interference may render them technically impractical.
- Since most security and surveillance applications contain sensitive information, there should be adequate firewalls and password protection to satisfy concerns relating to IP surveillance systems, particularly when the network may be open to the public.
- Network cameras with built-in motion detection can be programmed with more intelligent algorithms to recognize license plates, count people, etc. Intelligent video makes surveillance systems more reliable and effective than those that do not use motion activation.
- IP cameras are available with built-in memory that allows for some internal storage of video files and image assessment capabilities that maintain surveillance upon network failure.
- Zoom lens optics have become so advanced that the focal length can display the color of an individual's eye from 100 yards or more away. However, the problem with extreme zoom cameras is that they are expensive and offer no real advantage in normal surveillance situations.

Another major advantage offered by IP addressable cameras is that the equipment embraces open standards. Older analog video equipment manufacturers offered only proprietary systems that required the purchase of most, if not all, system components from the same original source. With the flexibility of an IP video system, the end user can integrate disparate systems. Cameras can be purchased from one company, the hardware that records the video can come from a different source and the client software from yet another source.

Most surveillance CCTV systems represent a huge investment if converting from analog (legacy) cameras but there are many options that will allow the existing systems to be carried over network cables instead of coax cable. This is accomplished by converting the output of the analog camera into an effective network camera by installing a video encoder/server. This device digitizes and compresses the video and converts it into a standard network appliance and carries it on an unshielded twisted pair - format. This method utilizes Cat 5e cable to transmit the video information in what is known as a hybrid (digital/analog) system. However, legacy cameras may not offer color or the same high-quality image as newer cameras.

The most limiting aspect of digital video security systems initially was that they had to be designed to function within the bandwidth limitations of available networks. Analog systems may be unencumbered by this limitation, but they require a point-to-point wire connection. For example, 100 analog cameras require 100 coax cables run from each camera to the control center.

IP cameras offer a multitude of benefits and features and more manufacturers are getting into this new technology, but there are some applications for which they may not be the best option. At one time, high-end IP cameras lacked the visual clarity of high-end analog cameras, particularly at zoom position, but that has all changed. Some IP cameras are more advanced than the other components in the surveillance network. Some of these cameras incorporate memory chips so, in the event of a line failure, the camera continues to store the imagery. In situations where there is a need to read license plates on vehicles at distances, it has been found that IP cameras that record with this level of clarity are much more expensive than analog cameras with the same capability. Weatherproof cameras were originally rare in the IP camera industry, because they are more susceptible to environmental conditions, but advances in this area have made that a non-factor these days.

Addressable IP cameras can be added to or moved on a network in much the same way as other peripherals such as printers or PCs. Standard servers can function as network video recorders providing storage capability far beyond that of an analog VCR or even a DVR ever could. And with the advent and rise of cloud computing, the ability to store virtually limitless amounts of data *somewhere*, via such service providers, has likewise increased capabilities.

A typical IP camera offers resolutions of 320 × 240 and 720 × 480 pixels. But, there are cameras that can present a display of up to 4,822 × 3,248, unheard of levels of detail that no analog camera can approach. A realistic IP camera is in the range of 1–5 megapixels and costs slightly more than high-resolution analog units. For surveillance applications, a camera that captures 40 pixels per foot is sufficient resolution to recognize a person or object in most situations. For forensic needs, 80 pixels per foot would be required. For extremely high detail like that needed in facial recognition, the requirement would be in the 120 pixels resolution range.

Automated license plate readers have been rapidly deployed as a source of revenue and speed enforcement across national roads and highways. They automatically capture all license plate numbers that come into view, along with the location, date, and time (Figure 6.1).

3. The resolution of IP cameras is far superior to that found in the most advanced analog cameras, but that is only one factor that makes IP-based video surveillance the technology of the future. There are other significant advantages of IP technology over the analog format, including the higher resolution and many programming features of IP cameras. Analog resolution cannot duplicate the higher resolution that an IP camera is capable of achieving. The time it took for the conversion of analog to IP systems was mainly due to the large number (millions) of analog units in use and the cost to replace such a large number of units.

FIGURE 6.1 New York, NY, United States – April 28, 2024: New York's congestion pricing program uses overhead booms with digital scanners that read license plates and bill drivers entering Manhattan below 60th Street.

Source: www.shutterstock.com.

4 HIGHER MEGAPIXEL VIDEO COMPRESSION

All digital video surveillance systems rely on some type of compression to carry the video signal to the recorder/monitor. Without the proper compression format, the video file would use all the available bandwidth in a relatively short time. Selecting the right compression is vital whether it be JPEG or MPEG or the formats of H.264 which was followed by the introduction of H.265 in 2017.

With the growing demand for higher video resolution and the limited bandwidths available, manufacturers are continuously developing expanded image enhancement compression formats. Soon, H.265 will likely become the most used format as older compression technology is phased out. When H.254 was first introduced in 2003, the most advanced IP digital cameras were VGA, which is at the lower end of the resolution spectrum and were not designed to accept the large file sizes like Ultra-HD and other higher resolution video imagery.

With the introduction of 4K resolution cameras along with the prevalence of five-megapixel video images now on the market, the H.265 will eventually supplant H.264 as the standard for IP video security. The H.265 is more cost-effective to deploy without significant additional infrastructure expenses. Over a decade ago (2003), the move from MJPEG to H.264 did not provide a great advantage particularly when compared to the savings in bandwidth and storage costs of the H.265 protocol.

5 ANALOG CAMERAS: A REBIRTH

In 2008, less than 15% of CCTV cameras were IP addressable but the industry became so captivated by its many advantages that it prepared for the eventual changeover to digital by installing Cat 5 wire for analog cameras in lieu of coax to support the future IP cameras. Some companies replaced DVR units as they failed with hybrid DVRs to support existing analog cameras as well as any newer digital cameras. The process of living with analog while introducing digital equipment was complicated but the trend reflected the belief that analog cameras represented obsolete technology. Yet with all the advantages of IP technology, the early demise of analog did not happen. Analog cameras with enhanced picture quality made a strong comeback after 2013.

In 2015, analog CCTV technology was still the most popular video format, capturing 60–75% of the marketplace, though evolving IP CCTV continues to supplant now antiquated devices. The advent of a technique known as Analog High Definition (AHD) – also described as HD-over-Coax justified continued use of analog cameras for some time. The enhanced analog technology allows surveillance system providers to enhance resolution limits into the megapixel domain. AHD users can rapidly adopt megapixel solutions by adding more cameras to existing installations. Even with the newly developed technology, the IP surveillance network is still the wave of the future, and the application of AHD systems has only slowed the inevitable change in video format.

This has rapidly changed in the last decade as IP networked video has seen widespread adoption. With the increase in remote workers, storage and bandwidth have become better and cheaper. With the rise of cloud computing and network connectivity – for many companies – security cameras will need to be integrated seamlessly with IT infrastructure. As a whole, as of 2023, it was estimated that the global video surveillance market was worth $53.7B US, with a projected value of $83.3B in 2028 (Source: MarketsandMarkets, www.marketsandmarkets.com). The US video surveillance market is currently $12.34B US (as of 2024) and is estimated to top $22B US by 2029 (Source: Mordor Intelligence, www.mordorintelligence.com).

6 INTELLIGENCE IP VIDEO

The use of computer algorithms to process digital video and identify specific object movement is a technology often deployed during some manufacturing processes to identify production defects. There is software available that accepts digitized video feeds that are then processed into separate objects of interest from the picture background. Some systems have embedded the analytic chip directly into cameras and servers. The system must be carefully programmed to identify objects of a specific size, motion characteristic, and type (human vs. vehicle). "Trip-wire" is the term that refers to a program that generates an alert when the specific object crosses a defined line or threshold in the camera image.

Car dealers were among the first to recognize the value of intelligent video to battle nighttime vehicle thefts. With this new video software, any vehicle movement would be detected by a change in imagery within the camera scene. The use of intelligent video will grow since it can replace the presence of a guard to protect a facility as long as the software is properly programmed.

6.1 Video Analytics

Video analytics is the term generally used to describe a form of an intelligent video specifically designed to detect changes in specific scenes within the camera image which activates a signal at the monitoring station.

The best book available on the latest specs for cameras, cabling, switches, storage, video analytics, and technology is the CRC Press/Taylor & Francis publication *Intelligent Network Video, Third Edition* (ISBN: 9781032534718), authored by Fredrik Nilsson and Axis Communications – market leaders in IP video hardware and software systems (Axis first introduced the first network-attached IP video surveillance cameras in 1996).

6.2 Video Smoke Detector

One of the many special-use cameras introduced to the commercial marketplace has been customized with unique software using algorithms to detect things such as "red" clothing and people walking too fast or remaining stationary for too long. The newest application of CCTV cameras is the video smoke detector (VISD), which can detect a plume of smoke rising from a fire before it is detected by an ionization smoke detector or sprinkler head. The conventional smoke detector requires some product of combustion that causes obscuration within the device while the VISD only requires a visual sighting of the fire before it rises to the ceiling-mounted smoke detectors and/or sprinkler heads. The problem is that the software must ignore dust clouds and particles from exhaust gases or cooking gas. The VISD will not replace the current methods of smoke detection but may have some practical use in large open areas such as domed stadiums, exhibit halls, and arenas.

6.3 Thermal Imagery

Upon their first introduction to the marketplace, low-light video cameras operated with uncooled thermal detectors which offered long-range, high-resolution performance but could cost as much as a car. Though effective for border security, industrial facilities, and other high-risk applications, low-light video cameras have limited add-on value for conventional CCTV surveillance systems found in 99% of all facilities. Basically, thermal cameras allow operators to see what their eyes cannot by detecting minute temperature differences between objects and converting the invisible heat radiation emitted into video images regardless of lighting conditions. These cameras see heat, not light, and are effective security tools in any environment, day or night, and in any weather conditions. Conventional analog and digital cameras, as well as the human eye, create images from reflected light which relates directly to the amount of light available. Under nighttime conditions, when there is limited light, the standard camera relies completely on starlight, moonlight, or artificial light. Insufficient light does not produce clear images, particularly if the object doesn't have a lot of contrast with the background.

(a) (b)

FIGURE 6.2 a and b, The Mobotix M16 Thermal Weatherproof thermographic Power over Ethernet (PoE) camera displayed at the in Lighting and Building Fair in Vilnius, Lithuania.

Source: www.shutterstock.com.

- Thermal cameras operate using two completely different operating principles, neither of which suffers from the basic limitations of visible-light imaging cameras. Thermal cameras capture images formed by their heat signature, having no need for reflected light energy used by conventional video imaging systems. Thermal radiation cameras generally create a better image at night than during the day since they do not rely on the amount of light to capture the picture. Capturing images in total darkness is a feature that is still difficult to achieve and different techniques have been tested to produce cameras that detect movement and/or to display imagery under poor light conditions or in total darkness (Figure 6.2).
- Two major technologies have proven relatively successful in capturing identifiable images in low-light situations. One requires an invisible light source from a light-emitting diode (LED) and the other relies on the available natural light. Both have limited applications but offer significant advances over the earlier nighttime systems which experienced short life cycles.

  LED Thermal Cameras use an illuminator that is invisible to the naked eye and was the first so-called thermal camera. These cameras capture a picture in the IR spectrum with the use of the LED that projects near-IR light waves directly onto the target area. The light is invisible to the human eye but allows the camera to capture scenes in the same resolution as found with regular cameras taking pictures under normal light conditions. There is a slight tendency for the images to bloom or for there to be some whiting out of the image. Both defects were common in earlier LED models, but have been mostly eliminated with newer technology. The illuminator can throw the light hundreds of feet but there is a loss of effective illumination as the distance increases. Another problem is that LED bulbs can fail or grow dimmer over a period of use. Some circuits can monitor the LED voltage output and maintain bulbs at a constant brightness throughout the life cycle of the bulb. Unlike old IR lights that burn out in a few years, newer IR LED lights last at least seven years, even if used 24/7.
- Cameras with built-in IR LEDs first came on the market in 2012 and currently compare favorably in price with standard video cameras while offering the low-light advantage.

  Thermal Imaging Cameras, which is the second but more popular thermal imaging device, was an older technology that allows cameras to capture images of objects that give off IR radiation (all objects and humans give off IR radiation). This technology operates under the same principle as the passive IR detector and therefore needs no artificial light source whatsoever. The "heat-source" essentially operates in the IR spectrum and detects thermal changes in the field of view. The human body gives off heat measuring around 100 degrees and thermal imaging cameras can translate the movement of heat as it passes within the field of view. The greater the difference in temperature between the camera and a person, the easier it is to create an image. Some very expensive thermal imaging cameras use cryogenic gas, but the cost is so exorbitant that their use is limited to the military. Commercially available thermal imaging cameras do not rely on cryogenic cooling but that limits the distance they can adequately cover. Thermal imaging cameras do not display clearly defined picture images but rather outlines of the heat source; as such, they are more useful to detect rather than identify. They may have initially been developed for darkness but can be useful in daylight conditions since the technology can detect humans hiding in bushes or shadows and cost a little less than LED thermal imaging systems.

The Continuous Wave Laser was developed to provide light for extra-long distances and relies on an active illuminating system as opposed to a passive system that uses body or building heat as the image source. The continuous wave relies on a laser for the light source and does not pulse like the LED units that operate in a similar way. Some test units can detect movement and capture the image up to 9,000 feet away, far exceeding any of the other nighttime systems. Some applications match the continuous wave with thermal units so that when movement is thermally detected, the continuous wave units can zoom in and focus their lasers on the objects or humans that require closer surveillance. These systems are expensive and probably more complicated than required for most commercial applications. They may be suitable for government use such as border patrol or other applications where long-distance nighttime surveillance is essential. One big shortcoming of this laser system is that the system can cause serious retinal damage if a person looks into the unit within 50 feet of the beams. The other shortcoming is that the first commercial models cost well over $100,000, though, as with all technology, the costs continue to come down.

7 COVERT SECURITY

Digital cameras with small, built-in memory cards have become so compact that they can interface with many common devices and provide surveillance without any indication that the unit is transmitting images. The cameras can be hidden in a wall or enclosed in what appears to be a clock or thermostat, with the most popular disguise appearing to be a motion detector or smoke detector. These covert cameras/detectors are either battery operated or use a low-voltage connection and, in many cases, there is no need for other interconnecting wires with current wireless technology. The memory card can store thousands of images, translating to hundreds of hours of use, since the camera is only activated if motion is detected within the field of view. They can be installed in minutes and removed just as quickly for use in other locations. The imagery can be viewed by transferring the memory card to a computer. The only problem is that many unions and other groups interested in privacy issues claim it is a violation of certain privacy rights. There are many situations in which the unit had to be removed because it appeared to be a smoke detector and the public was lulled into a false sense of security, believing the room was protected with an early warning fire-sensing device. When a room has a device that looks like a smoke detector on the wall, it had better be a real smoke detector, not a surveillance camera.

Cameras are found on street poles, in building lobbies and in every public institution, and now capture life in the nation's inner cities. Proponents argue that cameras deter crime, help catch criminals, and reassure the public in this age of terrorist threats. Privacy advocates are concerned that few safeguards protect citizens from excessive government scrutiny. Little detail is available about who watches our personal activities, how records are stored, and for how long.

TV or digital camera surveillance has been in general use since the early 1980s, but the technology has become much more powerful in the digital age. Previously, video footage would remain in the camera or on a tape on a shelf until someone decided to review it. Most footage was never analyzed at all, but, in this age of the Internet, the images are likely to be available on some website and accessible to a wide variety of public and private viewers. Older technology imposed practical limits on how much footage could be recorded and stored and for how long. Videotapes degrade over time, but digital video can be stored indefinitely, cheaply, and *en masse*, without degradation.

An argument used for complete surveillance is that a camera trained on a park bench records what a police officer would normally observe when patrolling the park. A counter-argument is that a permanent database is different from an officer on patrol because they are not mentally cataloging the associations and activities of each person in the park without probable cause. Gradually, the privacy abuse issue is becoming of less concern in this age of potential terrorist activities and the use of surveillance cameras will likely continue to expand.

Yet with all the criticism leveled at the nation's, and world's, acceptance of surveillance, the industry continues to be the first line of defense against catastrophic events, and sales in the tens of billions in systems and technology will only continue to grow.

8 COVERT TECHNOLOGY

Only in movies do people use special electronic devices and techniques to compromise cameras visible throughout the premises. Bank robbers realize that they are being recorded but don't seem to care since they know where the money is, and they are very aware that there is limited correlation between being caught on camera and being caught by the police. In certain situations, a video system may have made a thief think twice before stealing a wallet or pocketbook, but facilities concerned with security cannot rely solely on surveillance cameras to solve all their security problems. Studies conducted by several criminal justice agencies indicate that only a few criminal activities are effectively stopped with a complex video system, mainly car theft. Some crimes, like petty theft and drug abuse, also drop off significantly if the system is monitored by surveillance cameras but only if the activity is responded to promptly.

Once a surveillance system grows too large, capturing a crime in progress using video images serves no real purpose without a prompt response. In situations where the response is not prompt, the crime rate rises to its previous level within a few months. Murder, terrorism, and crimes of passion do not slow down regardless of the number of cameras capturing a particular event. A person committing a violent criminal act is completely indifferent to the number of images that are recorded. An unfortunate example is the 2005 bombing of a London train station that killed 52 commuters and injured 770 others, all caught on cameras scattered throughout the station terminal.

Since 9/11, there seems to be less public outcry regarding closed-circuit camera surveillance and privacy. While the obsession with video surveillance can be considered overkill in certain situations, it was proven to be very effective when the trains in London were attacked again a few weeks after the first bombs exploded. Suspects were seen entering the stations, attempting to detonate backpack bombs, fleeing when they failed, running from the station, being chased by members of the public, and running into nearby housing projects. Critics who claim the cameras represent an invasion of privacy have been silenced to a large degree since there is irrefutable evidence that major urban centers throughout the industrial world must consider the use of surveillance cameras just as important as any other building element.

- The crime, no matter how minor, must evoke some type of punishment. A person caught performing an illegal act must be afraid of the consequences of their actions.

- To instill a high expectation of being caught by a video surveillance system, the illegal act should provoke a swift, if not immediate, reaction.
- If the punishment is not quick, the purpose behind it is quickly forgotten. This is not helped by a slow-moving court system.

Video systems work best in controlled environments, such as where the same people report to work day in and day out. In that situation, a camera may pose as a deterrent to certain offenses because everyone is generally known to each other and by the security guard at the central monitoring station. However, in the average hospital, only 20% of the occupants at any one time are recognizable staff members. In this environment, with thousands of visitors moving freely throughout premises, the likelihood of theft is higher, with or without cameras. The familiarity of the population being observed is important because a visitor knows they will be miles away before anyone can respond to the misdeed.

9 LENS APPLICABILITY

Computer chips and megapixels are the backbones of digital photography and video surveillance, but it is the lens that separates cameras from one another. The **FISHEYE Lens**, an ultra-wide-angle lens that captures "curved panoramic" or "hemispherical" images, first appeared in security applications in 2005. Using sophisticated software, the distorted circular image can be reprogrammed to display the same images that would normally require four separate fixed cameras mounted on a wall at 90-degree separation. Figure 6.3 was captured from a single fisheye camera mounted on the ceiling in the middle of a facility. It can easily display the four sides of the room on separate screens by "de-warping," without any distortion of the images.

FIGURE 6.3 Fisheye lens cameras can be useful for certain security applications.

Section III

Caught on Surveillance

Jay Leno, a famous American comedian, suggested in a 2006 routine:

> Evolving countries seeking a democratic governing body like Iraq, are having trouble drafting a constitution. Why not give them ours which was written by a lot of smart guys, and it worked well for over 200 years – and we're not using it anymore.

The use of collective intelligence can easily monitor every movement and even the moods of individuals. It can be both a boon for security and forensic investigation, and a threat to privacy and civil liberties. The following chapters cover how surveillance cameras have become a way of life as they intrude on our privacy with little challenge compared to the pre-9/11 era.

1 VIDEO IMAGES

- Video data collections and their effect on quality of life> People who claim they have nothing to hide should realize that there is nothing that they can hide
- London, England has, what is estimated as, nearly 1 million cameras in the city – or roughly one for every ten people
- China collectively has millions of cameras in several major cities, with estimates that the country has over 700 million in the country (or nearly one for every two people)
- Cameras on buildings capture criminal suspects (they caught the Boston Marathon bombers)
- The typical US citizen is recorded an estimated average of 250 times a week, those in cities perhaps as much as 100 times per day
- Robocops with cameras attached to glasses

2 NATIONS UNDER SURVEILLANCE

- Traffic light cameras and adverse public response
- Police officers frequently have dash cams and body cameras
- Retail store cameras
- Schools with CCTV cameras
- Airport cameras

DOI: 10.4324/9781003091103-10

7 Video Images Worth a Thousand Arrests

1 BIG BROTHER IS WATCHING

Advanced CCTV surveillance systems act mainly as a deterrent and are designed to catch the bad guys. They've proven highly successful at monitoring the good, the bad, and the stupid. The bad guy simply wears a hat and sunglasses when they want to avoid the camera. But that doesn't suggest that closed-circuit cameras serve no practical purpose. There are many uses for this surveillance technology, such as notifying a guard that a suspicious package was left unattended, identifying a vehicle driving too quickly toward a building, or someone attempting building entry without pausing to identify themselves. Intelligent video is not ideally suited for every environment, but, much like its predecessors, it is increasingly becoming an affordable feature included in numerous cameras and network configurations.

The availability of these technical systems has raised concerns regarding privacy considerations. Administrators must balance potentially invading someone's privacy with what may result in only a slight improvement in the level of security. Simply put, CCTV surveillance systems provide living records and have limited deterrent effect, but they have been the genesis of a surveillance society. The concept of anonymity enjoyed by prior generations has all but disappeared. The need to protect Americans from the threat of terrorism has been used to justify the intrusion into their private lives but the debate over how far the government can and should go has grown in intensity.

A climate in which society is in a state of constant alert can lead to regularly treating others as potential terrorists. The public is inundated with reminders that if you "See Something, Say Something" and messages urging anyone to report suspicious activity to authorities. This ever-escalating skirmish between security agencies and a paranoid public has resulted in countless unfounded reports and false arrests. As was the case of a photographer who was arrested and his camera confiscated after being reported for photographing a government courthouse during a demonstration. The Federal Protective Service (FPS), an agency under the Department of Homeland Security responsible for guarding buildings and properties owned by or leased to the federal government, claimed the arrest was justified because the tourist was in violation of its guidelines which say, in part, that "persons entering in or on federal property may take photographs of building entrances, lobbies, foyers, corridors or auditoriums for news purposes." The guidelines prohibited tourists, or others, from taking pictures of these buildings for non-news-related purposes.

In the case referenced from 2010, the photographer engaged the New York Civil Liberties Union to sue the government for violation of his civil rights through his arrest and illegal seizure of his camera. The court ruled that the photographer had a First Amendment right to use a camera in public spaces without being harassed. The ruling also forced the FPS to inform its officers that the "public has a right to photograph the exterior of federal courthouses from publicly accessible spaces as there are no general regulations prohibiting exterior photography of federal buildings by individuals."

DOI: 10.4324/9781003091103-11

There appears to be a difference of opinion as to where surveillance pictures can be taken by government authorities in public areas and as to the public's right to take still photographs on smartphones or video cameras at any event. For better or worse, free societies are quickly entering the world of total surveillance and any rights to privacy are rapidly disappearing. The public's concern, or perhaps more aptly put will, toward protecting their privacy seemed to dissipate in the aftermath of 9/11. This has enabled the government to share a digital universe that records the public's every move at all times of the day or night.

This constant surveillance includes seemingly mundane things such as items purchased, what books were favored, Internet search history, and chat room participants.

Personal data mining has greatly contributed to the rapid erosion of privacy and is the inevitable result of a society under constant surveillance. Video data collected by government agencies isn't always used to help track down criminals. There are privacy advocates who warn that some of our liberties are being compromised without justification. Most Americans don't seem to be overly concerned about being watched as they believe they have nothing to hide. In actuality, the public should be concerned about being identified on the Internet. The major snoopers these days are not in government agencies but are in the private sector, busy collecting information about every person in America. Databases owned by private companies can process up to 3 billion records a day. They trawl the Internet, research public records and commercial transactions to create up-to-the-minute files on each citizen. The data ranges from name, age, and Social Security Number to how many times someone has been married, how many children they have and their purchases. All of this collected data consolidated in one place can be sold to eager marketers, and frequent data breaches – of even major corporations – mean that such vital personal information is stolen and used by hackers from around the world.

Incidentally, even seemingly mundane data can be captured and used, in court cases, for example, from vehicle airbags equipped with boxes that record the speed and braking force immediately before and after an airbag is deployed. Another instance of capturing random data points can be in printers – in which printers can embed paper run through them with invisible dots to provide details such as where and when a document was printed.

People who claim they have nothing to hide should realize that there is nothing that they* can *hide!

2 INADVERTENTLY SOLVING CRIMES

Cameras have multiplied on building fronts, light poles, and handheld smartphones, all contributing to new types of crime-fighting opportunities for law enforcement. Traffic cameras, while disliked by so many drivers for causing tickets being issued to red-light runners and speeders, can be useful in helping the police nab criminals:

- Police caught four suspects after they committed a violent crime by reviewing images from a red-light camera near the victims' home. The video showed the suspect's car travelling in the direction of the home and then away from the scene of the crime.
- A college student was dragged to her death when a drive-by purse snatcher caught the student's hands in the strap of her purse strings and the driver was traced through red-light camera footage.
- A drunken driver, blaming a crash on the actions of the driver he had hit and killed, was shown by a red-light camera to be lying.

Gambling resorts are the largest digital CCTV surveillance system users recording the hotel portion of the facility, the restaurants, as well as gambling floor and casino money cages. Images and data are logged, recording transaction amounts, the time each took place and the player's account number. Casino camera systems have been converted from analog to IP digital format that can be integrated with more sophisticated ID programs, including facial recognition that could not previously be recorded or duplicated using analog equipment.

In 2010, the NYC subway system installed CCTV cameras to record movement on every inch of a subway train. The cameras are not monitored in real time, but the footage is stored in a central database and used for more purposes than basic security surveillance. The cameras can assist the Municipal Transportation Authority in legal areas, where it has a history of paying millions in damages to injured passengers who claimed to have had accidents on the train which can only be verified by video. The primary value of CCTV as purely a security device is to have images in real time; but while that technology may be available, in some cases installing or maintaining streaming video on transit systems would be either cost- or bandwidth prohibitive.

Distinct IP addresses are assigned to each digital CCTV camera in order to locate it on the network, like other devices that are plugged into the network (i.e., printers and PCs). The introduction of digital video cameras in the mid-1990s met with resistance from the industry. At that time, the industry was supporting analog cameras and by 2006 it was estimated that only 80,000 IP cameras were in use. Most potential users thought it would take several decades before IP cameras would become a practical alternative to analog and digital devices were too costly. While acceptance was slow at first, both criticisms eventually proved to be inaccurate. The advantages of IP cameras were so significant that sales continued to increase each year with digitally operated cameras continuing to replace the older analog versions. Although analog cameras were less expensive than IP cameras, when the cost of running cables for each camera was balanced against the cost of a single "loop" cable for network cameras, the cost difference was insignificant. While analog cameras are still available, wired and wireless digital IP cameras have been widely adopted and have largely replaced such systems in most markets.

3 BRITISH CCTV SCRUTINY

Britain has become a "surveillance society," a place in which security cameras lurk at every corner. Databases that keep track of intimate personal details of its citizens have provided the government with extraordinary power, intruding into the daily lives of its entire population. The level of surveillance in London and other British cities has been far more intensive than in any other city or country, with the exception of China. The British increased surveillance in response to a series of bombing incidents in their transportation facilities and the identification of a growing number of communities near London that have been the homes of several terrorist cells. The coverage has become so thorough that Britain is now the most monitored society in the world, with 7.3 million surveillance cameras viewing the public-at-large, almost 20% of the world's closed-circuit cameras. There is one camera for every 11 people and the average Londoner is captured on one 900,000 cameras scattered around the city 300 times a day. Those images are not limited to terrorist or subversive activities. People are caught on buses and subways, in stores and office buildings, and are seen littering and committing parking violations. The British government maintains possibly the world's most comprehensive computer database on fingerprints and DNA to supplement their "Big-Brother" surveillance systems.

The CCTV system in London's Ring of Steel which captures license plates is capable of processing the records to up to 35 million vehicles a day. The British are generally tolerant of the cameras and view video surveillance as fact of life, whereas most Americans have privacy concerns which still limit the reach of camera systems. However, while the British may tolerate an unlimited number of cameras tracking their daily movements, they have successfully fought the government's attempt to require every person to carry a national identity card containing biometric information such as fingerprints or face and iris scans. eIDs are coming through and starting to be seen, and rolled out, in Europe.

The CCTV system can track the movement of suspected terrorists, but it can also be misapplied. For example, when used to track a woman suspected of nothing more than falsifying her address to get her daughter into the neighborhood school. Local officials began covert surveillance of the woman by obtaining her telephone billing records then followed her movements on the street with CCTV cameras. This procedure was legal under a regulation known as the Regulation of Investigatory Powers Act (RIPA) of 2000 which relates to surveillance powers. It permitted law enforcement agencies to use surveillance cameras and other means to follow residents secretly and has been used to investigate malfeasances like illegal dumping of industrial waste, loan-sharking, and false welfare benefits claims. The investigation fell apart when it turned out that the family had moved from an old address to a new one outside the school district, but they had met the residency requirements because at the time they were still living in the old house when school applications closed.

1. RIPA also applies to wiretapping and intrusive surveillance carried out by security services. It allows agencies to film with hidden cameras, trawl through communication traffic data such as phone calls and website visits, and enlist undercover agents to pose, for example, as teenagers who want to buy alcohol. It also specifically allows the investigation of reported noise polluters and people who do not pick up after their dogs. Local governments use the regulation to catch people who fail to recycle, put their trash out too early, sell fireworks without a license, and whose dogs bark too early, but not people who close their bars too late.

   The 2012 Summer Olympics was held in the London metropolitan district and the Ring of Steel surveillance system was fully tested and subsequently met and exceeded expectations during the event. The system had also been expanded and integrated with the Ground-to-Air-Missile system to shoot down any potential terrorists commandeering aircraft. The British Ministry of Defense had installed six missile launchers on high vantage points around various Olympic venues in London, including two on the roofs of private houses. It was, of course, never deployed, but the possibility was a little disconcerting and an indication of how far the British government would go to prevent another 9/11 type incident on its soil. The missile systems were part of a much wider layered defense system including radar facilities, Typhoon aircraft, and helicopters with snipers.

Regarding security for the 2012 Olympics, there were several concerns involving the deployment of a deadly missile to bring down a passenger plane, specifically the authority needed to actually shoot the missile. Other concerns revolved around the fact that the missiles were useless in poor weather as they relied on the operator being able to see the target. The missiles were laser-guided, steered toward their target by a soldier needing to have the aircraft in sight.

4 MARATHON IMAGES CAPTURED ON STORE TV

Large urban city retailers and many smaller food chain outlets have closed-circuit TV cameras that pan, tilt, and zoom to capture activities outside the store entrance, moving around to track people and incidents of interest. These all-observing surveillance cameras have captured hundreds of criminal acts and events as horrendous as the mass killing in Tucson (2011) where a camera captured a gunman entering a supermarket and placing plugs in his ears in preparation for his attempt to kill a congresswoman. The two terrorists who set of the bomb at the Boston Marathon in 2013 were clearly visible on Lord & Taylor's outdoor cameras, which helped law enforcement authorities to quickly identify the terrorists as they slowly walked away from the carnage (Figure 7.1).

FIGURE 7.1 The Boston Marathon Bombers caught on camera prior to the attack.

Without the CCTV cameras, the marathon bombers might not have been caught within days of the incident. Since the early 1990s, many other stores' cameras have helped in criminal investigations of incidents including child abductions, robberies, drug deals, and car thefts. Box stores like Costco and Walmart captured video of drive-by purse-snatchings in which victims were dragged and injured. Some stores go even further and have incorporated their own forensic labs where they analyze video and fingerprints related to crimes, often helping law enforcement with cases not necessarily related to the retailer's operation. While American cities may not have as many cameras as London's "Ring of Steel," it is estimated that an average commuter in one of America's more populated urban centers may expect to be captured on at least 150 cameras every day. Some civil liberty advocates are concerned as to this total lack of privacy and law enforcement officials believe the sacrifice is well worth it.

The Boston Marathon bombing was the eighth attack against a major marathon and the first in the United States, according to federal reports. Only two of the eight attacks resulted in the death of people attending a marathon. The two-man marathon attack in Boston in 2013 resulted in the deaths of two adult spectators, an eight-year-old child, one of the two suspects, a police responder, and the injury of an additional 180 spectators. The incident brought a major American city to an absolute standstill as it became a gigantic crime scene. The inner city was closed off to citizens, shops were locked down, streets were barren, trains did not run, and the state's major turnpike was as clear as a bowling lane. People were told to stay indoors, not answer doorbells, and call the police if they saw anything suspicious. All national media outlets, print, and broadcast continued showing the event and the search for the remaining bomber during the week following the attack.

The brothers responsible for the bombings were not skilled in the art of explosives and had simply purchased a copy of Inspire Magazine, published by al Qaeda, that included a chapter titled "Make A Bomb in the Kitchen of Your Mom." The instructions were easy to follow; the explosive material could be purchased at convenience stores and the detonator at an electronics store. The deadly bomb cost about $300 and used a toy's remote control as the detonator; the flammable powder came from dozens of firecrackers with the shrapnel made up mainly of nails.

The brothers who planted the two homemade bombs had attempted to create a national feeling of fear, insecurity, and vulnerability just as the nine terrorists attempted on 9/11. New York, Boston, London, Paris, Mumbai, and other cities that have experienced such attacks have shown bravery in response and resiliency in the face of such extreme acts that have been condemned the world over.

TABLE 7.1
Top 10 cities in China with the most estimated number of public surveillance cameras. Camera numbers are per 1,000 people

City	Number of Cameras
Taiyuan	117
Wuxi	90.5
Changsha	55
Beijing	55
Hangzhou	51
Qingdao	45.5
Kunming	44
Xiamen	39.5
Harbin	33.5
Suzhou	32

Source: Statista.com.

5 THE CHINA SYNDROME

Video surveillance is commonly used in major Western cities, but nowhere else is CCTV on public streets growing as explosively as in China. By the end of 2023, over half of the world's security cameras, an estimated 700 million, were watching streets, hotel lobbies, businesses, buses, major stores, schools, and even mosques and monasteries. To put it into perspective, that is roughly one camera for every two people in China!

The proliferation of surveillance cameras in China is mostly driven by the same rationale as found in Western nations. Police forces are stretched thin, there is rising crime, mushrooming traffic, and a need to protect against terrorism. Although in China, there is an added overriding concern of controlling dissent. Cameras are just one part of the surveillance infrastructure which includes Internet monitoring and censorship, as well as telecommunication and law enforcement databases. The growing use of surveillance technology can expand to any location since privacy safeguards in China are nonexistent, making the state entirely free to mobilize surveillance monitoring not only to fight crime but also for political purposes.

Chronological advances in CCTV could be summarized as follows:

- Advances in B&W resolution
- Motion detectors added to B&W CCTV cameras
- Analog color cameras were introduced in 1974, gradually replacing all B&W units
- VHS tape recorders captured pictures on two-hour tapes
- Disk drives replaced tapes
- Digital IP addressable cameras introduced
- VHS replaced by DVR disk recorders with hard drives to store images
- UBS memory disks record images in the IP camera's body
- Cameras monitored online

6 EYEGLASS COMPUTER SCREENS

Sixty years after the introduction of the remote control and 30 years since the rise of smartphones, the digital revolution has presented the world with wearable headgear that essentially turns an eyeglass lens into a computer screen. The wraparound glasses, which place a single lens above a person's right eye, display information, like maps and emails on the glass. Control of the device is done by voice or gesture and is not science fiction. Warehouse workers in Austria began wearing these futuristic eyeglasses in 2012. The lens gives visual direction through a Wi-Fi connection to the 1.4 million items stored in the warehouse. It also confirms that the worker made the right pick and frees up their hands.

The device was introduced with a retail price of $13,000 but that price is expected to significantly decrease as the number of users multiply and new applications are identified. The rise of smartphones and advances in display systems have accelerated the development of wearable augmented reality (AR) devices that continue to capture the imagination. In 2013, Google briefly entered the market with its Google Glass device which essentially turned an eyepiece lens into a computer screen. While Glass was discontinued soon thereafter in 2015, several others have entered the market hoping to convince skeptical consumers to wear unusual-looking gadgets from glasses that transmit directions so people won't have to look away from the road while driving, to displays that show skiers how fast they are traveling down the slopes. The devices will also be programmable to display email and text messages without the need for manual commands.

Eyeglasses with the built-in "computer vision" can recognize objects and faces using technology to retrieve information about objects displayed on the eyepiece and collected from the Internet. An American visiting another country could wear a translation app to overlay English on foreign-language billboards, street signs, and storefronts that fall within the field of vision. The Air Force has used this technology since 2009 to show weapons information in the helmets of fighter pilots. Some experts predict that AR glasses will be widely available in the coming years with AR market estimated to be worth $97B by 2028.

8 One Nation under Surveillance

1 CAMERAS ON MOTORIST

The number of US lives lost to terrorism and foreign wars since 2001 is fewer than 10,000. That figure represents a great number of what many people consider senseless fatalities, but it pales in comparison to the roughly 200,000 alcohol-related traffic deaths over the same time period. Since the introduction of car travel over a century ago, drunk driving, excessive speeding, and aggressive driving have caused fatalities. The nation seems willing to endure considerable inconveniences to address the potential threat of terrorism (airport security line delays, intrusive online tracking, searches, etc.) but is reluctant to accept the inconvenience of obeying traffic laws to reduce vehicular fatalities. As the POGO cartoon character once stated so prophetically, *We have met the enemy and he is us.*

Belts, Booze, and Speed. Since about 2006, new cars have been built with advanced electronic stability control systems to prevent skids, rearview cameras to prevent fender benders, and an increased number of airbags to protect occupants in collisions. In addition to all these safety features, automobile manufacturers have spent hundreds of millions of dollars on advertising campaigns to inform the public of the dangers of drunk driving, failing to use a seat belt and texting while driving. Yet with all these technical advances, the National Safety Council estimated that more than 42,000 people died in vehicle-related accidents in 2020, an 8% rise from the year before. Experts contend the increased death toll can be attributed to more lenient enforcement of seat belt use, drunk driving, and reluctance by lawmakers to pass more restrictive measures.

Only 18 states have laws requiring seat belts for both front and rear occupants and in those states not wearing a seat belt is considered a primary offense. In 15 states, failure to wear a seat belt in the front seat is a secondary offense. Yet, over half of all traffic fatalities involve unbelted occupants.

Universal use of surveillance cameras in non-crime situations prompted privacy rights advocates to challenge the installation of speed cameras as a violation of the Fourth Amendment. While the home is still sacrosanct, the courts have maintained that a citizen in a public place has no reasonable expectation of privacy. With that in mind, surveillance cameras have become a solution to every conceivable security problem, not just in large public facilities but in most office buildings.

The Department of Homeland Security granted New York City (NYC) over $1 billion beginning in 2006 to install thousands of cameras in the subways and some central monitoring stations in the nation's largest buildings resemble the layout of the tracking system in "Mission Control" but lacked the experienced manpower to adequately monitor these massive arrays of consoles and display panels.

A widely cited study from 2002 that is cited to this day indicated that, after 12 minutes of continuous video monitoring, an operator will often miss up to 45% of scene activity. After 22 minutes of viewing, up to 95% is overlooked (Ainsworth, T.: Buyer Beware. In: Security Oz, 19, pp.1826 (2002)).

DOI: 10.4324/9781003091103-12

In 2007, Virtual Fence brought surveillance into a new area in an attempt to prevent illegal border crossings. In a 28-mile section along the Arizona-Mexico border, nine 98-foot-tall towers were installed and mounted with CCTV cameras to replace metal wire mesh fencing. The test project, known as SBInet, was underwritten by the Department of Homeland Security. The video cameras and radar were designed to zoom in on people up to five miles away and detect automobiles as far away as 15 miles. The information collected was beamed to a command center and fleet of Border Patrol vehicles, alerting federal agents to border crossing attempts. The sensors would sound an alarm if anyone got too close to one of the structures and each tower was equipped with a "hailer horn" that can broadcast warnings at a volume equivalent to that of a jetliner taking off. The system was devised to be a deterrent but the problem-plagued and controversial system was cancelled by then Homeland Security Secretary Janet Napolitano after a huge expenditure in the $3B program.

Some experts in security surveillance not only see the unquestionable value in it but also understand that having an individual's privacy compromised won't help prevent terrorism. However, these same privacy advocates appear to ignore the fact that the original application for video recordings was in banks, where it served as more of a deterrent than a weapon against crime, and no one objected. Surveillance cameras spread rapidly to other sensitive areas where it had a significant effect on lowering the crime rate, particularly in hospitals, schools, and high-crime inner-city areas. Lower Manhattan in NYC employs over 4,000 mobile license-plate readers (LPRs), allowing patrolling officers to scan the license plates of dozens of parked and moving vehicles at a time and in a city that has over 5,400 cars reported stolen citywide each year. In this regard, the LPR system has proven very successful.

2 SPEEDING CAR SURVEILLANCE

The Insurance Institute for Highway Safety (IIHS) reported that in 2023 there were nearly 200 US communities in at least 19 states, plus the District of Columbia, using speed cameras, up from 100 in 2010. Some officials are not pleased with the photo enforcement of traffic violations and the topic remains highly controversial. AAA insists that the cameras are intended to inform motorists about traffic safety and make drivers aware that certain roads have strict enforcement policies.

The first state to deploy dozens of speed cameras on highways statewide was Arizona in mid-2008, and these cameras have since spread to highways in states throughout the United States. Supporters contend that the cameras slow drivers and reduce accidents but opponents argue that they are intrusive and are more about making money than improving safety. Some have shown their distaste for the cameras by covering them with boxes, sticky notes, and, in several cases, infuriated drivers intentionally set the cameras off by driving with covered license plates (and while wearing disguises).

Some simply ignore any ticket they receive knowing that law enforcement must prove it was received. The ticket, in some cities, becomes invalid if a violator is not served in person within three months of the violation. Overtaxed process servers cannot get to most of those people and many citations go unpaid. The companies that install and maintain the cameras work on a percentage of the fines collected, and in many cities, the income falls far short of the break-even point for the vendor.

While there is some question as to the value of these cameras in reducing traffic accidents, there is no question about the profitability of these systems since they are provided free to the municipality. The monitoring service receives either a fixed fee or a percentage of fines collected. In 2006, when there were only 50 cameras installed around the city of New York, 22,000 drivers were caught speeding and paid the $50 fine. In recent years, the number of cameras has increased tenfold, as have the fines collected.

A growing number of municipalities have responded to public concerns regarding payment for the camera monitoring service instead of paying a percentage of the fines collected. As a result, municipalities now pay monitoring companies a fixed monthly fee.

Many of the communities using traffic surveillance systems are having second thoughts. Opponents claim the cameras violate basic constitutional laws since they deny drivers the right to confront their accusers in court. They also argue that the system is simply a ploy by local governments to raise revenue, and the complaint usually includes details as to how monitoring companies are compensated.

The first speeding ticket issued to a motorist in the United States was in 1904. The driver was traveling 12 miles per hour where an 8-mile-per-hour limit was posted. The police officer, one of the nation's first traffic officers, was able to chase the speeding car by running alongside it.

3 RED-LIGHT CAMERAS

According to the IIHS, red-light cameras at street intersections are being used in over 338 communities and 23 states, plus the District of Columbia, and have at least one red-light camera. The cameras are not only used for safety reasons, but also as a highly profitable revenue-producing system, a rarity for security systems. Cameras are strategically placed at major intersections, each taking pictures of vehicle license plates as they pass or exceed the speed limit by more than 11 mph. The cameras are not only highly profitable but also highly controversial. Traditionally as well as legally, only police officers who witnessed traffic infractions have been authorized to write traffic tickets. Another factor that has added to the controversy surrounding the use of red-light cameras is that some studies indicate that while the number of side-impact crashes was reduced significantly, the number of rear-end crashes increased due to sudden braking of cars entering the intersection after spotting the camera. While there are many who consider red-light cameras an infringement of their basic rights, relatively few violators caught on camera have protested in traffic court and those who do almost always lose.

Many excuses are used in traffic court, including the driver being forced to pass during a red light because they were being chased by an ambulance and could not pull over (though no ambulance was seen on camera). These, and other poor excuses, have not proven to be very successful as evidenced by the Division of Motor Vehicles statistics. In cities with red-light cameras, approximately 100,000–300,000 summons are issued each year, on average and less than 5% of driver's appeal and almost 90% are found guilty. Some of the less creative excuses rejected by the court include (1) wife went into labor, (2) cell phone message indicating the babysitter can't find the child, and (3) the car was stolen (used only as a last resort).

RED-LIGHT FACT SHEET

Red-light cameras were installed to deter motorists from committing red-light violations.

Images were taken automatically by a camera connected to a traffic light which is only active when the light is red, not when the light is green or yellow.

Vehicles passing the stop line or entering the crosswalk after the light turns red are detected automatically. The camera takes two photographs, approximately one second apart, and the images show the date and exact time of the violation as well as the number of seconds since the light had turned red. A flash unit allows the camera to operate at night as well as in the day.

When a red-light offense is photographed, the registrant is mailed a Notice of Liability along with one or more images showing the violation. The registrant is liable for a fine but will not receive points on their driver's license. Under the law, the registrant can be held liable even if they were not the driver of the car photographed.

FIGURE 8.1 A red-light camera in Chicago, Illinois.

Source: Used per CC SA 3.0. User: Joe Ravi.

Under current programs, several cities, including NYC and Chicago, rotate the cameras among various intersections that have a history of accidents attributed to vehicles running red lights. The camera rotation has been helpful in large cities like New York which has 12,000 intersections and only 100–200 cameras. However, even in limited use, the cameras have proven far more successful than anticipated. Chicago brings in more than $50 million a year (Figure 8.1).

Some cities admit to having installed 200 dummy cameras as deterrents which are programmed to flash when a vehicle goes through a red light but do not trigger a ticket. The locations of the dummy cameras are known and a list of these alongside the location of real cameras appears on several popular websites.

Researchers found that violations dropped 36% following an extension of the yellow light period giving drivers more warning before a light turns red. This was not welcomed news for the camera installation companies that collect a portion of the revenue from the traffic fines, but it was good news for the elderly population:

- Most healthy adults up to the age of 65 average four feet per second when crossing the street, fast enough to cross the intersection when the red light stays on for a standard 25 seconds.
- Elderly residents manage only three feet per second, according to transportation experts. To improve the quality of life for seniors, major intersections in many cities have retimed their traffic light duration to allow pedestrians 29 seconds to cross, four more than the standard.

4 PUBLIC BACKLASH

Nationwide, red-light cameras have surged in popularity since the first of such cameras was initially used in the early 1990s. According to the IIHS, widespread adoption didn't come until the first decade of this century. The use of traffic cameras grew tenfold in that decade and includes a number of big cities, such as Chicago, New York, Atlanta, and Phoenix. During the ensuing two decades, millions of traffic tickets have been issued from remotely located central viewing stations thousands of miles from the intersections which house the cameras. The rapid growth slowed in 2010 as a growing number of municipalities began deactivating the cameras because they had not demonstrated a material decrease in property damage and bodily injury, in spite of the claims made by the monitoring companies who happen to receive a percentage of the fines paid by the violators.

The number of communities using red-light cameras had fallen to 508 by 2014 and – as referenced earlier – had dipped to 338 as of 2023, according to the IIHS. Nine states – Arkansas, Maine, Mississippi, Montana, Nevada, New Hampshire, South Carolina, West Virginia, and Wisconsin – prohibit red-light camera use, and several more states are considering such prohibitions. Yet, major cities in the United States continue to add more traffic cameras each year. This conflicting policy seems to indicate that cities with populations of less than 100,000 are discarding cameras while larger cities are adding them. Cities removing cameras from intersections cite statistics that cameras cause too many legal hassles as a result of dealing with erroneous tickets and complaints from drivers, many of whom perceive the cameras as invasive. Mostly, however, officials point to studies that claim the cameras do little to reduce accidents and, in some cases, may actually increase them.

Some insurance organizations contend that the cameras save lives while opponents of the lights counter that doing nothing is better than putting up cameras. Arguments on both sides of the controversy are convincing and, at the same time, questionable.

Proponents. Statistics from the National Highway Traffic Safety Administration (NHTSA) indicate that each year there are more than 850 deaths and over 170,000 injuries as a result of accidents caused by drivers running red lights. Statistics also show that there is a dramatic change in driver behavior where red-light cameras are installed.

- Red-light cameras at major intersections have proven to be effective in reducing the number of drivers who run through red lights or jump the green light prematurely with a 96% decrease in violations following installation.
- The IIHS quotes statistics which show that red-light cameras reduced the rate of fatal accidents due to red light running by 24%.

Opponents. The National Motorists Association, a drivers' rights group, contends that cameras increase the number of vehicle crashes in some areas and that other strategies are more effective in making intersections safer.

- Though they may appreciate the safety benefits and added revenue, many people are opposed to the intrusion on their privacy.
- Critics portray the cameras as Big Brother devices that allow cities to generate revenue from traffic offenses so minor that police officers may not have written them up if they had seen them with the naked eye.
- Constitutional concerns have arisen when violators are penalized before they have an opportunity to explain any extenuating circumstances and whether or not they were the person driving in car captured on the TV image.
- A Florida judge ruled that the state's camera law was unconstitutional. Florida law allows drivers to make rolling right turns on red without requiring drivers to come to a full stop, but many are picked up as a violation.

- Cities mail notices to vehicle owners, rather than to the driver, a process that can cause headaches for owners who weren't driving.

While the red-light cameras may appear to be too intrusive to some privacy advocates, the cameras may never be removed.

Given their position on potential abuse of records by police, the American Civil Liberties Union (ACLU) was probably pleased to learn that there are cases in which the camera may negatively impact police officers. A driver of a Mercedes-Benz was pulled over by traffic police for passing three red lights. While searching the vehicle, officers found 13 bogus credit cards and the driver was arrested for criminal possession of a forged instrument. The man was placed in jail for 15 days before posting $3,500 bail. The vehicle was seized and moved to the police pound for the duration of time the owner spent in jail.

After being released, the owner paid the red-light violation fine and retrieved the vehicle. Two weeks later, the owner received a red-light ticket in the mail that included a picture of the vehicle with a time stamp corresponding to a time when the owner was in jail. Someone (either police officer or civilian) had taken the vehicle out for a joy ride. While the owner hired a lawyer who was unable to prove damages, he at least got the police to pay the $200 fine.

5 MOTORIST REVENGE

At one time, motorists were being ticketed not only for being captured on camera for traffic violations but also for warning other drivers about upcoming speed traps and police vehicles. The flashing of headlights by motorists to oncoming traffic has been a common practice since cars appeared on roads in the early part of the last century, warning drivers to slow down. The police have tried to discourage the use of such signals and have resorted to pulling those drivers over and issuing tickets for improper flashing of high beams. A group of nearly 1,000 irate motorists who received these tickets filed 18 class-action lawsuits in Florida in 2011 against highway patrol and other state traffic-enforcement agencies. They sought an injunction barring law enforcement from issuing headlight flashing tickets, plus refunds and civil damages for cited motorists. Some of the lawsuits also contended that the tickets were unconstitutional because they forced drivers to prove their innocence rather than the government being obligated to prove their guilt. The court ruled in favor of the motorists who are now free to flash their headlights whenever they see fit, but by 2014 that ruling only applied to Florida drivers.

Police Defense: The International Union of Police Associations claims the Florida statute is intended to ban flashing lights to prevent people from impersonating a police officer. The flashing light allows a speeding driver time to slow down until they pass the awaiting police officer, after which they may speed up again. It can be argued that the headlight flasher is interfering with legitimate law enforcement activity.

Motorists' Responses: The First Amendment protects all types of non-verbal conduct and protects more than the spoken or printed word. The courts have found that a wide variety of actions, such as honking horns or flashing headlights, are protected forms of communication granted under the First Amendment.

Supreme Court Ruling: In 2014, the ACLU sued a Kansas City suburb for pulling over a driver for flashing their headlights at oncoming traffic to warn of an upcoming speed trap. The driver had violated a city code that bars interfering with radar and other speed-checking devices, but the ACLU alleged that the driver was "engaged in expressive conduct" and that the traffic stop had violated the First Amendment right to free speech and Fourth Amendment protection against unreasonable search and seizure. There were many court cases that involved the same constitutional principles. The dispute finally went to the Supreme Court that, in 2015, ruled in favor of motorists being allowed to give warning signs to incoming traffic.

In the century-old cat-and-mouse game of avoiding police radar, speeding drivers have a potent weapon: relying on each other in an automatic social network. Major radar detector manufacturers developed a feature that automatically shares the location of police, and photos of speed traps and or red-light cameras with all others on the network. And now, of course, many driving app services like Waze and others allow drivers to report police, cars, pulled over, construction, objects in the road, and many other such hazards. The system exemplifies how crowdsourcing, linking people to share mass information, can revolutionize transportation by providing motorists with real-time traffic data reported by fellow travelers.

6 HOT PURSUIT CAN BE FATAL

A police vehicle overtaking a vehicle leaving the scene of a crime or pursuing a stolen vehicle speeding through a red light can be dangerous, not only for the criminal but also for innocent bystanders and the police officers in hot pursuit. More and more municipalities and state law enforcement agencies have discouraged the hot pursuits due to the danger posed to civilians and expensive lawsuits that often result from the crashes. Police cars in pursuit are not permitted to violate traffic or other related laws, particularly when chasing a vehicle that was in violation of a minor infraction. If police officers drive in an unsafe manner, the municipality (or a state agency) is liable for property damages, injuries, or deaths that they cause.

DEADLY CHASES. According to an analysis by the San Francisco Chronicle, at least 3,336 people were killed in police vehicle pursuits from 2017 to 2022 across the United States, and at least 1,377 people died in 2020 and 2021 alone. People have been killed in police chases, with bystanders and anyone in the vehicles being pursued making up almost half of those fatalities. Nearly once a month, a police officer is killed in a high-speed chase. According to the NHTSA, in a typical five-year period, 1 in 100 pursuits ended in a police fatality and 55% of the deaths involved the driver of the pursued car and 42% of that total involved the deaths of about 360 innocent bystanders. Advocates of more restrictive chase policies claim that the number of deaths reported is lower than the actual toll because no mandatory reporting system exists for deaths in pursuits. There are some indications that the toll could actually be four times higher.

Police chases have killed nearly as many people as justifiable police shootings, according to government figures.

An important factor that has forced many of the nation's more than 18,000 police departments to reconsider their pursuit policies is that about 35–40% of all chases end in crashes. Police pursuits for minor traffic offenses are not worth the risk to people's lives and a growing number of law enforcement agencies have adopted a policy whereby a chase is justified only if there is probable cause that violent crime has occurred and not simply reasonable suspicion.

During the past three decades, 42% of police officers killed in vehicle crashes were not wearing seat belts or other safety restraints. A study by the NHTSA, which analyzed 733 crashes from 1980 to 2010, found that fatal traffic accidents were the leading cause of officer deaths for that 30-year period. There has been an average of roughly 50 motor vehicle fatalities per year between 2011 and 2020. Some of this, in past years especially, has been attributed to officers' continued resistance to wearing seat belts. The main excuse is that the belts slow their movement in and out of the vehicle while others complain that the straps get tangled in utility and gun belts.

Estimates by law enforcement agencies indicate that seat belt compliance is around 50% for officers versus the 86% rate of compliance among the general public. Some cities have instituted special training programs to change the culture and educate their police departments about the potential dangers of not wearing a seat belt, including:

- Police crash survivors speak to departments or make video presentations with the underlying message that seat belt usage saves lives.

- They form driving panels to study how to improve driving safety by learning from people in transportation businesses such as the United Parcel Service (UPS), who have developed successful programs.
- Departments encourage officers to report colleagues who don't comply.

A landmark unanimous decision by the U.S. Supreme Court in 2014 related to the actions of the police in a 100-mile hot pursuit chase that ended with police officers shooting 15 bullets into the vehicle, 12 of which hit the driver and passenger, killing both. The officers who had used deadly force to end the high-speed car chase were sued by the families of the deceased but the justices found that the police could not be sued in spite of evidence that the officers were never in immediate danger. The driver's family could not prove that the officers' conduct was unlawful. The case was a sequel to another court decision in which the court ruled against a man who was paralyzed when his car was rammed by a police vehicle during a chase.

Police officers shooting at speeding cars in an attempt to stop suspected criminals are standard fare in movies and TV, but, in real life, more law enforcement agencies are curbing the practice. US police departments are instructing officers not to shoot at vehicles and instead move out of the way whenever possible. Conventional thinking tells us that shooting at a moving vehicle is a bad idea for a variety of reasons, not the least of which is that others may be at risk of being struck by stray bullets. Back in 2014, the Miami Police Department prohibited officers from firing at any moving vehicles unless someone in the vehicle opens fire or displays a weapon. That change was due in large part to an incident that killed a motorist in which police officers fired 116 rounds, injuring four bystanders.

Some municipalities allow pursuits only for violent felonies and others permit chases for the simplest misdemeanors. The federal government has yet to pass legislation to standardize local hot pursuit laws. In the meantime, should pursuits be limited, criminals become aware of the policy and then know that they will not be pursued and will therefore flee as soon as they are confronted by the police. There must be some balance and municipalities will have to trust the judgment of the officer on the scene if criminals are to be caught and in many situations the only way to do that is to chase them down.

7 YELLOW DOT STICKERS SAVE LIVES

Yellow DOT is a free program designed to help first responders provide life-saving medical attention during the first "golden hour" after a crash or other emergency. Started in Connecticut, the program is slowly but steadily gaining interest in other state Transportation Departments and is currently adopted by 22 states. A yellow dot sticker is placed on the driver's side rear window to alert first responders that vital medical information is stored in the glove compartment. It provides a way to communicate with rescuers during the "golden hour" which is the first 60 minutes after a serious crash that can make the difference between life and death for the critically injured.

According to the NHTSA, there are over 40 million people in the United States aged 65 and older which is 13% of the population and by 2030 that percentage will be over 20%. About 32 million of this group have a current driver's license. The Yellow DOT alerts first responders to look for a container in the glove compartment that contains:

Yellow DOT decal
Yellow DOT folder with forms
Senior participant's name
Identifying photo
Emergency contact
Personal physician's name
Medical conditions
Recent surgeries allergies and medication

8 STICKY GLOBAL POSITIONING SYSTEMS

There is no way to catch a speeding vehicle without being involved in a hot pursuit unless the police vehicle is equipped with a technology introduced in 2014 which enables police vehicles to fire sticky projectiles containing small global positioning system (GPS) devices onto the suspects' vehicle. This unusual tagging system marks the target vehicle for tracking rather than chasing. Once the projectile takes hold, the adhesive is formulated to cope with metal, plastic, and even dirt, the tagged car can be tracked by a dispatcher, who can direct an arrest when the situation settles down and the suspect's vehicle is finally parked.

The compressed air launcher mounted behind the police vehicle grille holds two projectiles, each approximately 3.5 inches long, about the size of a small soda can, that are fired at sublethal speed up to a distance equal to two car lengths. Several police departments have installed the projectile system and combined it with automated LPRs. The reader will issue an alert when a stolen vehicle is detected and allow the police to tag the vehicle with the GPS device before they have to make a traffic stop.

9 LICENSE-PLATE READERS

In 2013, a revelation that the U.S. National Security Agency was capturing millions of email and phone records prompted debate about privacy, including the rapidly expanding police use of automatic license-plate readers (ALPRs). According to the Bureau of Justice Statistics (BJS), "93% of police departments in cities with populations of 1 million or more utilize ALPR systems. In cities with populations of 100,000 or more, 75% of police departments use ALPR systems" (Source: Robert Spinks, *What you need to know about automatic license plate readers*, October 20, 2023, American Police Beat). The cameras have become relatively inexpensive and are programmable to capture images of up to 1,800 plates per minute across four lanes of traffic and at vehicle speeds of up to 150 mph.

Police claim that the devices are highly effective at recovering stolen vehicles and reducing auto theft rates. In a shopping mall in a large West Coast city, ALPRs snapped pictures of about 3 million plates in 27 months, identifying 51 stolen cars. The police in the same city say that the ALPRs are also used to solve criminal cases, including homicides.

In the first nationwide evaluation of the ALPRs, the ACLU reported that the system may infringe on privacy rights, doesn't fall under any government regulation as to the proper use of the collected data, and its limitations may easily be abused by the police. The ACLU claims that, in 45 states, there are no laws on how long police can keep the records and some agencies hold on to data for years, or even indefinitely.

The cameras are often placed on moving police cars with video images streamed to a central station that compares the plate numbers to a vehicle database. The system has proven to be highly effective in locating stolen vehicles or those reported leaving the scene of a crime. But some civil liberty advocates believe it is a violation of the privacy guaranteed under the constitution.

Local police can track vehicles automatically by using cameras mounted on a patrol vehicle that record every nearby vehicle license plate, time, and location. As recently as 1995, more than half of license plate surveillance was unavailable or not in widespread use, but by 2012 more than a third of large police departments used automated plate-reading systems. In one medium-sized California city, the sheriff's 50 camera-equipped vehicles scanned 2 million unique plates in a two-year period. Until recently, the LPR was unavailable and/or in limited use. LPRs are now widely available and inexpensive. The U.S. Department of Homeland Security distributed more than $50 million in federal grants to law enforcement agencies to obtain LPRs.

Databases that contain license plate numbers and owner information (addresses, contact info, etc.) are supposedly protected by laws which limit public access to the information, but nevertheless, owner information can be obtained by government agencies, police departments, private investigators, insurers, researchers, private toll operators, and, in some states, journalists. Law enforcement officials claim the technology is used to track down stolen vehicles, collect unpaid tickets, and identify vehicles of suspected criminals. While this sounds good, there are a growing number of cases where the information can be compromised. In response to concerns regarding possible abuse, some states using the LPR system passed laws requiring that any police department employing the system must purge stored images of scanned plates after 21 days.

10 MINICAMERA FOR ROBOCOPS

In 2018, the BJS published a report on the use of body-worn cameras by law enforcement agencies in the United States in 2016. The report showed:

- 47% of law enforcement agencies had body-worn cameras
- Big city police departments approached 80% camera penetration
- About half of all agencies actually used the cameras they'd acquired
- The vast majority of agencies with cameras had a written video/camera policy
- Police departments not using body cameras cited cost as the main barrier

While there aren't more recent statistics available for most of these data points, BJS did report that the use of cameras had increased from 47 to 68% by 2020.

Police use tiny, wide-angle cameras on their uniforms to record a first-person view of interactions with the public. The resulting video can be fed into a smartphone to be shared with other investigators. Cameras have radically changed police responses to emergency calls and traffic stops by revealing officers' proper (or improper) behavior. In response to those worried that the video will violate privacy and possibly appear on YouTube like dashboard camera videos, manufacturers claimed that images from mini ear cameras used by police are transmitted to a secure site with no way for officers or others to tamper with them. The video can't be erased or edited. While the initial cost to supply a large number of police officers may be high, it can be argued that it is justified by averting potentially expensive litigation.

Mini cameras could be clipped onto a collar, helmet, or sunglasses and weighed a half ounce (about the size of a cigar stub). They were used mostly when police officers are away from their vehicles, capturing video on the surveillance unit and voice recording on the built-in microphone (Figure 8.2). The information was transferred to a small computer carried by the officer (about the size of a smartphone) and then to a docking station, eventually being stored on and managed by a cloud computing system. There was an electronic sleeve available that fit over the officer's arm which incorporated a built-in video display, camera, laser beam, and stun gun. One of the main purposes of these surveillance devices was to help defend against brutality complaints which, at the time, were the basis for over $2 billion a year in court settlements by police departments. There were units that could capture between eight and nine hours of continuous video powered by a 14-hour battery, and which cost about $1,000 in addition to a monthly data storage charge. Some police departments didn't request cloud storage due to fears of hacking and instead outsourced their evidence rooms.

The mini camera units can be programmed to operate in three modes:

- **Privacy mode**, when nothing is being recorded, and **Default buffering**, which records 30-second loops but does not store the data
- **Event mode** in which the device automatically saves whatever video is being recorded.

FIGURE 8.2 Close-up of police body camera.

Source: www.Shutterstock.com.

Police departments slowly began accepting mini cameras but were reluctant to use a system with infrared or other type of low-light capabilities, arguing that cameras with these features don't provide an accurate representation of what occurred or was seen during an incident. An officer may be involved in a shooting where what the officer perceived as a threat turned out to be a cell phone. If the camera was infrared, it would likely be able to differentiate a cell phone from a weapon as opposed to an officer's naked eye which may not, particularly in low-light situations.

Devices continue to be developed specifically for police officers on patrol, creating modern-day RoboCops. At some point, in addition to wearing cameras, they may also wear heart rate monitors to immediately alert station operators when an officer is in a high-stress situation. Another sensor may alert station operators that an officer released the strap on their pistol holster. All the information is stored for easy retrieval should there be any questions regarding the events in the future. Once the video evidence has been captured, date and time stamps are encoded and officers may review the tape but may not make copies or alter it. The tapes were previously stored for a year, but with the advent of cloud storage, images can now be stored for a much longer time.

As cameras become ubiquitous, digital video images are likely to become a go-to source of impartial evidence in much the same way that DNA did in the 1990s. Since 2014, it is estimated that nearly one-third of the 18,000 state and local police departments in the United States use body-worn cameras and authorities are trying to find ways in which to analyze, process, and store the huge amounts of data generated by each camera. Police departments will likely have to dedicate technicians to sorting through all the data captured.

People have the right to record anything they see in public places; a right protected by the First Amendment. It is normal for people photographing the police during intense situations to get flustered. They may forget to hit record or may not know how, or where, to upload a video. But there have been innumerable, prominent videos recording that have been publicized over the last five to seven years – not the least of which was the George Floyd murder – that have led many agencies and individuals to record any such police interactions. There have been cases in which police officers, sometimes in violation of the law, confiscate cameras or phones which may have been used to record encounters. As a result, since 2011, a growing number of apps have been developed to streamline the process of capturing and broadcasting videos of police interactions. *CopWatch*, an app launched in 2014, automatically begins recording when the owner taps its icon and automatically uploads the video to the Internet when the recording is stopped.

There continues to be a great deal of controversy surrounding the use of body-worn cameras by police officers. Often, police officers moonlighting at off-duty jobs in the civilian sector (hospitals, nightclubs, ballparks, etc.) don't wear the cameras, even in cities where cameras are mandatory for patrol shifts. An Associated Press survey of the 20 largest US cities conducted in 2017 found that nearly all have their officers wear body cameras, but only five had rules requiring them for uniformed officers working outside their regular hours.

Many police departments have only a limited number of body cameras; therefore, mandating off-duty use would require additional and perhaps unjustified costs for the department. However, some argue that, regardless of the cost, off-duty camera use should be required as long as police officers have guns with real bullets.

TRAFFIC RADAR. The military use of radar during World War II was highly successful and when the war ended, experts in the field converted the long-range detection device to capture traffic speeders. In 1947, the town of Glastonbury, CT, equipped its police officers with handheld radar guns, creating America's first speed trap. It was considered a technical marvel at the time, removing the possibility of human error. In 1955, one driver contested the ticket by hiring a lawyer whose main argument was that using a radar gun was a lack of fair play on the part of the police who were armed with high-powered wartime technology. The argument was dismissed, and the fine was paid.

Today, police in every jurisdiction use digital radar guns that are far more accurate than previous analog radar. The most important effect of radar gun use is not the collection of fines but the change in driver behavior when they think they're being watched. When the small town in Connecticut first parked their radar truck beside the highway, traffic slowed down, even though the officers were handing out "courtesy cards," not speeding tickets.

11 TICKETED CARS – SPIKES IN VIOLATIONS

The car insurance industry tends to set rates according to several factors including the verification of the claims made by motorists when applying for auto insurance, approximate distance the car will be driven, and where the vehicle will be garaged. All this information and more go into calculating insurance premiums. In a recent in-depth study by the Highway Loss Data Institute that involved about 2 million vehicles, it determined which make and model vehicles received the most moving violations. Overall, a vehicle is pulled over for a traffic violation once every 500,000 miles and data suggests that those who have expensive, high-performance vehicles are often unable to resist the temptation to test their vehicle's capabilities when on an open, low-traffic road.

According to the National Highway Traffic Safety Administration (NHTSA), the top five most dangerous holidays in order are:

1. Memorial Day
2. Labor Day
3. Independence Day
4. New Year's Eve
5. Thanksgiving

Running a red light has deadly consequences. According to the IIHS, 928 people were killed and 116,000 were injured as a result of red light running in 2020. Half of those killed were pedestrians and bicyclists and people in other vehicles hit by the red-light runners. A suggested alternative to red-light cameras would be to do away with intersections entirely and replace them counteract with roundabouts or traffic circles as neither would require traffic lights.

12 CROWDSOURCING – INFORMATION SHARING

As law enforcement continues to employ more advanced technology to prevent crime and traffic offenses, many citizens scramble to find ways to circumvent those efforts. Initially, seemingly primitive information sharing, such as flashing headlights to alert oncoming motorists of speed traps or truckers using CB radios for the same purpose, has been replaced by Internet communication networks. All these person-to-person communication networks were the forerunners of the computerized technology called "Crowdsourcing" one of the main motivating factors for each side to find more sophisticated technology to overcome the methods used by the other. As the first radar detectors arrived, police responded by utilizing different radar bands, which inspired the development of wide-band detectors which, in turn, prompted police to develop laser devices that are much tougher to detect and gave law enforcement a temporary advantage. However, the use of crowdsourcing information systems utilizing existing GPS systems switched the edge back to the motorist.

Early crowdsourcing systems consisted of an $80 power cord connected to a radar detector. The cord allowed the transmission of radar sightings to a smartphone with GPS which automatically transmitted the information to all others nearby on the network. Today, the radar detector and $80 power cord are no longer needed as apps like Waze have integrated reporting features which allow users to report travel issues (crashes, slowdowns, speed traps, construction, disabled vehicles, etc.) which will automatically be shared with anyone using the app.

13 CITIES FIGHT BACK AGAINST WAZE CONGESTION

Beginning in 2007, GPS apps like Google Maps, Waze (2009), and Apple Maps (2012) became the overwhelming choice of motorists because these offered ongoing suggestions of shortcuts when main thoroughfares become temporarily congested, especially during rush hour. When the app alerted motorists to an alternate route, the result often causes vehicles to venture into residential areas, unbeknownst to the towns. Reacting to the local gridlock, the small town of Leonia in New Jersey decided to take action. In 2018, the town adopted ordinances prohibiting non-residents from using its streets during morning and evening rush hours. To differentiate between residents and non-residents, yellow tags were issued for residents to hang in their vehicles, allowing non-residents passing through to be more easily identified. Any non-residents passing through during rush hour were issued summonses and faced fines of $200. The ordinance was careful to keep roadways within their city limits that are controlled by either the county or state from being exempt from the restriction.

The ordinance initially passed challenges in court but was ultimately struck down in 2020 by the state Superior Court in Hudson County.

14 DECOY LICENSE PLATES

In an attempt to level the playing field, some enterprising engineers developed a line of products specifically to hide license plates from pervasive surveillance cameras. Products sold on the Internet are made specifically to fool red-light camera systems. One such item, a Fresnel lens, is a clear plastic cover that snaps over the license plate and blurs the plate numbers when viewed from overhead light poles which house red-light cameras. The Fresnel lens works like a Venetian blind. Looking directly at a license plate, the numbers appear clearly; however, when looking from above, the "blinds" hide the numbers. It costs about $20 and can only be shipped to states that have not outlawed the device.

Another popular decoy device was a chemical spray squirted over the license plate leaving a transparent film which, though clear to the naked eye, becomes clouded when hit by the flash of a camera, blurring the plate numbers. The aerosol sold for around $15.

While some states have outlawed the sale of any material or device that covers the license plate, both items described can easily be obtained via the Internet and shipped to a nearby state where there is no legal restriction.

15 FLIPPING LICENSE PLATES

Smearing grease on a rear license plate, disguising the plate numbers, was a crude method of toll-dodging but was too noticeable to observant traffic officers to be effective. A much more sophisticated method was to make the license plate vanish as the vehicle enters the toll booth area and have it miraculously reappear seconds later. The "flip" is accomplished with little expense or effort on the part of the driver. Cameras photograph vehicles that don't pay and a vanishing license plate saves drivers the toll which can be expensive, particularly if the vehicle is used on a daily basis. A vanishing plate can save truckers with 12 tires rigs upward of $65 at some toll booths.

Inexpensive and commercially available kits allowing drivers to flip rear plates and to have it installed at a service station only a few more dollars. This electromechanical contraption includes a toggle switch that sends an electrical impulse to the plate causing it to flip over. The same switch reverses the movement and restores the plate to its proper position. Handmade models have been discovered which use cables attached to the plate that are activated at the driver's position. Even a fishing line can be rigged up to accomplish the "flip-effect."

16 OBSCENE LICENSE PLATES

Obscene language, drugs, and religion all have something in common. They all appear on the list of rejected license plate names scrutinized by the Department of Motor Vehicles (DMV) in almost every state. License plates cannot contain offensive or objectionable language and the restricted list in some states can number in the thousands. While many of the names are obviously obscene or insult certain groups of individuals, each state has its own rules and their motor vehicle agency reserves the right to reject any request for a name that it deems patently offensive to others. The rejection is automatic if the name is on the published banned list, or it can otherwise be rejected at the discretion of each agency's commissioner.

Every state employs a group of motor vehicle license plate censors who have a talent for solving word puzzles. Phonetically arranged characters that sound like four-letter words are the easiest to reject as are numbers associated with street gangs and atrocious historical figures. But, even to the trained eye, there are challenges which require the assistance of online slang references that have to be unscrambled. For example, the application for a plate with A221X was rejected in California because the number 221 is a police code for a "person with a gun" and X refers to a woman. The translation appears to read, "A woman with a gun," and any reference to a gun is considered hostile.

17 FREE SPEECH LICENSE PLATES

Texas DMV rejected a proposal in 2011 by the Sons of Confederate Veterans requesting a specialty plate with a logo of the Confederate battle flag. The decision was based on the assumption that many members of the public reasonably associated the Confederate flag with expressions of hatred and bias against minorities. The Veterans group sued, arguing the state was in violation of the group's

First Amendment right to free speech because the state had issued specialty plates to other organizations and was therefore guilty of what is known as a "permissible – viewpoint discrimination" and it was unfair to deny the group's request for the logo. A federal trial judge sided with Texas forcing the Veterans group to present their case to the U.S. Circuit Court of Appeals which found that the state couldn't selectively shield the public from speech because of perceived offensiveness. The controversy surrounding the choice of a logo on specialty license plates eventually went all the way to the U.S. Supreme Court.

In 2014, the highest court in the land rendered its decision in favor of the Texas DMV. The Justices voted 6–3 to allow states to reject certain logos and messages, given that such messages on government-issued plates could be considered sanctioned or endorsed by them. Thus, they should be subject to more regulation and private free speech under the First Amendment.

The last time, before the 2015 case, that the Supreme Court considered how the First Amendment related to license plate requests was in 1977 when it ruled that New Hampshire could not force a citizen to drive around with plates bearing the state's motto: "Live Free or die."

Another challenge that made its way to the Supreme Court related to Texas plates with the logo "*SONS OF CONFEDERATE VETERANS.*" No fewer than 92 of Texas's 385 specialty plates honor the US military and its veterans, but when the state proposed a 93rd patriotic plate commemorating Confederate soldiers, the Texas Bureau of Motor Vehicles refused to sanction the logo.

The contested Texas plate, the first one with the veteran wording to ever be turned down by a DMV, featured a Confederate battle along with the controversial logo. The proceeds were to go to a youth group where teenagers would learn how to fire muskets and cannons. The case was also brought before the Supreme Court and in 2015 the justices voted 5–4 to uphold the decision to reject the logo. All the other states with the same logo had to recall the plates and issue a less controversial one.

Every state in the nation offers specialty plates. Maryland offers a record of nearly 800 specialty plates, including one for the Sons of Confederate Veterans. The same plates became available in eight other states: Alabama, Georgia, Maryland, Louisiana, North Carolina, South Carolina, Tennessee, and Virginia.

The most controversial license plate on the market and the one that involved the most court cases bear the simple message, "Choose Life." It's available in 34 states and in the District of Columbia; in North Carolina, two federal courts have ruled the state cannot offer it without also allowing plates preferred by abortion rights advocates.

18 CAR THIEVES HAVE PRIORITIES

According to the Insurance Information Institute, statistics based on FBI data listed California, Texas, and Florida, as the top states in car theft in a recent year, with Vermont, Wyoming, and Alaska being the lowest as of 2022. Newer cars are harder to steal because of higher tech anti-theft devices but thieves' skills for obtaining keys with the electronic codes are improving over previous years.

According to the National Insurance Crime Bureau's (NICB), a nonprofit organization that tracks vehicle theft on behalf of insurance companies, 932,329 vehicles were reported stolen in 2021, which is a 6% increase over 2020 and 17% over 2019.

Newer vehicles, especially luxury vehicles, now include technology that – according to the Federal Bureau of Investigations – (at least initially) almost solved the problem of stolen vehicles. However, it is that same technology that has contributed to the rising number of stolen vehicle cases that look very different from thefts of the past. Thieves no longer need to put effort into breaking into vehicles; instead forgotten fobs and those left sitting in cup holders within the vehicle have made things easy.

- **CAR THEFT DAY.** New Year's Day continues to be the worst day of the year for auto thefts according to the NICB, which based its findings on data through 2018. President's Day, Halloween, and Labor Day followed closely behind. The slowest day is Christmas because apparently crooks take that day off to be with their families.
- **BUMPER PARTS THEFT**. There are specialized car part thieves who will not steal a car that can easily be tracked using GPS technology. Instead, these specialists steal body parts like bumpers, fenders, airbags, navigation systems, catalytic converters, and even entire engines. The removal work is often done in large, multistoried parking garages like those surrounding apartment complexes. To accomplish the theft, a truck is driven into the open garage, usually late at night, until they spot the exact bumper they're looking for. Within minutes, the bumper is professionally removed and hidden in the back of the truck. The specialists can find and remove as many as ten car components in one night in garages that contain over 500 vehicles. The hourly parking fee is paid on the way out, complete with a large tip for the unsuspecting ticket attendant. The value of the parts may be up to three times the value of a vehicle, there is less risk of being arrested because the parts are hard to identify, and GPS cannot be used to track bumpers as it can an entire vehicle.

  WHEEL and TIRE THEFT. The sight of partially dismembered vehicles left on cinder blocks was fairly common decades ago. This type of theft nearly disappeared only to experience a resurgence around the country beginning in 2011. There are two reasons for this, one is the introduction of anonymous online marketplaces that provided easy access to customers and the other is the advance in power tool technology that enabled thieves to remove tires and wheels with the speed of a NASCAR pit crew. In a matter of a minute or two, a vehicle is jacked up, the lug nuts are removed and then the wheels are taken. Alloy wheels are preferable, some of which can cost upward of $500. The wheels are then sold in the form of its individual components, tire, and rim.
- **ENTIRE ENGINES**. New vehicles parked in a garage are more vulnerable to engine theft than if the vehicle was parked in a public area or on the street. It is relatively easy to remove a car engine and replace it with a ten-year-old engine (>85,000 miles on the odometer) made for an entirely different year of the same model car. This feat takes about an hour and a half and must be done when there is little activity in the secluded section of the parking garage. More than likely thieves will select a mid-town garage in a theater district or near a sports arena and time the theft during a show or event when owners are not expected to return. Two skilled mechanics and a truck with a chain-lift are all that is needed to detach and lift the new engine and drop the old engine into place before the owners return and can be done in little over an hour. The unsuspecting owners will notice nothing strange since the vehicle appears undisturbed and will start and drive normally. The big surprise comes only when the vehicle is brought in for service.

According to the NHTSA, a motor vehicle was stolen every 39 seconds in the United States in 2020, more than three-quarters of a million vehicles costing owners more than $7 billion.

19 STEALING OLDER CARS

Old cars may be easier to steal and there are plenty of them on the road but they're not worth very much. Cars are typically stolen for their parts and, as the car gets older, parts become less valuable. Thieves have been forced to take the cars to salvage yards and sell them as scrap for just a few hundred dollars. The scrap yard owner, in many states, will accept the car without its title if it is more than eight years old. The law allows a car that old to be sold without a title and if it is worth less than $1,250. With fewer valuable stolen cars coming in, it has become less appealing to operate an illegal chop shop than in the old days. The model year appears to play a big role in the types of vehicles stolen. Honda products from the late 1990s and early 2000s were popular decades ago and that trend continues. In 2021, trucks were also popular with thieves and the NICB reported that the most commonly stolen Chevrolet and Ford trucks were 2004 and 2007, respectively. The NICB compiled data indicating that nine of the ten metropolitan areas with the highest auto theft rate are in California. Thousands of the West Coast vehicles end up in Mexico, with others found in container ports in Los Angeles destined for Asia.

The 2023 top 10 most stolen vehicles (by make and model), nationally as reported by Insurify and the NICB, are shown in Table 8.1.

TABLE 8.1
Most common vehicles stolen 2023

Make/Model	Number of Vehicles Stolen
Hyundai Elantra	48,445
Hyundai Sonata	42,813
Kia Optima	30,204
Chevy Silverado 1500	23,721
Honda Accord	20,895
Kia Soul	20,001
Honda Civic	19,858
Honda Civic	19,858
Ford F-150	15,852
Kia Sportage	15,749

Section IV

Controlled Access with Audit Trails

Closed-circuit TV surveillance and card reader access control systems have had the greatest impact on physical security over the last three decades and these two technologies continue to evolve. Card readers practically rendered the traditional locksmith obsolete, a trade that never accepted the movement away from mechanical locks to electronic control systems. More doors are now being monitored by sophisticated smart cards than by metal keys. Mechanical locking hardware is limited because it retains no memory or audit trail of use, nor can the conventional metal keylock cylinder be programmed to limit operation to specific time zones. That was true until 2004 when a memory cylinder (and key) was introduced.

The introduction of card reader technology came about with the Hollerith card which was adapted by IBM as a "punch card" used primarily for time and attendance and then by phone companies for billing purposes. Eventually, the technology was adopted by the access control industry and advanced over three decades to include features not initially deemed possible, such as being able to embed personal information onto smart cards. Basic ID cards went through several revolutionary changes from punch cards to magnetic strips to barcodes and finally in the early 1980s, to a proximity card that could open a door without the card touching or swiping through the reader's head.

In the 21st century, control systems are being integrated with broadband Internet connections and with biometrics that can read the size and characteristics of the iris or veins in the hand to ensure that the cardholder is the person to whom card was issued. Other advances include the introduction of open architectural platforms that allow differently formatted cards to be interchangeable. This is a change from the traditional use of proprietary card access systems which forced end-users to remain a captive of the original manufacturer even after an older system reached its saturation point, without the ability to add new readers or cardholders.

ENTERPRISE SYSTEMS. Probably half of all access control systems in the world consist of less than 35 card readers, but at large institutions (government agencies, hospitals, etc.), it is not unusual to have a system with 400 readers or more, falling into the enterprise category. Access control systems were first introduced to a marketplace that did not consider future expansion from their initial configurations. A system with an initial installation of 50 readers could, over a seven to eight-year period, expand to hundreds of readers and thousands of card users, beyond the capabilities of the basic system. Since 2005, more and more institutions have discovered that they have reached the saturation point and their system cannot add more readers or additional ID cards.

DOI: 10.4324/9781003091103-13

Conventional card access systems may have allowed for triple the original number of readers initially installed, but as buildings expanded and more doors were added, the total number of readers could very well exceed ten times the initial system which was well beyond its capability. Compounding the card reader issue was that the salesmen did not fully understand that a facility with 4,000 people would eventually require over 25,000 distinctly coded cards simply due to employee turnover (departures and new hires). Over about an eight-year period, the number of cards issued can reach saturation even if the facility population stays at 4,000.

Vendors have since adjusted to build in both scalability and obsolescence such that physical and software systems have become more adaptable and upgradable as technologies advance in sophistication.

9 Not Just Card Readers

1 SIZE DOES MATTER – SOMETIMES

Manufacturers of hard-wired card reader access control systems have experienced out-of-control growth since the first credit card reader was introduced. Dozens of firms entered the access control market during the 1960s and 1970s but most became obsolete almost overnight due to the rapid technological improvements, forcing 90% of them to leave the industry within five years. The first few access control manufacturers made systems that could only read about 50 coded cards and the system had to be shut down at night to allow the downloading of information onto a computer. Gradually, the number of cards and readers expanded but there were limits. Sometimes, the small number of facility codes allowed under the early technology forced manufacturers to be careful in assigning card-to-facility codes to avoid having one card activates a door reader in a nearby building. An attempt to overcome this problem was to assign available facility codes to different buildings within a certain neighborhood and then group the codes again in another neighborhood. This did not always work, especially if the user roamed around the city. By the mid-1990s, most of the problems were solved except for capacity, which rarely rose above 100 readers for a particular facility. It wasn't until the turn of this century that Enterprise systems allowed over 1,000 readers in a facility.

Card reader systems are now found in over 80% of commercial and government buildings, even though the newest technology allows more emphasis to be placed on larger, so-called Enterprise systems, most card access systems still operate with fewer than 50 readers.

SMALL: Up to 32 readers
MIDRANGE: Up to 512 readers
ENTERPRISE SYSTEM: Theoretically, it can control an unlimited number of readers.

2 SHORT LIFE CYCLE TECHNOLOGIES

"Legacy" systems are sometimes described as "vintage" security systems since they refer to any security device or system that has been in operation for over ten years. That life-cycle can also describe other technology systems found in computers, TVs, CCTV, TV video recorders, wireless devices, and even automobiles and appliances. Once a card access system operates for a decade, the electronic components are generally considered obsolete by the manufacturer who will no longer support the system with replacement parts (there is also a good chance that the manufacturer is no longer even in business). An even more serious problem than the lack of maintenance is that the system eventually reaches a saturation level at which point it can no longer allow additional ID cards or readers to be added to the system. This is a scenario that describes probably 35–40% of all existing card reader systems because the salesmen who sold the initial system did not understand that expansion is inevitable and could outgrow the limited capacity of the proprietary system.

Expanding or upgrading any type of existing security is an exercise in frustration, particularly when it involves card access control systems. Unlike CCTV systems where cameras, monitors, and recorders from different manufacturers can be connected to an existing system without significant compatibility issues, attempts to expand card reader systems pose unique and serious industry-wide problems. Manufacturers produce "proprietary" software and/or hardware that will not operate

DOI: 10.4324/9781003091103-14

with components from other manufacturers. An initial installation can easily grow to five times the original size over a three- to seven-year period and, under the present situation, facilities are forced to purchase new components from the original manufacturer regardless of poor performance or lack of technical support. Another consideration with older systems is that new communication cables may be necessary to modernize the network and improve security of communications.

An effective access control security system possesses the capability to expand in the future. Its installer should be factory-trained and authorized to maintain the system. Most major manufacturers have policies to support and protect a limited number of local installation firms, particularly if they have earned a good reputation from end users and were the firms that sold and installed the original systems. The facility will have to purchase additional components for expansion from the same installation firm unless a letter of complaint from the security director or engineering department is received by the manufacturer, requesting a new installation firm due to a lack of service and maintenance from the original installer.

Most access control systems use proprietary field panels. In the event the customer decides to change the operating system, they will need to invest in new field panels, sometimes new cables, as well as installation labor. Very few facilities can justify this expense while most manufacturers continue to protect the technology used in their panels to recoup their investment cost. "Interoperability" is a dirty word in most sectors of the security industry.

Laws Relating to ID Systems

- HSPD-12 requires a common ID credential for all federal government employees who need access to federally controlled facilities and information systems. (Explained in more detail in later sections of this chapter.)
- The Federal Information Security Management Act is another law that requires federal agencies to provide periodic training in computer security awareness and accepted computer security practices.
- Graham-Leach-Bliley Act regulates the sharing of personal information belonging to individuals who obtain financial products or services.
- Sarbanes-Oxley requires IP officials to take responsible action to ensure the security, accuracy, and reliability of information systems.

Health Insurance Portability Accounting Act is of particular interest to the healthcare industry since this federal statute mandates that hospital administrators comply with security standards to protect all information that is maintained or transmitted electronically. Each facility must assess its systems for potential risk and vulnerabilities to the health information records of its patients.

3 BADGING SOLUTION

1. **ID Badges** have become commonplace in most facilities and are considered the first line of defense in fulfilling the objective of upgrading safety and security in a highly cost-effective manner. Assigning badges is a relatively simple task that can be quickly implemented and requires little training. Badges also improve the visitor experience and can be applied to standalone systems or integrated with software-driven visitor management and security systems. From a management perspective, various levels of badging system technologies are available to best meet specific needs while providing powerful tools to track and monitor daily visitor traffic.

2. **Levels of Badging** solutions are available and the choice of level depends on the size of the facility and the frequency and number of visitors and temporary workers. In general, there are four levels of badging security, each with unique features:

   Level One: **Paper Label Badges:** Employees are given standard paper badges, which provide a minimum level of security and can be issued either manually or electronically. Paper label badges are generally intended for one-time use only and are considered a simple form of a name tag.

   Level Two: **Paper Badges with Color Photos**: Usually produced by scanning a picture from a driver's license or capturing an image via an onsite camera, these badges help to further identify visitors as they may contain color codes, symbols, and/or words.

   Level Three: **Electronic Badges** include proximity, magnetic, and barcode badges, as well as those containing biometric information. These semi-intelligent badges can be programmed for predetermined levels of access and can be integrated with access control and video surveillance systems to electronically monitor activity.

   Level Four: **Time-Expiring Badges** deliver instant verification of security status in addition to all the solutions and benefits described above. Time-expiring badges prohibit their unauthorized transfer and/or reuse.

4 BASIC SLIDE CARD READER TECHNOLOGY

The **Punch Card Tabulator** was the first machine to process information rather than material, a device operated by office clerks rather than factory hands. Invented in the 1880s by Herman Hollerith, the punch card tabulator was almost immediately sold to the US Census Bureau that had been unable to track the explosive population growth taking place in America. The 1870 census had recorded just five subjects; when the 1880 census was expanded to cover 215 subjects, the staff could not sort all the new data collected. The Bureau was very accepting of incorporating the punch card into their system for the 1890 census. It worked on the simple principle of storing information by punching holes in certain locations on a paper card. A hole in one place on the card indicated that the family had three children, while a hole in another place signified that the family lived in an apartment, etc. The card was placed onto an electrically charged plate in the Hollerith module and then a grid of metal needles was placed onto it. Wherever a hole was punched, a needle would pass through, completing a circuit and allowing data on the card to be recorded on a meter. It was the nation's first successful application of the binary system which anticipated, more or less, the binary operation of today's digital computers and card access control systems. The way in which the punch card tabulator company came to be sold set the pattern for the entire modern history of business computing.

During the Industrial Revolution, there was a pressing need to process more information than ever before. The Hollerith tabulator allowed big businesses to process information more quickly with fewer people and with greater accuracy than was possible prior to that period. In 1911, the Hollerith firm merged with other companies working with the industry and expanded. In 1924,

TABLE 9.1
Format of credential ID systems

Contact Cards	Contactless Cards
Hollerith/punch	Proximity
Magnetic stripe	Smart cards
Barcode	EZ Pass
Barium ferrite	RFID
Weigand	

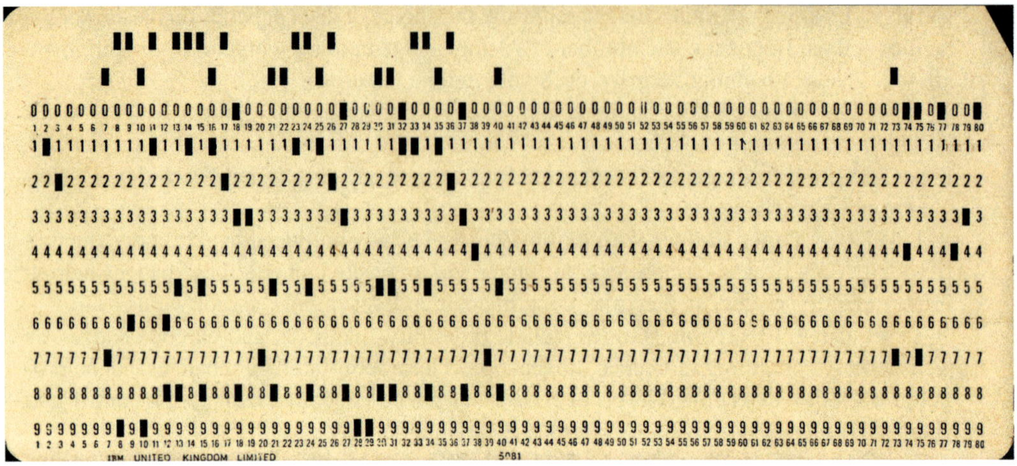

FIGURE 9.1 A 12-row/80-column IBM punched card from the mid-20th century (Used per CC Attribution 2.0 Generic; Author: Pete Birkinshaw, Manchester, UK).

Thomas Watson took over the company and changed its name to International Business Machine Corp. The data processing and information technology industry grew from the punch card system to the mainframe computer and on to the modern computer network.

ID cards can operate with different technologies and are regarded as secure if they resist duplication through forgery or use of a copy machine. The punch card has long been replaced with swipe, insertion, and proximity readers that were introduced in 1984. By 1990, these readers controlled 75% of the access control marketplace. They cannot detect a card's use by unauthorized personnel unless other identifiers are used such as a personal identification number or biometric features such as fingerprints, vein, or voice patterns, which identify characteristics uniquely associated with an individual. Succeeding generations of ID systems brought forth new advances in card reader technology that offered more features.

Microprocessors and computer technologies, contrary to current fashionable opinion, were not new forces unleashed upon an unprepared society. It was merely the latest installment of the continuous development of the access control revolution.

- **HOLLERITH** IBM Punched Hole card (a legacy technology) featured a uniform pattern of rectangle holes arranged in 80 columns and 12 rows holding up to 80 characters of information on each card (Figure 9.1). It was used as an inexpensive identification device where the punched holes (in different arrays) allowed light to pass through to identify each user. IBM then patented a "cardkey" (1941), which ushered in the era of using a card to verify a person's identity and gain access to a secured area. It was, for many years, the most popular "card" for billing purposes but has practically disappeared as a security card in favor of more secure ID systems.

Observation: In the 1950s, some people would punch random holes on the monthly Hollerith card that accompanied the telephone bill before returning it to the phone company. They would claim that they never received another bill from the phone company after the mutilation because it "tilted" the billing machine. This practice has never been confirmed, but it does point out that when the punched holes were not perfectly aligned with the reader, a false read would be generated (the Florida presidential election punch card election fiasco of 2,000 is a perfect example).

- The **MAGNETIC STRIP** was the first card technology specifically designed for access control system applications and is still widely used at ATM entrances and for ATM banking transactions. The information contained within the thin magnetic stripe is read by moving the card along a decoding track in the reader's head. Each card carries both the ID of the user and a facility code that is unique to the particular building location. While they are relatively inexpensive and have the advantage of allowing the end user to encode the card on site, they are high maintenance, and since duplication of magnetic stripe information is possible, their level of security is relatively low. One inherent shortcoming of early models was the risk of accidental magnetic erasure and the card characteristic that defines the level of resistance to erasure is known as the coercive force rating. This rating indicates the strength of a magnetic force required to erase the magnetic material. A card with a low coercive force rating is fairly easy to erase and, in general, plastic magnetic stripe cards possess a significantly higher coercive force rating than disposable cardboard cards. The higher the rating, the better the card is protected from extraneous magnetic fields.

  Observation: When magnetic strip cards were introduced to the security industry in the 1970s, they had such a low coercive force rating that almost any foreign magnetic source that came near the card would render it useless. For example, when individuals placed a card in their inside coat pocket and opened the refrigerator, the door would pass near the pocket with the card. The door has a magnetic latch mechanism to hold it closed and this small magnet would actually scramble the magnetic stripe code on the card. When an attempt was made to enter the office door the next day, the reader would not allow admittance. This is another example of problems with technology that was introduced to the security industry without proper testing. Issues associated with this older technology have not been corrected and the United States stands alone as a modern nation that continues to use magnetic-striped payment cards when almost all other industrial nations have abandoned them.
- **BARIUM FERRITE** cards were popular access control devices in the 1970s and 1980s when they were used in high-volume, high-turnover applications, such as parking lots. They use a pre-coded array of magnetic spots in the reader that react to magnetic material sandwiched between two layers of material in the card. They are slightly more secure than the older magnetic strip card systems since the magnetic barium ferrite codes are difficult to duplicate. These cards hold up especially well in hostile environments and harsh weather but can be distorted by outside magnetic fields and wear out over time.

  When first introduced, the same code array was used by all access cards issued in a facility. If the card was lost, the program cartridge and the ring cards had to be reissued. As the technology advanced, additional magnetic spots were added to the main array, forming unique ID numbers that could be interpreted by microprocessors. Newer "touch" plate models use an array of electronic sensor devices installed behind a touch plate to read magnetic spot patterns on the card.
- **WIEGAND** cards are the most secure of all contact swipe cards. These cards utilize short lengths of small-diameter, ferromagnetic wires subjected to a unique twisting process that imparts unique magnetic properties to the individual wires embedded in the card. When exposed to changes in magnetic flux, a current is induced in the wires generating a signal for the reader to pick up. These cards are very difficult to duplicate and provide far superior security than other contact cards.

5 BARCODE LABELS

The least expensive and easiest to use of all the reader systems is the barcode found on every type of item throughout the world (Figure 9.2). While initially developed exclusively for price marking in

978-1-032-25910-9

9 781032 259109

FIGURE 9.2 An example barcode from a CRC Press/Taylor & Francis forensic science book title.

grocery stores, barcodes have become a much more practical device for a variety of other applications; however, their primary use remains tagging merchandise. Barcodes are now scanned more than 10 billion times a day around the world for price marking and inventory control. The proper name for the barcode format is the universal product code (UPC), commercially developed by IBM engineers based on a known technology using 30 black and 29 white lines representing 12 bits of data in binary code. The 12 digits are printed in a pattern of thin black vertical lines to form a distinct number and address that can be stored in a database and can be "decoded" when read by an optical scanner.

History: In the late 1940s, two grad students at the Drexel Institute of Technology created a technology based on a printed series of wide and narrow striations that encoded consumer product information for optical scanning. The two received a patent in 1952. Their early ideas involved printing product information in fluorescent ink and reading it with ultraviolet light, but this soon became unworkable. They then considered applying some form of the Morse Code, with its limitless combinatorial potential to a graphic display, but ultimately elected to use wide lines and narrow rows instead of dots and dashes, calling it their "Classifying Apparatus and Method." The basic system depended on an immense scanner equipped with a 500-watt light, which was expensive and never went into production. The patent was sold to Philco for $15,000 which was all they made from the invention. Experts in the field at that time felt that "the numbers found beneath the barcode" were more important than the striped pattern and recommended that it changed to another font to be more readable by machines. In other words, they mistakenly believed that the bars were not as practical as the numbers and went so far as to predict that numbers would eventually supplant the striped pattern. Those experts were proven wrong because they could not see the practical application of devices outside the unrealistic world found so often in research laboratories. It was another of many situations where out-of-touch research engineers misunderstood the value of the innovation of their own devices that could be used in the access control security field.

Figure 9.3 includes some of the credential types, discussed here in this chapter, as well as the related technology they use.

The barcode was not an immediate success; between 1952 and the early 1970s, some manufacturers and retailers put their own product coding systems in place, but they were not uniform and one company's barcode was unintelligible to another's system. The major food product manufacturers were concerned that they would be facing conflicting demands from different stores requesting different code arrangements. Amid rising inflation in the 1970s, supermarkets wanted to cut labor costs by automating the way their merchandise was stocked, inventoried, and registered at checkout counters. After the original patents lapsed, a committee of supermarket executives formed the Uniform Code Council and examined submissions from more than a dozen technology companies before selecting a design of black-and-white vertical bars in 1973. It was a recreation of the original design developed by the engineers at IBM and inspired by two Drexel students, Norman Joseph Woodland and Barnard Silver, who developed the concept more than two decades prior in 1952 as grad students.

Credential Types	Related Technologies
Bar coded cards and objects	Light and dark patterns interpreted by optics
Barium ferrite cards	Magnetic pattern
Biometrics	Physiological patterns interpreted by various means
Hollerith cards	Holes that allow the passage of light or electrical current
Infrared cards	Light and dark patterns interpreted by optics
Keypads	Keyboard input
Magnetic stripe cards	Magnetic media
Mixed-technology	Combined technologies
Optical cards	Light and dark patterns interpreted by optics
Proximity cards and objects	Radio transmission and computer chips
Smart cards	Computer chips
Wiegand cards	Magnetic pattern

FIGURE 9.3 The credential types and corresponding, underpinning technology.

The real impetus for the barcode concept occurred two decades after the patent was issued when a food market executive purchased a ten-pack of Wrigley's Juicy Fruit chewing gum in 1974, watched it slide down a conveyor belt and past an optical scanner that beeped, the cash register faithfully ringing up 67 cents. That occurred in a supermarket in Troy, Ohio, and was the first product anywhere to be rung up using a barcode. That purchase of a package of chewing gum changed the world of retail marketing and personal identification (Figure 9.4).

Through its varying patterns of thick and thin bars, the barcode could efficiently represent the 11 digits needed to encode data about products like Wrigley's chewing gum. The UPC grew to the 12 digits in use today. Barcodes allow retailers to track product inventory and change pricing without needing to retag every item. Retail barcode numbers in the United States and 107 other countries are assigned by GSI, a nonprofit standards organization created in the early 1970s. A company applies to GSI for a barcode number specific to that company, then creates (or hires a firm to create) the barcode to match that number. GSI tests barcodes in its laboratories that are refilled with new and old flat-bed scanners, hand-held scanners, and overhead scanners. For a fee, GSI assigns each barcode a grade "A" through "F" – based on how consistently it scans.

Like many other innovative technologies that have been introduced in the security field since the early 1970s, the barcode revolution involved a 20-year gestation process. It took several years after first appearing on food items for the public to accept the accuracy of these small barcodes. At first, store owners were hamstrung by shoppers who refused to buy barcoded products, worrying that they might be cheated at the checkout counter without the backup of price-stamped labels. Today, almost every item in a supermarket is labeled with a barcode, and barcodes and versions of them are also used on airplane boarding passes, package tracking labels, glucose meters, calibration tools, and labels tracking the research statistics on the pollination of bees.

In 2009, barcodes began appearing on commercial television programs, allowing smartphone users to read sales information by pointing the phone straight at the TV screen. Barcodes are also

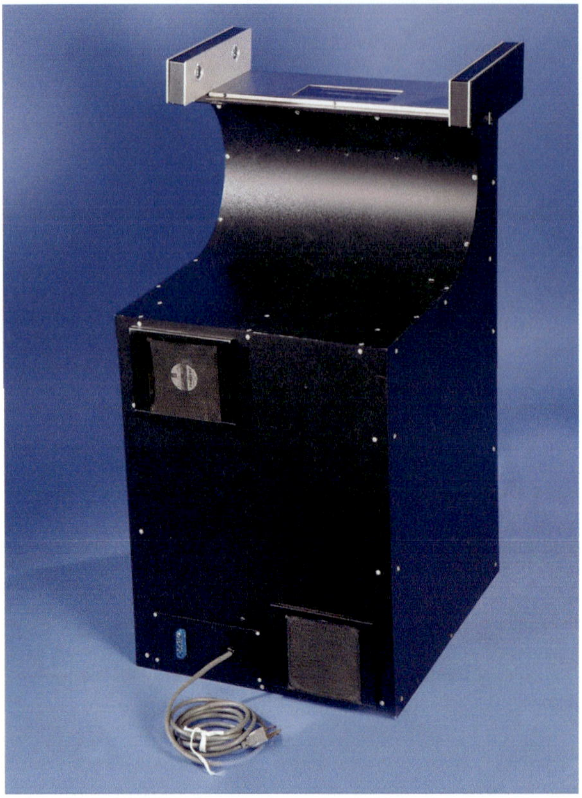

FIGURE 9.4 This spectral physics model, a price scanner, in the Smithsonian Museum, is one of those first ten scanners produced and used in a retail environment.

Source: Smithsonian Museum.

used in hospital nurseries. They are pasted on the standard patient ID wrist band of both mother and infant and a handheld reader scans the code to ensure their barcodes match. The same barcode is also applied to mothers' milk bottles before being placed in the refrigerator to indicate proper ownership.

- **Unorthodox Apps:** There are some special circumstances where barcodes are created randomly using an iPhone app that does not adhere to any international set of standards. The wine industry is one of these where individual store owners hire iPhone developers to create an app for them. Instead of paying for each unique barcode, the store owner simply reuses the codes that come with the bottles, reprogramming them using the unique database provided by the iPhone developer.
- **Vanity Barcode Symbols:** Packaging consumer products to attract customers relies, for the most part, on the design. Manufacturers have been known to spend more money on the package than on the product itself. The importance of package design has led some firms to take artistic license with the standard barcode image, attempting to disguise the staid image with colorful features without affecting the barcode's functionality (Figure 9.5). A handful of companies that specialize in making vanity barcodes have cropped up in recent years, though some companies create them in-house. On a barcode, there are certain things that can be changed and certain things that cannot, and then there is the gray area that allows some liberty in the final configuration of the lines.

FIGURE 9.5 A stylized barcode on Trader Joe's™ onion ring snacks.

For a scanner to read a barcode, it typically requires a blank space of about a half inch on either side and the line colors cannot be red, yellow, or orange as scanners are not able to detect them. Some vanity barcode designs go against GSI guidelines by using shorter lines than the organization recommends, though they do scan the product information accurately.

Barcode labels have many uses in the security field, but only limited use in high-security areas as they can be duplicated fairly easily. Since their introduction in 1994, QR codes have become one of the most popular types of two-dimensional barcodes gradually replacing the older UPS barcodes for certain applications.

6 QUICK RESPONSE CODES (QR TECHNOLOGY)

Developed by a Japanese firm in 1994, Quick Response (QR) technology was initially used in the automotive industry. While the inventor owns the patents and the trademark for this unique type of barcode, there are no licensing requirements to use the code. The familiar square imprint of individual black and white dots (or modules) comprises the data or message; the larger squares at the corners of the code, known as position detection patterns, allow the code to be scanned from multiple angles. The standard barcode reads in one dimension, left to right, with alternating bars of black and white, while the QR code is two-dimensional with black and white dots that read both left to right and up and down – allowing considerably more data to be encoded.

The size of the QR code image varies depending on the type of characters (alphanumeric, binary, etc.), the number of characters, and the level of error correction needed. If the code is going to get dirty or damaged, a higher ability to compensate for this can be built into the code. A single QR code can be broken down into smaller units, if necessary, to fit into the space allotted for printing. There are different versions of the QR code that go from a few characters to 1852 characters or more.

A QR code can be read by a variety of handheld or desktop scanning devices. Smartphones have been able to read them since 2010, and though they initially required an app, now the QR code can be read using a smartphone camera. The use of QR codes has steadily expanded, from in-store product shelf labels, coupons, magazine ads, and billboards to online ticketing, online repair manuals, wildlife refuge trail information, and tombstones. Hackers have been able to implement malware through QR codes, sometimes compromising the information on the user's phone by loading a malicious website through the code.

7 TURNSTILE – PROPER SWIPE TECHNIQUE

Turnstiles located in public transportation centers are swiped thousands of times each day; this many uses cause problems which are always frustrating to the traveling public. There is a proper way to use swipe cards and, as an example, the NYC Metropolitan Transit Authority has issued guidelines, which could conceivably apply to turnstiles all over the nation. The standard card, when swiped, must hit four sensors before allowing entry to the platforms and subway or train cars. The first sensor verifies that the user card is programmed for that particular site and not for other business locations. The second sensor ascertains that the card has enough money to cover the fare, the third deducts the fare, and the fourth opens the gate. If the card doesn't hit all four sensors, the turnstile will not turn. The card must be held in a level position throughout the process.

The speed of the card is also crucial – it can't be too fast or too slow. The optimal technique is a steady swipe motion like a punch to the gut but perfecting the swiping technique is no guarantee of success. The machines come under so much use that they must be carefully maintained. At high-volume stations, the heads (sensors) are removed and cleaned every ten days and station agents do "fingertip" maintenance daily by running an alcohol-soaked "cleaning card" through the turnstiles' sensors.

Experts on turnstile operations suggest that card users should head for the middle turnstiles because those turnstiles closest to the stairway entrance are the ones used first and are the dirtiest. When a turnstile malfunctions, a rather common occurrence, the unit will light up and inform the user to swipe again – and again – and again. It seems that if the frustrated passenger receives a "please swipe again" message, they are free to try their luck at a different turnstile, which is a natural reaction. However, if the alternate turnstile repeats the "please swipe again" message, it means the fare has been deducted and continuing to swipe will only continue to deplete the card. The next step usually involves jumping over the turnstile (or creeping under it) and then being arrested for this illegal maneuver. It is only after being fingerprinted and photographed at the police station that the prisoner can verify that the toll was paid and all the "swipes" were made. Sometimes these burdensome turnstiles are deliberately sabotaged for profit.

8 TAILGATE SNEAKS

Most of the early models of card reader's access control systems were incapable of preventing someone from closely following a card user from "tailgating." Since 2004, there are systems available that stop this practice to a great extent. If someone were to piggyback onto the person with the authorized card, they would most likely get through the opening. But since people generally walk a slight pace behind each other, there are detectors that react to the space between people, even those attempting to tailgate.

Nearly all tailgate detection systems use infrared sensor (PIR) beams to identify unauthorized entry attempts and notify staff of any violations. It is important that this add-on is always used to ensure that only one authorized person at a time gains entry. The PIR can be retrofitted to existing systems or installed with new access systems.

9 CARD READERS AND CREDIT CARDS

The first big change following the Hollerith card occurred in 1958 when two men made an agreement with an NYC restaurant, Major's Cabin Grill, allowing them to charge their meals using a piece of cardboard with their names and a number printed on it, the first "Diner's Club" credit card. The original idea was for the card to be used to pay for the restaurant bill within a month and the Diners Club would collect a small fee from the restaurant. Gradually, the credit cards were accepted by retail stores and then the banking industry entered the market and began charging interest to the cardholder plus the same small fee to the establishment that accepted the card. Currently, there are over 1.109 billion credit cards in circulation in the United States.

Credit card companies have become a major industry that discourages "deadbeats" – cardholders who pay off their balance each month, avoiding interest charges and fees (up to an exorbitant 30% in some cases). These companies consider ideal credit card holders, called "Revolvers," those who carry a balance to the maximum paying the exorbitant interest rate. Some states have usury laws that place a cap on the amount of interest that a credit card company can charge. Delaware and North Dakota, the two states where most credit card companies incorporate, have no such laws.

Credit Card companies collectively have taken pains to move the industry – and consumer – away from swipes, to the insertion of chip reader cards, to the chip "tap" feature in rapid succession. This is largely to combat the ubiquitous credit card skimming devices, scams, and thieves. Again, as is typical with security, when thieves find ways to capture personal data and information, the industry is often forced to innovate and devise new ways to thwart such criminal activity.

10 Contactless Card Access

1 PROXIMITY (PROX) CARDS

Traditional research and development processes usually involve laboratory testing, a pilot program, modifications, upgrading, and extensive field testing. The passive infrared motion detector that was previously described went through government-supervised laboratories prior to its introduction to the commercial marketplace 20 years later. The normal development cycle is rarely followed by most security product manufacturers who then sell systems without undergoing any standard or recognizable testing program. However, it could be said that the testing is done once installed on the customer's wall.

An example of how to get a product from its inception to its marketplace introduction is the proximity card access control system (PROX cards). This phenomenally successful technology came into existence accidentally and without being encumbered by conventional development stages. It was a revolutionary concept that exploded into the security industry, bypassing any formal procedures, but its initial marketing program was a disaster. Eventually, the PROX card became a leader in card access systems after overcoming being marketed as just another locking hardware arrangement rather than a revolutionary concept in access control identification.

PROXIMITY *refers to a card reader technology which used the first so-called contactless card when all other card reader technology required the card to be swiped across or touch the reader to be scanned. PROX (Proximity) cards have integrated coil circuit and antennas embedded in the cards, which have no batteries. They are powered from the electromagnetic field in the reader mounted on the wall which generates a signal, activating the card. It was the most widely used card access control system until the advent of the radio frequency identification (RFID) card in 2008.*

Premature exhilaration best describes the accidental birth of proximity technology. In 1979, the century-old Schlage Lock Company became interested in distributing a more advanced type of lock hardware and hired a two-man engineering firm to design an improved deadbolt-type lock. The project never left the drawing board because, during the early stages of development, one of the principals of the engineering firm had to leave his pet dog at home for many hours. He devised a small sensor tag encapsulating wire coil tuned to a specific frequency that was attached to his dog's collar. He then added an RF antenna on the door frame that, upon detecting the coil frequency, released the door hatch and the dog was able to exit the house to exercise in the fenced-off backyard. Schlage abandoned the lock project but bought the rights to the dog monitoring system, which appeared to be a significant advancement over conventional locking and card access control systems. It had no moving parts, and it did not require the access card to be in contact with the reader unlike other reader systems required.

The introduction of the proximity card evolved through a series of fortunate accidents and is another example of an intrusive and serendipitous discovery shared by other innovative products such as cellophane, Teflon, and scotch tape.

DOI: 10.4324/9781003091103-15

The wall-mounted receiver/transmitter is contained in a cabinet the size of a CD case and continually transmits an RF signal that sweeps through a specific range of frequencies out to about 4". The signal excites a tuned coil that is embedded in the PROX card along with a memory chip. The card then generates a pattern that possesses a unique code contained within the memory chip and is read by the wall-mounted receiver. It was considered revolutionary since it had no slots to jam or moving parts to break down and it did not have to be removed from a wallet or pocketbook.

Schlage obtained a patent for the PROX concept and formed the Schlage Electronics Division to produce the nation's first proximity card access control system. It was so advanced that the company had trouble selling it to the public. There was no comparable system posing a problem since the salesmen couldn't simply claim that the system was better than all other similar systems and there was no ready market since card access control systems were just evolving. The security industry has traditionally been slower to accept new technology than, say, the computer industry or the medical industry and security directors are reluctant to be the subject of beta tests or purchase new technology. They are not trained to understand how the system operates and are not awarded for being innovative and "early adopters"; in many situations, they are almost encouraged to purchase traditional systems rather than one that is state-of-the-art.

The Schlage Proximity Access Control System was initially distributed as a standalone system where each reader operated independently of other readers (this was before the advent of microprocessors and multi-reader interconnected systems). It was a very basic system since all the cards had the same facility code. The easy-to-recognize "blue PROX card" was slightly thicker than magnetic and barcode swipe cards since it contained a coil of wire that was tuned to the same frequency being transmitted by the flat wall-mounted antenna. Each tuned coil resonated at one of 256 preset RFs that allowed a limited separation of facilities. Since the card only had to be held in PROX*imity* of the antenna, it offered a significant advantage over all swiped readers in use at that time. While the manufacturer had over 100 years of experience in the lock industry, this strong mechanical background created two serious shortcomings: (1) the firm had no experience (or history) in the evolving electronic security industry and (2) their major client base was the 15,000 locksmiths who remained satisfied to work in the dark ages of mechanical locks where their skills were limited to cutting duplicate keys and not in electronics.

Although locksmiths offered a large captive group of established and skilled mechanics, they represented a collection of small proprietors who had rarely been accused of showing any interest in the evolving field of electronic technologies. That mindset along with an indifference to the trends in physical security has caused the century-old locksmith industry to remain stagnant while the rest of the industry exploded. Schlage had concentrated their initial market plan on utilizing their loyal client base along with the hardware distributor to sell what was basically a revolutionary "electronic" system. They believed that their established and loyal clients had been successful in locking doors long before the nation became obsessed with sophisticated electronic security. They wrongly assumed that the best marketing approach was to continue with the old client base. The company sold hundreds of demonstration kits to locksmiths who initially indicated some interest in expanding into the more complicated electronic security industry. While they may have had good intentions, they had neither the inspiration nor training. Schlage had sufficient financing, an outstanding reputation, and name recognition but was new to the access control industry, just like the locksmith. About a year after selling the demonstration kits to locksmiths, Schlage was surprised that no one had sold a system to a facility, and they were shocked when almost 100% of the demo kits were returned by the locksmiths when they came to the realization that they were skilled mechanics, not electronic technicians.

This proved to be an expensive lesson since it took a year to discover that they had overestimated the skills of the locksmith. Eventually, the company realized that the PROX reader was not a locking system but a unique electronic card access control system. The company was forced to restart their marketing program by reassessing their sales approach and focusing on the alarm integrator, reintroducing their PROX reader to engineering consultant firms. This more practical approach gradually made their proximity card reader a phenomenal success. Within five years, the company controlled almost 25% of the access control industry. By the time their patents ran out, they were the #1 manufacturer of card access systems in the United States. After a series of questionable technical decisions (their patent expired), the company left the card access control industry, selling their card access system assets to a conglomerate.

One of the reasons for the failure of PROX cards to maintain their share of the card access market was being too successful too soon. The first generation PROX coils were limited to the number of coded signals it could read: (1) facility codes up to a maximum of 256 distinct digits representing buildings and (2) 32,000 user codes. These numbers appeared sufficient during the first five years following their introduction to the marketplace, and the company was careful to avoid using the same facility code within the same zip code area. Building managers and owners started to overwhelmingly accept this revolutionary card access system, once they discovered that the card could be used without removing it from a wallet and the readers could not be vandalized like swipe card readers. This unanticipated success resulted in having all the facility codes reaching their limits in larger cities. Once the 256 codes were assigned in the same city, the company was forced to sell units with the same code to buildings within a small radius. An employee at one facility would be able to use their card to enter another building in the same city. Security was no longer assured; in fact, it was seriously compromised.

The situation at Schlage came to a dramatic conclusion, not with new insight but with insults and lawsuits. An executive of one facility entered a retail store that had tall, upright antennas as part of their Electronic Article Surveillance (EAS) system to deter shoplifting. The executive was carrying his PROX card for his building's entrance door, which was tuned to the same EAS frequency that was assigned to the retail store, and he set off the alarm. The executive was arrested for shoplifting and was insulted by the insensitive security guard. He sued the card access control manufacturer, and the entire fiasco ended up in court where a decision was made that the EAS companies would adhere to the FCC Part 15 rule "filter out unwanted frequencies." The executive received an apology and some money, but the incident affected the entire industry. (Before the legal dispute was adjudicated, employees with PROX cards were encouraged to wrap them in tin foil when they left their place of business.)

If the PROX card was not shielded, the guard stationed at the entrance of a store with the anti-shoplifting EAS device would ask the person to shield it inside a tin foil wrapper. This allowed the cardholder to shoplift all they wanted and walk out of the store holding the card in hand. When the system activated the audible alarm, the guard would wave the person through the barrier and out of the store, assuming that the PROX reader was causing a false signal, and of the tags attached to the seven dresses hidden inside the coat.

Sharing of frequencies prompted all firms dealing with RF technology to upgrade their systems and forced Schlage to design a system that would allow more facility codes, adding a third series of frequencies in their cards. They hired an engineering consulting firm (one of the nation's most prestigious) with so-called RF experts. The project became so complicated that the consultants were hired to run the company. It was the wrong decision and when the company's patent protection period ended, and nothing new had been developed to help the company retain its leadership in the industry, the company was sold, repackaged, and rebranded but was ultimately an inferior product. This scenario is another example illustrating that technology is not easily passed from one company to another and the end user is not always made aware of the fact that the system being installed has a new trade name but the underlying technology may be on the verge of becoming obsolete.

PROX cards were considered too thick when first produced but over time they have been flattened to approximately the same thickness of swipe ID cards. PROX cards are considered durable, secure, and only slightly more expensive than other card readers while providing more security. The card has become so advanced in its design that it is flat enough to fit inside a wallet.

PROX readers are made by HID Global, the largest provider of the proximity reader in the world and now a division of ASSA ABLOY, a foreign conglomerate.

2 RFID SMART CARDS

The RF system has proven to be very effective and has garnered significant interest due to the benefits it can provide across a wide variety of security-related applications. RF systems involve a method that can uniquely identify physical objects using wireless communication technology. Examples include automatic toll collection (EZPass), public transportation tracking, cashless payments, access control, and infant abduction security systems. Concerns have been raised by civil libertarians over the expanded application RFID technology has. They claim that computer technology using RFID could create a climate in which the public is acutely sensitive to the privacy issues it raises.

The RFID system uses interrogators (readers) and tags to create a unique type of information system. The reader is responsible for transmitting a message to all tags within the range of the reader and the "excited" tag then responds by transmitting individual identification numbers and other data pre-programmed and stored within the tag. Since RFID uses radio waves to transfer data between the reader and tags, it does not require physical contact or line of sight between the reader and the tag. In the security industry, this RF technology was first used in access control systems where the tags were held in proximity of the reader (e.g., the PROX card). Once the reader collects data from the tags, the data is passed to the information system where it is collected, processed, and transformed into a technical language that can be used to identify the user and initiate an event such as opening a door.

Contrary to most articles published on the subject, this is not a new technology. The first use of RFID was an aircraft transponder built in the late 1930s. RFID is defined as a technique that uses a radio signal to send identification information from an unverified device to a system that can interpret the coded information. As RFID entered more markets, it made the public a little wary of how rapidly it expanded into mainstream American life. The fears take on two distinct directions: a belief that it compromises privacy and a misunderstanding of its capabilities and limitations. RFID suffers from rumors that are more fantasy than reality caused by exaggerated embellishments and is a victim of the same technophobia that accompanied the introduction of ionization smoke detectors (radiation causes cancer) and the first commercial microwave ovens (food glows in the dark).

The first practical application of RFID can probably be attributed to the German military during WWII, when its engineers were exploring the use of "radio detection and ranging" (radar) as a practical method to track aircraft flying at high altitudes. At the time, radar operators encountered problems when aircraft entered the range of the radar signals. There was no way to discriminate between the blips on the screens caused by groups of friendly aircraft returning from a mission, and enemy bombers seeking to destroy their cities and factories. It is not known exactly how it was first discovered, but the ground controller at the radar screen was able to record the movement of the aircraft wings as it went up and down, known as a roll maneuver. This slight variation in flight pattern would change the radar signal to a unique and distinguishable display. When a device was installed in the aircraft that detected the radar signal from the ground, the pilot would initiate a roll maneuver to enable the radar operator to recognize the plane as friendly. This procedure allowed radar operators to dispatch fighter interceptors only when a non-German aircraft was detected on the screen.

British pilots began to notice that entire groups of German aircraft would occasionally simultaneously conduct a roll maneuver. British analysts detected a coded signal transmitted from the ground that always preceded this strange behavior and eventually understood that the aircraft was conducting a roll maneuver to signal the radar operator that they were German aircraft. With this knowledge, the British established a secret project in order to develop their own automated system which would provide their own radar operators with the capability of discriminating between friendly and unfriendly aircraft without requiring any observable maneuvering by the pilot. The result of the project was the development of Identify Friend or Foe (IFF), which was one of the first practical applications of RFID. This unique IFF system required that each aircraft be equipped with a powered transmitter and receiver pair which became known as a "transponder." This small unit was designed to either broadcast a coded identification signal continuously or transmit the signal only in response to a coded signal sent from the ground radar operator. From that military beginning, the IFF system is now standard equipment for all civilian and military aircraft. The British developed an "active system" that required a powered transmitter in the aircraft to send the signal back to the ground. The German technique is considered "passive" because the aircraft had no active power source and relied on the reflective energy transmitted from the radar signal.

The exponential growth of IFF technology occurred when the system was adapted to civilian applications as far back as the 1960s with the introduction of EAS. It detected shoplifting from retail stores by attaching a small, inexpensive, and passive one-bit RFID tag to a garment which could be detected when passed between two antennas. The antennas sent a signal to activate the tag, which responded with a coded signal that indicated its presence, a miniature, and practical application of the IFF system.

RFID systems can be compromised by four general categories of unauthorized attacks: Sniffing, Spoofing, Replay, and Denial of Service. These categories are not mutually exclusive.

- ***Sniffing*** is considered the greatest threat to any RFID system and is a common problem shared by all communication media. Anyone with an antenna can intercept legitimate communications between the RFID tag and the reader if the device is placed within range of the transmission. All the frequencies assigned by the FCC are standard and available through an Internet search. If the frequency is a proprietary frequency band, it can be obtained by a skilled individual using a spectrum analyzer. If someone could obtain an RFID passport or drive a vehicle with an RFID license plate, a terrorist could build an illegitimate reader that only detonates within close proximity of a bomb.
- ***Spoofing*** attacks require the programming of correctly encoded data so that the tag appears legitimate. The technique is mostly used to misidentify items in a point-of-sale situation, for example, removing the tag applied to a frozen lobster and retagging it with a tag meant for a pack of mints. Supermarkets tend to be the victims of this type of attack.
- ***Replay*** attacks are a combination of sniffing and spoofing and occur when someone can query a tag to obtain the information it sends, and then retransmits that information at a later time. As an example, suppose an employee carrying an RFID badge that allows access to a facility is at lunch and passes someone with a hidden reader in their briefcase. The attacker triggers the briefcase to send a query and records the response from any badge within its proximity. The attacker can then program a blank RFID badge and can enter the facility undetected.
- ***Denial of Service*** attacks affects the availability of a system's means. For example, by removing the tag from an item and hiding the item under a coat, it will not be detected by the EAS system. Another attack involves placing an item in a foil-lined bag (called a Faraday cage).

Regardless of the public's slow acceptance of this technology, it is now being used for payments in all forms of mass transit. Embedded RFID tags in automobile keys immobilize the vehicle until the correct key is presented, ski resorts use RFID for lift tickets, and hospitals are using RFID to track drugs and patient specimens.

3 MICROCHIP SURVEILLANCE

The introduction of RFID chips is often depicted in high-tech spy films, where RFID chips are inserted under the skin, completely invisible to the eye. Such futuristic devices were once considered a subject for science fiction but have been tested and approved for use on humans by the Food and Drug Administration as far back as 2004. Small, flat RF circuit chips were implanted in about 100 volunteers, including a member of the staff at the Harvard Medical School Technology Department. Advocates of the RFID implant feel it may be appropriate in many medically controlled fields, such as in patients who easily become lost or confused like those affected by Alzheimer's disease. There would be a need to obtain consent to the procedure early on when the patient still had the capacity to understand the ethical issues involved. It could also be useful for people who are adventurous (climbing mountains, etc.) and feel that a serious fall might render them unconscious. The chip would expose the person's entire medical history to the attending physician who would not be able to access the information otherwise. Some of these uses are many years away from full acceptance and, meanwhile, there are opponents who feel RFID is an affront to human dignity.

Under-the-skin implant chips were initially developed to be used to monitor the movement of cattle. However, they had been in use for many years before. The government implanted chips into employees, at one facility, to activate the access control system door reader without the use of any visible ID, strip card, or PROX card. It provided a foolproof and secure identification surpassing other biometric techniques – except that it was found to be easily compromised. All someone would have to do is stand nearby with an RFID reader, surreptitiously pick up the chip's data, and then make a duplicate chip that could be used to break into the system.

Implanted RFID chips have proven to be effective in tracking inventory and in recent years have been adapted for many other uses. In 2009, there was an outbreak of cactus plant thefts in an area of the Southwest United States that was under the control of the National Park Service Office of Special Investigation. This federal agency started a program to embed microchips in the cactus plants to protect them from thieves who ripped them from the desert and sold them to landscapers, nurseries, and homeowners. The cacti are very valuable, can grow to 50 feet high, and weigh several tons. Some take 50 years to flower and 70 years to sprout an arm and there are 1.3 million just in one park area in Arizona known as Saguaro Park. Mature 6-ft-tall plants have been poached and sold for $1,000. Each chip is uniquely encoded and can be read by waving a wand within about a foot of the plant. The chips cost $4.50 each and the wand costs about $1,000.

MEDICAL MARVEL. A novel microchip implant was introduced in 2012 for medical use that may very well replace the need for injections in about ten years. A tiny microchip is implanted inside patients who require daily injections to treat illnesses that is programmed to operate as a drug-delivery system when activated by radio signal. In test trials, surgeons implanted the microchips in the abdomens of women ages 65–70 during an in-office procedure lasting approximately 30 minutes. The implants received signals to release a drug from a microscopic reservoir etched in the chip for up to three weeks. The ultimate goal was to develop a chip that can deliver daily medication doses for an entire year, something that is still undergoing testing and consideration to administer a variety of medications.

4 MERCURY PLATFORM OPEN STANDARDS

Connecting security products and systems made by different manufacturers into a single integrated system is known as an "open standard" platform. Having a standard platform ensures that software vendors design their particular security system to support various brands of network products. It could be considered a revolutionary concept in an industry that prides itself on having never allowed much interchangeability between products. Traditionally, once a card reader system was installed, the facility was forced to purchase the components from only the manufacturers of the system.

Technological changes in the access control industry were slow to evolve before 1998, a year that witnessed a radical change in the card reader access control system industry with the appearance of the Mercury Platform. This unique software/control panel package was an attempt by a circuit board manufacturer to allow proprietary card access control systems from different providers to be interchanged with units made by other firms. Traditionally, once the initial card reader system was installed throughout the premises, it had to be maintained and expanded by a factory (OEM) authorized local integrator until one of the following events occurred:

- The building was torn down
- The access control system couldn't be expanded as the number of readers had reached capacity

- The number of ID cards exceeded the capacity of the basic system and new employees could not be issued additional cards
- The system was considered a legacy system, no longer made, or replacement components were no longer available
- The manufacturer went out of business or was sold to a conglomerate and the new company had abandoned the specific product line
- There were no factory-trained technicians or installation companies in the area capable of servicing, repairing, or programming the system
- Organizations using a card access system often involved several locations, and each building had a different system, requiring different ID cards and readers. Building owners required one system to be shared by their facilities in various locations

There is an industry-wide belief that over half of existing card access systems presently in operation are not being supported effectively, nor can they be expanded, forcing many building owners to seek replacement systems. Most facilities rushed to convert their locking hardware to card readers in the mid-1990s, discovering to their astonishment ten years later that there was no convenient way to replace the system even though it was almost impossible to maintain and could not be expanded. Added to this built-in obsolescence was the concern that the next system would exhibit the same defects and shortcomings in future years. The access control system industry has become the biggest sector of the security industry but has not attempted to standardize or make systems that are interchangeable.

The introduction of the Mercury Board offered an alternative to the industry-wide practice of delivering only proprietary systems. The Mercury hardware was developed to offer a practical solution that would satisfy the main complaints associated with the card access control industry. It is a "platform" consisting of a printed circuit control panel that contains all the features of conventional access control systems and is sold to manufacturers as a component – not as a complete system. Mercury Platform technology allows the interchangeability of card readers and software by the facility owner.

Maintaining hardware and firmware is very costly. Most access control manufacturers do not have large volumes, so they try to retain older designs as long as possible to maximize the return on investment. The industry-wide practice is the systemic upgrade to the operational features of the control system. Not using a Mercury Platform would require new field panels and, in some cases, all new cable when changing from one manufacturer to another. In essence, an end user can change their access control system from one Mercury OEM to another with a simple change in software. Since all Mercury OEMs have access to the complete set, ideas of one developer are shared by all Mercury OEMs. (Mercury has developed and released new hardware to support the increased memory and processing speed required to meet the government's Homeland Security Presidential Directive 12 (HSPD-12) requirements).

The choice of the Microsoft or Linux operating system is made by the manufacturer and end user. Linux has primarily been a platform used by developers on systems that require a limited user interface as well as reliability.

Y2K PHENOMENON – Man-Made Problem – The year 2000 had a designated abbreviation of Y2K which defined the infamous "millennium bug" that caused a panic throughout the security industry as the new century approached. There was a worldwide fear that security systems relying on computer programs, particularly card access and remote door locking systems, contained a software glitch that threatened to melt down millions of computers when their internal clock attempted to roll over on January 1, 2000. The only country that had enough software programmers to adjust these millions of computers before they went haywire at a reasonable price was India.

The Y2K episode turned out to be a false premise but by the time it was exposed as such, governments, businesses, and ordinary citizens joined together to combat a complex, man-made problem that threatened to wreak global havoc. Many unscrupulous firms made millions solving a problem that, as it turned out, never actually existed nor manifested – while giving birth to the now common practice of outsourcing.

5 SMART CARD

Smart cards are not an entirely new technology. First introduced in the early 1980s, smart cards were initially limited to bank cards, phone cards, and some federal agency employee ID cards. It wasn't until 2005 that they began to be widely accepted and were expected to be the wave of the future in card technology. Smart cards differ from conventional credit or credential (security access) cards since they are the only cards that have a memory chip embedded within the card. These chips are similar to those used in computers and in microwave ovens. The main difference between the smart card and electrical devices with chip technology is that there is very little interest in stealing information on how long it took to heat a bagel, but there are many people interested in stealing personal information such as health records, bank account information, or other sensitive data.

A benefit of the smart card is that it makes biometric access control a practical security option. Before smart cards, biometric systems required large, relatively expensive databases holding biometric files for staff members. Door access hardware included a card reader and a biometric reader. When an employee presented a card to a reader or tapped a personal identification number (PIN) onto the keyboard, the next step was to present a biometric in the form of a fingerprint, handprint, or iris scan. If the system accepted the card, it would then compare the biometric offered at the reader to the biometrics stored in an access database. If a match was found, the door was opened. This two-step process took too much time to compare the biometric data, find the match, and then signal to open the door. A smart card can store encrypted data capable of recreating a biometric image or template. A match-on-reader reads the biometric data in the card along with other data and opens the door after a successful match.

One of the most common uses of RFID technology is the EZPass tag that allows drivers to zip through a tollbooth without having to stop to pay cash. Today, rather than swiping a card through a machine and signing a receipt, customers only have to wave the card in front of a sensor, a great convenience that has attracted the attention of computer-savvy thieves. Thieves purchase RFID readers for as little as $50 allowing them to read the card from short distances – and steal private information right from the unsuspecting victim's wallet. Some illegal and so-called underground readers have been developed to meet the demand from the criminal element to accurately read a card in a mailbox. The extent of this criminal activity is still unclear, but it is estimated that the number is growing every day. Anyone using a no-swipe card that is placed in a back pocket or pocketbook is just as vulnerable to information theft as if their credit card number were printed across the back of their coat.

The standard, so-called credential card only sends a coded number and functions more-or-less like a conventional metal key for a restricted entrance. Smart cards send as well as receive data and are more secure than any other ID card since they use encryption and the internal computer power of small chip technology. Smart cards hold promise to eventually replace the pure proximity card since an embedded chip can be programmed to hold much more information than any other card-based access control system. Once these cards become nationally accepted, a hospital visit won't be as perilous as a patient's complete medical record can be encapsulated within the card, including allergies and the reason for the visit. Surgeons can swipe or touch the card while the patient is on the operating table, ensuring that they are operating on the correct spot. Emergency rooms can save lives since the card provides instant medical records which are essential for unconscious patients and accident victims. In 2006, the first hospital to give patients this card was a Manhattan facility which issued 100,000 cards to patients requiring multiple hospital visits – so-called frequent fliers. The hospital expanded the cards to 45 metropolitan area hospitals in a cooperative venture with the manufacturer, who intended to issue these cards across the nation by 2011, creating the first healthcare databank in the nation. It is already in use throughout France and Germany.

As with most technology, there is some concern about the user's personal traits being stored in a smart card, but almost everything in daily life has become a challenge to our privacy rights.

Though not yet universally accepted, the smart card has grown in popularity. Colleges alone use close to 1 million smart cards representing one in 17 students. Cards are programmed to ID each student, grant access to computer networks, serve as access to banking funds, pay for dining services, check out library books, etc.

6 MULTIPLE APPLICATIONS

The smart card had become one of the fastest-growing segments of the ID industry. In fact, the global smart card market is currently $9.8B US and projected to reach $13.8B by 2023. This growth has been facilitated and encouraged by the federal government. It has become an integral part of the Windows platform because smart cards provide new and desirable features that might be as revolutionary as the mouse and compact disc were to the computer industry. The major advantage of smart cards over other credentials is that they allow multiple applications on a single card:

- Staff identification and authentication
- Physical security
- Premise security
- Storage of biometric information
- Secure access to the Internet

The China Syndrome. In one of China's largest cities with a population of over 12 million, residency cards fitted with powerful computer chips are programmed to include data that include the citizen's name and address along with other personal information such as work history, education, religion, ethnicity, police record, medical insurance status, and landlord's phone number. The program is being expanded to incorporate credit history and allow payments for subway travel and other small purchases. It is the world's largest effort to meld cutting-edge computer technology with police work to track the activities of an entire population in an effort to fight crime. It may spread to every major city in the free world unless privacy advocates discourage its use in the United States, but if the nation remains focused on terrorist threats, this card, along with video surveillance systems, physical searches in public areas, eavesdropping on email messages and phone calls without a warrant, the "big brother is watching" concept may become an acceptable part of the nation's quality of life.

- **HSPD-12** is the federal mandate signed into law in 2004 requiring all federal employees to possess "a secure and reliable form of identification." This credential allows both physical access to federally controlled facilities and logical access to federally controlled information systems. It is intended to standardize interagency identification and access control protocol. Government agencies, including the Departments of Defense, Interior, State, Justice, and Treasury, must accept a compatible smart card device to merge physical and logical access control onto a single card. This Directive required some agencies to address issues that were handled by separate functional groups in the past. Most had to overhaul "legacy" physical access control systems that relied on proximity, Wiegand, or magnetic stripe technologies which didn't offer the same security features that were available with smart card technology. HSPD-12 requires federal identity credentials to be secure and reliable. Smart cards are to be:
  - Based on criteria that verify the individual's identity
  - Resistant to identity fraud, tampering, counterfeiting, and exploitation by terrorists
  - Authenticated electronically and rapidly
  - Issued by providers who underwent an accreditation process

The Federal Information Processing Standard 201 (FIPS-201) issued by the National Institute for Standards and Technology introduced a standard that establishes the minimum requirements of all federal identification and verification systems associated with the HSPD-12 directive. FIPS-201 provides the implementation standards for a more secure credential process. The entire process is designed to substantially improve homeland security by creating better access control, reducing identity theft, and ultimately establishing a three-factor, biometric authentication system as the standard for anyone working for the federal government. FIPS-201 has become the driving forcc that has helped to introduce the smart card to the many firms that will get involved in the potential $8 billion industry created by HSPD-12.

Personal identification verification (**PIV**) specifies a single smart card to be used for both physical and logical access and operates with both contact and contactless interfaces. PIV cards should not be confused with RFID tags since contactless smart cards are small computers that include microprocessors, memory, operating software, and applications. These unique features found in smart cards will secure, store, and provide access data on the card while performing complex computer-to-computer security functions.

The HSPD-12 changes affect over 6 million employees and contractors. Most cards and badges used for physical access are not standardized enough to be used at multiple sites, and many facilities choose to issue their own unique cards. The issuance of multiple cards wastes time and money and increases the chances of a security breach: the more cards an employee possesses, the more likely one will be misplaced.

There were a number of key security weaknesses found in the older legacy access control systems which were based only on what the user knows (PIN or password) or possesses (key or token). Both methods have proven to be inadequate and ineffective. Tokens and PINs alone provide poor security as both can be easily lost, stolen, or shared. All legacy systems are essentially flawed as there is no way of verifying that the person in possession of the key, card, or badge is the true owner. The PIV card uses biometrics and by authenticating based on a person's unchanging personal characteristics, such as fingerprints or iris, is far superior to any conventional access control system. Unlike most legacy systems that rely on traditional methods of authentication, biometric identifiers are inextricably linked to one specific individual and therefore cannot be forgotten, lost, or stolen. The cards must function in two ways, with a non-contact reader and a contact reader, both of which must meet International Organization for Standardization standards. Another condition requires federal employees to receive a favorable background investigation, including FBI fingerprint checks, a practice that costs $50 million, or $25 per person.

Older access card reader installations are known as legacy systems and have been the cause of serious problems since the law requires the replacing of thousands of existing systems which, for the most part, cannot support the memory required for the new smart cards. Also, most legacy proximity readers function on a frequency that does not meet the requirements of FIPS-201. There are so many issues involved in converting legacy systems to use the FIPS-201 encryption algorithm, the concept that was the basis for initiating HSPD-12, that the entire effort will never be fully implemented. Furthermore, the FIPS-201 algorithm has been cracked which makes a justification for the project highly questionable.

Biometrics, as outlined in federal guidelines, will enable the federal government to achieve a higher level of agency-wide trust among geographically dispersed facilities and to virtually eliminate the threat of someone assuming an employee's identity. Of course, a threat still exists if the employee or contractor turns out to be a terrorist sympathizer, or lone wolf actor, with no indication of this on the background check.

7 TRANSPORTATION WORKER IDENTIFICATION CREDENTIAL

The TWIC (Transportation Worker Identification Credential) program is part of the Maritime Security Act of 2002, which aims to protect US ports and waterways from the possibility of terrorist attacks. TWIC ID cards were issued with the purpose of ensuring that anyone who has unescorted access to secure areas of US port facilities and vessels receives a thorough background check and is not a known security threat. Employees and vendors are required to obtain a tamper-resistant TWIC card that encapsulates a picture, a thumbprint, and personal ID information. TWIC is the newest entry in the vast array of federal agencies involved in some phase of the nation's growing determination to expand ID systems in all aspects of public commerce. The TWIC program crosses over into other identification programs previously established at various ports. It is currently managed by the Transportation Security Administration (TSA) and the US Coast Guard.

Probably the most recognizable agency within the Department of Homeland Security (DHS) is TSA which manages security at commercial airports, but its responsibility also includes the safety of the nation's ports and shipping lanes. This is a less publicized area of vulnerability that may actually be the most important and most difficult to control since there are no personnel guarding entrances or metal detectors to search every person working on the cargo docks. The problem at ports on the waterways around our nation is more than 1.5 million workers who handle the loading and unloading of cargo whose background and loyalties are often difficult to establish. In response to this problem, the Department of Homeland Security issued a set of rules under a TWIC program that requires longshoremen, truck drivers, and other dock workers to be issued a tamperproof smart card with embedded personal information and a thumbprint as a first step in making it more difficult for terrorists to sneak into secure port areas as part of the labor force.

Each applicant will have to prove that they are legally working in the United States and cannot have been convicted of any terrorism-related charges, crimes like spying, sedition and treason, or a capital offense. The TWIC ID system was fully implemented in 2009 and maintains roughly 2.2 million people enrolled in the program.

Background checks, which became mandatory for all port workers in 2009, rely on FBI and state criminal records that are often incomplete and out of date. The errors have been catastrophic for port workers across the country according to a report by the National Employment Law Project, an advocacy group for workers. The federal government had mistakenly denied credentials to tens of thousands of people who had lost their jobs while awaiting clearance or gone without a paycheck for months until the denials were finally reversed on appeal. Federal lawmakers erred by not focusing the 2002 port security law exclusively on offenses that relate to terrorism, like possession of explosives or espionage. Instead, they sometimes denied credentials to people who committed minor drug offenses and property crimes. The original intent was to deny credentials primarily to felons with convictions less than seven years old, but rejections have been issued to people who had been arrested even though the arrest was dismissed.

8 REAL ID COMPLIANCE

Congress passed the Real ID law in 2005 in response to 9/11 with the intention of making it more difficult for terrorists to hijack airplanes and use them as flying bombs. The program generated by the law, Secure Flight, came under the TSA and relies primarily on authentication of driver's licenses presented by passengers to board a plane or enter federal buildings. The law attempts to force states to issue national identification cards to their residents, specifically driver's licenses. It encourages states to use sources like birth certificates and national immigration databases to verify that those applying for or renewing driver's licenses are American citizens or legal permanent residents. The effectiveness of Secure Fight in preventing hijacking depends largely on the identity confirmation process employed by state agencies issuing the driver's licenses. Defenders of Secure

Flight believe that some of the 9/11 hijackers had driver's licenses which may not have been issued if the Real ID law had been enacted prior to that date.

Real ID requirements for commercial air flight were delayed by COVID-19 and May 7, 2025, was the new enforcement deadline according to the Department of Homeland Security's website. All 50 states must meet the strong security standards required under the Real ID – Secure Flight – program.

Many states were resistant to the implementation of REAL ID and the US government had to require it as the only form of accepted identification on military bases and in nuclear facilities in order to ultimately get all states to comply.

CRC Press/Taylor & Francis has a fascinating book on the topic of personal identification as a security measure that features the REAL ID requirements and 12-plus year implementation process (Chapter 7): David J. Haas's ***Personal Identification: Modern Development and Security Implications, Second Edition*** (ISBN: 9781032523712).

11 Innovative and Programmable Access Control Systems

1 CARDLESS CONTROL SYSTEMS

Access control systems are traditionally considered those that use some form of formatted card to ID someone seeking entry through a monitored door. Each time the card is swiped in the reader unit on the door lock, or tapped if a proximity card and device, the event is recorded and an audit trail is created on a centrally located computer terminal (Figure 11.1). Since thousands of employees and visitors use the reader every day, the only reason the event is recorded is for use in perhaps a future investigation and to reject an unauthorized card. No one actually sits in front of a screen and watches user names fly by.

Mag-stripe access cards have increasingly been replaced by proximity cards and some companies, like CyberLock™, have developed locking systems in which both the key and locking system have microprocessors, the cylinder being able to read the programmable key when inserted. For installation, traditional cylinder locking systems only need to be replaced with CyberLock hardware – a cheaper solution, where implementable, to running wiring and lines. In this way, there is no power or battery in the cylinder – all power is contained in the small, computerized key and

FIGURE 11.1 Proximity cards have increasingly replacing swipe card readers for security access within the industry.

Source: Shutterstock.com.

DOI: 10.4324/9781003091103-16

the ID of the user is stored in both the key and the cylinder as well as the audit trail – a remarkable and inexpensive alternative to card reader systems.

2 EMBEDDED CHIP CARDS

The last decade has seen the movement from mag-strip credit cards to chip-embedded ones. By 2015, more than 80 countries around the world use chip-based card technology while only 1% of the credit cards in America had contained the embedded chip. Initially, the transition away from mag-strip systems was seen as cost-prohibitive for some (taxis, restaurants, and small businesses) which contributed to the delay in adopting chip cards. In late 2013, the retail chain target suffered, at the time, the largest and most sophisticated cyber-attack in modern times when 40 million credit and debit card account numbers were breached. Following that and other high-profile breaches of other companies, the United States began transitioning to chip cards in earnest starting in October 2015.

It was up to the merchants to decide whether to upgrade their payment terminals, but the banks and the card networks such as Visa and Mastercard had set the rules for liability. Since card issuing companies will no longer pay for a fraudulent transaction like using a lost or stolen mag-stripe card was a strong incentive to purchase the new terminals. Chip cards have retained the magnetic stripe on the back of the card so they can be used with both old and new terminals.

A chip card is not swiped but, instead, is inserted into the payment terminal that can identify the characteristics of the user. The information is available on the embedded microchip that is infinitely more difficult to counterfeit than a mag-swipe which uses the same data for each transaction. If the chip-based card is stolen, it is useless to the thief because it was only relevant for the previous transaction (the code constantly changes).

With the advent of skimming devices being installed in bank teller machines, in grocery store checkouts, and at gas stations by thieves, retailers – and bank and credit card merchants – are looking at implementing more contactless point of sales readers and strongly encourage usage by consumers to combat this crime and trend. In these, the chip is not inserted and, thus, the information cannot be "skimmed" but the device is inserted into the terminal by the criminals.

There are two types of chip-based cards, and only one of them uses radio frequency identification (RFID).

One type requires that the card physically touch the card reader to transfer data. It is called a "chip and PIN card" because the user must both swipe the card and enter a personal identification number (PIN). These cards were developed in a joint effort by Mastercard and Visa. The phase-in period for converting to EMV card readers was extended to 2020 (and again extended for fuel retailers to 2021) and, in the interim, magnetic stripe card readers continued to be used.

The other type of chip-based card doesn't require physical contact between the card and the card reader. It uses RFID radio technology to send data short distances through the air and has been commercially available for many years. The problem with all RFID cards is that, unless the card is inside a protective covering, it can be read from a few inches away by someone who has a portable RFID reader. Metal foil is the best protection to prevent data theft and a variety of RFID protection items are now sold (wallets, card holders, etc.) but the degree of protection is far from uniform. If the RFID cards use security codes that automatically change after every use, information stolen from the card can only be used for a single fraudulent transaction.

Again, by using the credit card with the embedded chip on the face of the card, any fraudulent transaction is covered by the card issuing company as has been the custom for decades. Using the little back magnetic stripe on the rear of the card will place the loss to the merchant until the terminal is upgraded.

TABLE 11.1
List of the smartphone technology devices and communication links

Manufacturer	Communication Technique
CyberLock	Flash
Latch	Bluetooth
KISI	Cloud base Wi-Fi source
August	Bluetooth
ASSA ABLOY/Lockitron	Wi-Fi
Lockitron	Bluetooth and Wi-Fi
Kevo/Kwikset	Bluetooth
Danalock	Wi-Fi
Best smart lock	Bluetooth
Goji	Bluetooth
Haven	Bluetooth
Noke	Bluetooth
Master lock	Wi-Fi

3 SMARTPHONE ENTRY

Another advanced technology that has also altered conventional access control systems is the Wi-Fi or Bluetooth entry system that relies on a smartphone instead of keys or cards. This system uses smartphones to communicate with the locking hardware to remotely unlock a door or a padlock. Smartphone key entry was introduced in 2015 and, within a few months, dozens of manufacturers almost simultaneously entered the market. Some rely on Bluetooth as the communication link between the door and padlock and others use Wi-Fi. FlashLock by CyberLock uses a unique type of flash technology (Table 11.1). All the techniques are capable of unlocking doors while recording the activities on smartphones and, as a result, have become the next generation of access control.

4 SMARTLOCK AND INTERNET SHOPPING SERVICE

Amazon revolutionized the standard package delivery service by having customers share the front door key (for a few seconds) with the person making the delivery. The major flaw in all deliveries is the persistent problem of missing packages left at the front door. This is a growing problem in the age of online purchases, especially for Amazon and their customers given that they are the world's largest mail-purchase company.

Packages that remain outside homes are easily stolen or damaged by rain. Furthermore, packages left on a doorstep for lengthy periods of time indicate the owner is not home – a tempting target for thieves. To combat this, under the delivery concept started by the Amazon Key system, Amazon offers delivery of packages and groceries into a garage, behind gates, and inside front door multi-unit buildings. Drivers who can't safely make a delivery will leave the package outside the front door.

The Amazon Key In-Garage Delivery costs $1.99 per delivery and can be used with the Ring app. If users can open or close their doors, using the myQ, Alladin Connect, or OHD Anywhere apps, then the garage is compatible with Amazon Key In-Garage Delivery.

Of course, many people living in cities will know that the swipe cards, that had been traditionally issued and used for public transit systems, are being phased out or discontinued to be replaced with riders being able to wave a smartphone or certain credit cards at the turnstile (Figure 11.2).

FIGURE 11.2 NEW YORK CITY. MetroCard swipes at subway station.

Source: Shutterstock.com.

12 Biometric Identifiers

No longer the province of security services and science fiction films, biometric technology has gained a place in almost every aspect of American life. Facebook uses facial-recognition software so its members can automatically put nametags on friends when they upload photos. Apple has incorporated voice recognition to use Siri on its devices, and some theme parks take digital fingerprints to help recognize season pass holders. Hospitals and school districts are using palm vein pattern recognition to identify and efficiently manage their patients or students – in effect, turning a palm into an EZPass tag.

1 PHYSIOLOGICAL AND BEHAVIOR BIOMETRICS

Conventional card reader-based access control systems have replaced the lock and key hardware with swipe cards using many different formats like barcodes, magnetic strips, Wiegand, and proximity card technologies. These card readers can be programmed to identify and admit certain individuals to specified doors at pre-selected time periods. Since about the mid-1980s, access control systems have become the most widely used security systems to control room and building entrances as well as other areas that formally relied on lock and key hardware. They will continue their rapid growth since the card ID system eliminates the need to carry numerous keys, fobs, and badges. Also, this system can record entries and selectively limit activity at a designated door.

Flat cards can be programmed to grant or deny access to whomever is in possession of the card (often called a token). These cards not only replace keys but also have practically eliminated the push button (digipad) units that open doors through punching in a Personal Identification Number. Basic card reader systems, while more secure than conventional key locks, can't assess the identity of the card user – a serious shortcoming when a higher level of security is required. The user can't be authenticated without an upgraded access control system, for instance, one measuring unique human biological data. Iris and palm scans and acquired traits like voice stress and walking gait are examples.

IDENTIFICATION is not a reliable method to determine who is holding the card or pushing the digipad buttons. It is very similar to determining who a person is by viewing a photograph that has no label.

AUTHENTICATION is a far superior method of identifying someone since it involves some biometric proof that the person attempting entry is not an imposter. Biometrics come in various forms and use different technologies, which can be classified into two basic types:

- **Physiological**: Based on data derived from direct measurement of a part of the human body with generally unchanging characteristics. This means an injury, such as a cut on a finger, will not affect the end user's validation. Although as with any application there are exceptions, including the loss of a body part like an eye or a severely burned hand. Fingerprints, iris scans, hand geometry, and facial recognition are leading physiological biometrics.
- **Behavioral biometrics:** Refers to access control systems that incorporate algorithms that vary in how the end user utilizes the application. It is based on actions taken by the individual and can recognize voice patterns, keystroke-scan, and signature-scan. These so-called performance tasks also include walking and gait analysis, which can be observed and stored by biometric devices in a database that can be used for future comparison – very

DOI: 10.4324/9781003091103-17

much like a bank having a representative signature on file for an account holder. If the end user's action is different each time, an attempt is made to authenticate a biometric characteristic, the results of gaining entry will be restricted or denied.

Passwords can be hacked, and smart cards and tokens can be stolen. But the human body is not so easily compromised. Security researchers in academia, industry, and the military are devising ways to use different parts of the body to protect computer systems and sensitive data.

Biometrics eliminates the weaknesses found in the basic access control systems that use programmable cards or metal keys or programmable keys. Authentication of each individual can be made after it is verified by any of the dozen biometric devices that are incorporated into the access control system. Biometrics is the technology that disallows unauthorized individuals from entering a secured area using stolen information or duplicated card/keys. Measuring human characteristics is the definition of biometrics, a word derived from the Greek bio (self) and metric (measure). The original use of biometric technology can be traced back to ancient Egypt and China, but in modern times, this method of authorization can be traced to the early 1960s when researchers began investigating voice recognition as a means to separate individuals by unique characteristics stored within data banks. By the mid-1980s, the biometric industry was beginning to be recognized for its value and established the first such systems for general distribution. It has had a rapid growth in recent years due to the heightened need for security after a number of disturbing events around the world and, of course, 9/11 in particular. (Source: biometricupdate.com)

Early biometric devices were limited mainly to government agencies that required a security system that did more than simply identify the supposed user. It wasn't a popular alternative to conventional card reader systems due primarily to the high cost of the early biometric devices. In recent years, the advent of inexpensive microprocessors and advanced imaging electronics has dramatically reduced the cost and increased the accuracy of biometric devices. Also, the public's image of biometric devices had been limited to James Bond movies and was considered too exotic for public facilities. That attitude has radically changed with the public's desire for more security. Biometrics, in all its forms, has become a very popular identification and authentication device.

Biometric technology continues to change and improve. The more popular biometric technologies are those that are easy to administer, user-friendly, nonintrusive to the user, free of errors, and rapid in operation. There was some concern when early versions of these systems were installed in large facilities that caused slowdowns at entrances and parking lot gates resulting in problems for the staff and visitors. Another factor that initially limited its acceptance was a general belief that biometrics was not a reliable identification system, that there were high false acceptance rates, and that it required too much service.

Those concerns were justified and most of the shortcomings have been corrected in areas such as processing speed, improved algorithms, better accuracy, and faster scan time. Biometric access has become a serious addition to the security marketplace and is current valued at $36.5 billion US dollars (2024); it is expected to reach over $113B by 2032. A strong impetus that caused a sudden growth in the use of biometrics was 9/11 and the Homeland Security Presidential Directive 12 (HSPD-12), which mandated biometric devices at military and government facilities and the Transportation Worker Identification Program (TWIC) program and requires a biometric device at ports and ship terminals.

With all the advantages of a biometric verification system, it should not be considered to be a device that solves all security problems and may not be applicable for use across entire building complexes. Instead, biometrics is commonly used in specific areas of a facility, such as laboratories, pharmacies, psychiatric wards, or other rooms where confidential records are kept. When the level of security needed is relatively high compared to the rest of the building, then biometric systems should be considered to supplement card access control systems.

Advantages

- Ease of use and the biometric profile can't be misplaced like a card or key
- Lower cost due to competition by vendors eager to enter this growing market
- Need for improved security methods following a breakdown in basic locking systems
- The user cannot be duplicated (except by identical twins in some situations)
- Best used with multiple applications such as a key card and digital video surveillance

Disadvantages

- Slower than access card readers
- Higher cost in general compared to traditional cardkey systems
- Considered by many to be too intrusive

Along with the flexibility and higher level of security provided with biometric technology, some concerns have been raised regarding the information collected by private organizations which could be used in ways that would violate the legal right of privacy. Some biometric techniques gather information that might be more revealing than necessary. The data stored that is necessary for biometric purposes can potentially include sensitive information that is highly personal. Some biometric technologies can double as medical diagnostic tests, such as iris technology where changes can be traced to direct physical damage or certain diseases.

Unfortunately, there are limited laws on the books and in place to provide guidelines and guardrails on biometric usage. The National Biometric Information Privacy Act (NBIPA) of 2020, though introduced on August 3, 2020, by Senators Jeff Merkley (D-OR) and Bernie Sanders (I-VT), never received a vote and – was thus – never enacted. While several states have enacted their own privacy laws, the NBIPA would have regulated the collection and usage of biometric information at a national level. Any collecting agency or entity would provide individuals a right of action to seek a remedy, if information was misused, and would require individuals' consent to collect such biometric data.

A person's gait, for example, could be an indicator of a neurological episode. Biometric information must be protected, especially if it violates HIPAA or other health information statutes that were issued to protect information on individuals' health conditions.

Samples of the most popular biometric techniques offer some idea of the different concepts that are available. Some facilities have installed multi-factor authentication where two or more techniques are provided; for example, fingerprint readers at facility entrances and iris scanners at specific room entrances requiring higher security are offered.

- **HUMAN PHYSIOLOGICAL APPEARANCE**
  1. Retina scanning (no longer in use)
  2. Iris technology records characteristics of the eye
  3. Fingerprint readers (ridge and valley patterns)
  4. Hand and palm scanners
  5. Vascular features (back of the hand)
  6. Photograph

7. Facial recognition
8. Nose biometrics

- PHYSICAL PATTERNS

9. Signature authentication
10. Graphology
11. Voice/speech patterns
12. Larynx analyzer
13. Keystroke patterns
14. Walking gait
15. Gripping pattern
16. Linguistic analysis

- REALM OF FACT AND FANTASY

17. Skin spectroscopy
18. Salinity
19. Facial thermography
20. Ear and ear lobe shape
21. Brainwave (EEG, MEGA, PET)
22. Footwear
23. Body temperature
24. Body scanner
25. Heart rhythm
26. DNA signature
27. Body odor (dogs and most other animals have been using it for centuries)
28. Laughter recognition
29. Finger bones
30. Inner ear bone
31. Lip shape (called the Angelina Jolie Effect)
32. Physiological screening (Future Attribute Screening Technology) medical exam detector

Some of these systems, particularly iris technology scanners and facial patterns, are not just for super-spies in James Bond movies or Star Trek reruns. Systems using seemingly exotic technologies have been operational for 30 years or longer in thousands of businesses around the world. These special effects devices are especially useful to control movement in certain sensitive areas like secret military laboratories – or to enter a computer terminal or server closets containing data that involves privacy issues protected by law.

1.1 Retina Scanning

This technology began with much fanfare in the mid-1990s and was met with high expectations, but during its initial introduction phase, a misconception stifled its growth. The public wrongly believed that the scanning involved a laser, which sounded dangerous, and health officials also raised concerns regarding safety (such as contracting an eye disease by sharing the device with previous users). Both criticisms were based on early models that scanned and measured the veins in the back of the retina, which were compromised as the person aged or developed an eye disease. This technology had a short life cycle and rapidly faded into obscurity by 2001. It was soon replaced by a newer eye scanner called iris technology.

1.2 Iris Technology

The public is still confusing retina scanners and iris technology even though they rely on two completely different biometric measuring technologies. Iris technology uses a more reliable and less intrusive technique since the reader does not have to touch the eye; it does not involve an actual scan or a laser as the retina scanner unit does. The iris technology system uses software algorithms and harmless infrared light sensor to analyze minute textures in the eye to determine whether the person is alive or dead. It is less dangerous than posing for a picture on a digital camera. The system operates with a digital video camera to capture images and the color of the iris and registers the unique patterns found in each person's iris in a centralized data file. Iris patterns are more unique than snowflakes. In simple mathematics, the odds of being misidentified by an iris scan range from one in 1.2 million to 1.44 trillion if both eyes are scanned, according to IRIS ID, a manufacturer of such systems. The iris scanner can detect 235 unique details and differentiate between right and left eyes and those of identical twins. (A fingerprint has about 70 unique details).

Irises aren't affected by age, Lasik eye surgery, or disease. In recent years, there has been a new wave of interest and some experts in the field of biometrics believe that this technology may well become as popular as fingerprinting devices. One company seems to provide most of these devices and over 2,500 police agencies in 27 states are taking pictures of eyes and storing the information in databases that can be searched to identify a missing person within seconds, as compared with weeks for fingerprints and months for DNA. They also use iris scanners to identify sex offenders, runaways, abducted children, and wandering Alzheimer's patients.

While its growth has been slow, iris scanning received great interest in 2007 when the US military began taking iris scans of thousands of Iraqis to track suspected militants. The technology was used in about 20 US airports from 2005 to 2008 to identify passengers in the Registered Traveler program who could skip to the front of security lines. Finance companies hoped the scans could stop identity fraud, and in 2010, the Department of Homeland Security began to test a newer generation that allowed cameras to capture images from up to six feet away. This longer-range system was initially used at a Border Patrol station in Texas for speed and ease of use. A unit can identify people who simply walk up to a box and look at the camera.

- Once the image is recorded, it's stored on a hard drive, much like a digital photo. The saved image is good for the life of the subject, as irises don't change after a person's first birthday.
- When a person scans his or her eyes to enter a building a similar video camera compares the image to those in the facility base. The process takes a little longer to record the reading than other biometric devices and maintenance can be higher than fingerprint and hand geometry systems.
- The iris scanning technique is much more expensive than fingerprint readers ($3,000–$51,000) but appears to be far more reliable. Newer systems feature two optics to scan both eyes instead of the single eyepiece that has been associated with iris scanners since they first appeared on the market.
- Civil liberties lawyers fear that the camera could be used to covertly identify individuals at a distance without their knowledge, meaning that a person could be tracked anywhere there's a camera and Internet access.

  The eye of a deceased person would deteriorate too quickly to be useful, so no extra precautions have to be taken with iris scans to be sure the user is a living human being.

Advantages

- Very high accuracy
- Verification time is generally less than five seconds

Disadvantages

- Intrusive
- A lot of memory for the stored data
- Very expensive

1.3 Fingerprint vs. Genetic Profile

Fingerprint matching is considered the oldest biometric technology, and for a century, the fingerprint was the most widely used identification device, the predecessor of DNA profiling. According to recorded history, the Babylonians were the first known civilization to use fingerprints to identify criminals almost 4,000 years ago. It is still a relatively inexpensive biometric device when used for access control systems but has lost its share of the access control market each year since the turn of this century to DNA profiling and other, more reliable biometric methods. Fingerprint readers are easy to use and rely on the pattern of ridges, whorls, loops, arches, and furrows that are recognized as distinctive to each human's fingerprint. Fingerprints may change in size as a person grows, but the structure of the ridges and other features don't change as people age. Fingerprinting is also considered less intrusive than other biometric measures, and almost everyone has had their fingerprints recorded at some point during their lifetime.

There are two different types of fingerprint biometrics; one is used by government agencies and law enforcement for immigration and police investigations and is very sophisticated since it is used as reliable evidence to identify criminals. The other is a very different class of fingerprint readers used for commercial purposes that simply open doors or activate computer keyboards and go for a much lower cost.

Modern fingerprint technology started in 1880 when a British physician discovered that powder could make the fingerprint visible, a detail that could possibly link the fingerprint from the crime scene to the perpetrator. Scotland Yard quickly upgraded the discovery by developing the first fingerprint database and accumulated a large collection. That collection gave birth to the forensic science of fingerprints, which were permitted in court as evidence. A milestone occurred in 1903 when two convicts with the same name and same physical characteristics were imprisoned in the same federal penitentiary and the only way to clearly distinguished one from the other was through their fingerprint.

The acceptance of fingerprints as a crime-fighting technique grew slowly with only a few law enforcement agencies in the US collecting them and creating databases, until People v. Jennings in 1911 when a landmark decision was made and a print of the fingers was accepted by the Illinois Supreme Court. It became the first American court to allow an evaluation of the controversial methodology. The case involved an intruder who had entered a house and shot and killed the owner. A suspect was apprehended wearing bloodstained clothing and carrying a revolver that was matched to the unused cartridges from the residence. In addition, four fingerprints were found in fresh paint on a fence near a window the intruder had used. Four experts, claiming to be familiar with the new fingerprint concept, testified that the prisoner on trial was the source of the telling print evidence.

That opinion provided almost a century ago by so-called forensic experts led to a guilty verdict by the jury. An appeal was made challenging the admissibility of fingerprint evidence since there was no scientific data indicating their uniqueness. The appeal was heard by a higher court and was rejected, allowing fingerprint evidence to be accepted and the court sanctioned the testimony of fingerprint expert witnesses, "who demonstrated a peculiar knowledge or experience not common to the world, which renders their opinions foundered on such knowledge of experience, an aid to the court or jury in determining questions at issue."

In other words, the original verdict stood based on the evaluation and testimony of experts, and the defendant was hanged shortly thereafter.

In 2004, the FBI arrested a man living in the Northwest named Brandon Mayfield, claiming his fingerprint matched those obtained at the scene of a Madrid terrorist attack. That "perfect match" turned out to be flawed. The innocent man sued the government and was awarded a $2 million judgment. (NY Time 6/5/2004, *Spain and Us at odds over Mistaken Terror Arrest.)* The case became a prominent fingerprint misidentification and a challenge to the common thought that fingerprints were as reliable as DNA evidence and matches.

There is no agreement by forensic scientists on how many points of comparison are necessary to determine a conclusive fingerprint match, though an increasing amount of research and studies have been done to try to both professionalize, and bring reliability, to the human analysis process – from an investigative and forensic standpoint. Fingerprint security applications are different than in criminal cases. 10-point is commonly used as a benchmark and standard.

When used as a biometric access control device far fewer points of recognition are used and the result is not always acceptable, especially in high-security situations. Fingerprint readers are not the most reliable of all available access control devices since they produce too many false positive readings, rendering fingerprints unreliable. The inexpensive fingerprint access control system reader used in the commercial marketplace differs greatly from the more sophisticated fingerprint reader used by law enforcement agencies. The law enforcement version takes images of ten fingers instead of the one or two generally done commercially.

The use of the fingerprint reader as a biometric entrance door device began in the 1970s and quickly became the most popular such device on the market with its success peaking in the 1990s. Since the turn of this century, it has gradually been replaced by more sophisticated and more reliable biometric technologies and its share of the market has gone down from about 80% in 1995 to less than 20% in 2014. Most end users who accept fingerprint technology do not realize that the fingerprint image is scanned and delivered as a mathematical representation. This code is literally a piece of data – not the actual fingerprint.

The validation process is familiar to most people where the user presses one or several fingers onto the reader card to be examined to confirm authorization. The reader analyzes the finger's unique ridges and patterns and compares that information against a limited database, held either on file or on the smart card that is often used in conjunction with the biometric reader. The fingerprint, while easy to use and cost effective, can produce false readings caused by environmental factors such as contamination, moisture, and temperature and should not be used outdoors. When used in critical areas where a higher sensitivity is required, the false acceptance and false rejection rate can become troublesome and compromise its value as a security device.

- While fingerprint readers may be struggling to retain some level of popularity, criminals and illegal immigrants are still going to extremes by burning their fingertips to hide their identities. Efforts to erase or alter fingerprints go back to the early 1930s, when infamous gangsters and bank robbers, like John Dillinger, tried to burn their prints with corrosive acid. Some alteration is accomplished by expensive plastic surgery where the skin is overlapped to create new prints.
- Beating the system by using a fake finger made from a cast of an authorized user that includes print minutia was a serious weakness that has been solved using a newer technology that was developed to detect the difference between a real finger and one made of silicon or gelatin. The new system uses a pair of electrodes built into the sensor to send weak electrical currents that measure the conductivity of the material placed on the sensor, allowing it to tell whether the finger is human skin or a plastic compound. These days, most fingerprint devices are able to detect vital signs – for this reason.
- With the new analytical technology being developed, a fingerprint can reveal much more than the identity of a person. It can also identify what the person has been in contact with, including drugs, explosives, and poisons. A tiny sample of electrically charged liquid is sprayed on the fingerprint and the droplets dissolve compounds in the fingerprints and splash them off the surface into an analyzer. The liquid is heated and evaporates; then the electrical charge is transferred to the fingerprint molecules, which are identified on a mass spectrometer costing several hundred thousand dollars. While this process

is not yet adaptable to the mass market, researchers have used the chemical signature process successfully for many years and called the technique "desorption electrospray ionization." Once in mass distribution, a fingerprint could be analyzed not only for ID purposes but also to detect tiny amounts of various chemicals, including cocaine and explosives.

FINGERPRINTS 101: Fingers, thumbs, hands, and toes have ridged patterns that share a biological purpose from which prints are taken. They provide assistance when gripping objects and probably amplify the tactile vibrations (notifying a person) that are triggered when these appendages contact uneven surfaces. Finger and thumbprints are the most common patterns collected with the fingerprint, referred to the "pad" as the area above the last joint of the thumb and the fingers.

- When fingerprint examiners compare two fingerprints, they look for common characteristics and matching patterns. Such matching characteristics that are consistent across fingerprints are called points, referred to in such terms as ten- and 12-point matching. The ridges on fingers contain perspiration, oils, and amino acids, which are left on objects touched by a person. Most of the time, these prints are practically invisible and cannot be seen by the naked eye. These are called latent prints, while visible prints are those that are discernible when the finger comes in contact with something like paint, ink, or blood and then touches a surface.

Fingerprints are made up of friction ridges, which are raised strips of skin. There are three main types of fingerprint patterns: **loops, whorls, and arches**. Loops are found in 65% of the population, while 30% have whorls and 5% have arches. A fingerprint is identified as a *loop* when the ridges begin and end on the same side of the fingertip they started on. On the *whorl* fingerprint, the ridges form a spiral that does not seem to begin or end on either side and usually form a circular pattern around a central point on the finger. The ridges in an *arch* pattern rise in the center and begin on one side of the fingertip and end on the opposite side.

- Fingerprints can vary according to factors such as hormonal changes, weight gain or loss, and aging. Fingerprints naturally produce a glandular secretion from within the skin's friction ridges that can be used when pressed against a surface. Since it is important to make faster imprints, a more permanent method is to place the finger on an ink pad prior to pressing on the paper to collect the pattern image.

Once the print is collected, the next step is to identify the unique characteristics of each ridge pattern. In the days before computers, an army of people would visually compare the fingerprint against all the questionable fingerprints on record. Once computers arrived, scientific calculations replaced the inefficient human eye with a faster and more reliable comparison. These calculations were originally made up of 12 points, but some of the more sophisticated systems use additional points for a more positive identification. The fingerprint reader and associated software program are used in three different ways:

- Optical is where a light source illuminates the surface of the scanned area, and a charge-coupled device collects an image on the illuminated surface.
- Thermoelectric detects a localized heat source emitted by fingers and reads the electrical variances on the surface to acquire an image of the fingerprint.
- Ultrasound is rarely used because of the cost. It collects ultrasound images of the fingerprints by bouncing very high-frequency sound waves off the three-dimensional structures of the prints and records a 3D model in a database.

Only about one in 50 people in the world lack fingerprints. Some of the causes are occupational and some health-related, including:

- Hand-foot syndrome, a condition caused by some chemotherapy medication
- Work as a bricklayer
- Work with dry cleaning solutions

- Rare inherited conditions called Naegeli syndrome and dermatopathia pigmentosa reticularis.

  Live Scan technology replaced the familiar ink and paper method of obtaining fingerprints by digitizing the fingerprint minutiae (the characteristic that makes fingerprints unique). Traditional ink cards had to be sent to state law enforcement agencies to determine if a suspect had a criminal record, a process that could take up to three weeks. If a suspect was released while awaiting the results, the department had to then use its manpower to track down the suspect if there were outstanding warrants. In addition to the length of time involved, the ink cards were sometimes deemed unreadable if too much or too little ink had been used. It was estimated that 10% of ink cards are unusable. Live Scan systems do not have that problem since they are inkless and digitized images are sent directly to a databank, usually at the Department of Justice, that stores millions of images. The response time to an inquiry from a police department regarding a suspect's identity, criminal record, or outstanding arrest warrants is sometimes completed within an hour, while the suspect is still in custody.

  Embedded fingerprint technology is gaining some attention from the security marketplace. It places the fingerprint directly on the access control ID card itself without any need for a database or servers to hold the personal biometric data. No computer is involved to register the fingerprint. This self-contained card is slightly thicker than the conventional prox card and contains software, a rechargeable battery, and an indentation for the fingerprint sensor. It is compatible with existing proximity readers so there is no extra hardware involved. Once enrolled, the card is held in the hand with the thumb resting on the spot designed for the finger. The prints of several other fingers can be stored for multiple access use. For example, the right thumb will open interior doors, the right index finger lets the user log on to their computer, and the left thumb allows access to another door. For increased security, the use of both the thumb and index finger of the right-hand forefinger may be used for accessing another door, etc.
- There are also some prototype cards that hold the embedded fingerprints that are read on special readers connected to a databank. This type of smart card is in use at all seaports under the TWIC program.
- Another card has an indentation that reads the user's thumbprint while holding the card in a normal position before placing it near the prox reader.

Advantages	Disadvantages
• High accuracy method • Most economical biometric • Authentication • Being a familiar device and considered • Nonintrusive • Easy to use and is standardized • Less intrusive than other physiological biometrics	• Criticized because of its stigma as a criminal ID • False readings: dryness of the skin or dirty finger technique or age • Not appropriate for children: fingerprint changes quickly

A simple touch of a finger unlocks a smartphone or computer – no password required but research has shown that smartphones can be fooled by fake fingerprints digitally composed of many common features found in human prints. While full human fingerprints are difficult to falsify, the finger scanner on phones is so small that they read only partial fingerprints. The typical setup procedure requires eight to ten images of a finger to make it easier to make a match and since a finger swipe has to match only one stored image to unlock a digital device, the system is vulnerable to false matches. It's a similar situation to using 30 passwords and the attacker only has to match one to enter the system.

It's estimated that roughly 55% of smartphone users use a biometric authentication factor (face or fingerprint), and the majority are thought to be Gen Z and Millennials. Time is of the essence when police try to use a dead person's fingerprint since the fingerprint function stops working after 48 hours after the last time it was used to unlock with the fingerprint. After that time lapse, the user is required to tap in their password. Some fingerprints are difficult to reproduce as living tissue and more sophisticated systems use capacitive scanners to detect the electrical properties of the human skin as part of the measurement. After death, the conductive properties of the skin are quickly lost.

HAND GEOMETRY – PALM SCANNERS: This technique is often used to log timing and attendance and as a payroll recorder. It matches the topology of the hand – size, length, and width – for verification against a previously stored database of the individual's handprint at a particular facility. The technology has several advantages over fingerprints and iris scanners, mainly cost, low maintenance, and ease of operation. It is also one of the fastest biometric devices, taking about ten seconds to confirm authorization, which is a great advantage where large groups of people need access at any given time. Its main drawback is that hands do not give unique biometric prints, which can cause a higher percentage of rejections or false acceptance rates, whereby an unauthorized person is granted access by mistake. One problem that has caused some reluctance from potential customers is the possibility of contamination by placing the palm on a platen previously used by unknown others.

Palm scanning is losing ground to hand vein scanning but is still frequently used by law enforcement to provide a more complete analysis of crime scenes. About 20% of the prints taken from crime scenes are palm prints, though fingerprints are still the prominent biometric used by authorities.

Advantages

- Easily integrated into other devices and systems
- No public attitude problems as associated with other biometric systems
- Amount of data required to identify the user is the smallest of other ID systems

Disadvantages

- Expensive
- Considerable size
- Not valid for arthritic or injured person unable to place hand on scanner properly

1.4 Vascular – Hand Vein Pattern

A recent addition to the array of biometric systems relies on hand-vein patterns that appear just below the skin surface on a finger or on part of the hand. The device recognizes and records certain distinct characteristics of veins in the wrist, palm, and dorsal surfaces of the hand (and even on the surface of the finger). It captures an individual's vascular pattern, usually on the back of the hand or the palm. The main dorsal blood vessels have higher temperatures than surrounding tissue, so they appear brighter in the infrared image. The reader separates the vein patterns from the background skin surface and must be able to adjust to changes. Since blood vessels grow as people grow, the system only records the shape and distribution of the veins. It won't work in an environment where the hands get dirty.

The veins throughout the body have a similar pattern to those of other people, but there are enough differences to determine the identity of the person at a higher acceptance rate than other hand-based technology. An image of the vein structure is taken by shining an infrared light into the hand, which makes the red blood in the veins appear black, while the surrounding areas appear white. The high contrast image of the vein structure makes it suitable for comparison and subsequent authentication.

After a slow start in the biometric industry, vein technology found a niche market where conventional fingerprint devices were unacceptable for privacy reasons. Some organizations (labor unions in particular) do not feel comfortable knowing that employees' fingerprints are on file with the law enforcement agencies. Since there is no government database on vein patterns, these recognition devices became more acceptable in certain sensitive situations. Also, in rough environments where there is a risk of getting fingers oily or dirty, identifying veins in the finger has not been met with the same level of success as when devices were developed to identify veins on the back-of-the-hand. Studies and experiences indicate that vein patterns are unique like finger and palm prints, and the images lend themselves to analysis better than other surface methods.

The vein technology was used in about 2002 when a researcher for a major digital developer recognized that the veins in the hand appeared as unique as a fingerprint and could be photographed. A severed finger or hand will likely fool a fingerprint reader, but that same finger or hand will not produce a good vein image, so it can't be used to gain access. Veins on a severed hand will not produce the "infrared heat" that is needed to scan them. Palm scanners are installed in hundreds of schools and just as many hospitals and are also used at ATMs. Hygiene is not a concern because users hold their hands above the device to register a scan. To date, vein pattern recognition has been more widely adopted and utilized in Asia than other parts of the world.

1.5 Photographs

This common biometric card was first used as a recognition ID system in state motor vehicle agencies when the picture on the license was used to match the picture from the website of "America's Most Wanted." It is now the most common biometric measure (facial recognition), used at airport terminal security desks and, of course, passports. The more sophisticated computerized facial-recognition systems are gaining popularity and one state has a database of 9 million digital license photographs and can find an exact match in a matter of minutes. About six states are working on the same biometric technology where their databases are being combined with increasingly accurate facial-recognition features to be a practical resource for the law enforcement community. There may be other more reliable biometric databases, such as DNA and fingerprint information, but these biometric patterns are not commonly collected from the general public. This computerized facial-recognition matching system has discovered hundreds of cases of fraud each year in states that have adopted this technology. Some suspects make it easy for investigators by wearing the same clothes when captured as they did when seeking licenses under different names.

Four states require "neutral" expressions to avert fraud when posing for the license picture, which is no laughing matter. Arkansas, Indiana, Nevada, and Virginia insist on serious poses at the Department of Motor Vehicles (DMV), so the applicant can't smile very much. These states installed high-tech software that compares a new license photo with others that have already been shot and stored in the database and will expose someone who doesn't have the same facial features as the real person. Problems arise if the person grins since it can give a false identification. Dull expressions make the comparison process more accurate, justifying the no-smile policy. Individual states may allow slight smiles but reject an attempt to grin. A total of 31 states have been users of computerized matching of driver's license photos and most of them do not follow the no-smile policy. One state reported that in a ten-year period since the software was installed over 6,000 people were caught trying to obtain a fraudulent license.

Other practical uses are not hard to imagine and may involve the coroner's laboratory, which compares photographs of unidentified dead people to the DMV's. The process may be considered low-tech since it is not as accurate as other commonly used biometrics. The picture image on the license is used with algorithms that take into account about 8,000 facial data points – allowing the computer to narrow the field of potential matches. The mortuary staff can then review the picture for a perfect match by looking at lips, noses, ears, scars, and tattoos.

Critics of the technology are worried about privacy and contend that a license is for driving a car, not to be used as an identification arm of law enforcement. The initial goal of searching for facial characteristics on the license was to stop people who try to get a second license after their first had been suspended, not identifying criminal suspects.

1.6 Facial Feature Recognition

130 million American adults were in facial-recognition networks used by government-controlled law enforcement agencies for various criminal investigations. That number is surely higher today.

Many law enforcement agencies began using facial recognition to find suspects caught on surveillance cameras and compared the image against databases of driver's licenses and mug shots. This type of biometric system operates by using the camera to take a photo of a person and compares it with an image stored on the database. All a thief would need to fool the system would be to hold a photo of the face of the real person in front of the camera. The advantages of a facial-recognition system are that it could conceivably identify a known group of suspected terrorists (or the common criminal) by programming it to scrutinize large crowds of people and spot the face of a specific person. A template is made of the shape of the face – determined by distances between the eyes, between the eyes and nose, and other distinctive facial characteristics and features proven to be unique.

When recorded on surveillance cameras, images are matched against the template, but people moving in crowded areas or wearing glasses or a hat proved to be difficult to positively identify. Facial recognition offers great potential, but it is not ready to be competitive against other, more reliable biometrics. It is, however, being accepted as a valuable ID method when used to identify and grant access to a computer terminal, for example.

Passwords and fingerprint biometrics have been used on many computer terminals since about 2002, but facial recognition is the newest technology for this application and has many advantages over most other ID sensors. Facial-recognition devices check computer operators by using a small video camera in the computer, a concept that received a great deal of attention when Apple introduced the iPhone X in 2017 which relies on facial recognition as the password to unlock it.

Facial-recognition technology quick went mainstream and is no longer found just in police departments. Facebook employs it to let its users more easily identify their friends in photos. The iPhone and Android cameras can identify Facebook friends by simply snapping pictures of them. Yet with all their potential as security devices, facial recognition and iris recognition technologies still make up only a small percentage of the biometric industry that has been dominated by fingerprint technology.

All facial-recognition systems require software programmed to capture full-face images and an axis on the screen helps users position their face so as to record the strongest image. The continuous facial authorization functionality checks computer users' faces every ten seconds. When the operator steps away from the computer, the system locks. Upon their return, users see an image of their face on the computer screen as it is matched to the database, activating the system. The system must be carefully programmed since it can inadvertently turn off the computer if the user turns away and faces the side for more than a few seconds – or moves too far back from the screen.

Facial-recognition systems used for crowd surveillance and recognition first appeared on the market in 2011 and received a great deal of fanfare due to their potential for law enforcement agencies, even while raising important questions about privacy and civil liberties. The system was initially sold for use by law enforcement and expanded to the healthcare and financial industries soon after. The device attaches to the back of an iPhone and allows an officer to snap a picture of a face from up to five feet away or scan a person's irises from up to six inches away. It can immediately search to see if there is a match in a database of people with criminal records.

A fundamental question is whether or not using the device in certain ways would constitute a "search" that requires a warrant. The issue has not, as of yet, been challenged in court. It is generally legal for anyone with a camera, including the police, to take pictures of people freely passing through a public space. However, once a law enforcement officer stops or detains someone, a different standard of the law might apply. The US Supreme Court ruled that there must be "reasonable suspicion" to force individuals to be fingerprinted, but facial and iris recognition technologies haven't been put to a similar legal test. A warrant might be required to force people to open their eyes since iris scanning must be done at close range, which could be considered a "search."

With little publicity and only a few complaints, the American military began recording biometric identifying information on a vast number of the population in Afghanistan and Iraq in early 2011 (particularly men of fighting age). By 2012, biometrics on over 1.5 million Afghans had been placed in databases operated by the United States, NATO, and local forces. That number represents the equivalent of one out of every six males aged between 15 and 64. In Iraq, the number was one out of four males of fighting age. To develop the data, soldiers and police officers take digital scans of eyes and photographs of the face and fingerprints.

There have been attempts by some research labs to build systems that could match faces in a crowd with the names on a terrorist watch list. The system could also be used to identify criminal suspects at high-profile events like a presidential inaugural parade, look for fugitives in places like Times Square, or even identify known card cheats in crowded casinos. While both the government and the private sector have made significant strides in facial-recognition devices, there is still no reliable system that can separate people in a crowd by using facial features.

The Department of Homeland Security tested a crowd-scanning project called the Biometric Optical Surveillance System or BOSS. The system cost over $5.2 million and consisted of two towers holding robotic camera structures with infrared and distance sensors. They take pictures of the same subject from slightly different angles. A computer processes the images in a 3D signature created from data that records the ratio between various points on someone's face to be compared against data about faces stored in a watch-list database. The system had not perfected the surveillance of crowds from a distance in which lighting and shadows vary, and faces tend to be partially obscured or pointed in random directions. These environmental conditions have created obstacles that have not been effectively addressed as yet. The perfect facial-recognition system will not be available without further technical advances and testing.

A growing number of privacy advocates hope the BOSS system will never be available unless the government establishes oversight rules and limits on how it will someday be used. While this technology helps law enforcement agencies, it can easily be abused if authorities use the images to identify antiwar protesters, people at other legal gatherings, or even track the public movement of ex-lovers or rivals. The Department of Homeland Security is not the only agency developing facial-surveillance capabilities (Figure 12.1). The FBI has spent more than a billion dollars on its Next Generation Identification program, which contains over 30 million searchable photos, some of which were supplied by seven states from their database of driver's license photos. Defenders of this practice claim that it is not a violation of any law since no one has a legitimate expectation of privacy when it comes to public information, including the use and distribution of a license photo.

There is a vast difference between a law enforcement officer sifting through thousands of hours of video footage in search of a person of interest and using software to instantly locate that person anywhere at any time. Prior to the introduction of facial recognition's sophisticated technology, time and effort created effective barriers to surveillance abuse, but those barriers are quickly disappearing. In order to uphold the nation's privacy protection laws, it is felt by many that the law enforcement agencies should be given access to crowd surveillance images only after a court grants a warrant. Authorities should be forced to report what database is being mined and provide aggregate numbers on how often they are used. Furthermore, the facial-recognition database should be populated only with known terrorists and convicted felons. Driver's license photos and any image

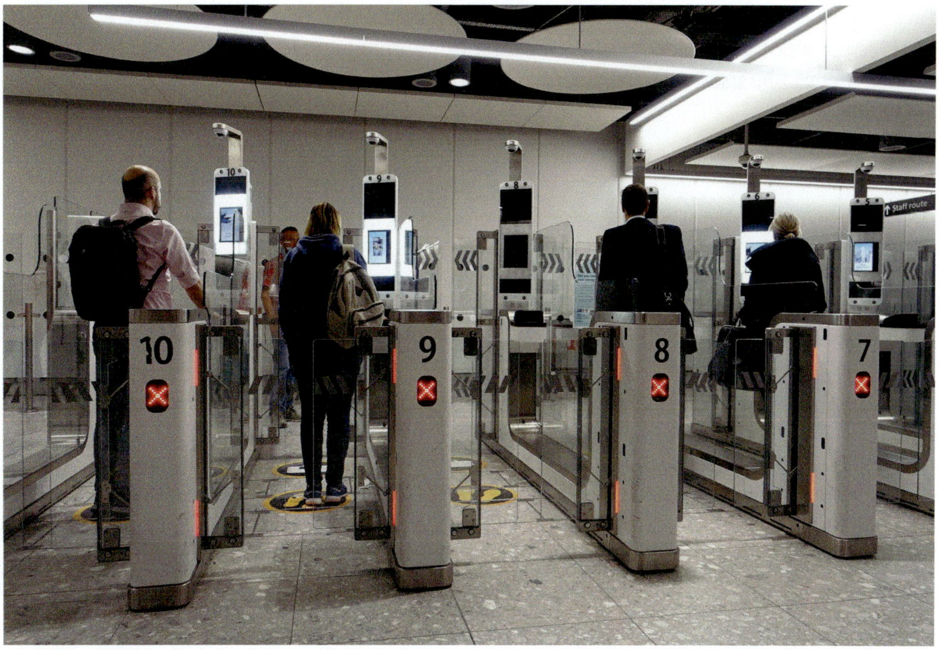

FIGURE 12.1 Air travelers pass through automated passport border control e-gates at Heathrow Airport. Heathrow and Britain's Border Force use facial-recognition technology.

Source: Shutterstock.com.

of ordinary people should not be included in the database without the knowledge and consent of the public.

- FACIAL EXPRESSIONS. Computer software has been developed that uses frame-by-frame video analysis to read subtle muscular changes that flash across a person's face in milliseconds, signaling emotions like happiness, sadness, and even disgust. The technology may be new, but the concept has long been recognized by scientists who, since Darwin, have systematically analyzed facial expressions, finding that many of them are universal. People often reveal their private feelings in fleeting facial expressions, visible only to a best friend or to a skilled poker player. Humans are remarkably consistent in the way their noses wrinkle or their eyebrows move as they experience certain emotional reflexes. People can be trained to spot changes in facial muscles, learning to distinguish common expressions by studying photographs and video. Computers have been programmed to make those distinctions practical and reliable since the mid-1990s.

Software has been available for many years that uses an algorithm to analyze subtle expressions. It has shown a great deal of potential in both the commercial marketplace and in the security community.

Facial-recognition and profile technologies appear to work best when people have normal facial expressions, since a smile or a frowning face can confuse the recognition software. The stored algorithm works best with no expression which led the New Jersey Department of Motor Vehicles to prohibit people from smiling when being photographed for their driver's license. No one noticed the new smile-ban until a lady stormed out of a motor vehicle office after refusing to pose with a sad face since it was in violation of her First Amendment rights. She had the last laugh when the state waived the rule and issued the license with a smiling face. Facial expressions can be acceptable in certain applications like dating services, as long as people agree in advance to have webcams watch and analyze the emotions reflected in their faces. But without consent, it raises serious privacy concerns since it could be used at airport screening areas without a person's knowledge.

Advantages

- Nonintrusive
- Many uses

Disadvantages

- Affected by changes in lighting, the person's hair, age, and if the subject wears glasses
- Requires camera equipment and is not likely to become popular until most PCs have better cameras

The practical applications of using a computer programmed to read facial expressions as an indication of someone lying would include the replacement of lie detectors. The program could also be installed at border crossings and at airport security checkpoints; it can also serve as a diagnostic aid to doctors. Researchers have written software that not only detects whether a person's face reveals genuine or fake pain but does so far more accurately than human observers. People generally excel at using nonverbal cues, including facial expressions, to deceive others (the so-called poker face). They are also good at mimicking pain, instinctively knowing how to contort their features to convey physical discomfort. The medical staff success rate in determining if a patient is really in pain or is faking it is effective 50% of the time, meaning that detecting fake symptoms is like guessing.

Short Comings Software developed by CERT (computer expression recognition toolbox) can measure the presence, absence, and frequency of 20 facial muscle movements and is accurate over 85% of the time in detecting fake pain and work is continuing in this area. A Georgetown University conducted a 2017 study that found face recognition less than acceptable in identifying (1) African Americans, (2) women, and (3) young people.

To this point, several cases – and articles published in recent years – have highlighted how racial bias can be inherent in the datasets used to train artificial intelligence. Facial recognition is a thing, but also bias in it. (*Unmasking the bias in facial recognition algorithms | MIT Sloan.* (2023, December 13). MIT Sloan. https://mitsloan.mit.edu/ideas-made-to-matter/unmasking-bias-facial-recognition-algorithms)

1.7 Nose Recognition

The nose is a facial feature that could be ideally suited to biometric identification. It is almost impossible to hide, isn't altered by facial expression changes, and relatively few people undergo rhinoplasty (a nose job). Detractors of this biometric technology claim that, besides surgery, the nose shape can change due to an accident or normal age-related changes. Wearing glasses that rest on the nose could also affect the reliability of this ID technique. Testing continues at several research universities using a three-dimensional image of the nose, similar to the method used to capture human facial features. The initial efforts to concentrate on this single feature have defined different nose types: turned up, Roman, Greek, snub, hawk, and Nubian.

The recorded images focus primarily on the nose bridge and the tip as well as the nasion, the depressed space between the eyes that is right at the bridge of the nose. The ridgeline is the feature that has had some success as a one-dimensional measurement that distinguishes one nose from another. Experts in the field recognize that the nose will never be as reliable as iris pattern recognition, but it could possibly be used in conjunction with other biometrics. Entrances that have secured turnstiles could use a photo of the nasal shape in conjunction with other ID devices, such as a swipe card or fingerprint pad, to speed along the movement of people through checkpoints.

2 HUMAN-DEVELOPED FEATURES

Nearly everybody lies but relying on hunches to spot a lie is unacceptable for many reasons – it takes science to reveal the truth. There are five channels of emotional expression in humans including body gestures, choice of words, voice pitch, and the way the person speaks, but facial expression is the most reliable source of analysis. A poker face is not necessarily an indication that someone is lying. To lie, a person needs to work on more facial muscles and even experts in the field have trouble separating a pathological liar, or if the lie is just a situational phenomenon. Handwriting is the result of interactions between various structures and circuits inside the brain, and an analysis can be a more reliable prevarication detection method.

2.1 Handwriting Analysis

When learning how to write, people are taught to form letters by copying examples in a rote way so that at a young age handwriting looks a lot like everyone else's. As people get older, they start to develop their own unique writing styles. Eventually, writing becomes subconscious so that no one even thinks about how the letters are formed. Some people mix cursive with print writing; others embellish their writing with fancy swirls; some write letters in a word close together while others spread them apart. Then there are those who slant their letters to the right or to the left. It is a combination of these factors that make a person's handwriting unique and recognizable by experts in the specialized field of graphology.

Signing another person's name is the most common type of forgery. There are three ways a person may forge a name. A traced forgery is done when a real signature is traced onto a different document. In a simulated forgery, the forger carefully writes the name freehand by looking at the real signature. Blind forgery is when the forger uses his or her own handwriting to sign someone else's name without any effort to disguise the handwriting and sometimes not even knowing what the genuine signature looks like.

AUTHENTICATION/VERIFICATION devices capture six handwritten sample signatures using a graphic tablet to create a template for future comparison. The system does not rely on the physical appearance of the signature but instead focuses on the manner in which the signature was written. It captures special algorithms to measure a signer's rhythm, speed, pressure, and acceleration against a referenced database. While it is easy to duplicate the visual appearance of any signature (on a copy machine), it is very difficult to duplicate the behavioral characteristics of the signer. Some independent tests indicate more reliability and fewer intrusions than other biometric techniques, but signature verification has not become a market leader. Mostly used in banking transactions, it is slowly getting accepted by many security applications where the speed of identification is not essential. This ID technique has improved substantially since the 1970s when IBM produced the first signature recognition system. Those early devices did not provide something that was trouble free, and it wasn't overwhelmingly accepted when compared to more reliable fingerprint and hand geometry tests. One of the drawbacks of those older signature biometric systems was the need for massive storage and the high-cost computer terminals. With lower costs and the availability of expanded storage capacity, signature authentication technology can provide a precise biometric profile trait that is as unique as a fingerprint while also being cost-effective.

2.2 Two Types of Signature Verification

- Dynamic/biometric requires comparing a captured signature against the sample biometric signature previously stored in the database. Biometric handwriting characteristics are unique and virtually impossible to duplicate and remain one of the more reliable of human identifiers.
- Static signature verification is as accurate as the dynamic methods and is far superior to visual verification. It is a procedure that is widely used in banks to detect forgeries and with the proper software, it can easily be deployed worldwide. It is also finding a niche in authenticating absentee ballots.

KEYSTROKE PATTERNS measure and identify a user by the habitual "typing dynamic" patterns that are unique to each individual just like handwriting and thus can be used biometrically. A device is programmed to capture certain characteristics such as pressure, speed (the time it takes for the user to type particular words), dwell time (the time a person holds down a key), and flight time (amount of pause time between keys). Keystroke pattern systems are not very popular, yet they are the easiest biometric system to implement because they are completely software-based and can be installed on a computer. The main drawback is that it does not replace usernames and passwords. Its advantage is that it enhances the security of existing username/password-based systems. The system is losing favor since typing is not as deeply ingrained as walking or writing for most people, but studies have shown that such factors as inter-character timing and the dwell time can yield a very accurate identification of the typist. The potential is very limited since people are leaving typing behind and spending more time on text messaging – and there is no work being done to use texting dynamics as a biometric tool.

Before the computer keyboard replaced the typewriter, document examiners were skilled in analyzing text from a typewriter, which was much easier to identify than the keystroke pattern. Typewriters become old and worn and begin to make unique marks when printing. The FBI maintained a file of the makes and models of typewriters and could compare known types to identify the possible source.

In the 1880s, telegraph operators had personal styles of tapping out dashes and dots, and the same is true for modern-day computer users. This idea of people possessing unique typing patterns has led scientists and companies to experiment with computer security systems. Based on "keystroke dynamics," or subtle variations in typing styles, researchers at several universities developed a keyboard that captures the electrical charge emanating from human skin. The software is then able to differentiate users from one another by their typing patterns. Some experts believe that the electronic typing patterns individuals produce are the equivalent of a fingerprint in their uniqueness and consistency.

The system requires a keyboard to be coated with a layer of electrode material that conveys electrons from a nonmetal substance, like the human fingers. The flow of electricity creates a record of both the speed and intensity of each keystroke, plus the manner and rhythm, typing habit, finger size, and applied typing force. These factors are correlated to produce a detailed electronic profile for each user. The technology became available in 2017.

SPEECH RECOGNITION or voice recognition combines both physiological uniqueness (larynx) and behavioral characteristics (word pronunciation). Either characteristic by itself could produce a reasonably unique biometric measure, so combining the two produces a surprisingly accurate biometric. When acquaintances call on the phone, their voice and speech pattern make it easy to identify them. Based on this fact, systems have been developed that collect a database by requiring each individual to enroll by repeating set phrases. For example, "the worm can be used for bait" is used to enroll the individuals into the database. Speech recognition systems that attempt to know how a person speaks, no matter what is being said, are far too complicated and will use way more computer disk space than necessary.

GAIT-LEG RECOGNITION is based on the way people walk since the way the body moves is distinctive and can be measured. Someone walking in the distance can often be recognized well before the person's face is in view. Studies have shown that gait is a measure that is both behavioral and physiological because the way we walk is a learned trait. Gait is somewhat dependent on physical characteristics such as weight, length of legs, and foot size. In anti-terrorist applications, being able to determine whether a target is among a group of people walking around a campsite from a half-mile away is another advantage of the yet to be fully developed technology of gait biometrics. Part of the appeal is that analysis doesn't require the subject's consent and even blurry images can still be useful. A study conducted in 2018 found that a person's unique gait is set by the age of 7 and will rarely change throughout the life cycle. (IEEE Internet of Things Journal, 2018; Birch, I., Nirenberg, M., Birch, M., & Vernon, W. (2020). *Forensic GAIT analysis: Principles and Practice.* CRC Press.)

HAND-GRIP PRESSURE has been investigated since 1999 under a study conducted by a state college (and sponsored by the state of New Jersey) to study biometric techniques to prevent children from using a gun that is left on the premises due to parental irresponsibility. Locking the gun in a box or using a trigger lock was determined to be inherently unacceptable since the key could be stolen. Fingerprint readers proved ineffective with false readings due to dirty and sweaty hands. Prox cards could be carried by someone other than the authorized user. A conventional trigger lock was not acceptable since the key would have to always be in a safe place, which wasn't a feasible solution since keys can be stolen. The study focused on a dynamic biometric technique that measures the way and the amount of pressure is applied to a gun handle by the authorized owner (Table 12.1).

TABLE 12.1
Comparison chart

Biometric Identifier	**Operation**	**Comments**
		Rate Scale 10 – best; 1 – worst
Fingerprint recognition	A digital scan of the fingertip is taken and whirls, arches, and loops are recorded. The units are inexpensive copies of the models used by the FBI and not as reliable.	Used with card readers at doors and time and attendance terminals. Found to be very effective, require little maintenance. Failure rate acceptable. **Rated 6**
Hand geometry	Person's hand is measured and analyzed, including shape and size of the hand, length of fingers, and ridges in the skin.	Same general application as the fingerprint device, but appears more effective. Because of higher maintenance. **Rated 7**
Facial recognition	Attributes of a person's face are measured and analyzed, including the shape of the face, distance between eyes, nose, mouth, and jaw.	Very new and is being tested as a distance surveillance device for perimeters or parking areas. **Unrated** – too new
Voice recognition	Scans of voice using a selected word or phrase are recorded. The idea is as old as the telephone.	Not used as an access control reader. Probably more suited or financial services. **Rated 5 –** some potential
Blood vessel recognition	Registers a user with a passive IR image of their blood vessel pattern. Stores it in a template database.	Vascular skin devices should become more reliable than voice or iris devices. Not yet a replacement for a card reader. Still considered a little weird. **Rated 6.5** and getting better
Iris technology	Can analyze more than 200 points of the iris including rings and furrows.	One of the more popular devices and used in certain areas of a hospital or airport where very high security is required, for example. **Rated 9**

(Continued)

TABLE 12.1 (Continued)

Biometric Identifier	Operation	Comments
Signature recognition	Signatures are unique and banks have been using this characteristic for decades. Starting in 2018 major credit card companies stopped requiring holders to sign their signature for a transaction. The chip embedded in credit cards provide more fraud protection to the than signature could provide.	Can be used to control documents and acknowledgment of service received. **Rated 5** and going lower

LINGUISTIC ANALYSIS is not very practical, but studies of email messages have indicated that it may be possible, in certain situations, to actually predict human relationships based on the language displayed in a large sample of the original email messages. The basic idea is to identify the person who wrote something by analyzing the characteristic way they structured the language, examining: how are the sentences are put together? What words are frequently used? How are the ideas presented? The messages can then be biometrically evaluated against a pre-programmed database to determine if the writer is disloyal and might leak confidential information.

It should be noted that the current and future advances in AI capabilities are going to impact many of these areas. This includes both the analysis of linguistics and the potential to obfuscate what was written by a human. This is evidenced by as the increase in usage of AI in students writing college essays and professionals generating marketing copy, posing the question: what writing is original versus AI generated.

13 Realm of Fact and Fantasy

1 DEVELOPING TECHNOLOGY

SKIN SPECTROSCOPY recognizes the unique optical properties of the skin. A special sensor illuminates a small patch of skin with multiple wavelengths of visible and near-infrared light. After being scattered in the skin, the light is reflected back and measured for each wavelength. While it may sound like science fiction, the process actually analyzes the reflectance variability of various light frequencies as the light passes through the skin. Skin spectroscopy may never be used at building entrances since it processes data too slowly and, in cold weather, each person would have to remove their gloves. Although it has its limitations as a standalone system, it may be useful as a secondary or redundant system to back up a fingerprint scanner. It can sense body temperature at a great distance and can determine if a person is deceased. First responders use the device to make quick decisions in dangerous mass-injury and rescue situations.

SALINITY is a rarely used device but is worth noting since it is the only biometric system that measures the natural level of salt in the human body. The newer systems use an electric field and salt's natural conductivity to measure tiny electrical currents as they pass through the body. This technique was never demonstrated at any security trade show, so it is suspected that very few such devices have ever been purchased for commercial application.

FACIAL THERMOGRAPHY generally refers to the pattern of heat in the face caused by blood flow under the skin. Special infrared cameras, while very expensive, offer advantages over other biometric devices that may encourage additional research to make this technology competitive with other facial methods of identification. Facial thermography can capture this heat interface to produce a thermal pattern and since every person has a different vein network, the infrared (IR) thermal pattern is distinctive. Of course, if a person drinks too much, the thermal signature on the face could be altered. None of the security directors in the various New York City facilities who contributed information used in this book have ever witnessed this system in operation.

BODY ODOR or scent analysis is usually associated with bloodhounds who follow the odor of the clothes worn by the person being sought. Converting this practice into a technology-based concept is understandable since it is an accepted fact that everyone smells a little different. Each unique human smell is made up of volatilized chemical compounds, the analysis of which can confirm a person's identity. While it may never be used as an access control system, since it is too slow to decode, it may have some use in identifying missing persons.

Even the most pleasantly odor-free person leaves behind traces that trained dogs can identify and follow. Agencies and police departments work with electronic "noses" to identify not only explosives but also any traces left behind by humans. Another factor that limits its future acceptance as a practical biometric technique is that, socially, it's not particularly acceptable to tell people that you would like to identify them by using their body odor. On the other hand, it may be possible to collect "smell evidence" presented as proof that someone was present at a crime scene, allowing the prosecutor to state to the jury that "something about this case stinks…."

EAR SHAPE is not as viable or practical a technique as facial imaging, but it is considered superior to facial thermography, which changes with each expression. The ear's stability over time doesn't change significantly and provides a better-behaved biometric than the face; however, collecting ear images can be compromised due to their being hidden by hairstyles, hats, or cosmetic surgery.

EAR CAVITY measurement technology was developed in 2016 and commercially available in 2018. This technique uses the ear canal as a means of identification and authentication. Since the

DOI: 10.4324/9781003091103-18

shape and size of all ears are unique, research has found that the acoustic features of the ear are just as unique and can be highly effective in identifying individuals. The technology uses an earbud to measure individually unique acoustic characteristics by inserting an earphone with a built-in microphone to generate a few 100 milliseconds of acoustic signals from the earphone speaker and to receive the signals transmitted within the ear through the microphone. The average of the waveforms of the multiple signals received is used to eliminate noise from the received signals. It then calculates how the sound resonates within the ear. Tests have indicated that the accuracy rate for recognizing an individual is greater than 99%. This biometric method of authentication places only a slight burden on users who are already likely to be comfortable with using earbuds. Authorization can be continuously confirmed while users are on the move without being confined to the range of a camera.

BRAINWAVE identification research has had limited success in identifying unique differences in humans, but some labs have continued to investigate this method. The technology, however, has limited use in real-life situations and is not a realistic method to rapidly and reliably identify people. Yet, brainwave technology is not as far-fetched as it first appears since the human nervous system is electrochemical in nature and creates a minute electric field that can be measured. The fields can be mapped using medical equipment such as electroencephalography (EEG) and magnetoencephalography (MEG) machines.

EEG devices measure activity created by electrical impulses passing through electrodes placed on the scalp, generally requiring direct contact with the skin.

MEG is a relatively new technology that uses superconducting coils cooled to -269 degrees Celsius that are placed near the skull to detect electrical fields in the brain.

PET is Positive Emission Tomography that measures brain consumption and activity and can map out structures in the brain.

FOOTWEAR databases can be developed that will store images of footprints at crime scenes that police can check against the markings of common shoes to identify a suspect's footwear.

BODY TEMPERATURE has not been used in biometric technology but is highly effective in Passive Infrared Detection, which simply measures body infrared heat, not distinguishing between individuals.

HEART RECOGNITION uses a bracelet and an algorithm to authenticate the wearer's unique cardiac rhythm against a stored record to grant or block location access or block use of specific electronic devices.

Deoxyribonucleic Acid (DNA) signature is an excellent biometric measurement since comparing one person's DNA to another is more or less foolproof and is not prone to errors that far too often occur in most of the image-based biometrics. To accomplish a DNA comparison, cells must be collected from the person being identified or authenticated and sent to a lab for testing, which can take weeks.

DNA has gained acceptance by the criminal justice system as a foolproof method of personal identification and is considered a frontier in the evolving world of forensic science since it has successfully been used to both identify suspects and exonerate the wrongfully convicted.

A recent development surrounding DNA testing is when it is used in what is called "familial searching." This technique of Forensic Investigative Genetic Genealogy has been used by investigators to search criminal databases to identify someone likely to be a close relative of the person who may have committed a crime. It is less a piece of evidence than it is a lead that can help solve a crime when there is no database profile for the primary suspect. In other words, the suspect can be traced by finding a cousin who is in the database and shares the same DNA. The cousin can then identify the suspect when confronted by the police. This crime-solving technique is becoming more common and attractive to criminal investigators despite questions and the controversial implications for privacy and the investigative technique's efficacy.

DRUNK DETECTION was introduced into the marketplace by the ride-hailing company Uber in 2018. The system can identify if a passenger is impaired after ingesting too much alcohol by analyzing a number of factors: (1) stumbling out of a bar and holding a smartphone at a strange angle, (2) how the person is walking, and (3) the speed at which the user switches through pages and types directions. These characteristics are pre-recorded on the app and compared with other times it was used. This system will likely never be commercially available because of too many variables.

2 PUBLIC ACCEPTANCE OR REJECTION

The most critical factor in the success of a biometric system is "user acceptance" of any of the numerous biometric devices. Whatever the device or method, it should cause no discomfort or concern to a user other than possibly a small inconvenience. When people are concerned about the technology, they usually will not use it properly and that often results in not being granted access even though the individual was authorized. The user must understand how the system operates and that the biometric measurement technology is harmless to the user (as in measuring veins in the iris will not cause glaucoma). Another factor to be considered before selecting a device is that it must be user-friendly. Any complicated device will not be readily accepted; most people prefer things that are simple and intuitive. By example, people become frustrated if there are no indications on card readers as to which direction should be used when swiping the card.

The last and most important factor in the selection process is that whatever device is selected must work correctly with a minimum chance of false identification. For a biometric device to work properly, it must accomplish two things:

- Reject unauthorized personnel attempting to circumvent the system.
- Allow an authorized user to enter the building or system.

People generally treat biometric devices as intrusive, but they have come to recognize that these systems are for everyone's protection. The acceptance rate by users often depends on how fast the system admits authorized individuals and how often the system breaks down or fails to admit an authorized individual. Frustration can set in quickly regardless of the type of biometric system and none of the systems will last longer than the five-year cycle that most undergo before being replaced by more advanced technology. It is expected that the global biometric industry will grow from a $42.9B industry in 2022 to an $82.9B industry by 2027.

3 PRACTICAL APPLICATION

In one hospital, fingerprint readers were tested in the emergency department with 40 readers installed around the hospital and another 20 readers in the administrative office. In the emergency room (ER), where several physicians and nurses shared the same PC, it was very time-consuming for individual users to log on and off every time they needed to pull a patient record or other information. Without some type of ID system, records could be viewed by people who didn't have any proper authorization – a potential violation of federal law (HIPAA). After fingerprint readers were installed in combination with identification badges, the result was a significant improvement over the previous method. When a doctor or nurse approached the station, the computer recognized their badge and automatically logged the person in, who then verified their identity using the fingerprint reader. If the user leaves the computer, the PC senses it and automatically logs them out. If any files are left on the screen, the machine remembers and opens them automatically the next time the same person logs on.

The old system took five minutes to access a particular record, but with the biometric reader, that time was cut down to about a minute or less. The failure rate was minimal, which was more than acceptable considering that no security device ever made will operate perfectly all the time.

It is always desirable that any biometric device has a high rate of reliability, which is rated in two operating categories that describe "error." Two errors that could compromise the value of a system that relies completely on human characteristics for identification are defined as a device that allows unauthorized individuals to enter and one that prevents authorized individuals from entering. These two errors are what separate one biometric device from another. They are specified as the False Accept and False Reject error rate.

3.1 False Accept Rate

The probability of authenticating any user registered or unregistered to another person's stored template is known as the False Accept Rate (FAR). It is important to keep in mind that the only way a false acceptance can occur is if someone tries to use the device. In order to compare different devices, the FAR for each device supplied by the manufacturer should also include the number of unauthorized attempts in order to determine the actual number of possible occurrences.

3.2 False Reject Rate

Also called "Insult Rate," False Reject Rate (FRR) is the probability that the biometric does not recognize an authorized user and therefore denies them access. The FRRs vary from .00066 to 1.0%. An effective way to determine the number of times that an FRR occurs is to list the number of times people attempt to gain access in a given day. A hospital could have a biometric device on the door leading to one floor with a staff of 100 people, each of whom uses the door four times per day. This would yield 400 transactions per day by authorized employees. A false rejection rate of 1.0% would predict that each day four authorized people (1% of 400) would be denied access. Over the course of a five-day work period, that rate would result in 20 problems. Reducing the FRR to 0.1% would result in only two problems during the same one-week period. Acceptable levels of a false accept and FRR depend on the application.

The growth potential for biometric devices is almost unlimited since public facilities throughout the nation often have some form of a card access system, but only about 8% have added a biometric device for security purposes. Usually, the first venture into this field is a fingerprint device used with employee attendance systems before expanding biometrics to the entrances of the buildings. The National Biometric Security Project is a biometric-based test organization, but its test results are never conclusive due to many factors out of their control. Technology is changing so rapidly that establishing testing standards is impractical. With the federal government sponsoring programs and directives issued by Homeland Security, there may be more reliability surrounding the various biometric technologies in the near future.

The public's reluctance to replace keys with a biometric reader has slowed industry growth, but there are a growing number of proponents who insist that this type of technology offers increased security and efficiency without slowing down movement throughout a facility. Typically, a person's hand, fingerprint, or eyeball is measured once when he or she is enrolled in the system and that measurement is stored in a computer database or on a smart ID card, and movement through the system is just as rapid as having staff members swipe their ID cards at the door. Scanning body parts is far more secure than asking users for passwords that can be forgotten or stolen. In psychiatric wards, where patients can steal the card, that same patient wouldn't be able to get through a biometric device. A biometric measurement must be: (a) unique for each individual, (b) must not change, and (c) must be easily measured. Cost and various technical obstacles have slowed mass purchasing by healthcare facilities, possibly due to concerns that these systems leave an electronic paper trail – open to abuse or theft – of the activities of the medical and support staff.

Biometric systems may not be infallible, but they are far more reliable and effective than the highest level of conventional key-locking hardware devices. Biometrics suffer from some potentially serious flaws depending on the complexity of the technology. Retina scanners were so troublesome that when a user had droopy eyelids, they were forced to open their eyes wider than normal; then it was discovered that aging changed the vein pattern on the retina. These units are no longer being made. Facial recognition systems can be faulted by photographs of an individual held up to the camera and, in some cases, are compromised by a drawing. All these shortcomings illustrate the importance of conducting a survey of the available systems before selecting one. Even the time-tested fingerprint ID method used at crime scenes has its limitations. The well-publicized study conducted by the National Academy of Sciences (NAS) in 2009 found that when police match a crime scene print, showing unique ridges and whorls on the skin surface, they frequently rely on only a smudged or partial fingerprint which makes a comparison very difficult, especially when the investigation is performed by a minimally trained person.

The NAS study also found that bite marks that link a suspect's dental molds to impressions and photos taken of a bite wound are highly questionable. Teeth alignment changes and skin shifts when bitten; the visual comparison is considered as much opinion as science.

4 PRIVACY ISSUES

Even when used carefully, biometric devices can easily compromise an individual's privacy, an area that always seems to cause a great deal of controversy, particularly if used to identify people without their consent. People desire security and convenience but often ignore the possibility it can turn their lives into a video library. Hand geometry systems seem to cause the least apprehension because such measurements are not commonly used off the premises, so they would have little or no adverse repercussions if the biometric pattern or unique code were to leak out. That doesn't eliminate the privacy issue since the trail of use, along with the confidential database, can be considered an invasion of privacy.

Employee records are required as part of an access control/biometric system and in some situations, the data is protected by legal statutes. In a hospital environment, staffs have sometimes compromised the privacy of patients by selling their data to the media. Hospitals and many government agencies must establish a policy that treats all biometric information as confidential and should resist any effort to obtain copies through a state's public records law. It is still not clear how the collected data should be treated in the event of a court-ordered subpoena, but generally, the information should not be distributed without a staff member's permission. It should also be noted that cooperation with legal "authorities" is not always in the best interest of the hospital. Whenever there is a problem with releasing patient information, or if a subpoena is served, a facility staff member should seek advice from the legal department before releasing any personal data.

Section V

Analog to Digital Revolution

Digital technology has completely altered how images are recorded, stored, and transferred via the Internet. Digital technology has revolutionized the security industry and offers distinct advantages over analog and traditional security systems. Internet Technology (IT) departments have been given greater control over the operation of security department administrators. This has significantly changed the function of security departments, which must now appeal to IT before purchasing security systems that operate using an IP building-wide network.

Understanding the terminology and operation of IP Addressable Surveillance Cameras and Video Image Compression are two subjects that have become of major importance to the average security director, who often bases the selection of new surveillance systems on often incorrect and/or easily misunderstood information provided by overzealous salespeople. While in many large companies, municipalities, and networks, IP cameras have replaced the existing analog units (called legacy systems), that transition took some time and was not an all-at-once conversion, as will occur with a new technology. Most organizations had invested heavily in legacy analog equipment and while Internet protocol video surveillance was unquestionably the wave of the future, the transition has been costly. Manufacturers claimed that unless the inevitable transition was immediate, the existing surveillance systems are doomed to certain failure – a completely unsubstantiated conclusion.

Analog Video has been the industry standard for many years and its success can be attributed to its relatively low cost and reliability. It is a proven technology that doesn't confuse the buying public with issues like multiple scan rates, different sync levels, frames per second, and compression rates that relate to IP devices. Based on the high level of picture quality produced by existing analog cameras, the cost to replace a system with IP devices was not always justified, depending on the scale of the application and usage. When expanding existing systems or installing new video systems, the components now must of course be digital compatible.

There is no debate about the advantages offered by IP cameras. IP video security is vastly superior to analog systems since it exposes the user to certain features that can't be duplicated by an analog camera, which is limited to displaying and storing images. IP can perform searches far faster than any analog system can ever achieve, especially if the facility still relies on VHS tapes to store images rather than DVDs.

DOI: 10.4324/9781003091103-19

Multiply Video-Compression Schemes are available that allow many cameras to be connected to a single cable rather than running the old coax from each camera to the central office where the monitors are located.

Digital Video Recorders (DVRs) became so successful that the older tape recorders are no longer manufactured and nearly all facilities have either replaced the tape units or will be forced to do so in a few years. DVRs are capable of storing the existing analog imagery with provisions to also store IP images in the future. The choice of the proper DVR is still based on factors not easily understood by some consumers, but units are getting less expensive and offer additional features each year.

14 Digital Communication in Place of Personal Contact

1 IN THE BEGINNING

In 1969, humans first walked on the moon. That same year, a less publicized event took place when the government's Defense Advanced Research Projects Agency funded and installed the first computer network that allowed scientists, engineers, and senior officials to communicate with each other. Public attention to the moon project overshadowed this small, inauspicious computer network project, but while the moon project came to an end in 1972, the computer network, initially known as ARPANET, eventually grew into something bigger even than the moon program – the INTERNET (Figure 14.1).

The Internet is the classic example of how a significant technical advancement, one that was initially misunderstood, changed the world. Spawned by the US government, it has been a phenomenal success simply because no one owns or controls it. Built on the backbone of common carrier telephone companies, it became commercialized into a worldwide information highway. The Internet is the largest network in the world, expanded by over 250 million pages of information every day. There are more than 4 billion internet users, located in virtually every country with no centralized management; instead, it is a collection of hundreds of millions of individual networks and organizations, each of which is a series of high-speed links between major supercomputer sites worldwide, with the main concentration in the United States. While it started as a small research experiment, by the 1990s, there were so many networks in use that the public came to realize that the Internet was not limited to research and educational purposes. In the last two decades, the Internet has flourished without burdensome federal regulations, becoming an integral part of American society and regular attempts at government intervention have not been successful. In other words, leaving the government's policy of "network neutrality" in place is the same mindset as "if it's not broken, don't fix it."

Between the early 1990s and into the second decade of the 21st century, the Internet experienced dramatic growth, influenced by three important technological changes:

- During the initial growth period, the major online services like America Online (AOL), CompuServe, etc., provided email, which caught the general public's attention, but the early programs were limited only to customers of the same service.
- Once these independent services connected to the Internet for email exchange, the Internet became a global switching center. An AOL member could finally send mail to a CompuServe member and so on. The Internet brought the world together for electronic mail and was the forerunner of the digital revolution.
- With the development of Web browsers such as Netscape Navigator, Opera, and Microsoft's Internet Explorer, the World Wide Web experienced explosive growth. Delphi was the first proprietary online service to offer Web access and was quickly followed by many new Internet service providers into a booming market.

2 AREA NETWORKS

The advantages offered by digital technology are almost infinite, and they will eventually replace all the legacy analog systems used to operate conventional communication and security systems: (a) data can be stored with minimal space requirements, (b) that same data can be retrieved quickly

DOI: 10.4324/9781003091103-20

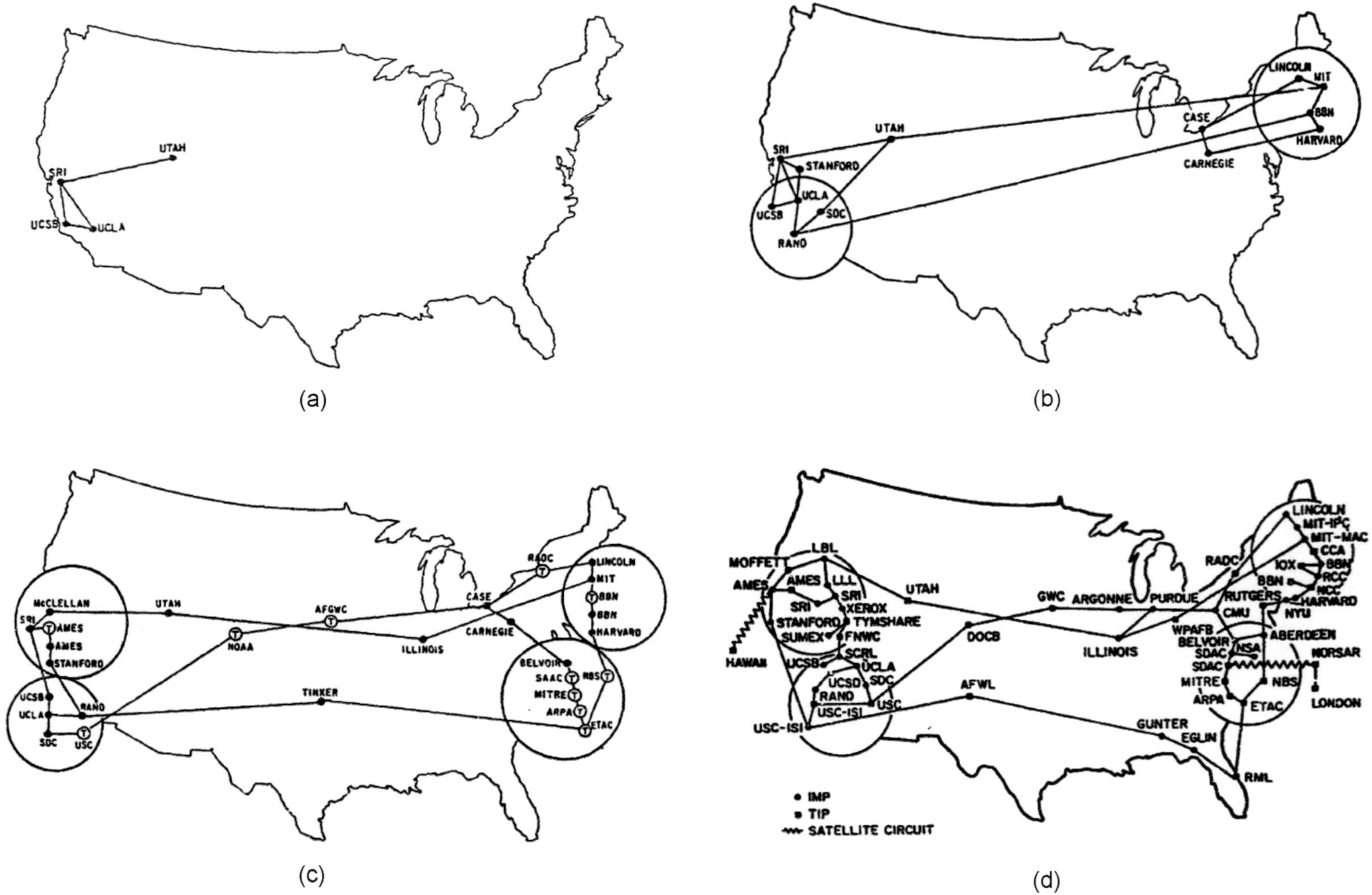

FIGURE 14.1 The early development of ARPANET from an experimental network to an operational Department of Defense (DoD) run operational network (a) December 1969), (b) 1970, (c) 1972, (d) 1975.

and efficiently, (c) data can travel great distances without signal degradation and loss of video quality, and (d) when installed during new construction, digital systems will cost about a third of the cost of security systems using conventional wiring such as coax cable. Once data is collected in a digital format, it can be programmed to provide a converged, single integrated application. It can also interface with different databases to produce information like photos, street addresses, and fingerprints that a secured facility requires. Probably one of the most unique features of all the information-gathering techniques is that a single distributed database can be presented on one or multiple PCs in networks or websites of any configuration. With the advent of cloud computing, such data can be housed, shared, and accessed from virtually anywhere in the world.

Each network cooperates with other networks to direct Internet traffic so that information can be passed between them. Together, these networks and organizations make up the wired (and wireless) world of the Internet. Users have access to unpublished data and journals on an unlimited variety of subjects. In 2005, the European Union and other countries tried to assert international control over the Internet. The Internet lacks a central authority – there is no "Internet Inc." that controls it. Beyond the various government boards that work to establish policies and standards, the Internet is bound by few rules and answers to no single organization. In 1996, the Communications Decency Act was enacted, which provides criminal penalties for those who post or transmit "indecent" material via the Internet. The law was challenged by those who felt it would unfairly prohibit many legitimate uses of the Internet and was ruled unconstitutional.

"Network" is an overused word that simply refers to a connection between two or more computers that form a collection of devices into a "network." Computers can share resources (such as wireless printers, switches, routers, and storage), exchange files, or allow electronic communications. The three basic types of networks are Local Area Network (**LAN**), Metropolitan Area Network (**MAN**), and Wide Area Network (**WAN**).

LAN is confined to the immediate area, usually in the same building or on one floor of a facility. In a typical LAN configuration, one computer is designated as the file server that stores the software that controls the network.

MAN can cover large geographic areas such as entire cities or school districts.

WAN connects larger geographic areas, such as New York, the United States, or the world. A large WAN facility in New York City can communicate with similar types of facilities in Tokyo in a matter of minutes without paying enormous phone bills.

Surging growth in the amount of data transmitted across the Internet, particularly moving images, is cluttering the global network at a growing and daunting rate. The need to carry more digital bits of information requires more and more space – called bandwidth. Computer and big box stores are now selling external hard drives for the home capable of storing terabytes of data, a concept unimaginable 40 years ago. A single terabyte equals 1,000 GB and can store about 1,000 copies of the Encyclopedia Britannica. A petabyte is 1,000 times as large as a terabyte and can store more than 300 billion pages of text. The next generation of digital technology will have to contend with even more storage capacity to handle the changes in the way data is controlled and transmitted. It is a sobering thought that an estimated one-third of the Internet is running on Amazon's AWS service.

3 ETHERNET AND NETWORKS

The method of networking a computer in a LAN configuration goes through the Ethernet, which can be thought of like the cable connecting the computers, but that is not the complete meaning of the word. The Ethernet Protocol is the most widely used "protocol" and is a system where each computer listens to the attached cable before sending anything through the network. Sometimes two computers attempt to transmit at the same instant, called a collision, where each computer on the Ethernet then backs off and waits a random amount of time before attempting to retransmit (Figure 14.2).

FIGURE 14.2 Network cables connected to switches and routers, mounted on racks in a data center.

Source: Shutterstock.com.

PDF is the Portable Document Format created in 1993 to allow document exchanges and was a proprietary program owned by Adobe Systems. The firm continued to release additional versions up to 2008, when it became an "open" standard; there is no monetary obligation for transmitting PDF documents. This popular electronic document format is not easily understood but is essential in transmitting important material in exactly the same format that a document started in. A good example is the creation of a résumé on Microsoft Word that must be sent over the Internet. Most résumés are carefully constructed with margins just right, selecting the perfect font and spacing the lines just so, etc. If the document is to be sent via email to someone who has a different version of Microsoft Word, the carefully drafted document may be distorted in the transition between versions. To overcome this recurring problem, the document is saved as a PDF to send via email or a hard copy is mailed.

A PDF is not a word processing file; it is more like a photograph of that file. The recipient receives exactly what was created, regardless of operating system or software. Virtually all computers now come with built-in PDF viewer software installed.

IT is the common designation for the **I**nformation **T**echnology department that manages and controls a facility's digital network, which carries information on the network, including the alarm signals and camera images, throughout a building. IT allows the security alarms to use their cables but manages how much bandwidth can be assigned to security devices without interfering with other communication signals being transmitted on the same cables. These cables are sized under a category program, with cables referred to as Cat 3, Cat 4, Cat 5, etc. If the cameras take up too much bandwidth to transmit their images, they may not be allowed on the network, forcing the security department to use coax cable between the cameras and the monitor, a method that is largely being phased out of existence. The ideal situation would be to have an IP cable installed that is dedicated specifically for the security department's use. Until that unlikely event occurs, the security directors should be advised to remain on friendly terms with the IT department as they may very well be working for that department in the future. In the meantime, the IT and physical security worlds are forced to coexist, whether they like it or not.

The main contention between IT and security departments is usually regarding the use of network cabling for video transmission without taking up too much bandwidth. Another problem is the relative cost of video storage and whether or not the images will be accessed infrequently or constantly. Most important is the need to compromise when requesting bandwidth for video systems in cases where it may not be necessary to record everything on each camera at a higher rate or frequency.

4 BASIC DESCRIPTIONS FOR THE UNINITIATED

- **IP** stands for **I**nternet **P**rotocol, an obvious abbreviation known to all security professionals, but IP may be unfamiliar to architects who are bewildered by the difference between IP and analog cameras. The IP address is a unique identifier for a computer or camera or other device on a network, be it on a LAN or a WAN. This protocol provides the capability to transfer packets of data to and from IP-addressable cameras, including audio, email, or any other data that needs to be sent (while connected to either the internal network or the Internet). Each device (camera, computer terminal, etc.) is assigned a unique IP address (number) that no other device shares. Once IP video is connected to the Ethernet infrastructure, the streaming images can be viewed anywhere that has Internet access.

  IP's application to a facility security system has completely changed the concept of having to wire each device back to the command center. Newer versions of IP identify an excess of 4 billion unique addresses, known as Internet version 4. An upgraded version 6 of the IP-based technology would increase the number to a 3 with 39 zeros after it (IPv6). The reason such a virtually unlimited set of IP addresses is desirable is that, when this protocol was conceived, it was thought that 4 billion addresses would be enough since they were directed to computer terminal addresses only. In the future, it is predicted that for every human being on earth, there will be hundreds, if not thousands, of devices communicating via wireless networks, each needing its own IP address.
- **PROTOCOL** is a set of rules that provide standardized communication between computers. It is simply the guidelines that regulate how data is transferred from one computer to another. Typically, networks utilize an IP/TCP configuration, which is a switch that describes how information is received and understood by the receiving computer. (IP/TCP stands for Internet/Transport Control Protocol).
- **FILE TRANSFER PROTOCOL (FTP)** is a standard network protocol used to copy a file from one host to another over the Internet. FTP users can connect anonymously if the server is configured to allow it.
- **WEB BROWSER or Internet Browser** is primarily intended to bring information resources to the user. This process begins when the end user inputs a Uniform Resource Locator (URL), for example, http://en.wikipedia.org/, into the browser. The most commonly used URL starts with *http://* and identifies a resource to be retrieved over the Hypertext Transfer Protocol. When the concept was first applied in the late 1980s, it did not attract too much attention until Netscape Navigator was released in 1994 and quickly exposed the value of a Web browser. Netscape became the world's most popular browser, accounting for 90% of all Web use at its peak. In 1995, Microsoft responded with its own browser, Internet Explorer, initiating the first browser war. By bundling Internet Explorer

with Windows, Microsoft was able to leverage its dominance in the operating system market to take over the Web browser market and by 2002 it dominated 895% of the market. Other browsers appeared during the first decade of the millennium, including Mozilla Foundation (Netscape's attempt to win back some of the market), FoxFire, Opera, Apple's Safari, and Google Chrome (released in 2008).

- **HUB** is a phonebook-sized box that is an older version of a "switch" and is less efficient and will eventually become obsolete. The hub forwards packets of data to available network segments instead of just the final destination. This causes the network to be slower and offers no advantage to a faster switch unit.
- **SWITCHES** are best described as the wires that come out of the back of each computer or network device that allow data to be organized and distributed through the LAN or WAN network. In more specific terms, it is a multiport bridging device that connects together the different devices on the same network. A switch repeats and forwards a packet of data only to the network segment where the final destination device is located, making for more efficient network trafficking.
- **ROUTER** is the next step in the chain of transmission and is essentially a computer with no storage capacity dedicated to routing and forwarding data packets from one network to another. Routers take digital signals from the hub or switch and selectively transmit data according to an IP address to another computer, switch, or other IP device. They ensure that all data is sent to its proper destination. Routers operate as the traffic cops of the Internet and are essentially computers dedicated to routing and forwarding packets of data by selecting the best path to route the message (based on the destination address and origin). They use the IP addresses assigned to each appliance (terminals, cameras, card readers, etc.) in the network. When it becomes necessary to connect the LAN to the Internet, the router serves as the translator between the information on the LAN and the Internet. Routers can listen to the entire network to determine which sections are busiest and can redirect data around those sections until they clear up.

  Usually, when smartphones, computers, and tablets are one, two, or five years old, there is a tendency to replace them, but routers are generally held onto for longer periods of time. Routers should be replaced at least every five years to enhance the operation of existing communication devices and for optimal Internet functioning (faster speeds and stronger signal to be received at greater distances).
- **NETWORK-ATTACHED STORAGE (NAS)** is a standalone hard disk storage device usually used within a specific area or facility to store and share computer-driven information used by all authorized individuals within an organization.
- **WORKSTATIONS** are those computers connected to the file server on a network. A typical workstation is a computer that is configured with a network interface card. Networking software does not necessarily need drives, a hard drive, or otherwise, since the files are saved on the file server that all workstations feed off.
- **USB (Universal Serial Bus)** is a convenient and easy way to connect computers with nearly any peripheral devices, such as digital cameras or printers.
- **REPEATERS** make up for the loss of signal strength as it passes along a cable. The repeaters can boost the signal by electrically amplifying it as it receives it and then rebroadcasting it along a network.
- **MODEM** is the connection between the computer terminal and a phone line that allows each terminal to dial into an Internet service provider like AOL (along with its roughly 1.5 million subscribers, down from a peak of 23 million back in 2000). The Internet contains thousands of modems that accept dial-ins from the subscriber computers in order to communicate with other modems on the Internet. It is possible that the modem might get a busy signal because the receiving computer might be handling too many subscriber calls (a common occurrence). Two types of modems are in use. External modems are connected

to the computer via a serial or USB port while internal modems are attached directly to the computer motherboard. Almost all the new computers have an internal unit, which can be used in a **wireless** environment. The operational format consists of five steps: Modems are controlled by built-in software. When a message is sent, it connects to the phone line on the premises and dials a pre-programmed number to make the connection.

- After dialing the number, the modem at the other end answers the call, after which the originating modem sends a tone to notify the receiving modem that the sender is trying to make a connection.
- Once the two modems connect, they exchange information about how they are going to send data to each other (a handshake, like with a fax). The data is exchanged in digital language.
- The telephone operates as an analog stream of electrical current that varies in frequency and strength, while the computer talks in digital binary bits of data. Therefore, the digital signal is changed to analog before it can be sent over the telephone line.
- The modem takes digital information and changes it – modulates it – into an analog signal. This information travels along the telephone system as analog data until the receiving modem demodulates it – changes the analog data back to digital data. This modulating and demodulating of data gives the modem its name: **MO**dulating/**DEM**odulating.
- **FLASH DRIVE** is a simple device that can interface with any computer that has a USB port (which can be found on every computer produced since the late 1980s). Due to its compact size, it was given its generic term, thumb drive. It also has an enormous and expanding storage capacity that makes it an ideal device for covertly copying data or uploading malicious software onto computer systems.

IBM was the first company to sell flash drives in the United States. The first model had only 8 MB but improved manufacturing technologies increased flash drive capacities while decreasing their cost. Some of the more popular flash drives store 32 GB and cost about $20 and there are drives that hold over one terabyte of data or well over 100 million pages of documents. This model can transfer the data at speeds of 160–240 MB per second. It is the kind of model that was used by the National Security Agency (NSA) employee Edward Snowden in 2013 to steal thousands of pages of secrets that exposed the various programs used to eavesdrop and monitor communication messages between all American citizens.

5 SERVER

In more general terms, the server is basically a bulked-up desktop computer, minus the screen and keyboard that contains chips and a special software package to provide a service to computers on the network. Any computer can potentially be used as a server as long as it has enough hard drive space and a fast-enough processor to handle all of the requests for information from other connected computers. Servers are essential in applications where there are hundreds of computers on a network. Every computer manufacturer produces computers assigned specifically to act as a server. It may not be necessary to install a server if the network contains only a few computer terminals or if all the computers on a network are capable of storing the files and accessing the programs needed for all the computers on the network. On large premises with thousands of computers, there may be many dozens of servers scattered throughout the building, all requiring special security measures to keep these scattered "server closets" secure.

Data centers are all over the world and likely number in tens of millions. They house servers to support the overall explosion of digital information. Stupendous amounts of data are set in motion each day as people download movies from iTunes, send emails with file attachments, buy products on Amazon, post on X (formerly Twitter), check credit card balances, or read newspapers online. All this streaming information consumes vast amounts of energy in a wasteful manner.

Online companies typically run their facilities at maximum capacity around the clock, whatever the demand. As a result, data centers can waste 90% or more of the electricity they pull off the grid.

The exact number of data centers continues to grow to meet the unending need for information and exchanging of data. The number of federal data centers has grown from 32 in 1998 to over 4,000 in 2012 and it is estimated that there are more than 3 million data centers worldwide. These power-hungry digital warehouses use about 30 billion watts of electricity, roughly the equivalent of 30 nuclear power plants every day. The United States accounts for one-quarter to one-third of that load. A single data center can take more power than a medium-sized town and on average these centers are using only 6–12% of their servers to perform computations. The rest of the power is essentially used to keep servers idling and ready in case of a surge in activity that could slow or crash their operations. For security reasons, companies typically do not reveal the locations of their data centers, which are housed in anonymous buildings and vigilantly protected. They must be built near power sources and have to be backed up with diesel generators and batteries to prevent a loss of power for even a millisecond.

6 CLOUD STORAGE SERVICE

The loosely defined "cloud server" may have been available for some computer users well before 2009, but that was the year it began to gather momentum and the attention of Internet subscribers both large and small. It has become a major technology for off-site data storage in which individuals and companies (and almost every federal agency) rent storage space in remote site servers that exist on an unseen Internet-based cloud service. The sharing of remote servers in the cloud has replaced the need to buy additional hard drive storage devices, like NAS, memory sticks, or remote hard drives that collect and store excess data in-house. It is simply a backup of the hard drive that stores data at a remote site – in the cloud. The cloud is not an abstract idea but is a very real and extremely large collection of computer servers, storage devices, network wires, and communication technologies. Cloud data centers are typically vast collections of computer servers and supporting equipment run by businesses that specialize in wholesale data processing (Table 14.1). Individual software developers, large companies, and governments use these cloud services to run their operations at a fraction of the cost of buying and maintaining their own on-site machines.

The unseen servers, located anywhere in the world, can have data retrieved from any location in the world. Users pay the service provider a monthly fee to store recorded data in the cloud and

TABLE 14.1
As of Q1 of 2024, a list of the top global cloud infrastructure service providers; Amazon and Microsoft are far and away the two largest of the bunch

Company	Share of Market (%)
AWS (Amazon)	31
Microsoft Azure	25
Google Cloud	11
Alibaba Cloud	4
Salesforce, Inc.	3
IBM Cloud	2
Oracle	2
Tencent Cloud	
Q1 2024: Total cloud infrastructure service revenues: $76B	

Data source: Synergy Research Group, Statista.

maintain system components remotely, eliminating the need for on-site storage hardware as well as in-house technical expertise to manage, troubleshoot, upgrade, and service the system. A hosted surveillance system or an access control operating system can easily tie into a monitoring service, enabling an alarm operator to visually verify an event before calling responders to the facility.

The only major downside to this off-site backup is that in the event of a major regional disaster affecting many customers, there may be a delay in recovery due to the simultaneous demand for backup restoration.

Most administrators and IT people had only a vague notion of what "the cloud" meant when several major computer software providers entered the cloud marketplace with a vengeance around 2009. At the time, only 40% of Americans understood the function and advantages of the cloud service and even fewer – 9% – actually used it, according to a survey conducted by a market research group. The firms that control a large share of this emerging market stand to be the leaders of the IT industry for years to come. Like the mainframe room-sized computer terminals of yesteryear – and their replacement, the personal computer – cloud software and services are merging to become the computer industry's third major platform.

In cloud computing, dozens or even thousands of computer servers are joined to create a giant machine capable of handling many tasks at once, from storing data to running Websites and mobile apps to tackling complicated analytical problems (Figure 14.3). The main reason cloud services have grown so rapidly is due, in part, to the fact that they provide a dynamic, cost-effective way to link all the office communication gadgets without any additional hardware. The concept of the cloud is as old and simple as the Internet itself. The cloud is about accessing storage or software remotely and is a modern twist on an old concept of time sharing on giant mainframe computers dating back to the 1960s. A more up-to-date off-site storage program is the operation of the TurboTax program, the Internet-based tax preparation service. Log on. Crunch numbers. File from TurboTax. That process uses a cloud service. Another example is the many places a consumer visits on the Internet to access such things as Google's Gmail, which is web-based and far different from

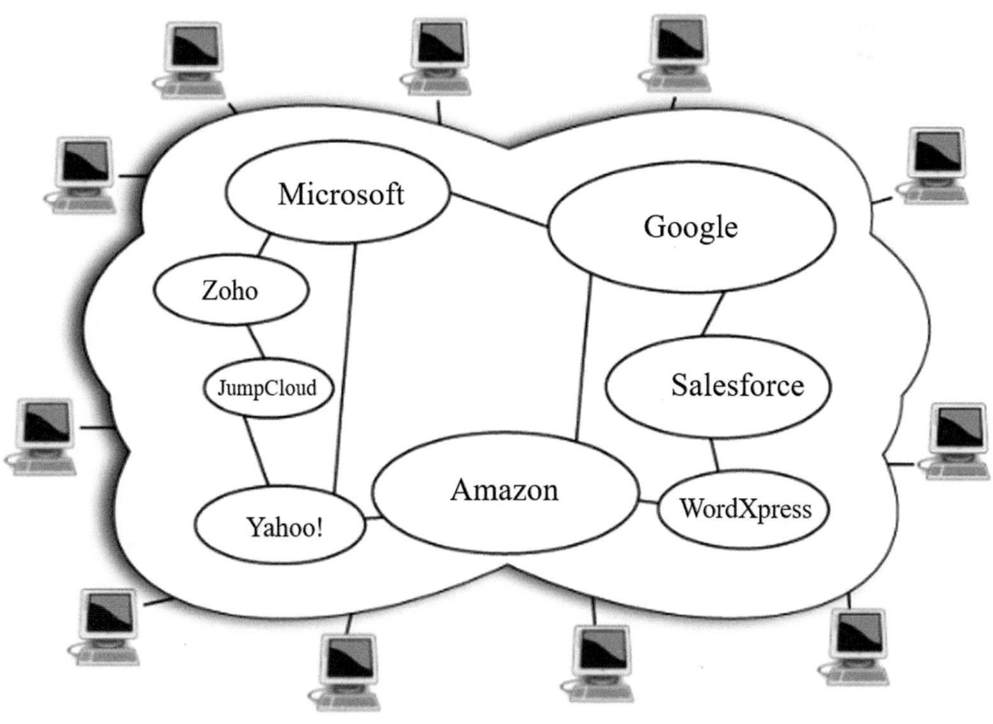

FIGURE 14.3 An illustration of cloud computing services and some of the major providers.

using Microsoft Outlook, which saved the information on the user's hard drive, that is, until it went online in 2011. Cloud computing is in the process of reshaping entire industries so long as security concerns continue to be properly addressed. Much of corporate America has been in the process of adopting and using this technology over the last several years.

For corporations around the world, and since its advent, cloud computing has completely altered the landscape for access to, and storage of, data and programs.

Facebook has become a member of the growing family of cloud storage companies with many photos and comments that are uploaded via smartphones and computers that remain nestled in its cloud or access one of its armies of storage servers. Social network Internet activities are shuttled to massive data centers; nothing is stored on an individual's device. With the cloud, the information doesn't fry or disappear when the PC hard drives crash. Microsoft introduced a cloud version of its popular Office applications in 2011 and Google entered the field at about the same time. Microsoft's main competition in the cloud data storage industry may actually be itself. Nearly nine of every ten computers run on one of the 14 versions of Office that Microsoft has released since that popular program was launched in 1989. The task is to get the 1 billion users to switch to Office in the cloud without disrupting the legacy versions, which are actually financing their own replacement.

Most people want to be able to use mobile devices at any location and the cloud allows them to use any connected environment as their home office. Until the introduction of the cloud concept, the only way to have a more flexible way of storing data was to get the IT department to coordinate a growing array of on-site hardware, which sometimes cost hundreds of thousands of dollars.

Data privacy laws in Europe forbid the transfer of information about individuals outside the European Union bloc of 27 nations and have prevented many companies on the Continent from moving to the cloud concept. Cloud data can possibly be stored on remote servers in Asia, the United States, or elsewhere and with the privacy concerns, cloud computing has become the exception, not the rule, in Europe. That will change as the European Union amends the data protection laws that govern how personal information travels within and outside it. More and more countries have started to follow the trend and accept the outsourcing remote server concept for their nation's use, despite the complexities of the European data protection laws. The cloud trend has become the primary method for data storage throughout the world.

7 FEDERAL CLOUD SKEPTICISM

Contractors like Amazon, Microsoft, and Google market their cloud services as a way for the private sector and government-controlled agencies to avoid having to build and manage costly new data centers as they expand their computer data use and capabilities. Agencies can use the cloud to help change the size of a project, without having to add or subtract from their computing infrastructure, at a minimal cost through this new storage network. Those agencies dealing with less confidential information have been quick to adopt the cloud model, yet cloud computing is often more secure than traditional computing because companies with deep pockets like Google and Amazon can often attract and retain cybersecurity personnel of higher quality or experience than many governmental agencies. The Agriculture Department was one of the first major government agencies to accept this matchless technology and moved about 120,000 new accounts into the cloud environment in 2011. The Departments of Defense and State discouraged those agencies with extreme security concerns from moving too quickly into cloud computing by continuing to keep confidential data within the military's advanced security systems. Defense officials are understandably concerned with any approaches that may introduce new security risks. The State Department, on the other hand, has introduced cloud computing, but only with low-risk projects like the website for its Office of the Historian, which offers public information about the history of American diplomacy, a not-so-sensitive collection of general data. Each agency has balanced its risk against the many advantages of cloud computing before following agencies such as the General Services Administration, which placed its entire email system on the Google cloud service network in 2010.

In 2024, the US government spent $74 billion on non-DoD included information technology, making it the largest consumer of technology projects in the world. At the same time, the government has also developed a reputation of wastefulness and for pouring money into projects that grow in scope over time without delivering significant results and for building immense hardware systems that are unnecessarily duplicated. The reluctance to accept cloud servers using security concerns as the excuse may have overstated the threat and much of that skepticism has been applied by those in the technology industry who have a business interest in maintaining the status quo.*

**Note: Starting with the 2022 federal budget, the Department of Defense or other classified IT spending is not included in the federal agency budget that is released. Undoubtedly, IT vendors and services providers remain some of the largest government contracts for military spending.*

8 DATA CENTERS

There may be over half a million data centers worldwide, some of which are contained within a single facility as large as 4.6 M square feet (Meta Platforms – i.e., Facebook) in Prineville, Oregon. About 5,800 of the giant data centers are located in the United States, all filled with computers that operate the Internet, process email, preserve medical records, and perform countless other tasks for corporations, governments, and other organizations. The number of data centers will continue to expand well into the third decade of this century. Despite serious concerns of a global economic slowdown, the companies that construct and operate data centers storing vast amounts of corporate and government data have grown at a rate of 20% every year since 2008. These centers represent complexes that require high-security protection because inside are the engines that keep smartphones smart, businesses connected, and social networks humming. Some data centers have installed traps that isolate intrusions by unauthorized individuals; technology that weighs people as they enter and sounds an alarm if that weight is different when they depart. Bulletproof walls and blast-proof doors are commonly found in the major centers. High security is essential since these facilities are critical to the world financial systems and the overall function of the Internet. The added cost to provide adequate security has raised the average price per square foot to build a data center from $600 to $1100, while the cost to build a normal industrial building is less than $85–$140 a square foot. The amount of data in the world doubles every couple of years and people and businesses are willing to pay for it to be stored. Some experts contend that far too much emphasis is placed on security hardware when the real security breach of data centers takes place on the other side of the world in cities like Moscow, Russia, where computer programmers sit in office buildings from 9 am to 5 pm hacking into databases all over the world, completely ignoring the security in buildings housing the vulnerable servers.

Giant communication companies such as Google, Facebook, Apple, Yahoo, and Amazon often build data centers in rural areas that are just as vulnerable to hackers as any other location. Apple opened a 500,000-square-foot, $1 billion facility in a rural North Carolina county that had no signs to indicate what the building holds. But everybody knew what it was. Nearby, data centers were opened for Facebook and there were others to follow. Although the rapid growth of data centers has diminished the ability to hide in plain sight, many building owners are secretive about the locations. These are two rules that, in general, relate to the center's location. Rule No. 1 is don't talk about data centers and Rule No. 2 is that you do not talk about data centers. Not every data center is a fortress. One center owned by a medium-sized city in Florida was a former water tank located next to city hall. It was selected because previous buildings holding servers were impacted by three hurricanes, forcing the servers to stop operation until each hurricane passed. The water tank did a good job holding water, so the city knew it would keep water out.

One of the biggest growing concerns in the Internet community is not hackers or intruders; it's the need to find a source for the huge amounts of electricity required to power the vast number of servers, which can number in the thousands at some sites. Data centers use 100–150 watts per square foot, in contrast to the 3–5 W for each square foot generally needed for an office building. In order to ensure continuous power, some centers have installed upwards of 65 generators and collect data at backup centers in case of a power loss.

IT departments are usually overloaded with responsibilities and the addition of IP-based security systems adds more conflicting goals between the traditional security and IT departments. Since less hardware infrastructure is required onsite when using cloud services, IT departments are relieved of many responsibilities associated with on-premise installations. While IT might be involved to a certain degree due to network connectivity requirements, cloud services minimize the impact on their available bandwidth.

RAID stands for a redundant array of inexpensive disks that are managed by a controller embedded on a motherboard. If properly configured, RAID systems can provide some level of data protection because if one of the hard drives fails, the data on the system will not be lost. Some security directors strongly suggest that using multiple hard drives provides the perfect method for data protection, especially if the situation requires digital video material for use in litigation or criminal investigation. All data should be backed up on a regular basis.

15 Advanced Communication Technology

1 VOIP (VOICE OVER INTERNET PROTOCOL)

Communication technology is constantly changing and the introduction of VoIP at the beginning of this century is a perfect example of how the communication industry adapted to a method that provided an alternative to the traditional telephone and cell phone networks. VoIP, or Voice over Internet Protocol, operates by routing phone calls over a computer network (much like email messages). It was touted to be the wave of the future for many technical reasons as well as the fact that VoIP has no long-distance charges or hefty federal access taxes levied on hard-wired-based phone calls. VoIP is a practical and inexpensive system that allows voice calls to anywhere in the world for pennies – no old-fashioned phone company operator required or the need to maintain cell towers.

Voice data is sent in packets rather than by the traditional communication circuits that are required for conventional phone communication. The method has proven to be more reliable and less vulnerable to the most primitive forms of communication, like the Plain Old Telephone Service (POTS), which runs over a network called the Public Switched Telephone Network, a method that has been around since Mr. Bell invented the system. Over the years, companies with 15 or more employees who needed a telephone acquired their own telephone system, all using some form of the hard-wired model as their baseline. Four non-VoIP telephony system models went through their short cycle of acceptance:

- **POTS**. This model does not need to consider any complicated switching network, and each customer has a phone with a line to the carrier. It is the oldest of the conventional telephony model – sometimes called the *wireline model*. An example where POTS may be beneficial is when used to communicate in elevator entrapments that are monitored by a security command center. Phones on the VoIP system have been added to elevators, but during a blackout, VoIP is useless, while POTS still allows communication.
- **KTS** Key Telephone System is a modified key system that reduces a company's dependence on the total number of POTS lines. It provides, at no extra cost, many of the traditional call features.
- **CENTREX** service is owned and operated by the carrier and provides essentially what is the physical equivalent of a POTS line. The lines run from the carrier's switching equipment to each telephone station. Centrex is a practical solution for startups or companies unsure of their strategic plans because they can gain all the usual features of a POTS plan. When the company outgrows the Centrex, they can easily convert to a newer telephony system.

  In the current cloud computing era, there are several services now that offer cloud-based mobile services, as opposed to the traditional wired Centrex systems. Such calls and data get routed to the provider's central servers, where they are stored, connected, and processed digitally.
- **PBX** is better known as the in-house Private Branch Exchange or Premises Business Exchange and is the more expensive of the models described above. The PBX delivers the most value since it can use dedicated high-bandwidth lines out to the carrier or to other locations on the company's network. It is often used to provide full-motion video conferencing.

DOI: 10.4324/9781003091103-21

VoIP services became increasingly popular with people looking to save money on long-distance and international calls. Many VoIP services (Nextiva, RingCentral, Zoom's service, and Vonage, for example) allow users to call another VoIP-enabled device anywhere in the world free of charge or call a standard telephone or a cell phone at a deeply discounted rate. The universal attraction of the VoIP service means even criminals and terrorists can use this communication technique. This easy-to-subscribe-to-communication network is a trend that has worried law enforcement and intelligence agencies since the mid-1990s. One UN report highlighted the growing interest of Islamic militants in using VoIP and other internet-based technologies.

A new array of difficulties for intelligence and law enforcement services evolved with the widespread use of VoIP services in that the old-fashioned, but reliable, phone tapping techniques for traditional interception don't work with VoIP. A federal agency using conventional tracing techniques to track a call from a landline or cell phone to a VoIP subscriber would be able to get only as far as the switching station that converts the voice call into Internet data. The switch, usually owned and operated by the company providing the VoIP service, could be located on a different continent from the subscriber. Unlike a local landline phone with its location clearly noted in the area code, a VoIP user would reveal no information about the location. Back in the 2008 Mumbai attack, handlers were in Pakistan, but the calls went through services in New Jersey. To determine the location of a VoIP caller, an investigating agency must access a database kept by the service provider. The IP address of a caller is a unique numerical identifier, and it can be in the form of a microphone, a special VoIP phone, or even a cell phone with software that routes calls over the Internet using wireless connections as opposed to cellular signals. IP addresses also change frequently and are less tied to a specific location than phone numbers.

Pinpointing the location and IP address of the VoIP caller is a huge challenge for any intelligence agency since doing so may take days longer than tracing a standard phone would (especially if the service providers are in a foreign country). This challenge has been solved since VoIP calls leave a far richer data trail for investigators to mine than someone calling from an old-fashioned payphone. If the Mumbai attack had taken place in the 1990s, the attackers would probably not have had the capacity to make calls to their handlers during the course of their attacks, depriving investigators of vital clues to their identities.

2 CENTRAL STATION MONITORING

One of several solvable problems that hindered the advancement of VoIP circuits as a security network is that it wreaked havoc on the hundreds of existing central alarm monitoring stations throughout the nation that oversee 30 million alarm systems (20 million of which are in private residences). The central station industry had, for almost a century, relied on old-fashioned direct connection telephone lines to record a security alarm condition or a fire detection system on the client's premises. Some of the larger monitoring services supervise over 75,000 subscribers through the use of digital communicators or radio frequency transmission networks. Besides monitoring fire alarm sensors and motion detection systems that are used at many sensitive facilities, they also handle the hold-up button installed at the main console on the premises' central station. The need for a hold-up button occurs when a protected building reduces its night staff to a single individual and the security control center (reception desk, etc.) is in an exposed room or open area.

When the connection is converted from a conventional telephone line connection to a VoIP circuit, the remotely located central station has to change much of their equipment to match the requirements of the new communication technology. It seemed that one of the problems with the new cable system was that it did not have redundant computer capabilities. As a result, when routine computer upgrades were required, the entire system was taken down for various periods of time. If the alarm communication network at the protected premises relies on the VoIP technology and an emergency occurs while the facility is performing routine maintenance, there would be (1) no way to call 911 and (2) if a security situation occurs during the period of maintenance, the alarm signal would not be sent to the central station. Eventually, the central stations would rise to the occasion and adjust to the new technology as they have done in the past. Fifty years ago, the central station used a direct connect phone line between the protected premises and the central station. Then they relied on a party line or McCullough Loop and standalone tape dialers before the universal acceptance of automatic digital dialers. The central station industry, comprised of an innovative and resilient group of entrepreneurs, certainly adjusted and – with the advances in technology – bandwidth issues have become a thing of the past.

Central station monitors sit and talk in front of the computer either via a headset or into a microphone, using special software at both ends of the conversation. The computer can be moved to any location, which is both an advantage and a drawback. While the user can call from any computer anywhere, including making an emergency call to 911, the answering service doesn't necessarily know where your computer is and can't route an emergency team to the caller's location.

3 SKYPE, MICROSOFT TEAMS, AND ZOOM

The software application known as Skype allows sharing of visual communication between subscribers over the Internet. It was developed by programmers in Estonia in 2003 and experienced rapid growth as a peer-to-peer service for messaging and making calls since its introduction. It used a proprietary Internet (VoIP) protocol, and, by 2009, it had become the largest international voice carrier, with a reported 21 million users.

The company was acquired by eBay in 2005 for $2.6 billion and spun off as an independent company in 2010. In 2011, Microsoft bought it for a whopping $8.5 billion. Microsoft, which still supports Skype but to a lesser degree than its other product options – especially Teams – said Skype still had 40 million subscribers as of 2020, a number that had slipped to 36 million by 2023.

By comparison, Microsoft's own developed proprietary Teams app, launched in March 2017 as a direct competitor to Slack, has risen from 250 million users in 2021 to over 300 million in 2023. Obviously, Microsoft has expanded on its Teams app, since it has been incorporated as a standard part of its Windows operating system and the Windows 365 family of products.

Founded in April 2011 in San Jose, California as Saasbee, Inc., with its name changing to Zoom in May 2012, Zoom was founded by Eric Yuan – who had previously headed up Cisco Webex (previously Cisco Systems). The company IPOed (initial public offering) and went public in April of 2019.

With the beginning of the COVID pandemic in 2020, which marked a meteoric rise in, and the ubiquity of, Zoom and Zoom calls – for remote schools and schooling, businesses and remote work, and personal calls due to social distancing protocols – dramatically impacted the business of video telephony, chat, and communication technology.

Partly as a result of COVID, but also due to some of the fallout from the strain put on medical and health systems – and a lack of availability of services, tele-health and psychiatry using video communications have also greatly increased. By using video calls, providers can reach more patients in hospitals, prisons, veterans' health care facilities, and rural clinics. It makes online medical and private psychiatry sessions accessible to a wider group of patients, including those who shun office treatment or who simply like the convenience of therapy without commuting to the 40-minute sessions. One of the many third-party online therapy sites signed up 900 psychiatrists, psychologists, counselors, and coaches to offer web therapy, which has, in some cases, gradually replaced face-to-face therapy. Several studies have concluded that patients are treated just as well as if they were on the couch in the doctor's office. Some insurance companies reimburse for the virtual session, as long as a local provider has some oversight.

4 MESH NETWORK: ONE ROUTER AT A TIME

This evolving technology radically changed the way the public connects to the Internet since it offers an alternative to relying on conventional Internet cellular service providers like AT&T and

Verizon for router Wi-Fi connections. It can best be described as a community-owned local network that provides the same high-speed Wi-Fi broadband service for a fraction of the cost charged by the giant cellular providers. So-called mesh networks are not the same as connecting directly to the Internet; rather, it is a technology where users are connected to a mesh network and share an Internet connection. However, joining the mesh doesn't necessarily mean that the members can eliminate the cellular provider altogether.

Mesh networks are not a new concept since many foreign countries with limited Internet access developed their own free, collaborative networks. It is only since about 2010 that the United States has shown much interest in expanding and sharing an alternative to the conventional Internet technology. In New York City, for the few neighborhoods that had use of mesh networks, it proved to be a lifesaver during Hurricane Sandy in 2012, which crippled low-level sections of the city that lost their Internet connection. Several of the flooded neighborhoods that had the mesh network were able to restore communication between users instantly. The mesh forms its own network where subscribers can still communicate with each other via Wi-Fi under the mesh network umbrella.

Traditional networks rely on a series of hard-wired access points or wireless hotspots to connect to their many users. As an alternative to that familiar configuration, the wireless mesh network connection is spread out among dozens or even hundreds of wireless mesh "nodes" that talk to each other to share the network connection across a wide area. These small radio transmitters (nodes) function the same way as a wireless router that people install in their homes to provide a local wireless Wi-Fi service. The nodes also allow mesh users to communicate with each other. Wireless mesh networks are truly wireless, while the wired or fixed wireless networks everyone is familiar with require a modem in each contained area (Figure 15.1).

Only one node needs to be physically wired to a cellular network like an Internet modem. That one wired node then shares its Internet connection wirelessly with other nodes closest to it. That sharing process is repeated and the more nodes, the further the connection spreads, creating a

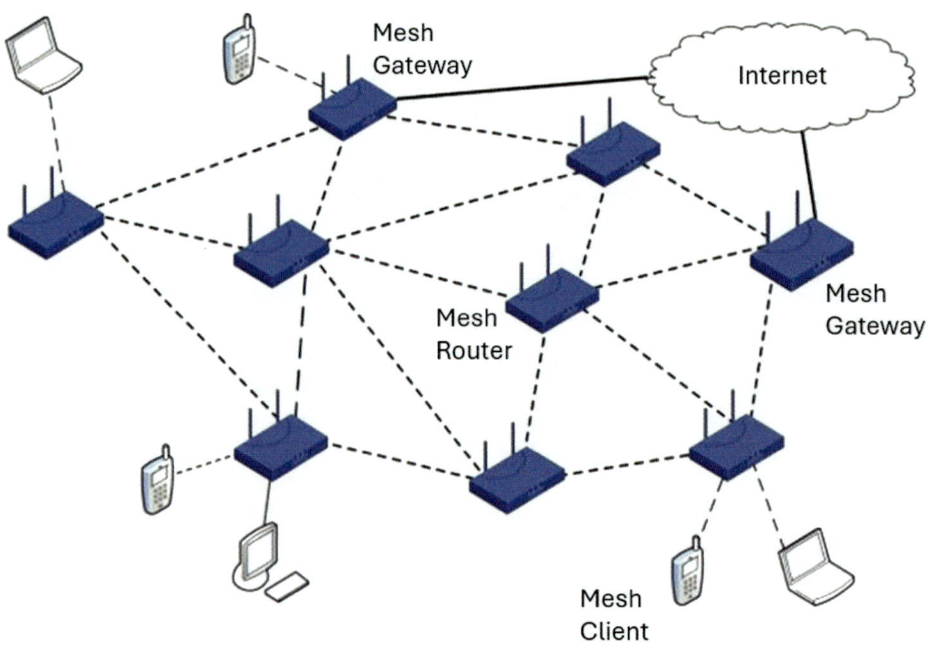

FIGURE 15.1 A diagram of how a typical mesh network is configured and operates.

"cloud connection" that can serve a small office to a city of millions. Mesh networks have many advantages:

- The more nodes installed, the bigger and faster the network becomes.
- Useful for non-line-of-sight signals from cellular towers that are intermittently blocked.
- They automatically find the faster reliable path to send data.
- Signals don't have to travel back to a central server.
- Access to the network is free.

5 PHONE NUMBERS AND AREA CODE LIMITATIONS

Until rules for area codes changed in 1995, there were only 160 possible combinations available in the United States and its territories, Canada, and parts of the Caribbean. After that date, area codes were expanded to 800 but even that increase may not be adequate long term. The problem is not that there aren't enough numbers; it's the way they're allocated.

An example of this potential problem is the situation in the state of Idaho. With only 1.7 million residents using a single area code, that could accommodate as many as 8 million seven-digit telephone numbers. Even though only 43% had been assigned, the supply of new numbers would have run out within a year, so a new area code (986) was added in 2017. This was done because of a shortage of available numbers but it was necessary due to the result of a billing and routing system that caused surpluses of numbers to accumulate in places where they are not needed.

Phone numbers are assigned in this part of the world according to the North American Numbering Plan (Nanpa) which was started by AT&T in 1947 to allow callers to dial long distance without an operator. The first three digits are area codes assigned a geographic region.... The next three are a prefix assigned to a switch within the same code area.... the final four numbers are a suffix attached to a landline or mobile phone. Each prefix accommodates 10,000 different suffixes, and the unused numbers accumulate when carriers have more numbers than customers.

The format is NXX-NXX-XXXX where N is any digit from two through nine, and the X is any digit from zero through nine.

The rapid increase in cell phone usage and ownership means that the numbering system necessitated an increase in area codes and the overall numbers available.

Area code exhaustion does not occur when all 8 million numbers are assigned to an end user. It will occur when all 800 prefixes are assigned or are no longer available. An example of this complicated code assignment is the situation in many cities that share a single area code where less than 30% of the numbers are assigned, but only 40 or fewer prefixes are left. With the release of four to seven new area codes annually, with 225 unassigned, it is forecast that it has enough numbers to last more than 30 years.

By 2018, there were 6.4 billion possible ten-digit combinations available. Some numbers, such as 911, are reserved. Nearly 800 million numbers are assigned, with an additional 86 million unassigned in carrier inventories. Those numbers should be adequate for at least 30 years; after that time, it may be practical to hang up the ten-digit telephone numbering systems.

16 Bandwidth, Broadband, and Single Compression

1 TRANSMITTING TOO MUCH DATA

The most important feature within the network configuration is the size of the bandwidth. When a facility shares the existing network for a security system, the broadband has limits that can restrict the amount of information gathered and transmitted. Only so many bits of data can be transmitted every second through the transmission medium (commonly referred to as "bps"). All these movements of bit-rate information can possibly fill the bandwidth to its defined capacity. The bandwidth size (or capacity) depends on the size of the physical wire, the devices on the network, and the function of the device. The limitation is best described by comparing it to a wastewater drainpipe system. The pipe diameter (bandwidth) can support only so many sinks, toilets, and other fixtures to effectively drain waste. The drainpipe selected must, therefore, be wide enough to allow drainage when all the devices are at full capacity and working simultaneously. If the pipe is too small, the drain will back up or stop draining. In the case of the digital network, overload can cause security devices to slow down, shut off, or reduce the bandwidth, causing signal degradation and data loss. This limitation is the reason why the operation of the digital security system is controlled by the technicians who monitor the bandwidth on the building network (they are not from the security department).

Before selecting any digitally based component that travels over the Internet, it would be advisable for every security director to become familiar with the terminology used in the new age of digital transmission and the proliferation of microprocessor-related devices. The alarm system salesperson isn't going to take the time to explain the meaning of the different phrases, and, in all likelihood, the salesperson may not be entirely sure of the meaning either. The world of bandwidths, compressions, CAT cables, etc., has taken over all aspects of security and all security professionals should become familiar with the new digital language.

Today, research might be conducted on the Internet while enjoying $6 lattes and listening to Starbucks's popular song loop in the background. The new lifestyle engenders certain sacrifices: spelling skills lost to automatic spell-check and entire generations of "digitized people" who no longer know the phone numbers of their immediate family because their cell phones do for them. Electronic calendars instead of memory keep track of appointments and anniversaries these days. Bandwidth limitations will become a serious problem in the not-too-distant future, along with the current limitations on energy use. Here is a list of the top ten countries with the fastest Internet (as of 2024):

1. Singapore – 286.66 Mbps
2. Chile – 272.01 Mbps
3. Hong Kong – 270.72 Mbps
4. Iceland – 259.72 Mbps
5. United Arab Emirates – 257.14 Mbps
6. United States – 242.38 Mbps
7. France – 240.36 Mbps
8. China – 238.17 Mbps

DOI: 10.4324/9781003091103-22

9. Monaco – 236.00 Mbps
10. Thailand – 226.36 Mbps

(Source: Based on the Speedtest Global Index – www.speedtest.net)

A family of four likely spends several hundred dollars a month on cell phones, cable or satellite television, streaming services, and Internet connections, which is about what is spent on gas and heating oil. Just as the industrial revolution depended on oil and other energy sources, the information revolution is fueled by bandwidth and, in future years, may be controlled by a few enterprises in the same manner as found in the oil industry. An engine can't operate without gas, and a cell phone can't communicate without bandwidth. Both are resources controlled by a tight group of producers, whether oil companies, the Organization of the Petroleum Exporting Countries (OPEC) nations or communications companies like AT&T, Xfinity, and Verizon.

In the United States and in most of the world, a monopoly controls the pipes that supply homes with information. These companies, primarily phone and cable organizations, have a natural interest in controlling supply to maintain price levels and extract maximum profit from their investments – similar to how OPEC sets production quotas to guarantee high prices or can force the price of oil down as it did in 2014–15. Americans are as addicted to bandwidth as oil and until alternate sources can be developed to offer more choices in the use of Internet bandwidth, there will be many adjustments that will have to be made by the public if it is to avoid a similar crisis that the world has experienced with the oil/gas situation since the turn of the 21st century.

High-speed broadband connections provide greater bandwidth – more data transferred per server – than connections over the regular phone line. There are two popular broadband Internet connections:

- Digital Subscriber Line is a special phone line provided by the local phone company.
- Cable Internet is provided by the cable TV company using the same cable that displays 250 brain-numbing channels on the home TV.

  Wavelet compression is a very important phenomenon and has revolutionized the entire field of digital technology. By compressing a signal (communication or video), more camera images can be transmitted on the same CAT cable. It's like compacting dozens of cars and squeezing the smaller models through the Holland Tunnel. Without the "compression," only a few normal-sized cars could get through the opening. Once the cars exit the tunnel, they are automatically expanded back to their original shape. Or you could leave the squeezed cars in the same shape and store them in a small room until you need them for transportation. Video signals can also be compressed, transmitted through the cable using only a small amount of bandwidth until they are received at the Digital Video Recorder (DVR), where they remain in the compressed stage until they have to be retrieved. DVR units can compress video as well in order to increase video storage capacity. The higher the compression ratio – the smaller the amount of bandwidth that is required, but the more compression – the higher the degradation of the image once uncompressed for viewing.

There are many reasons to compress video images:

- Reduces the amount of storage space required to store video – so that more hours of video can be stored on any of the various media devices (hard disk drives, solid state memory, etc.).
- Reduces the bandwidth required to transmit the video across digital connectors, including LAN, WAN, and Internet.
- There are limitations to bandwidth, which poses a problem in compressing the video because digital video files are very large, requiring a big section of the available bandwidth.

Resolution is a primary component of a digital video signal and is defined as pixels – the tiny dots that comprise the image, which is further defined in terms of the number of pixels. A picture identified as 640 × 480 would be a picture with 640 horizontal pixels and 480 vertical pixels. The higher the number of pixels, the sharper the digital display or resolution. The factors that make up the picture have a direct impact on the bandwidth required to transmit images over the network. A TV with a resolution of 8K is considered top-of-the-line and costs the most. The 8K refers to 7,680 horizontal pixels (or roughly 8,000 – i.e., 8K) and 4,320 vertical pixels. While this totals a staggering 33 million pixels, in reality, super-high-resolution pictures usually can't be distinguished from lower-resolution images in normal environments.

Video compression has become the most important consideration in the selection of a format to transmit the signal over the network. When this technology is used in the security surveillance industry, it allows for the addition of hundreds of cameras to a system that involves so much data that it would have been almost impossible to transmit without some form of network compression. The selection must be balanced against other considerations such as the image resolution, size of the image, and the number of images recorded or reproduced per second. The method used to transmit the video signal is to encode the signal and send it over the cable in a compressed form and then decode it at the storage or display unit. Compression technology records every video frame and compresses each frame by combining pixels into large blocks which are transported over the network and then reconstructed when the video is displayed.

The most effective way to transmit video pictures over the network and maintain some image resolution is not to share the bandwidth with other devices on the network. A networked video surveillance system would ideally not be part of a multi-purpose network because, in some cases, there might too much information being processed and not enough bandwidth to support video operations. Plus, the security system may not be considered as a high priority by the IT Department and sharing security with other building devices places the security department at a disadvantage.

Video is compressed using two types of frames, such as I frame, also known as the index or key-frame, which contains the whole image. The P frame only contains the information that is different from the previous frame. One of the more common forms of compression technology is called JPEG or M-JPEG which stands for Joint Photographic Experts Group or Motion-JPEG and uses I frame. JPEG can reduce the size of the image to about 5% of its normal size. The development of newer and better compression formats has allowed more video to be transmitted using less bandwidth without noticeably affecting resolution (picture quality). M-JPEG-4 and H.264 use a combination of I and P frames and consume considerably less bandwidth than M-JPEG.

1.1 Listing of Compression and Transmission Formats

JPEG – Frame-based formats refer to storing individual frames but is also used for streaming video as Motion JPEG.

MPEG-1 – Developed in 1993 to store video on CD, this captures certain frames that change from one frame to the next (a person entering the camera's field of vision). It is considered a low-quality video compression format. It was the first agreed-upon standard created originally for the entertainment industry and consists of a sequence of digital still pictures that have been compressed using the JPEG compression format. It is suited for applications where the recording can be set to a lower frame rate, and it doesn't require larger amounts of bandwidth for transmission, an important advantage offered by this format.

MPEG-2 – Developed in 1994, this provides higher quality video compared to MPEG-1 at the same frame rate. Used for DVD discs and high-definition TV.

MPEG-4 – An advanced version of the MPEG standard that offers more processing power and is capable of producing much smaller video files than MPEG-2, reducing the bandwidth by using algorithms to define objects and offering high video quality compared to older formats still in use. With high-volume processing power, it became standard for IP video after 2006.

H.264 AVC (Advanced Video Compression) is the most advanced compression technology used in video surveillance and is an extension of the MPEG-4 protocol. The H.264 format requires up to 50% less bandwidth than MPEG-4 to transmit the same quality image; therefore, it is the chosen compression standard for the highest-performance IP cameras. H.264 has been the compression standard since 2015.

2 CIF – COMMON INTERMEDIATE FORMAT

The image format is a very important factor when retrieving video from the recorder or directly from the IP camera and is a misunderstood subject by many CCTV system designers in the security surveillance industry. Basically, the size of the video picture on the computer screen is called CIF. It comes in four different formats. 1 CIF presents a picture about the size of a Post-it Note, approximately 4 inches by 3 inches high, while the more popular and largest 4 CIF format is the size of a full computer screen. There is an advantage to smaller CIF pictures since they take up less disk space or bandwidth – always an important consideration. Also, when the picture is blown up to 4 CIF, it can become blurry and distorted. Another format is QCIF, which is a picture size one-fourth the CIF standard size but is rarely used.

Frames per second (FPS) refers to how many video frames are sent from the camera or are recorded on the DVR hard drive. In the past, the videotape recorded 3–7.5 FPS. In order to understand how advanced digital technology has changed the way images are received, it should be noted that a TV picture transmits at 30 FPS, and the human eye perceives full motion at 16 FPS. (The lower the frame rates, the less the bandwidth and the less disk space utilized).

BITS OR BYTES is the measure used to describe the size of digital capacity:

- Storage capacity in the DVR is measured in bytes, kilobytes, megabytes, gigabytes, and terabytes.
- Bandwidth size can be bits per second (bits or baud), kilobits per second, megabits per second, gigabits per second, and terabits per second.

Byte and bit are related but not interchangeable. A byte is made up of 8 bits. As an example, the compact flash card can store 1MB (one megabyte) or 8Mb (eight megabits) of data. A 100-megabit Ethernet network can carry either 1MBps (100 megabits per second) or 12.75 Mbps of data.

Example: If a camera produces ten images per second where each image is 15 kB, and the system must carry several cameras but is limited to an Ethernet connection bandwidth of 100 Kbps, your camera capacity will be limited. Furthermore, the hard disk might hold, say, 20 GB before it is filled up and begins to record over itself. If the DVR salesperson can explain how many hours of storage are available, it is recommended that the DVR be purchased from him.

Note that Bits per second rates are commonly used when describing networks
***Bytes** are used when referring to **storage capacity**.*

2.1 Byte Scales

1 MEGABYTE: 873 pages of 1,200-character plaintext

1,000 MEGABYTES is a GIGABYTE: 4.5 thousand 200-page books.

1,000 GIGABYTES is a TERABYTE: 4.5 million 200-page books – the Library of Congress contains about 10 Terabytes of printed material.

1,000 TERABYTES is a PETABYTE: 4.7 billion 200-page books or 358 million 3MB digital photographs.

1,000 EXABYTES is a ZETTABYTE: about the capacity of 250 billion DVDs.

Measuring Internet data consumption of individual activities on a home computer:

1,020 MB One hour of watching a Netflix movie on TV 792 MB One hour of YouTube on a laptop

533 MB Watching a National College Athletic Association (NCAA) basketball game on TV

It takes up to three gigabytes to stream an hour of high-definition video and a household could stream roughly 50 movies a month before hitting 300 gigabytes which many of the popular service providers consider the monthly limit and charge $10 for every 50 gigabytes over that limit. It is estimated that if a household watched 5 1/2 hours of high-definition streaming a day – the average time an adult spends in front of the TV set – it would consume about 75 gigabytes a month.

PIXELS, as previously presented, is the shorthand term for a screen resolution that is used by the television and computer industries for the number of dots (pixels) on the screen – more pixels and the picture has more definition. A TV with 1,080p displays a superior image than a screen with a resolution of 720p. The resolution will continue to improve, just like digital cameras that take pictures with more megapixels or PCs that have ever larger capacity hard drives. Spending money on the newest fad is not always justified as seen with cameras that come with ten megapixels when five and six megapixels are more than satisfactory for everyday use. The same "overkill" mindset exists with monitors, where screens will be available to display 1,440p, and even higher, but these extremely high resolutions are not worth the additional cost, especially when used for general security surveillance purposes. There is a vast difference between a home theater and a surveillance console and the basic point to consider is that in console screens with a diagonal measurement of less than 60 inches, it is hard to tell the difference between monitors of 720p and 1,080p. There are three specific goals that must be addressed before a selection is made for a surveillance video system.

- Category #1: The most utilized since it provides a satisfactory picture for most situations and will show the movement of cars, for example, but not necessarily the license plate. It can capture the crowd and the number of people in the scene but not recognize the faces.
- Category #2: This is a more detailed picture that is better suited for forensic applications and projects a clear picture of the license plate and recognizes faces.
- Category #3: This has the highest detail image that not only reads the license plate but also the model of the vehicle and the face of a driver or, if in a bank, can clearly see the denomination of cash being handled.

In older analog CCTV surveillance systems, the resolution was defined by Total Video Lines but that has been abandoned with today's digital technology. When an image is listed as 640 × 480, that is actually 640 horizontal pixels and 480 vertical pixels which is a total of 307,200 pixels when multiplied together. That is the same resolution as 0.31 of a megapixel (million pixels).

3 THE CABLE PUZZLE

There is a great deal of confusion in selecting the proper wire for the various security system applications. This subject is often overlooked, but it is one of the most important factors in the design of a security detection system. There are a few sources that explain the significant differences in function and applications of wire/cable conduit even though wire material has been around since the introduction of the DC light bulb and AC electricity. Conducting electricity is done with nonferrous metal, available in a great variety of thicknesses and types. The rating of the insulation covering is another factor often misunderstood. Most wire or cables are usually circular in cross-section and more often than not, the wire is made of copper even though there are better conductive metals such as gold (thc most conductive metal), with silver being a close second. There is also tin, but due to its low cost, copper has become the base material for almost all cables and wires in use in the electrical and electronic industries throughout the world.

There are two basic types of wire in use today: the *single strand* and the more flexible *stranded* wire. The solid core (single strand) consists of one piece of nonporous metal wire, produced in different diameter sizes (gauge), encapsulated within a covering or jacket for insulation. Solid wire has a great many uses since it is the least expensive to manufacture (provides greater protection against the elements) but it is limited to applications where there is no flexing. Stranded wire is composed of many smaller gauge wires than are found in the solid core wire and the combined wires are bundled together to create the electrical characteristics of a larger wire – it goes up in cost as more strands are added. There is a trade-off that makes the higher cost for stranded wire acceptable since this type of cable is more pliable and will not break if bent or flexed continually. If one strand in the bundle breaks, it will have a negligible effect on the current-carrying capability of the wire.

Wire insulation is either the poly-vinyl (PVC) or the superior Teflon insulation material that will not stretch as much as other coatings when pulled with excessive force. Teflon is naturally more slippery and will slide through tight spaces more easily than PVC and since it is also plenum rated, it can be used in above-ceiling situations where an air plenum is included in the building construction. Shielded cable is used primarily to absorb the magnetic energy that is emitted from other electrical devices. This so-called magnetic flux can interfere with the signals carried in the cable. The metal shielding is installed beneath the outer jacket.

Wire gauge is measured under the American Wire Gauge system and sets the cross-sectional area of each gauge of solid and round electronically conductive wire. The gauge is the most important factor in determining the current-carrying capacity of the wire. A 12-gauge wire usually found in house wiring is.0808" in diameter and an 18-gauge lamp cord wire is.0403" in diameter. The greater the total thickness of the wire, the more current gets carried because there is less electrical resistance over a given length. Therefore, thicker wire is required for longer wire runs than for shorter runs. An electrical cable contains one or more conductors that can be individually insulated or bundled.

Alarm systems are designed to incorporate low-voltage components (under 50 volts), which can be installed by anyone with a ladder and staple gun. If the components operate at a higher voltage, most municipal codes require that the installer be a Licensed Master Electrician or supervised by such an individual. That requirement would probably triple the cost of installing a standard security system. Most cables and wires used to connect components in a conventional alarm system have been thin pairs of wires wrapped inside a COAX cable when used with Analog Closed-Circuit TV Surveillance Cameras. With the introduction of IP addressable devices connected through computer-based networks, the wire now has to consider such factors as bandwidth, which requires the use of communication-type cables that are popular in the telephone industry. COAX cables are slowly being replaced with CAT wire and cables and even wireless systems in some cases.

4 GENERAL USE CABLE TYPES

Cables are available in several types and functions, including twisted pair, coaxial cable, multi-core cable, and ribbon cable, among others. The following description may untangle some of the mysteries surrounding the cables used in the communication networks of most security systems. It's completely up-to-date – until the tech industry adds another confusing format to the list.

COAXIAL CABLE is familiar to the general public since it has been the universally accepted cable for CCTV analog surveillance systems, though it has been replaced with the introduction of digital technology. It was invented in the 1920s and has been in commercial use since the 1940s and is considered the granddaddy of cables. Medium-thick and either white or black and up to about 1990 was the most popular cable used in the security industry, especially for CCTV applications and is the standard for home use when connecting the cable bus to the TV and to the modem. Monitors purchased after 2009 use a much superior CAT wire or fiber optics cable that will eventually make coaxial cables a relic of the past. The coax cable transmits signals (not power) through a single copper core molded into a dielectric material and shielded with the copper braid to form a shield that surrounds the core to act as a ground. The coax appears as a tube-like cable with only one pin connector (Bayonet Neill-Concelman (BNC) type) that connects the camera to the video equipment.

Common	RG59 up to 750 feet
Standard of	RG66/U for distances up to 750 feet
Coax cable:	RG6/U for distances up to 1500 feet

Distance can vary if improper installation methods are used or if multiple splices or tight bends are required in the connection between the camera and monitor. Some manufacturers offer video amplifiers extending the cable length, and most installers ignore these limitations by exceeding the distance limitation with the improper use of the BNC connector. All of which results in carrying a noisy video signal because of poor installation practices.

The coaxial cable is the least expensive method for transporting video signals for distances of about 2,000 feet and does not require any signal boosters if the distance remains within its stated limits. The cable is also susceptible to electromagnetic interference (EMI) and the ground loop effect even though it is a shielded cable. It is a cable that is no longer a major factor in the security industry and will disappear as the digital age converts analog images to digitalized signal transmission over all the communication networks.

TWISTED PAIR is a popular form of wiring and has been available for many years. The twisted pair consists of two wires twisted at a given number of twists per inch. For the most part, it can outdistance coaxial cable by twofold and is ideally suited for any analog or IP camera system. It cancels out external sources of EMI and is commonly used for the transmission of video. It is an inexpensive substitute for coax except that the system requires a transmitter on the source side and a receiver on the receiving end to modify the basic video signal to run on the twisted pair. The rated distances are similar to or greater than those of coax; with heavy enough wire and a bigger power supply, distances up to 4,000 feet are obtainable. This wire is more forgiving than coax, but it probably has a limited future use in the digital era.

COMPOSITE is far more versatile than coaxial in terms of signal quality and is the cable of choice for connecting DVD players and other video equipment to older analog TV sets. Not much of a future for security system applications.

S-VIDEO is considered superior to composite but not as good as component cable (the "s" refers to separate). Cables do not carry audio nor do they support high-definition video and in general, S-Video has been superseded by component connections for the home. In a few instances, it is still used on PCs and is popular for linking DVD players, game consoles, computers, and older video equipment to the TV set.

COMPONENT is a three-headed cable that supports HDTV signals by dividing video data into three separate signals for a better-quality image (no audio signals are carried). Connects DVD players and other video equipment to newer HDTV sets. Component cable, during its lifespan, was the video cable of choice until HDMI was introduced.

HDMI (High-Definition Multimedia Interface) is the standard cable for high-end AV audiovisual equipment and is an all-digital connection that carries both audio and video over one thin cable.

VGA (Video Graphics Array) has been around since the late 1980s and is considered the lowest common denominator for video output on many PC systems. Some game consoles and video projectors connect by way of VGA with advanced variations that include Super VGA and XGA (Extended Graphics Array).

DVI (Digital Visual Interface) and HDMI are very similar since both can handle an all-digital signal, but DVI cables are video only, which appears a little confusing. There are different types of DVI cables. Somc arc analog only and some are digital DVI-I cables that can handle both digital and analog signals. Technical standards require DVI cables to maintain signals up to 16 feet as image quality can degrade if the cable is too long. DVI cables are mostly used to connect computers to LCD monitors and digital projectors. Mini DVI connectors were introduced by Apple but were phased out with the introduction of the Mini Display Port. Both connectors connect various Apple computers to display devices like TVs and external monitors.

USB (Universal Serial Bus) is the most common way to attach just about any type of computer peripheral: hard drives, MP3 players, printers, etc., to the computer and it can provide power to devices.

PLENUM WIRE is also called FireWire which is available in all sizes and refers to a fire-retardant covering of the wire that can be run exposed in the area over the hung ceiling (an area known as the plenum). The cover will not give off the same level of toxic fumes that commonly used wire will emit. Plenum wire is expensive, so it is avoided whenever possible during the installation of low-voltage (under 50 volts) alarm systems. Before the acceptance of plenum wire, all cabling had to be run inside the rigid galvanized pipe. It allowed for high-voltage plenum wire connections when used with fire alarm systems or when electromagnetic locking systems are connected to the fire alarm system. To determine if the wire is in code compliance, a knowledge of national electrical codes and local municipal fire codes is essential.

5 CATEGORY (ETHERNET) CABLE

Ethernet cables look much like a telephone cable, but the connectors are a bit wider than the cords that are used in phones. Ethernet jacks are standard on desktops and laptops and can also connect game consoles and TiVos to home networks. This technology required that alarm systems use "Category Program" wires which offer approximately six different categories, each one with different applications. The public was confused and considered the higher number category as the recommended size, making the lower numbers obsolete. This is not true since each category suits a particular application. The biggest-selling Category Cable is CAT 3 even though there are higher CAT cables (up to CAT 7) being manufactured. A basic description of each category is listed for reference purposes only and is not a recommendation for any particular system.

CATEGORY 1 cables are used very much and meet only the basic requirements for older analog voice and low-speed transmission of data.

CAT 2 cables carry the same amount of communication data as CAT 3 and carry voice and high-speed transmission data.

CAT 3 is the most popular size used in the telecommunications industry and has been accepted for LAN networks. A pair of unshielded/twisted for transmission characteristics up to 16 MHz.

CAT 4 cables are unshielded/twisted pairs with similar characteristics to CAT 3 (up to 20 MHz). CAT 5 is the same as the above, but the pair is for transmission characteristics (up to 100 MHz). CAT 6 exceeds the standards met by CAT 5 and CAT 5e and up to 250 MHz.

CAT 7 cables are able to achieve higher performance than preceding Ethernet standards such as CAT 5e, CAT 5, and CAT 6 by requiring each of the twisted pairs to be fully shielded, which eliminates crosstalk and significantly improves noise resistance. It costs more than other cables but has a 15-year life cycle compared to the ten-year life cycle of the other categories of cable.

5.1 Fiber Optics

While the fiber cable is far superior to other communication cables, it has been slow in gaining acceptance from the security industry. The first models introduced into the commercial marketplace were a little difficult to install (particularly when they required a cable splice for longer runs). This cable is made up of glass filaments instead of copper, allowing light to be transmitted over each filament, creating a stream of bright and dark spots. When measured over time, it results in an on/off code that can be interpreted to form clear video images. Fiber optic cable is gaining in popularity mainly due to the growth of computer networks since fiber optics are totally free from electrical interference and can transmit high-quality video data further without signal degradation. It covers more than 11 times the maximum distance of coaxial and 15 times the distance of twisted-pair cable before requiring signal boosters. Fiber optics is more secure than copper since an attempt to intercept the signal requires a break in the glass fiber, which is not an easy task. When first introduced in the mid-1990s, it required skilled personnel to install the fiber connectors. Since the cable is glass, splicing or cutting the cable had to be done accurately to minimize light loss (signal loss).

By 2003, most manufacturers had corrected this problem by providing simple-to-use tools. With the universal use of network IP connection, it is becoming the most popular wire in use (it can be used with analog cameras and, unlike coax, it can be used with digital devices). The only drawback is that it is more expensive than any other type of network cable commercially available.

Fiber optic cabling is often used selectively alongside a cheaper form of cabling to create the backbone of a network. The CAT 5 or 6 cables are then used to support the flow of data from any single device into the fiber network which is then used to connect all the switches and routers together since it has a large transmission capacity. The ideal configuration for a complete security network is a fiber backbone with the CAT 5 or CAT 6 cable connecting each device.

Section VI

Video Recording and Storage

The introduction of IP video technology changed the CCTV surveillance industry because of the effect it had on the legacy analog cameras that were the mainstay of the security surveillance sector prior to the advent of the more effective IP conceptions.

The topics cover digital surveillance cameras and their many significant advantages over analog cameras since they not only record events but can also look for suspicious or threatening behavior and alert security personnel to intervene. They can be programmed to identify packages left unattended for a specific period of time, detect when a new person moves into the scene, and send an alert or alarm when a piece of equipment suddenly disappears from the scene. They can also be programmed to make the camera follow the movement of individuals or vehicles captured on the display. These are important characteristics offered by digital formats since most studies indicate that a security officer can effectively watch 15–20 cameras for 20 minutes before fatigue sets in. With that short a time span, officers are at risk of missing an event. Being able to program the system to look for suspicious movement is an outstanding advantage over analog cameras.

This section will also cover camera quality and resolution advantages and storage capacity of the more advanced IP technology that includes a concise history of newer technology, from the floppy disc and removable storage devices to Cloud storage's revolutionary effect on how modern-day surveillance is used. Details on video compression are presented in an easy-to-understand format and much of the information has never been discussed in such detail. Technical subjects are described using terminology that is directed toward both the engineer and the security practitioners as well as students taking criminal justice and security courses.

A tape recording of camera imaging has long been the term to describe the process of capturing video on an analog time-lapse recorder. With the advent of IP-addressable cameras, the term "recorder" is a misnomer since digital video recorders are more accurately described as digital processors that store the image, not record the image.

Manufacturers have developed software that is used with a commercially available computer terminal and its hard drive to store digital images without the use of a digital video recorder (DVR). They provide only the special software instead of a cabinet with electronics and hard drives.

DOI: 10.4324/9781003091103-23

17 CCTV Recorders and Digital Conversion

1 BACK TO THE FUTURE

The security industry has long relied on the older analog technology that used video cassette recorders (VCRs) to record camera images onto bulky video home system (VHS) tape. Since 2009, technology has been available to upgrade CCTV surveillance equipment from analog to digital operation. Through the use of a relatively inexpensive hybrid approach, the cost was reduced by keeping the existing analog cameras but feeding the signals to a digital video recorder (DVR). This is accomplished by using special software that converts analog signals to a digital format without rewiring the circuit. There are an untold number of advantages that justify the replacement of the older analog tape recorders. One limitation of analog is that it can take days to search through a tape to discover an image of a reported crime, but the picture is so fuzzy that identification is all but impossible and the image cannot be used in court.

Analog CCTV surveillance systems required various types of complex "switchers," not required in digital format configurations and no longer manufactured, to handle different numbers of cameras. Some features are described below for historical reference purposes only:

- Sequential Switcher – automatically switches images from multiple cameras.
- Four Position Switcher – simply pushing a button on the unit causes one of four camera images to appear on the screen.
- Dwell Time – the pre-set time to display one camera image before switching to the next camera.
- Quad Splitter – displays four images on a single screen simultaneously. Works on the same principle as a sequential switcher.
- Multiplex Switcher – allows the video recorder to store more information per camera.
  - can be either analog or digital and organizes multiple groups of inputs and outputs
  - programmable to play video streams separately or in a quad format.

For almost three decades and up to the introduction of the DVR, VCRs had been the recording workhorse in CCTV surveillance systems. It relied on recording electronic signals via coax cable onto two-hour VHS tape cartridges that were placed on racks in the security room closet, cluttering the room since the common practice was to store them for more than a month before reuse.

2 BRIEF HISTORY OF VCR

VCR is an electromechanical video storage device that records and plays back full-motion audio-visual programming on magnetic tape cassettes. It was introduced to the marketplace in the 1960s at a cost of $50,000 and is now collecting dust in some long-forgotten storage area. The last manufacturer ceased producing units in mid-2016.

The VCR life cycle started 60 years ago when the first cassette recorders were purchased for $50,000 after Columbia Broadcasting System (CBS) executives were amazed to have witnessed "the first practical videotape recorder" at a demonstration of this high-quality, instantaneous replay device. The first VCRs for homes were released in the 1960s and became widely available in the 1970s when Sony's Betamax and JVC's VHS formats began to compete with VHS. Far outselling

DOI: 10.4324/9781003091103-24

the Beta units, there were two VHS formats: NTSC in the United States and PAL in Europe. All the cassettes were limited to recording one video stream per three-hour tape at 25/30 frames per second (real time), forcing the institutions to create large storage closets for the hundreds of cassettes needed if the user wanted to keep the images for over a month.

By the end of 1984, there were 15 million homes with the machining, triple the 5 million in 1982. The popular home unit sold for about $600, but in 1995, the introduction of the DVD sounded the death knell of the JVC. Less than a decade after the DVD began its reign, streaming video became the rage of the video storage industry.

The VCR went through several development stages as more and more institutions began to install increasingly elaborate CCTV surveillance cameras. In order to keep the number of stored tapes to a minimum, VCR used four or eight tracks on a single tape – called a multiplexer, which dedicated each track to a single camera. That allowed several streams of video to be recorded onto the same tape and separated out into discrete, viewable streams on replay. The time-lapse VCR came later which dropped frames and in doing so permitted a three-hour tape to be used over much longer periods, but this caused a loss of information. The unit could play back one camera at a time by being programmed to view only the desired camera. Another method to save recording space was to record only after being triggered by an input device such as a holdup button in a bank or a motion detector in a hospital.

While all these image-sharing features became available and reduced the number of VCRs needed, the images still took up a lot of storage room. Another significant drawback was that when a scene was to be played back to view a particular incident, it required the tape to be run for long periods of time to find the scene. The VCR became impractical for commercial use in the age of digital technology and higher security requirements. They were replaced with DVRs as fast as budgets allowed. VCR tapes can only be converted with devices specially made for those purposes. Once converted, scenes can be displayed on a computer and stored digitally.

3 DIGITAL FORMAT

In comparing features of analog tape recorders and DVRs, it is important to note that a reason there is so much confusion in selecting a newer DVR unit is that there is no industry standard to which all DVRs must conform. Not all DVRs are created equal; in fact, their differences can fill half the pages of this textbook. The older analog video is based on standards developed in 1953 for broadcast systems, but a comparable standard doesn't exist for digital video, so hundreds of digital manufacturers seem to follow their own design goals, formats, and decompression techniques. This incompatibility between DVRs means that when a facility purchases a proprietary system, it will probably have to keep buying additional units from the same manufacturer.

The adaptability, flexibility, and many technical features of digital video storage make it far more superior than analog tape recorders in every single category. Digital security systems provide the capability to adjust a multitude of settings designed to manage the finite amount of hard drive space to which the digital video images are recorded. A digital signal can be compressed and transmitted over a communication cable (CAT5, CAT6, etc.) and reformatted by the recorder, allowing many video signals to be carried in a single communication cable as long as there is sufficient bandwidth in the cable. In contrast, signals from analog cameras are carried by a separate coax cable for each camera and retrieving a scene in analog units takes hours while in a DVR it takes seconds.

Digital spy cameras can instantly pick out people in a crowd, as depicted on the television show "24" but most video systems are not quite that sophisticated, and most of the cameras are analog, retrieving sometimes blurry images from the VCR tape recorder. Stopping an incident in progress is nearly impossible unless the security officer happens to be watching the right monitor at the exact time of the incident. The technology available today can identify faces and license plates and software can be programmed to move the camera to follow a single person or a moving vehicle. Whenever a face is captured on the digital camera from a different angle or in a different light, the system is able to create a series of mathematical models. Once the profile reaches a certain critical level of detail, it becomes fairly simple to search the "motion event" to discover the exact location where someone has been. With analog technology, a security guard had to remember a face months after the event in question took place. By adding intelligent video to the recording device, it can "remember" images for many months. This technology may not be appropriate for some situations because CCTV works best where the lighting is constant and only a few people pass in front of the camera at a time. Cameras are more effective in controlled areas like parking lots, waiting rooms, and lobbies.

4 CHANGING TECHNOLOGY

IP video is more flexible and cost-effective than DVR-based systems, which makes it ideal for enterprise-level video imagery systems that connect multiple satellite offices or citywide surveillance systems. Almost all enterprise installations with more than 100 cameras are IP systems. While quite a few systems in the world have more than 10,000 cameras integrated into a single surveillance network, there are even a few with more than 100,000 cameras. The level of scalability can only be possible with network technology.

Analog signals cannot be compressed like digital signals and video images have to be carried on the existing dedicated coax cables (one camera = one coax cable), which adds a significant additional cost to the installation. A digital CCTV surveillance network can cover dozens of cameras on a single cable that is installed around the facility campus. This eliminates most of the installation cost with the individual coax cables required for each analog CCTV system. The network cable does have a limited bandwidth capacity, so as each camera uses a certain bandwidth, it will narrow the bandwidth set aside for all the information signals used in a typical network.

With the fast growth of IP systems, many companies have sprung up with very little experience in the technology. Separating the knowledgeable from the opportunists is the first obstacle to overcome before selecting a surveillance system that will last a decade.

5 DVR WITH IP CAMERAS

The tape recording of camera imaging has long been the term used to describe the process of capturing video on an analog time-lapse recorder. With the advent of IP-addressable cameras, the term "recorder" is a misnomer since DVRs are more accurately described as digital processors that store images, not record images. Some manufacturers have developed software that is used with commercially available computer terminals and their hard drives to store digital images without the use of a DVR. They provide only software instead of a cabinet of electronics, hard drives, and software. The warranty is supported by the computer manufacturer, not by the software provider.

Remote Image Storage was introduced to the marketplace in 2006, where the storage is accomplished at a remote site controlled by a vendor. When a facility needs to retrieve images, it simply searches the web for the information. The off-site vendor leases the data storage space, charging a fee for each camera in the system. Offsite storage is necessary since any disaster at a home facility (power failures, floods, etc.) might cause a loss of stored data and the backup at a remote site ensures that the information is never lost. The only problem with this process is that following a major disaster that affects the entire city (9/11 for example), any subscribers to the off-site service may make requests for data retrieval at the same time and overwhelm the system. It literally took several days before the data was able to be transmitted to customer sites. What many facilities do to avoid this problem is to establish their own off-site storage facility rather than utilizing a private vendor. This allows the facility to control all aspects of the memory storage network.

Surveillance cameras installed in hospitals usually do not have to produce very high-resolution images as long as the images are recognizable (even if a little fuzzy). There is no need to capture finer details like the color of a person's eyes or license plate numbers. To gain significantly higher resolution would require an additional cost which may not be justified. Most of the surveillance cameras in hospital corridors are able to identify common thieves or incidents, not terrorists.

6 CAMERA QUALITY AND RESOLUTION

A DVR will record images that fall between 30 and 60 FPS (frames per second) for each camera. This speed requires a large section of bandwidth. The human eye perceives 16 FPS as full motion even though 32 FPS is actual normal motion.

The FPS capacity of a DVR depends on the number of cameras supported multiplied by the number of FPS per camera. "Real-time" movement is often captured at 30 FPS (streaming video), but this level of resolution is optional and depends on the facility's requirements. If a specific area has very low movement, it can take a lower setting and be programmed to sound an alarm if there is a change in the picture.

Very slow motion is a great feature of digital communication since it adds intelligence and analytics to the conventional video image. It can also help combat fatigue in the security guard who may have to watch many monitors that show little movement. Digital technology can be set up to look at one particular section of the screen, identify items missing or out of place, or identify items that have been in one place too long, like a suspicious suitcase left behind. Many professionals in the field contend that a careful evaluation of all the features offered by a DVR must be made if the unit is to be cost-effective and still provide adequate storage capacity.

There are so many DVR manufacturers that have entered this field since the turn of this century that it has become almost impossible to separate the features of one unit from the others. Without a strong technical background, the average Security Director or purchasing agent is prone to making the same mistakes that occurred when the facility purchased, in the early 1990s, the card access control and surveillance systems. It is essential that anyone responsible for the selection of a DVR system become knowledgeable with the many technical features that separate one unit from another before selecting the final product. Most important is that since the technology can connect to the existing network infrastructure, the selection of a system should be a combined effort between the security and IT departments.

Security department administrators know the characteristics of the areas they need to monitor and what information they want to see in a security system, while the IT department will know the capacity of the current network and understand the different technology features available in digital cameras and recorders. Another consideration is that the video compression algorithms may be proprietary which could cause problems if equipment from one manufacturer is used across the board. The type of facility being monitored will also impact the selection process, with a luxury retail shop or a critical infrastructure facility seeking a higher-end solution and financial and/or correctional institutions a mid-range solution. Healthcare facilities usually seek out a mid-range or lower-end solution, depending on the particular hospital or clinic within the hospital.

7 MATH 101

To further understand the size-to-storage capacity of a DVR, multiply the 30 FPS by 16 camera inputs to set the 480 FPS (30 × 16) storage capacity of the DVR. Since each DVR offers different capacity models at different costs, the less expensive units will provide the least amount of storage and, thus, more would need to be purchased.

Without trying to complicate an already complicated subject, the end user should first select the image FPS, for example, 15. Then, if there were 16 cameras, the size of storage would be 16 × 15 = 240 FPS. In some situations, where there is very little movement (closets, storage rooms, etc.), an even lower FPS could work and require a smaller amount of storage.

Another often used method to extend the length of storage capacity is to program the system to analyze an image for pixel changes (motion detection). Recording a corridor that stands empty is a waste of storage space. Once motion appears on the screen, the recording can be increased to 30 FPS for real-time movement, then reduced when the corridor is empty. Too many empty screens can put an untrained security guard to sleep, and most studies indicate that after 12 minutes of continuous video monitoring, an operator will miss up to 45% of scene activity. After 22 minutes of viewing, up to 95% is overlooked. Trained guards, on the other hand, actually increase in monitoring effectiveness over time, seeing more after the first hour. Most video analytic software is designed to address these problems, supplementing conventional monitoring with algorithm-based monitoring. The basic premise of video monitoring involves algorithms that analyze an image for pixel changes. Upon the detection of pixel changes, the program responds to alarms, alerts, or any number of other inputs.

Digital Image Compression relates to bandwidth limitations of the IP network, dictating how many cameras, or actually, how many pixels, can be accommodated as the digitized video signal travels through the network CAT cable. Compression is a very important characteristic and may be one of the most critical factors in assigning "space" on the network and it also affects the DVR storage size.

Spatial Compression is used on a single image which is a completely independent entity in relation to other frames.

Temporal Compression stores change between subsequent images rather than isolating the frame (spatial).

A popular analogy that digital equipment providers use to explain bandwidth capacity is to compare it to the Holland Tunnel in New York City which has, more or less, the same limitations and the same roadblocks found in bandwidth capacity. There are a finite number of vehicles that can enter and use the tunnel at any one time, just as there are a finite number of cameras that can be carried on

> the network. If the vehicles were smaller, many more cars could drive through the tunnel. This simple analogy is a way of explaining the capacity of a typical network. The greater the compression of the video signal, the smaller the image, allowing more cameras to get through to the available digital recording unit. Another way to define the "compression" factor is that it is the number of video images that can be transmitted and stored without compromising the clarity of the transmitted images.

8 CATEGORIES OF DVRS

Many manufacturers entered the market with DVRs that had similar basic models, each with a choice of options (storage capacity, resolution, etc.). Unique features are used to determine the category of each system but in most circumstances, it is difficult to tell the difference between categories, especially when used in a hospital environment. Some members of the International Association of Healthcare Security and Safety local New York City chapter claim that in their experience working with DVR salespeople, 50% of the salespeople exaggerated the capabilities of their system and the other 50% did not understand how their own systems operated. When selecting a DVR, seven of the most important operating features must be considered before selecting a specific unit:

8.1 Category

High-end systems are suitable for a luxurious store or a nuclear storage facility. **Medium-sized** systems are suitable for banks or prisons and some large-scale hospitals. **Low-end** systems are adequate for most healthcare facilities.

- Wavelet Compression is difficult to understand. Seek help from the IT department.
- **Storage capacity**. Can the unit be expanded? How many weeks of video can be stored? What camera speeds are available?

- **Management system**. If there are 10 units with 16 cameras each, can all 160 be viewed at once? This is sometimes referred to as a management system.
- **Frames per second** choice ranges from real time (30 FPS) to slow motion (3.5 FPS).
- **Networkable** accessible over a web browser.
- **Multi-screen mode** with playback/record/live view simultaneously.

DVR #1	DVR #2	DVR #3
1 TB internal storage capacity for up to three months of continuous recording Compression Algorithm – MPEG-4	120 IPS recording at CIF resolution, 60 IPS at 2CIF	External Storage Firewall via (CD – RW and HDD up to 20 250 GB) Optional USB Thumb drive- or Firewire CD-RW Drive for data recording
Triplex in multi-screen mode Triplex [Display/Recoding/playback Easy Windows interface	PSTN, ISDN ADSL,	
	Cable Modem, Ethernet. Integrated facility maps.	
	Resolution 352 × 240,	
	740 × 240, 704 × 480	
Leverage IP and CCTV cameras	Partitioned to isolate save-disk specified event-triggered	DVRA Enterprise TCP/IP network
Private – IP, Port/DDNS		
CD-R/RW, Removable HDD, Network Integrated facility maps Resolution-352 × 240, 740 × 240, 704 × 480	Multi-screen mode 2 DDH	1 Terabyte of storage Pre-Alarm feature

While there is no attempt to detail the features offered by the more popular recorders, some comments about the confusing technical characteristics indicate why the average security director is still at the mercy of factory representatives. The following terminology is abstracted directly from several magazine ads for three different DVRs.

Our DVR is 16 channel design at 1 IPS at 360X240 (CFI) resolution at average image quality [Q4]. With wavelet compression and 1 Terabyte internal storage capacity to record at a higher resolution for clearer images or faster fps.

9 STORAGE CAPACITY AND VIDEO RECORDING TECHNIQUES

Two of the most important characteristics in a DVR are often defined by (a) the number of frames per second being collected and stored and (b) the storage capacity, which is the number of days that video can be retained for future retrieval. With that number in mind, a set of factors must be considered to determine the storage size:

- Video camera resolution (megapixels, etc.)
- Quality of cameras capturing the image
- Frames per second captured

In comparing analog and digital video, some of the discussion seems to ignore the more important subjects of video recording and data storage. With data retention and file size requirements increasing, the fundamental difference between recording and storage has been misunderstood by many security professionals. Understanding the difference is essential in communicating surveillance storage requirements to IT departments. "Recording" has been the term used to describe the process of capturing video in an analog surveillance environment. With the introduction of IP cameras, the method of processing, transmitting, and collecting the images means that they are never actually recorded; they are more accurately described as being stored in the digital image processor.

Storing digital data, instead of recording analog images, represents both an evolutionary and a revolutionary change for the security industry. The marketplace transitioned from VCRs to DVRs and the term "recorder" has moved out of the security vocabulary.

There are four major methods by which DVRs store video images:

- DVR used with existing analog cameras using coaxial cameras already in place and converts the analog signal to a digital format at the DVR.
- DVR connected to IP addressable digital cameras through the Internet on communication cable (CAT 6). The signal is digital before it enters the DVR.
- Network Video Recorder (NVR) replaces DVR in large facilities.
- Cloud Storage may replace all on-site recorders in the future.

DVRs will replace obsolete, time-lapse tape machines in every facility once it is recognized that there is a need to combine existing analog cameras with DVRs. The DVR can receive electrical signals that are transmitted over the coaxial cable from the analog cameras, and with a built-in digital signal processor (similar to the circuit in IP addressable cameras), they convert the analog signal into digital data (binary ones and zeros). The images are compressed by the software embedded in the DVR and stored on a hard drive for replay and management.

Manufacturers use a proprietary compression algorithm, forcing the end-user to use a device that utilizes the same format for viewing the data. More often than not, this proprietary compression algorithm is limited to the original manufacturer, and once installed, the facility must continue using the same device. To change the manufacturer or upgrade to new technology, the facility would have to replace the system entirely, a very expensive and very difficult proposition to justify, so purchasing the initial system is the beginning of a long relationship.

DVRs represent a major advancement in video surveillance technology by offering far more functions than the bulky VCR did. The DVR stores data on a digital disk that can be viewed on the computer terminal as opposed to the VCR tape that requires a device for playback. The DVR not only duplicates every feature of the VCR, but it can also skip to any part of the program in an instant rather than being forced to rewind or fast forward to review a VCR data stream. If a computer were stolen during the night, five or six two-hour tapes would have to be laboriously played back on the VCR to determine when and who took the device. With a DVR, the actual spot where the computer was originally placed would be retrieved and the screen where it disappeared would be on display in mere seconds.

- DVD – Digital Versatile Disc or Digital Video Disc

Digital Optical Disc Storage Format Comparison. DVD+R can be written once and read many times, whereas DVD-R is a different, competing DVD format.

Since almost all DVD writers sold today can record to both formats, this is not an issue for most users. When creating DVDs for distribution where the player unit is unknown or older, using the DVD-R format is preferable because some standalone video players and DVD-ROM drives aren't fully compatible with the DVD+R format.

- DVD-ROM – Digital Versatile Disc – Read Only Memory

This memory format was first introduced in 1996 in Japan and came to the United States in 1997. Commonly used for permanently storing large software applications that cannot be changed. Similar to a compact disc's read-only memory (CD-ROM) but has a larger capacity. DVD-ROM stores about 4.7 GB for a single-layer disk, or 8.5 GB for a double-layer disk, as compared to the 650 MB data storage contained in a CD-ROM.

NVRs are significantly different from DVRs in that they store digital images directly from the network, whereas the DVR compresses analog video feeds and stores them on a hard drive. The DVR must be located near the analog feeds, in contrast to the NVR that stores the compressed data streams from IP cameras and then indexes the data stream. There is no video connection anywhere on an NVR; its input and output are IP data comprising compressed and encoded video. This will typically be in either the MPEG or H.264 formats which have enjoyed widespread adoption in the CCTV industry due largely to their efficiency. More demanding applications may require a rack-mounted, standalone NVR unit that offers removable hard disk drives.

Recordings can easily be mirrored to multiple NVRs around the network. NVRs possess features that can't be duplicated by DVR:

Hot-swappable disks
Built-in diagnostics
Protection of files against deletion (accidental or otherwise)
Firewall for the protection of data against unauthorized access
Fully redundant power supplies and network connections
Provides uninterrupted operation in event of power supply or network failure

- The **SD (SECURE DIGITAL) MEMORY CARD** is a very important addition to surveillance that could never have been incorporated into an analog camera and is another example of the advanced technology offered by digital systems. The SD card is an integral part of many devices, including IP cameras, phones, MP3 players, and game consoles. The cards are highly efficient and programmable for internal storage that can be accessed remotely over a network. The card eliminates the cost of an on-site server, DVR, NVR, or PC to house the recorded video imagery. Basic SD cards range in capacity from 64 MB to 4 GB. Higher storage capacity cards go from 8 GB to 32 GB, and since 2010, there have been cards that provide storage as high as 2 TB. Instead of consuming bandwidth by streaming video to an external storage device, the onboard storage can be accessed remotely over the network just as if it were on the cloud or on the DVD.
- **USB FLASH DRIVE:** These small, flat, removable storage devices equipped with flash memory capabilities and a USB connector are compatible with any system that provides a USB port. Most flash drives or devices under the names of jump drives, thumb drives, or pen drives have all become invaluable devices for storing and transporting data by connecting to any computer through the USB interface outlet. USB 3.0 and USB 4.0 have high data transfer speeds and have replaced the clumsier floppy disks that were prone to failure. Flash drives are small and light, plus they plug and unplug easily and can be rewritten thousands, if not millions, of times.

The early version of the flash memory device was developed in the 1980s. USB flash drives were developed in the late 1990s and most experts in the field credit the Sandisk company for their mass distribution and the fact that the unit did not need special software. IBM sold the earliest 8 MB version in 2001 as the Memory Key and quickly upgraded it to 16 MB and then to the 128 MB version that cost $30 when first introduced. The storage capacity continues to increase and now 2TB thumb drives are available.

The companies claiming to have originated the USB flash drives have been unable to protect their intellectual property, and there are now numerous versions made all over the world. The USB flash drives have become as much a part of computing as the keyboard.

9.1 Floppy Disk

This storage device will be remembered by veteran computer users as the data storage medium they grew up with and is no longer used as a recording format. The first floppy disks were introduced in 1969, the same year as the Internet. The disks were initially eight inches in diameter and stored data magnetically like a hard disk, but they were not really floppy. However, when the actual disk was removed from the protective casing, it was found to be flexible. These disks were inserted into computer terminal slots and were read-only, like the CD-ROM, and held only 80 KB of data. Later versions held as much as 800 KB and in 1987, the 3.5-inch high-density floppy disk was released. It held 1.44 MB and for the next decade, this flat, jacketed disk became the standard means of distributing commercial software titles and for backing up personal data.

In 1997, CD-ROMs began to replace floppy disks as the standard means of distributing software. A few years later, consumers began migrating to recordable CDs for backing up their data. Apple's original iMac, released in 1998, was the first mainstream computer to not include a floppy disk drive. Several years later, PC manufacturers eventually followed suit. Some software was distributed on CDs and DVDs and data was backed up on either recordable CDs or USB flash drives. Cloud storage and the ability to download and install programs from the Web have all but replaced these now antiquated formats.

9.2 Video Management Software

Video management software systems are more expandable and flexible than DVR-based systems and range from the basic to the sophisticated, with major differences, including reliability features and the number of cameras and locations supported.

10 CLOUD STORAGE

Internet-based "cloud storage" may be the wave of the future as a centralized form of data and was growing faster in the private sector than in commercial communities that had been much slower to adapt. This technology can accumulate an unlimited amount of data at a remote server site – somewhere in the clouds or in the mountains of Utah. The location of the storage facility is commonly a highly secure data center operated by a third party under a monthly subscription service. Multiple companies share in the real estate of the storage facility instead of maintaining their own in-house storage on an NVR or DVR, and each company is provided with a unique web portal that only allows them to retrieve their proprietary information 24/7.

Storing video in the cloud is referred to as hosted video or video as a service and allows users to access their video no matter where they are located – whether at the office, at home, on the road, or even on vacation. By logging onto the web-based portal, all the information becomes available since it is on the remote cloud, not on the devices in their business facility. Other advantages of cloud storage:

- Lower up-front capital investment since it eliminates the need for on-site storage devices like NVR.
- No worries about DVR failures.
- No mega-terabytes of server storage to purchase or find space for the unit and no maintenance by using cloud service; the only upfront costs are for the cameras and an Internet connection.

The service provider supplies system updates and upgrades, integrates new features, troubleshoots, and performs regular maintenance – rather than placing the burden on the customer, who may not have an in-house technical staff to complete these tasks. It offers a perfect solution to a multi-site retail store, monitoring from central or multiple locations.

Using video-in-the-cloud surveillance at retail store chains or remote corporate facilities provides clients with a means to centrally oversee the entire operation and leverage the aggregate information to improve individual locations. This allows senior executives to remotely view video from all the facilities through a single portal and identify and monitor the operation of the entire network, eliminating the need to physically visit each site. There is no longer a need for standalone DVR-based systems.

Simple IT infrastructure best describes network cameras and streaming images to the cloud since there is no need for an expert in networking technology to operate cloud storage. Staff members who use email or a shared printer can easily adapt to remote storage technology. Another advantage to cloud storage is that institutions with low camera count branch offices that originally chose standalone DVR-based systems at various locations can turn to cloud storage as a better operational model. Instead of being restricted by the port capacity of a DVR, a cloud-based solution allows the facility to add cameras one at a time as needed. The hosted video provides local branch managers, district and regional managers, and corporate executives remote access to live and recorded video at each site to analyze ways to improve and grow the company-wide surveillance system.

Security became a concern after the worldwide hacking of giant retail store data storage systems in 2014 and 2015. The number of large hacks and breaches has only increased, including household names, well-known corporations, with hundreds of thousands of customer data being compromised at one time. Giving data to a third party can add to that concern. A certified cloud storage facility maintains the highest classification of data center security by complying with numerous regulations and industry guidelines that include extensive reliance at every layer to ensure consistent, uninterrupted power supply to the storage centers. Network cameras can add a digital watermark and time stamp to all recorded material to mitigate image tampering while creating an audit trail of who has seen what images and whether they've made any edits to those recordings. The **bandwidth** used in video cloud recording is often wrongly accused of monopolizing the bandwidth needed for other data communication traffic. This is a misconception since most network video cameras can have their frame rates and resolution adjusted to accommodate available bandwidth. This can be accomplished through the use of advanced compression techniques (like H.264), where the amount of data flowing through the pipeline can be reduced by more than half while maintaining excellent image quality for playback. Another method of minimizing the issue of bandwidth is to store lower-resolution, lower-frame-rate images in the cloud and higher-resolution, higher-frame-rate images on an SD card in the camera or on a local network device (NAS), which also provides redundancy.

Video Compression is used before recording video and TV signals onto a computer hard drive. It is necessary to compress the video because raw video footage requires lots of space. Video files would quickly eat up gigabytes of hard drive space resulting in only a short amount of video and TV signal that could fit onto the hard drive. In simple terms, the compression of the signal allows the storage of much more TV and video on the computer than if the video files were not compressed.

11 FIREWALL SECURITY AND CYBER HACKING

The digital communication world is constantly moving large amounts of data over the Internet, which has to be protected from unfriendly eyes and ingenious hackers. The method of choice has been to use a filter between the individual computer terminal and the Internet called a firewall. A filter allows information to leave and enter the server, but nothing else is permitted. The firewalls do not specifically prevent hacking – a practice that has proven to be the biggest threat to any organization that handles sensitive information through its network. Hackers from all over the world target the nation's infrastructure, military data, intellectual property, and personal identity information.

Thousands of foreign computer "bots" churn out thousands of hack attempts per millisecond looking to gain millions of dollars' worth of information and sooner or later the targeted network will open up. Once that occurs, either the network or application will crash or the data will be compromised. It doesn't matter if the target is a doctor's office, a Certified Public Accountant (CPA)'s office, or a government agency with secret data. Each of the millions of cyber hack attempts can also congest the network bandwidth until the bandwidth is so full of hack attempts that normal business cannot be transacted. An entire industry surrounding anti-hacking firewalls is constantly evolving. In 2015, the FBI created a sub-agency – the Internet Crime Complaint Center or IC3 – to combat the growing problem of computer-based crime. The National Security Agency (NSA), Government

Accountability Office (GAO), and Homeland Security each have their own cybersecurity divisions. Meanwhile, the problem only grows.

12 FLAT SCREEN MONITORS

Most commercial TV and surveillance monitors in production since 2009 are flat-screen panels; they have become a status symbol for central security surveillance stations. The shift to flat-screen units was inevitable since the surveillance industry has accepted the flat screen as the standard. As most units are almost identical in construction and offer the same choices of resolution, the final selection should be made based on size, pixels, and whether it is an LCD or Plasma. But, for all practical purposes, there isn't that much difference between the screen technologies, and any of the familiar manufacturers produce a reliable unit.

13 SIM CARD AND CDMA OPERATING CELL PHONES

In the world of cellular mobile phones, there are two primary phone types that are in use throughout the world. While these two systems have nothing to do with video recording or retrieval, they are an important sector of communication storage devices and are used in millions of phones worldwide:

>GMS (Global System for Mobile) cellular phones are in use worldwide and rely on a SIM (Subscriber Identity Module) card to tap into a mobile network. The card holds all the critical information needed for a wireless two-way conversation and is issued primarily in Europe. The SIM card offers convenience when it comes to switching phones since the necessary data is on the card itself, which can be plugged into a different phone to automatically transfer the data. To accomplish this, both phones should be in compliance with GSM standards. The major drawback is that if the card itself is somehow lost or damaged, any crucial data is lost forever.

>CDMA (Code Division Multiple Access) systems do not use a SIM card; instead, the carrier saves the phone number and other identifying information in the handset itself, which are tracked by their electronic serial number. CDMA has proven to be convenient for users and has always been the most common cell phone operating system in the United States. These handsets are usually heavily subsidized, and users may have less incentive to switch phones frequently.

Section VII

Wireless Communication

As of 2020, more than 5.4 billion people worldwide use the Internet (Figure S7.1).

That is an amazing number and almost unimaginable when compared to past communication networks. It took more than 25 years to get radio into 50 million homes, 14 years for television to do the same, but it took only 9 months to get 50 million subscribers on YouTube. Apple had sold 250 million iPhones, 70% outside the United States, and 70 million iPads. 350,000 applications are available across both devices.

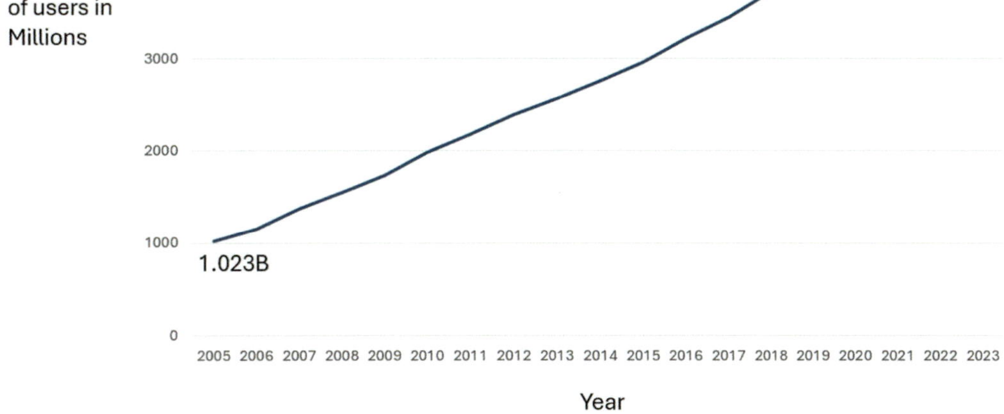

FIGURE S7.1 The number of Internet users, and growth annually, from 2005 to 2023 worldwide.

Source: Statista, 2024.

DOI: 10.4324/9781003091103-25

These numbers will continue to grow. Currently, there are 15 billion IoT devices connected worldwide and the number is expected to double by 2030.

Computer History: The personal computer is credited as being an invention from several sources, such as Xerox Research Center, Apple, and IBM. However, according to a technology historian at the Smithsonian Institute in a book titled *A History of Modern Computing*, credit should be given to a small company called MITS Altair which attracted the attention of a Harvard undergraduate and two Honeywell employees (Bill Gates and Jim Allen) who went to work at this new company, wrote a version of a basic programming language that could run on the machine made by Altair, and called it Microsoft Basic. It was the beginning of what would become the world's largest software company and would make its founders billionaires. Meanwhile, scientists working at the Xerox's Research Center in the early 1970s came up with a radical idea that was unthinkable at that time: one person, one computer.

Some experts now express a belief that the personal computer may soon vanish. Single computers once filled entire rooms, then sat in closets, then on desks, then on a laptop – to now where most handheld devices and photos can be carried almost anywhere and now, nestled in pockets. Soon, the computer may become invisible, hiding away in everyday objects. It will be called the "ubiquitous computer" and reappear embedded in all the objects that make up daily life. Pens, pads of paper, cars, and watches indeed virtually everything people use are becoming computer smart.

18 Laptop Computers Exposed

1 OPTIONS AND FEATURES

In 2015, an epic battle between Apple, Google, and Microsoft ensued over smartphone operating systems. These undisputed tech leaders knew smartphones were fast becoming the dominant computing device on the planet. A lot of controversy surrounds the public's choice for a second digital device. Apple will be the leader in the second category for the foreseeable future with its iPad and Mac, while Google produces the Android tablet and Chromebook laptops. Microsoft, mainly a software company, was the last to enter the competition with its Surface Pro 4, a tablet-PC hybrid introduced in 2012. Apple then came out with an iPad Pro that was close to the Surface Pro 4 in its features. With emerging touchscreen capability, many laptops can be flipped and used as de facto tablet devices.

Microsoft continues upgrading their software programs along with computer manufacturers, who also seem to be upgrading their models every year to meet changing software technology features requested by the nontechnically oriented public. Basically, most websites that offer suggestions on the most important features of laptops include the processor speed, amount of memory, brand, and the model of the graphics card. There is no need to seek out a computer to get 3-D graphics or one that is capable of managing the air traffic control system at LaGuardia Airport. A laptop should be lightweight, have an adequate screen size, have Internet access, be able to stream videos/movies, and provide access to email and documents, including PowerPoint presentations.

WEIGHT. Anything over six pounds is too heavy for students or professionals who may need to transport their laptops frequently. There are many lightweight laptops available, all the way down to two and a half pounds that are only ½ inches thick and are a few hundred dollars to $2,000 depending on the make and specs.

SCREEN SIZE. There are displays as small as ten inches, measured diagonally, which can be too small. The 11.5-inch screen is perfect for a laptop that can be carried while traveling, big enough but still portable enough to be thrown in a bag. The 15-inch models provide more versatility for gaming or working on for longer periods of time. 17- and 18-inch models have some advantages, but they often weigh too much and the battery life is affected.

PROCESSOR. Processor speed is always improving, so as long as the processor is current as of purchase, and gaming is not on the table, the processing speed is really not as big a factor in choosing a laptop as it used to be. For the average user, most processors are acceptable, and it's not worth the extra cost to get an incremental speed boost.

BATTERY LIFE. A search on the website about battery life specs from Best Buy shows that battery life varies on the product configuration and model, applications loaded, power management settings, and product features used by the owner. The specs mean little, and it's best to carry the power cord at all times.

MEMORY. Unlike the processor, RAM (random access memory) *is* quite important, and most have 8–32 gigabytes of RAM, more for high-end systems.

STORAGE. Most laptops these days come with 256GB of storage or more. 5120GB of storage can hold up to 1,200 standard videos, 48 hours of low-resolution video, or 8,300 hours of music. An application like the popular Microsoft Office requires 4–6GB on a PC, up to 10GB for a Mac. At one time, the storage capacity mattered, but with the rise of streaming services and cloud computing, the amount of storage has become less important. Even if files are stored on the hard drive so they are readily available, and there is no online connection, services like Dropbox can allow the selection of specific files for local use while keeping copies in the cloud.

DOI: 10.4324/9781003091103-26

GRAPHIC CARD. Almost any processor will be perfectly acceptable to most users, apart from those who are active gamers or Twitch streamers.

WIRELESS. Some laptops are Wi-Fi only, and most have built-in cellular chips so the laptop can get on 4G or 5G networks. The minor additional cost for a network model is worth it.

PRICE. Spending $500–$800 will get a very versatile unit with all the features that an average user will require. Spending twice as much will get a fancier machine that is thinner and lighter. There are units called Ultrabooks that are very thin and provide generally good battery life.

Before buying any laptop, it is best to try it out by going to a brick-and-mortar store and handling it to see if the keyboard is too cramped or if the screen and speakers are acceptable. Lower prices may be found online but holding the unit is important and the stores have technical help available, which is always essential.

MAC OR PC. This is an age-old dilemma but has become less important in the last decade. Most of the tasks performed on a laptop are web-based, and Macs may cost a little more than the competing Windows machines, but the gap has shrunk considerably. In the Ultrabook category, Macs can be the least expensive. People who have been using one or the other will continue to do so when they purchase the next generation of laptops.

2 COMMUNICATION REVOLUTION

What follows are terms with several meanings and some involving multiple definitions:

Cellular: A service in which the spectrum allotted to a carrier is broken up into geographic areas called cells. Cells next to each other use different frequencies. Cell phone towers are revolutionizing communication networks.

Wireless: Uses radio waves instead of hard wires or fiber optics to carry electrical signals.

Cordless: Operates in an unlicensed spectrum and is restricted to a small coverage area within a defined area (home or office).

Mobile: A wireless system that can be used over a wide distance as people move around on foot or in cars and private transportation.

VoIP: Conversations carried over the Internet as opposed to a phone line.

Digital Assistants: In 2016, major tech companies offered communication devices with voice-activated search capabilities.

- **Siri** was the platform on iPhones and the first digital voice on the market (2010).
- **Viv** was a follow-up to Siri and introduced in 2016; it has since been discontinued and integrated into Samsung Galaxy's Bixby, as had S-Voice, which preceded Bixby.
- **Google** for Android.
- **Cortana was** used by Microsoft but has since been replaced by Copilot.
- **Alexa** is the cloud-based voice-activated service by Amazon – used with Echo speakers.

These popular voice-activated applications each have a different profile.

Although some of these digital assistants sound almost human, they are far from equipped to grapple with the range of human conversations. The ability to handle responses to distress situations varies a great deal. Translating "I am having a heart attack" or "I want to commit suicide" does not always yield a response. Siri can dial 911 and help people find a hospital or emergency hotline.

Wireless connections, along with the information recorded and cataloged on the Internet, allow anyone to become a publisher. Even the most basic cell phones contain a camera, creating whole armies of amateur paparazzi. Uploading videos on YouTube permits the creation of new filmmakers. This explosion in information technology makes everyone, potentially, a public figure and the personal life of each citizen has become transparent. People now conduct their personal and business lives as if they were members of a gigantic family and everything they do is visible for all to see, leaving nowhere to hide skeletons of past foibles or indiscretions (as many politicians have, much to their dismay, discovered). Since the turn of this century, much of what is said, written, or performed will be preserved online forever. Before an employer gets a chance to read an applicant's résumé, they'll Google the person or review their Facebook profile.

While this newly acquired power to instantly circulate information that is read by thousands of Internet subscribers may be revolutionary to an older generation, it can cause a serious reaction when the wording on Facebook, Instagram, or Twitter targets a particular subject or person. In the past, if a person verbally criticized a business while standing around a bar, it went no further than the sound of the person's voice. In the Internet age of mass communication, there's a permanent record of the sentence as soon as the "publish" or "post" button is pressed. One of the growing number of freelance writers was incensed by having his car illegally towed by a private towing company and vented his justifiable outrage by attacking the towing company on Facebook. The towing company then filed a defamation suit against the writer by claiming that the comments were hurting their business and sought $700,000 in damages. It was not the first lawsuit in response to an online comment. Many lawyers consider a lawsuit of this type an example of a decade-old legal maneuver known as a *strategic lawsuit against public participation,* or "SLAPP." The label has traditionally referred to meritless defamation suits filed by businesses or government officials against citizens who speak out against them. The plaintiffs rarely expect to succeed in filing a frivolous suit but rather use the suit to intimidate critics who are inclined to back down with the prospect of a long, expensive court battle.

In 1992, California came up with a solution: an "anti-SLAPP law" that made it easier for defendants to seek early dismissal of such suits at no cost since it mandates that the plaintiff has to pay for the defendant's legal fees. Twenty-six other states followed, though only a few laws in these other states are quite as strong. Without some protection against these frivolous lawsuits, the average person's savings would be wiped out in legal fees before the case is half done. The anti-SLAPP laws do not eliminate abusive cases, but they make them far less frequent.

The Internet has revolutionized "revolutions" as they effectively can be used to stir people up about almost anything, almost anywhere. Facebook and Twitter pulled the Internet trigger that turned out hundreds of thousands of protesters in the Middle East during the many government takeovers in the early months of 2011, which had been dubbed the "Arab Spring." This organized assembly of people was made possible only because of the huge Facebook and Twitter followings. The number of followers in the hundreds of millions far outpaces the old-fashioned communication networks that rely on the 46 million US newspaper subscribers and the average prime-time daily viewers of the top four non-cable networks (ABC, CBS, NBC, and Fox). Information that used to take weeks to be distributed to the masses now takes seconds. Interactive social media has empowered anyone with a cell phone or Internet connection to access news and allows the world to share information without censorship regardless of the consequences.

3 CELLULAR UPRISING

Push-button phones replaced rotary phones in almost all households and with the acceptance of cellular phones, corded phones are practically obsolete. Over 3/4 or 75% of the population in the United States relies completely on their cellular phone (having disconnected the household wall-mounted hard-wired "landline phone").

Portable phones were truly revolutionary and, on rare occasions, today's smartphones are actually used to make phone calls. Digital communication networks have heralded in a new era of mobile phone applications (apps), especially since the introduction of touchscreen technology and faster network response. There are newsfeeds from CNN, language lessons, and musical instrument simulations, and by holding the phone up to a radio, the title and artist of a song being played can easily be identified. Among the most popular apps is one where users scan the barcode of a product and the wireless provider searches for the best price online, and the details are then provided to the cell phone user.

The history of the wireless handheld phone can be traced to the bulky walkie-talkies of World War II (WWII) and the Vietnam War, which led to marine/shipboard and car phones that required 30-pound batteries. Then, in the early 1970s, Motorola introduced a brick-sized phone with a smaller battery – containing 30 printed circuit boards and weighing over 2 ½ pounds – that allowed a person to make calls anywhere in the world (Figure 18.1). It was accepted as an executive toy since it took ten hours to charge the huge battery, which yielded about 85 minutes of airtime between charges, and there were more dead zones than live reception zones. It became a symbol of yuppie excess, and it would be another decade before anyone could offer to purchase the unit. Those early models were the playthings of Wall Street tycoons like Gordon Gekko, who, in the movie "Wall Street," carried the brick-sized, two-pound Motorola DynaTAC that had cost him $4,000.

Over 1.8 billion cell phones are now sold each year worldwide.

The cell phone has become the world's most ubiquitous computer, with more than 18 billion in use around the globe, carrying every conceivable type of personal information. The cell phone is not just a replacement for hard-wired telephones of years gone by; it is a device that has changed lifestyles and methods of mass communication. It is used to find friends, restaurants, the nearest gas station, and even the location of police cars hiding at speed traps.

Smartphones were introduced with great fanfare in 1987 when only 1.2 million people owned a cell phone. In the intervening years, the nation has become addicted to the various forms of this handheld communication device and the many social networks it has given birth to. Smartphones

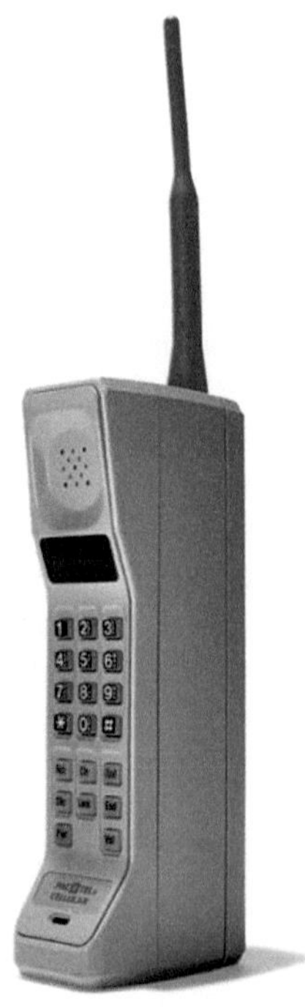

FIGURE 18.1 The original Motorola satellite phone model.

now offer a range of features that make them essentially equivalent to the desktop computers of a few years ago. Most of these devices are so amazingly versatile that many of the rafts of features they provide are rarely used because they are too complicated or impractical.

These smartphones are a great convenience and can enhance productivity but, like other technologies, they pose great security risks.

Searches can be traced to individual devices through the captured IP address (the unique coded identifying number assigned to all computers that access the Web). Internet search companies have stored detailed profiles of each person's personal interests and taste, causing some privacy advocates to claim that the data could easily be misused and/or sold to anyone willing to pay for the information… like a publication's mailing list.

There was a time in the not-too-distant past when people actually communicated face-to-face and our grandparents actually wrote letters to grandchildren from time to time. The teenager delivering messages between offices on a bicycle has disappeared along with the FAX machine. With the advent of email, purchasing a stamp has become a rare event since the formerly simple task of determining the proper value of a first-class stamp is no longer an easy task with the regular increases instituted by the Postal Service and no one wants to leave the computer terminal to mail a letter [assuming they can find a mailbox]. The Internet has changed the world and people under 50 now spend more time in front of the computer screen than they do conducting conversations with other human beings. Those same people have forgotten that there was a time when the only way to change TV channels was to walk all the way across a room to the TV set and manually turn it on or change the station/channel.

4 CELLULAR PROFUSION AND CONFUSION

Bluetooth and Wi-Fi are different methods of digital communication. Bluetooth has a comparably shorter range – around 30 feet. Also, it doesn't depend on any type of central device like a router. It is almost always used to connect two devices in close proximity.

One example is that of a wireless speaker and smartphone, where the phone and speaker talk directly to each other over Bluetooth, which beams the music from the phone to the speaker without the need for any third device or wide network. That's why it doesn't matter whether there's Wi-Fi in the room.

Other common examples of Bluetooth scenarios are wireless headsets for making phone calls or wireless keyboards and mice for computers and tablets. Vehicles are now made with Bluetooth installed.

Because Bluetooth is a direct device-to-device technology and is used for so many different things, it typically requires that you first "pair" the devices being linked, which sometimes involves typing a number generated by one device into the other. Wi-Fi has no such pairing requirement, though you'll need a password to access a private Wi-Fi network.

Wi-Fi can also act like Bluetooth, connecting two devices directly over a short range. A version called Wi-Fi Direct does this. It can transfer photos and files between nearby devices, just like Bluetooth, which is a short-range wireless solution for pairing one device with another, whereas Wi-Fi has a greater range and works with multiple devices.

Voice-controlled Devices have been around for decades but initially had limited use, like turning on the light, etc. In 2014, Amazon introduced a more sophisticated voice-controlled device called "Echo" which became a surprise hit, and, later, the Dot. The devices provided the weather, traffic information and could play songs from streaming services using only a voice command. It functions as a wireless speaker that connects with phones and the Internet through Bluetooth and Wi-Fi.

5 BIG BROTHER IS TRACKING

Cell phones have long been a fixture of life, becoming more powerful and more "aware" of their locations every year and, it seems, the law has remained behind each advance in wireless technology. This readily accepted use of cell phones has created ongoing legal issues concerning privacy and how far the government can go in gathering information about the whereabouts of cell phone users. The government wants relatively free access to "location" information, while civil liberties groups claim such information should be handed over only in response to a search warrant. These conflicting and opposing attitudes continue popping up in courts, but no definitive policies have been made to satisfy all parties.

Most people have no idea how much data cell phones collect about users, including the fact that phones report back to the carriers on where the users are at any given time – in some cases even when the phone is not in use. Carrying a cell phone is, in effect, the same as carrying a tracking device. Police agencies can request that a phone company monitor the movements of suspects in real time or can obtain stored records on a customer's movements in the past weeks, months, or even years. Information about a person's movement is, by nature, extremely private since it reveals personal details (attending church or protests may reveal religious and/or political affiliations, for example). It can also provide evidence of marital infidelity. If the courts allow the government to obtain these records without probable cause, the impact on ordinary people's freedom will be substantial.

In 2013, Edward Snowden, working for the highly secretive National Security Agency, left with a flash drive that contained information about many of the programs used by employees to track cell phone movement and listen in on conversations over the Internet through a project that became known by the government as PRISM. Some citizens became outraged because the program violated privacy guarantees, while others believed that privacy had to be compromised to ensure the safety and security of the nation.

CELL-TRACKING: A test case has surfaced that challenges the meaning and interpretation of the laws involving "entrapment."

Tracking is an intrusion on an employee's right to privacy since movements are recorded even when the employee isn't supposed to be at work. According to Repo Buzz, as of 2020, most states adhere to federal guidelines on GPS tracking and require consent before a GPS device can track a person or property. Most vehicle lessees agree to accept a GPS device as part of their loan agreement. The Teamsters Union negotiated a policy limiting the use of GPS data to discipline United Parcel Service drivers.

It's a double-edged sword. As GPS makes it more difficult to get lost, new cell phone services make it harder to hide. This power to minutely track others creates privacy risks, and, at the same time, effectively pursuing criminal organizations offsets the loss of privacy. Also, families may desire to track each other's whereabouts. If someone was expecting a friend to attend a meeting or a shared dinner, they can discover that the other person was stuck in traffic 40 miles away simply by viewing the cell phone instead of waiting at the meeting place simply because each person can arrange their schedule to arrive at the same time.

STOLEN iPHONE: An estimated 7.35 million iPhones were stolen in 2023, according to Allstate Insurance, causing many police departments throughout the country to use the Find My iPhone application to track them down. But obviously, they couldn't give the task the same priority as bank robberies or homicides. When a complainant reports the theft of an Apple product, the device must have access to Apple's iCloud and previously activated location services on the device. Additionally, the stolen phone must be turned on for Find My iPhone to work. Police use the app to prompt the stolen iPhone to play an audible alert, lock the phone, and/or permanently erase its data without actually having a search warrant and trying to retrieve the phone. With apps available to locate stolen iPhones, many victims take desperate actions to recover their devices and that action could prove to be dangerous. An unintended result of the proliferation of GPS tracking apps has caused worries among law enforcement officials that people are putting themselves in danger, taking disproportionate risks for the sake of an easily replaceable item.

CELL-STALKING (Spousal Abuse): Sophisticated wireless technology aids domestic abusers. GPS advances have unexpectedly made it easier for stalkers to track their victims using common cell services. On the other hand, GPS is now being used to effectively monitor restraining order compliance. A study commissioned by the Justice Department indicated that GPS usage decreased violations and rearrests.

While calling 911 might seem a refuge if targeted with abuse – a cell phone may make you feel safer, but you can't always call 911 when under attack. According to a sciencedaily.com article, cell towers leak location data since they use it to provide effective service. If you know a target's cell phone number, they can be tracked. The code and methodology are available through open-source projects, so just a computer and some cell network basics are all that's needed to hack your whereabouts.

ANIMAL TRACKING: The Animal Identification System is conducted under the authority of the Department of Agriculture, which had intended to make it a law but was met with so much opposition from the farm community that it was finally issued as a volunteer program. Many farmers felt that the government was starting to track items in someone's backyard and for some, a farm is their backyard only larger. About a quarter of the eligible farms have voluntarily registered for this free program where each farm is assigned a seven-digit ID code and the animals are given a 15-digit number and tags. The idea of tagging animals originated in 2001 following the outbreak of foot-and-mouth disease in Europe, which gave rise to a fear of agricultural terrorist attacks on the US food supply, which grew even more serious after 9/11. While the program remains voluntary, there was an outcry when it became known that a farmer had to be enrolled to receive state grants.

CAR TRACKING devices have been available for many years but remain controversial in the United States. Law enforcement agencies claim that tracking is essential in criminal investigations, while the public feels it is an abuse of privacy. The early acceptance of such a device in the Netherlands occurred without any complaints from their populace, but the concept would not be readily accepted by the American public because drivers are reluctant to allow traffic authorities to know where they are going or where they have been. More threatening is (a) the fear that the traffic information is being recorded in real time, and (b) anyone could follow a car's movements online simply by entering a license plate number.

The acceptance of GPS for auto leasing did not come from any law enforcement authorities. The insurance industry began offering discounts for voluntarily tracking driving data – produced by the onboard computers in every modern car. This information is streamed back to the company. By matching a "good driver" profile, participants in the program could reduce their insurance premiums by 10–50%. Not many drivers took advantage of the offer because many Americans still consider their cars to be one of the last truly private places they can go. Of course, there are red lights to contend with, speed cameras, etc., but for the most part, the occupants of a car are anonymous to the outside world… until now.

The technology can keep the driver under continuous surveillance, measuring the way a left turn is made, if travel is at night, how hard the driver brakes or accelerates, if the car goes over 80 mph – all factors that can affect the odds for accidents. This newfound source of driver habits could add important data to the traditional way insurers set premiums, which is to fit drivers into broad categories – age, gender, residence, history of accidents, or traffic violations. Learning how an individual actually drives allows companies to fine-tune a driver's risk profile and charge accordingly. The same way financial institutions use credit scores to set loan rates.

There is no need to carry a beeping handheld receiver to track the transmitter like the actors do in the movies; instead, there is a Web application that accomplishes the same feat. Trucking companies use them to manage their fleet of vans but there is a more user-friendly unit that became available that uses a transmitter the size of those older pagers that can be placed in a car's glove compartment in case it's stolen, on a pet's collar, on a child's backpack, or even on the lawn equipment that the neighbor borrowed and never returned. Apple AirTags are a popular device geared largely to the average consumer for such purposes.

6 TRACKING STOLEN PILLBOXES

Pharmacies in 33 states have caught painkiller thieves by using decoy pillboxes containing GPS devices. The growing number of pharmacy robberies perpetrated by drug addicts created a new industry around decoy pillboxes made up as popular prescription painkillers. The New York City Police Department admitted to the practice, but most other cities choose not to publicize the fakery, fearing robbers might target pharmacy workers if aware of this deceit.

The bottle or box is designed to be weighted and to rattle when shaken, so the thief does not instantly realize that the container does not contain the sought-after pills. Each pillbox sits atop a special base, and when the bottle is lifted off the base, it begins to emit a unique GPS tracking signal. The police can follow the getaway car by using GPS and apprehend the suspect.

Additionally, these devices can also be used to indicate the closest theaters and restaurants and can even be programmed to automatically notify someone in the event of an accident. There is also a buddy system available that shows on a displayed map the location of up to 25 friends – if all are carrying a similar smartphone.

The most unusual application of GPS is used by the Nepalese government to catch "Hima-liars" who make false summit claims when they attempt to scale Mount Everest. Climbers who reach the world's tallest peak produce photos of themselves at the very top as proof that they achieved this sought-after goal. This type of evidence has encouraged many to produce easy-to-fake photos to substantiate their questionable accomplishment, a practice no longer considered valid. In 2017, Nepal began providing GPS tracking devices to climbers to prevent them from making false summit claims. The data from the tracking devices are checked once climbers return to the base station.

When GPS devices first emerged, they made map-reading skills unnecessary if not obsolete. Relying on cell phones with apps like Google Maps or Waze has but replaced the traditional car GPS systems in allowing users to not only map routes but also be directed to the shortest or quickest route using real-time traffic information reported by other app users. Dash cams, primarily used for record keeping in case of an accident, can also be connected to a dashboard GPS device to automatically send a text message if there's a crash, while the average smartphone cannot duplicate that function.

7 ILLEGAL GPS TRACKING

Until 2012, GPS technology had been used extensively by law enforcement officials to collect evidence to prove that a suspect was in the vicinity of a crime by pinpointing their movements and analyzing stored data from the GPS. The technology can no longer be used for this purpose without a search warrant.

In a landmark case in 2012, United States v. Jones, the Supreme Court ruled that using a GPS device to track a vehicle constituted a search under the law. Jones, a suspected drug dealer, was the target of a police investigation. After obtaining a warrant, police attached a GPS device to his car. By a slim majority, the court ruled the warrant flawed. Police can't search without evidence indicating probable cause. Mere suspicion isn't enough. The Jones decision follows a history of complicated interactions between the law and technology:

- 1967 – Established the Fourth Amendment standard that people are entitled to a "reasonable expectation of privacy" against electronic surveillance; the court ruled that the use of an electronic listening device attached to the outside of a public telephone booth to gather information from a private conversation inside constituted a "search" requiring a warrant.

- 1983 – Finding that a person riding in a car opublic streets has no reasonable expectation of privacy in any movements from one place to another, the court permitted police use of a beeper to follow a car on public roads.
- 1984 – Declaring that the use of a beeper to track information into a private residence – the device was used to track a can of ether – violating the Fourth Amendment.
- 2001 – Holding that officers' use of thermal imaging technology to detect heat radiating from within a home (where marijuana was being grown) was a "search" subject to the Fourth Amendment; that decision emphasized the sanctity of the home.
- 2012 – A decision made it illegal to attach a GPS to a suspect's car without a search warrant or the owner's knowledge. The practice was deemed unconstitutional by a unanimous US Supreme Court decision which marked the justices' first-ever review of high-tech tracking and how its misuse was in violation of the Fourth Amendment protection against unreasonable searches and seizures. That constitutionally based decision will not allow the use of a GPS device, smartphones, or other devices with GPS capabilities for law enforcement surveillance without first obtaining a warrant from the courts.

Despite the Jones decision, the average driver is already tracked with little regard for Fourth Amendment rights. Ubiquitous video monitoring and toll roads with automatic collection systems all create a precise movement record.

8 JAMMING GPS SIGNALS

The Global Positioning System is owned by the federal government as a natural resource and was originally developed in the early 1970s by the US Department of Defense to overcome shortcomings of previously available navigation systems. In 1983, a commercial passenger plane was shot down when it accidentally strayed into the USSR airspace, which caused the US president to make GPS freely available for civilian use. Before that decision, the GPS signal was for military use only, and when releasing the technology for civilian use, the signal was intentionally degraded – a practice known as selective availability.

Criminals and would-be terrorists have discovered that since GPS signals are faint, it is relatively easy for a jamming device to overwhelm the receiving hardware circuitry so it cannot detect the GPS signal. In 2014, the Federal Communications Commission (FCC) recognized this weakness and issued a $34.9 million fine, the largest FCC fine allowed by law, against a major Chinese electronics manufacturer. That firm had marketed 280 models of signal jamming devices in the United States for over two years. The jamming devices were found to have endangered life and property by preventing individuals from making 911 or other emergency calls and disrupting the basic communications essential to aviation and maritime safety.

Jamming technology does not discriminate between modes of communication and prevents cell phones from making or receiving calls, Wi-Fi-enabled devices from connecting to the Internet, GPS devices from receiving location positioning signals, and first responders from locating someone needing emergency assistance. The jammers have been used for business purposes by sending inaccurate location data to competitors.

Jamming devices are sold on the Internet under differing descriptions. Jammers have military, commercial, and criminal purposes:

- South Korea discovered that the North Korean government often used GPS jammers that interfered with military and civilian receivers on land and at sea.
- Criminals in the United States have hijacked trucks filled with high-value goods by jamming GPS and mobile phones so that drivers cannot specify their location and call for help during the attack.

9 POLICE DETECT THROUGH WALLS – WITHOUT A WARRANT

In 2015, a wall-penetrating radar device readily available on the Internet caused a great deal of controversy when it was discovered to have been used by the US Marshals Service and other law enforcement agencies since 2012. The device can penetrate walls in private dwellings, which critics contend is in violation of the Fourth Amendment unless a search warrant was issued. Police and federal authorities defended its use because it allows law enforcement officers to detect movement and location when trying to apprehend suspects after obtaining a warrant, not as a ruse to conduct a warrantless search for evidence.

While it doesn't produce a picture, the use of radar devices without a warrant is much like pointing a thermal imaging gun at a house to detect heat from the lights or using a drug dog to sniff out the same house for the smell of marijuana. The court ruled that both techniques were illegal and in violation of the Fourth Amendment. Penetrating radars were first designed for use in Iraq and Afghanistan and at the time represented the latest battlefield technology, finding a use in civilian life like a great many earlier detection systems such as the passive infrared (PIR) and microwave sensors that were introduced in the 1980s from military research.

10 DIGITAL COMMUNICATION TECHNOLOGY

BANK TRANSFERS: Web-enabled cell phones opened up new horizons for money transactions never before imagined when first introduced in the early 1990s. Some banks allowed their customers to view account information, transfer funds, and even pay bills by phone. It may not be a universally accepted method for paying bills because it's a little too complex for some people and often requires having bills on hand – unlike checking balances or transferring money. As mobile banking becomes more popular, cell phones have become even more of a target for thieves.

INCOME TAX apps for smartphones allow taxpayers to file taxes on their smartphones; for many who don't have complex tax situations, it's now a common and convenient option, thanks to dedicated mobile tax apps and mobile-friendly tax software websites. As with any online option for convenience that deals with personal information, there are always criminals looking to hijack information or otherwise trick consumers into identity theft schemes.

TYPOSQUATTERS: Internet surfers who mistype an address can arrive at sites that appear similar to their intended destination but are actually the domain of "typosquatters" who collect a fee as low as pennies for each inquiry, which can add up to big money over time. The practice is discouraged by law, but it is still being done, with some firms establishing websites with an identical name to a popular site but a dot, colon, comma, or exclamation mark away.

BRAND-JACKING is the illegal use of trademarks online and poses a serious problem since most companies cannot afford the time or money to eliminate the spread of phony sites that rely on confusing the public by establishing a name similar to that of a legitimate one. Some of the lists can remain undetected for many years before being taken offline, enough time to cause serious damage to the original brand and unfair high profitability to the domain counterfeiters.

CYBERSQUATTING is the most prevalent form of brand abuse online, with almost 450,000 cybersquatting sites active in recent years. Search engines seeking a brand in response to an inquiry will often automatically link to a fake, unauthorized brand name instead of the real one, particularly in the apparel field and the automobile industry.

PROTECTING CELL PHONE DATA had not attracted the same level of attention that was directed to wireless computers since snoops and hackers had initially posed much less of a threat to cell phones than to PCs. But that low priority to cell phone data protection has changed over the last decade, where there has been a stronger need for software to preserve and shield a phone's data. Protection systems for mobile phones became available in 2010, but they did not work when a family or business had multiple types of cell phones on various wireless networks. Furthermore, mobile security has historically been considered a solution in search of a problem, but that has slowly changed due to the overwhelmingly increased use of mobile phones, especially when used to shop and conduct banking. As cell phones are increasingly used to do more transactions, the more valuable a target they become. Corporations have even installed data backup software so that they can remotely erase a phone if an employee leaves it in the cab, etc.

REMOTE CODING is one of the most remarkable uses for wireless phone technology since it can transmit remote cryptic signals from a base station unit to remote cell phones anywhere in the world as long as each unit has a built-in LED that carries a coded signal between each unit. The key is a miniature computer that stores the ID code to match the stored code in the cylinder (that contains a microprocessor). By holding a key next to the base station cell phone and the remote key to another cell phone, a coded signal can be transmitted via the Internet from one phone to the other, changing the code on the remote key and permitting the holder to access a door thousands of miles away.

E-TIPS is the name given to websites created by law enforcement agencies in major cities that allow good Samaritans to submit information, often accompanied by photos and/or video, on criminal acts in progress directly to the police. Tipsters can go directly to Internet sites where they are prompted to fill out a questionnaire and upload any images that depict a suspected criminal act. These sites are secure, anonymous, and available in several languages. Those who submit e-tips receive an identification number that can be used to track the case via a hotline set up by the local police department. Some e-tip reports include pictures of a fleeing car's license plate involved in a hit-and-run incident. This is an additional tool for local "Crimestoppers" programs used by a growing number of municipalities, which have been instrumental in solving thousands of violent crimes, including homicides. Some organizations offer rewards for transmitting the information if it leads to an indictment.

Another less discussed subject is the use of cell phones when traveling overseas and the different methods used by the company providing the service. A majority of the world's cell phone subscribers use GSM (Global System for Mobile Communications) technology, while others use CDMA (Code-Division Multiple Access), the technology found in North America as well as in some Asian countries (but not in Europe and South Africa). CDMA can be used in just 26 countries, while GSM is perfectly suitable in over a hundred countries.

11 LIFELESS ROBOTIC ANSWERING SERVICES

Life is better in the digital age, not only because of the advanced electronic system technology, but also because the digital age has given mankind new ways to communicate… social networks and, of course, the ever-annoying voice on the telephone answering machines that the industry refers to as the robotic "Interactive Voice Response." It may well be the worst invention of the last 50 years. The service, which should be called "anti-retentive device," is used by every company in existence. Major organizations rely completely on these prerecorded repetitive messages to direct the caller to hold for the next available operator – or request the caller to state the problem by saying or touching a corresponding number representing the reason for the call on the zero to nine buttons on the cell phone, a ceremony that seems to last between 15 minutes and 2 days.

The use of Interactive Voice Response saves millions of dollars for companies that utilize such systems since the voice on the other end is a recording; there are no sick days, no unions, or anything else that goes along with employing a person. Placing people on hold for 30 minutes is the antithesis of the age-old philosophy that the customer is king; in effect, robotic answering services tell customers that the company's time is more important than their own. RUDENESS IS DEFENSIBLE.

12 ISSUES OF PHONE ETIQUETTE

Voice-activated technology in smartphones first appeared in 2008 when mobile phones running Google's Android operating system and other software began offering basic voice commands to conduct Web searches and other tasks. It permitted people to order their phones around like digital tutors, commanding them to dictate text messages, add appointments to their calendars, and search for nearby restaurants. Apple took this concept to a new and more sophisticated level when it introduced "Siri" in 2011. This voice-activated program, better known as a virtual assistant, responds to natural-sounding phrases like, "What's the weather looking like?" and "Wake me up at 8 a.m." There is a great deal of difference in a phone conversation between two people as compared to a conversation with Siri.

13 EMERGENCY 911 CALLS

By 2020, private companies represented roughly 30% of all ambulance services in the nation, according to the National Association of State EMS Officials.

The various cellular service centers across the country receive over 240 million 911 calls every year, and if the call is from a cell phone, the exact location of the caller cannot be identified as easily as a call from a hardwired phone. This can pose a problem for the millions of Americans who gave up their landlines in favor of cell phones. Dialing 911 may no longer mean a quick response and can lead to misrouted calls or delayed information about the caller's location, serious shortcoming in emergency situations. Even the most advanced 911 systems do not allow a dispatcher to get a specific street address for a wireless call, and the signal can only be traced to a cell tower, then to a multistoried building, but the caller's exact location cannot be separated from the other potentially 350+ apartments in the building.

The nation's very successful 911 emergency response system, built in 1967, was based on the expectation that calls for help would come from an easily identifiable landline telephone. With the increasing reliance on cell phones, calls that bounce from tower to tower pose significant technical challenges, and the public can no longer assume that they are going to be located in a short period of time – if at all.

Over 90% of the nation's 911 centers have technology that allows dispatchers to immediately see the caller's number and the location of the cell tower that picks up the call, but the dispatcher must request the caller's GPS coordinates from the wireless carrier that operates the tower. A caller's location is determined through a combination of methods. Calls are generally routed to the nearest 911 center, with the location determined either by the cell tower that receives the call, or, more often via the phone's GPS or other such location services.

14 IPHONE AND ANXIETY

There have been numerous studies conducted that have found that people suffer from a strange case of iPhone withdrawal that can have serious psychological and physiological symptoms, defined as "separation anxiety." It affects people who rely on their phone for just about all their daily activities and admit to being mentally tethered to it. The study involved participants completing word search puzzles while their phone was in their possession, and another time when it was taken away from them. Researchers found that respondents had higher heart rates, higher blood pressure levels, and lower cognitive abilities when they weren't allowed to answer their ringing phones. The problem occurred more often in younger people who had been exposed to the technology for a longer period of time. The separation anxiety feeling shared by most people is due to the fact that they have had their entire lives and social interactions built around the smartphone. Until the late 1980s, it was assumed that individuals memorized their closest circle of friends' phone numbers, but by the second decade of the 21st century, the loss of an iPhone was as devastating as losing a loved one, since all the numbers are stored in the phones' memory.

People who realize they've left their wallets or office keys at home might find another way to survive the day, but leaving a smartphone means doing without a critical accessory. The pains required to retrieve it – distance travelled and added time – are a second thought. There is no cure for the malady save for medical phone implantation, preferably on a body part easily accessed when it rings.

15 HEADPHONES AND SILENT CARS

Almost 20% of teenagers in America exhibit what many medical associations call a slight hearing loss, which means they can't make out consonants like T and K, or the plinking of raindrops. The number of teenagers with hearing loss – from slight to severe – has jumped significantly over the last several decades. During such time, the rise of personal media players like the iPod, MP3 players, and smartphones went up a hundred-fold and many researchers attribute this widespread hearing loss to exposure to sound played loudly and regularly through headphones attached to one of these devices. Not only are young people losing the art of person-to-person conversation since they're always on some antisocial wireless communication device… but they're slowly going deaf. Protecting children's hearing is not just as important as protecting their brains; it *is* protecting their brains.

Headphones were invented at the beginning of the 20th century and were the brainchild of a tinkerer who grew frustrated when he couldn't hear sermons over the noise in the vast Salt Lake Tabernacle. He designed an amplifier with two sound receivers that eventually were placed inside earphones. Not many consumers owned headphones outside military and commercial pilots until after WWII. The 1950s led to another invention, stereo headphones, designed explicitly for personal music listening, which led to the introduction of Sony's Walkman in 1979. Headphones became more popular than the boom boxes of the 1960s, and now there is less ambient noise in everyday life simply because no one is talking anymore; they're just listening to music even while they're driving. It is important that car drivers not only be forced to concentrate on controlling the car but also be forced to hear the traffic around them. This has become a challenge, particularly if the traffic involves silent battery-powered cars.

People have traveled billions of miles over highways and byways for a hundred years, always aware of the sound of internal combustion engines around them. With the introduction of battery-operated and fully electric vehicles, automakers have eliminated the familiar engine noise and replaced it with silence. This was originally considered one of the advantages of electric-driven cars and automakers spent millions on insulation and sound-damping technology to make cars quieter, but it turned out lack of noise is a defect. The government has enacted legislation that requires electric cars to emit some type of noise at low speeds when their battery-driven motors usually run silently. New government rules will make the vehicle safer for vision impaired pedestrians and others who rely on aural cues to avoid being hit by an oncoming silent car.

16 CELL PHONE RADIATION

There was a time when nobody realized that cigarettes and tanning caused (possibly deadly) health issues. This scenario may be repeated in future years when the public looks back to another time when nobody knew how dangerous it was to hold a phone right next to their head and chat away for hours. Different phone models emit anywhere from 0.2 W/kg of body tissue up to 1.6 watts, the legal limit. The amount of radio frequency energy seeping into the body and brain is measured by a unit called the Specific Absorption Rate. Some have worried about the emission of radio waves by cell phones, which is a form of nonionizing, low-frequency radiation, but studies consistently show that the amount of radiation is too weak to damage DNA. Other types of radiation, such as that produced by an atomic bomb, are known to cause cancer, but that's an entirely different story.

The FCC set a standard as far back in 1997, requiring that all cell phones be rated at 1.6 W/kg or lower. That standard has never been updated with all the new information available and with the proliferation of cell phone use, a great number of unknowns remain concerning the harmful effects of having the phone practically attached to the ear for up to eight hours a day – or longer.

It has been suggested by some medical authorities that it would be safer to choose a phone that falls on the lower end of the radiation spectrum scale to minimize potential health issues. Until there is some conclusive evidence that radiation at these low wattages causes cancer, the public will continue to choose the most fashionable and cost-efficient models offering the most features. Some governmental authorities have considered action to require cell phones to carry warnings that they can cause brain cancer. The devices carry such warnings in some countries, though no governmental authority in the United States requires it.

A Swedish study of young people using cell phones as teenagers for at least ten years described a 400% increase in brain tumors, though other studies conflict with that finding.

17 COMPUTER VISION SYNDROME

In the age of digital communication, there are other devices, aside from cell phones, that can pose a risk to our health. Computers have been found to cause computer vision syndrome, which affects as many as 60 million people worldwide who use computers more than three hours a day.

This syndrome is typically a problem shared by professionals – accountants, architects, bankers, engineers, flight controllers, graphic artists, journalists, academicians, secretaries, and students – all of whom are practically attached to a computer. The main victims include millions of children and adolescents who spend many hours a day playing computer games. The effects of prolonged computer use are not just vision-related but include neurological issues, such as headaches, and musculoskeletal issues, such as neck and back pain.

The most common computer-related complaint involves the eyes, which can develop blurred or double vision as well as burning, itching, and redness, all of which can interfere with work performance. Causes vary, but experts in the field seem to agree that it may be due to the way in which words are presented on a computer screen. Unlike words on the printed page that have sharply defined edges, electronic characters, which are made up of pixels, have blurred edges, making it difficult for the eyes to focus. There is also the effect of less blinking, which results in dry, irritated eyes. The normal rate of blinks is 17 or more a minute, but when working on a computer, the blink rate is reduced to about 12 blinks. Steps can be taken to lessen the effects of working on a computer:

- The distance between the screen and the head should provide a comfortable focusing distance.
- The screen should be 20–26 inches away from the face.
- The eyes should be at the level of the top of the monitor to minimize dryness.
- Contrast is best achieved when the screen is brighter than the ambient light.

There are ways to reduce or eliminate computer vision syndrome by following the "20-20-20" rule. Every 20 minutes, take a 20-second break and look at something 20 feet away.

18 DIGITAL PRIVACY – THE FOURTH AMENDMENT

The 2018 Supreme Court ruling 5-4 in Carpenter v. United States on privacy law under the Fourth Amendment ruled, and agreed, that data and cell phone location collected by cell phone companies can't be turned over to authorities without a search warrant. In the decision, the Court agreed that there should be a higher standard for accessing location records due to their intrusive nature. Technology companies (Apple, Facebook, and Google) argued that the Fourth Amendment protects consumers from the risk of warrantless government surveillance simply by using available technology. Government officials claim that it is legal for prosecutors to rely on months of records obtained from a wireless carrier to prove their case. Prosecutors can use the location of cell phones to prove that a suspect was near a crime scene simply by viewing the location of calls made on the cell phone. According to research, the average smartphone carries from 30 up to as many as 80 apps, including banking and email, which can be used to paint a revealing portrait of the user's life and everyday habits.

The case brought before the court involved data collected on a suspect that located the cell phone 12,898 times over 127 days, an average 100+ each day. The records were obtained under the Stored Communications Act of 1986, which allows phone companies to turn over records if the government has "reasonable" grounds. A search warrant would require a tougher "probable cause" standard. The government relied on the *third-party doctrine,* which allows such searches without a warrant. As cell phones have come to hold almost every piece of information about an individual, the case law surrounding digital privacy only grows.

19 Traffic Hazards

1 DISTRACTED DRIVER FATALITIES

Decades ago, lawmakers fretted over whether windshield wipers and car radios would throw off the focus of drivers. Many states later banned eating, reading, and grooming while behind the wheel. More recently, multiple states ordered drivers to stop texting and get off the phone. In 2012, Rhode Island attempted to pass a law that would have banned having dogs sit on a driver's lap because they obviously distract the driver… and dogs don't wear seatbelts. Other states, such as California, Maryland, and New York, passed laws banning texting and the use of handheld cell phones while driving, but only a few have considered laws banning dogs from driver's seats; only Hawaii has a law on the books prohibiting drivers from having dogs on their laps.

Distracted driving is a growing menace on the roads and has only gotten worse due to car manufacturers installing new technology without a full examination of whether it should be in the car in the first place. In 2012, the Federal Department of Transportation issued voluntary guidelines that auto manufacturers were urged to implement in future vehicles, which included devices to disable text messaging, Internet browsing (including social media), navigation destination entry, and phone dialing unless the vehicle is stopped and the gear shift is in "park," but implementation has been lax. Increasingly, popular voice-activated, in-car technology that allows drivers to text, talk on the phone, or even use Facebook while driving causes dangerous distractions, according to a comprehensive 2017 and 2018 study conducted by the AAA Foundation for Traffic Safety. The study found that as mental workload and distractions increase, reaction time slows, brain function is compromised, and drivers scan the road less and miss visual clues. This could potentially result in drivers being unable to see things right in front of them, such as stop signs and/or pedestrians. In 2019, the AAA Foundation for Traffic Safety then looked at the effectiveness of *countermeasures* for distracted driving. National statistics by the National Highway Traffic Safety Administration (NHTSA) indicate that distracted driving claims, on average, about 3,000 lives a year.

Government standards for airbags are intended to maximize protection for unbelted drivers, a holdover from years ago when very few drivers buckled up. Several years back, there had been dozens of documented instances in which airbags, originally meant to protect average-sized individuals, had killed children and small adults during minor crashes. The government reacted by allowing manufacturers to depower airbags so they would no longer deploy in low-speed crashes, and when they did inflate, it was with less force than the previously designed airbags, which burst out at speeds of 150–200 mph. By 2010, carmakers began to install advanced technologies to assist airbag systems in distinguishing between large and small passengers, belted or unbelted. Studies are ongoing.

The government required automakers to install side door/window safety features by the model year 2018 to prevent occupants from being ejected from cars and trucks during rollover crashes. The deadliest of all vehicle crash types, rollovers, resulted in over 8,000 deaths in 2010. The 2018 rule was intended to prevent the equivalent of an unbelted adult from moving more than four inches past the side window opening in a crash.

2 CAR ACCIDENTS AND CELL PHONE TEXTING

The first mobile telephone call using early radio technology took place in 1946 at a time when the phone was connected to a car battery. Because of the expense and limits of early technology, wireless phones were used mainly by truckers and for marine use (phones weighed about ten pounds and were the size of a brick). By the mid-1960s, New York City had 2,000 customers with these mobile phones, and they typically waited 30 minutes for a call to get through to the land-based

DOI: 10.4324/9781003091103-27

marine operator. Those first-generation models cost upward of $2,800 (approximately $28,000 today). An engineer who helped develop the first modern type portable phone in the early 1960s recommended that there be a lock on the dial to prevent its use while driving. That was long before cell phones became common, but many industry pioneers were aware of the risks of multitasking behind the wheel but could not predict the astonishing growth that would eventually define the cell phone industry. In 1985, wireless providers had 340,000 customers. That number has grown to over 330 million today. About 97% of all Americans currently own a cell phone, with those aged 65 and over the least likely to own them.

In a perfect world, no one would use a cell phone while driving a car, except in emergencies. In a nation with more cell phones than vehicles, that's not going to happen anytime soon.

Early warnings about talking or texting while driving were validated by many studies showing the dangers of both. Despite the mounting evidence of accidents caused by driving and communicating, the wireless communication industry built itself into an over $500 billion business in the United States, largely by winning over a crucial segment of the population – drivers. Originally called car phones, they were extolled as useful status symbols in ads and other promotional material. Wireless manufacturers moved to design small handheld devices to keep people talking in their cars – an effort that, for the most part, succeeded beyond expectations.

Text messaging can be a powerful tool under certain circumstances as when a kidnapping victim managed to text authorities from a car trunk or an earthquake victim in Haiti trapped under a grocery store texted a plea for help; both were rescued. Those are just two examples of the 135 billion texts per month by US cell phone users of all ages that may have saved lives. Distracted driving is not limited to texting; it also includes such activities as talking on cell phones, conversations with passengers, and eating. Texting while driving is the most obvious menace because it requires looking away from the road.

Safety advocates and families of crash victims believe the cellular industry has to do more than placing warnings in their manuals or messages on the screens of cell phones. Not only is the industry lax in this area, but a new generation of technology that has large display panels that support driving and other smartphone apps has significantly increased the ways in which a driver can be distracted.

The National Transporation Safety Board (NTSB) was the first federal agency to adopt an outright ban on wireless devices while driving. The ban prohibits all their employees from using any wireless device behind the wheel while on duty. The restriction applies whether the device is hands-free or not. This agency, which investigates accidents but has no regulatory authority, has issued several recommendations to state and federal agencies, calling for restrictions on wireless devices for motor vehicle drivers and train operators. Numerous studies have shown that it may be difficult for people to ignore the ping of an incoming text or call – for psychological and physiological reasons. People may fear missing an important call from a friend or boss or get excited at the prospect of receiving interesting news.

Policymakers are eager to find a new way to attack distracted driving because, for all their efforts, driver multitasking is still on the rise. According to a 2019 Center for Disease Control study, individuals between the ages of 15 and 20 are most likely to be involved in a distracted driving accident. Part of the lure of smartphones is that they randomly dispense valuable information and people feel compelled to check for messages all the time. The unpredictability makes the phone incredibly irresistible and a hard habit to break.

This analysis offers hope to safety advocates who would obviously rather not battle an irresistible behavior impulse. It can be compared to someone who continues to smoke when many stopped after it became known that the habit could kill them, suggesting that even though nicotine is addictive, people could (and did) avoid it. Even addicted smokers do not light up in theaters or churches. The same scenario can happen with distracted driving, where people will automatically place their cell phones in the trunk of the car before getting into the driver's seat.

Previous efforts to change societal views about drunk driving – and to increase compliance with seat belts, infant car seats, and motorcycle helmet requirements – took root over the course of many years following a three-pronged approach of tough laws, enforcement, and education. The same sequence is expected to be followed as more and more states pass texting and cell phone restrictions.

At present, Missouri and Montana are the only states that *don't* ban texting while driving, whether the device is a unit held in one hand or a hands-free Bluetooth type device. Any use of a cell phone will disrupt the concentration of the driver, which can prove to be fatal. Currently, 29 US states (plus the District of Columbia) have a ban on using handheld devices for all drivers anytime they are behind the wheel. While statistics are available on the number of texting citations given to motorists by state police, there is practically no data offered as to the number of tickets given by local and county police agencies, but it is assumed the rate of ticketing is about the same for all law enforcement agencies.

3 TALKING CARS HELP OLDER DRIVERS

Senior citizens who hold a valid driver's license may remember the TV programs *My Mother the Car* (1965–66) or *Knight Rider* (1982–86) in which the protagonists talk to their cars. Talking vehicles might be the future in attempts to prevent collisions or accidents. Smart technology programmed into cars now has backup sensors, lane drift sensors, and more. Cars – with the increasing ubiquity of smart technologies – will someday soon likely have the widespread ability to "speak" to each other as well as speak to road and traffic signs. The ability of vehicles could provide visual and audio warnings in advance of events that could thus prevent a collision, such as a car approaching too fast on the other side of a curve or someone a few cars ahead breaking suddenly.

The "talking car" could be of particular help for drivers over the age of 70, who are of great concern to motor vehicle agencies and insurance companies. Many states have instituted eye tests for seniors prior to renewing their licenses, and a few states require a driver education course after a certain age. Today's vehicle sensors and lasers have an improved and increasing capability to detect potential dangers.

4 DROWSY DETECTION SYSTEMS

According to the NHTSA, microsleep, a brief state of drowsy unconsciousness (sometimes as little as four or five seconds), caused over 800 deaths annually from 2010 to 2020. Microsleep is a common condition shared by millions of drivers at one time or another that can occur even if the driver's eyes are opened. Manufacturers have developed algorithms that indicate when breathing changes to reflect patterns that are typical of someone who is about to fall asleep and give a warning before that person actually feels tired or sleepy. Sensors can be used to monitor vehicle movement, such as steering wheel angle, lane deviation, and time-driven, as well as body posture, heart rate, head and eye movement, and body temperature which can be programmed to sense the level of drowsiness. These specially designed systems, when installed, can allow for limited autonomous driving before the vehicle takes over once drossiness is detected. The data will be used to stop the movement of the car or pull the car off the road.

Regardless of the advances in technology in detecting drowsiness, fighting off sleep is futile because sleep is a biological need. The best solution for drivers is still a low-tech solution. Pull over and take a nap.

5 TEENAGE DRIVERS

The age group involved in more car crashes than any other age group is, not surprisingly, teenagers. According to the Insurance Institute for Highway Safety, the four deadliest months for teen crashes have consistently been the months of August, July, May, and June (in that order), and on an average day in the United States, eight teens die in crashes. Many parents think drunk driving is the main threat to teen drivers, but driver error, speeding, and distractions are bigger problems. There are laws that have been passed in a growing number of states, usually following a fatal accident involving teenagers, but the trend in car crash fatalities has not noticeably abated. These laws are difficult to enforce, and the best way to get teenagers to act more responsibly when in a car with peers is by having conversations about vehicle safety with their parents.

In the United States, it is essentially an understood rite of passage that when teenagers get their driver's licenses, they celebrate by piling as many friends as possible into the car for a joy ride. This ritual was clearly depicted in popular movies of the 1960s and 1970s, like "Grease" and "American Graffiti." States have reacted to this practice by passing laws that restrict when, how, and with whom teenagers can get behind the wheel. Fifteen states and the District of Columbia now prohibit teenagers from driving with another teenager. All but seven states forbid teenagers from driving with more than one other teenager, and for good reason.

Lacey's Law was one of many state laws passed after a teenager was killed in a car full of other teenagers. This Pennsylvania law limits the number of passengers in a car driven by a teenager and also allows the police to stop a car if the passengers are not wearing seat belts. Other state laws are beginning to address the growing problem of teenage driving accidents. One state tried to limit teenage drivers to one passenger with an exception for household members, which resulted in cars driven by teenagers packed with their "cousins." The law was amended to disallow even siblings.

All states have graduated driver's licenses requiring periods of supervision and probationary driving before teenagers can get a full license. In a growing number of states, teenagers cannot drive after 6 pm in winter (8 pm in summer) and in Idaho, they are banned from sundown to sunup. New Jersey was the first state to mandate that those with driver's permits or probationary licenses – generally anyone under the age of 18 who has been licensed for less than a year – must place a decal on the corner of the license plate to indicate a teen driver to traffic police. Since the law went into effect, there has been a significant reduction in the number of teenage driver accidents.

Not all teens are menaces to the population, but when teenagers crash, it's often their passengers or those in the other vehicle(s) or on bicycles involved in the crash who end up dead. The Centers for Disease Control and Prevention confirms that teenagers have a tendency to overrate their driving skills and underrate the risks on the road (2014, Social Norms and Risk Perception: Predictors of Distracted Driving Behavior Among Novice Adolescent Drivers, Patrick M. Carter, et al). They also have more trouble multi-tasking – talking to friends, listening to the radio, and texting are particularly hazardous. Teenage drivers' risk of a crash increases 44% with one teenage passenger and quadruples with three or more. All these facts have attracted the attention of federal government agencies which have set up incentives for states to tighten restrictions on teen drivers, with particular encouragement to impose stricter limits on the number of passengers and the hours they can drive, to ban cell phone use, and to extend the restrictions in states where they end earlier.

6 SAFETY MEASURES

Technology has been developed that is specifically aimed at curbing dangerous distractions by temporarily interrupting service, short-circuiting the temptation to respond to the chime of their phones. One software-based approach was to use the phone's GPS to determine if the vehicle was moving and block all wireless communication when the car was going faster than 5 or 10 mph. Another available app is able to detect when the smartphone switches from one cell tower to another and temporarily suspends service. If the driver needs to use the phone, the block signal effect can be overridden in an emergency.

This technology did not receive an enthusiastic response from the public since it could not distinguish whether a phone belonged to the driver or to the passengers – or, for that matter, passengers on a bus or train.

The first test case involving a lawsuit against a cell phone manufacturer and wireless service provider for causing a car crash was filed in 2009 (Pamela Meyer, et al. v. Sprint Spectrum, LP, Jan 29, 2009). It faced steep challenges since it forced the courts to decide who was responsible for the driver being distracted prior to the accident. That suit, as well as all previous suits filed against only the service provider, was dismissed based on the unforeseeability of the accident and the absence of a legal relationship between the victim and the wireless provider. The court decision also ruled that the crash was caused by driver inattention, not by cell phones – and the person talking on the phone had done so even though there was a warning in the phone user manual. It is a misdemeanor to cause a fatality in a car crash due to the distraction caused by talking on the phone; it's not treated as seriously as drunken driving accidents.

> Seat belt laws have proven highly effective in lowering the death rate from car crashes and thirty states have" primary seat belt laws" giving the police the authority to ticket a driver solely for not buckling up. "Secondary seat belt laws" are enforced in most of the remaining states which allow tickets only during stops for another offense.

Focusing public attention on distracted driving has led experts in the field of vehicle accidents to claim that there should be some shifting of resources to draw attention to other important safety issues, including seat belt usage, speed enforcement, and overall safer automobiles.

7 REARVIEW CAMERAS

Automobile manufacturers gradually accepted new technology to make their vehicles safer. First came the automatic transmission, four-wheel drive, power steering, seat belts, airbags, child seats, etc. The latest addition is the rearview camera and an in-dash navigation screen that was offered in luxury cars. By 2012, 42% of vehicle manufacturers offered rearview cameras as standard equipment and they were an optional feature on another 23% of models. After more than a decade of limited use, all cars produced after 2014 were mandated to install the camera.

Government statistics issued by the US Department of Transportation regulations now require these cameras since they eliminate a wide blind spot behind a vehicle. According to a 2008 report by the NHTSA, roughly 280 people per year – of all ages, 44% of whom are under age five – had died in backover accidents involving passenger vehicles in the early 2000s. As of 2023, that number has dropped to about 210 fatalities and about 7,000 people per year are injured in such accidents. In terms of absolute numbers of lives, it isn't the highest of all vehicular deaths, but in terms of emotional tragedy, backover deaths are the worst imaginable, especially when a rear camera could, for example, have prevented a parent from running over a child.

8 INFANT SEATS AND HOT CARS

The NHTSA rules had previously covered only how well car seats protect children in crashes from the front. Under a law passed in 2014, the seats must now pass a test that simulates a typical side impact vehicle collision at up to 30 miles per hour. Car seats are one of the most essential tools for keeping young children up to 40 pounds safe from head trauma and other injuries in vehicular accidents. The redesigned seat to protect against side impacts had to be mounted in all cars by 2017. Accidental injury is the leading cause of child death in the United States and auto crashes are high on the list of the causes of those injuries.

In a side crash, a vehicle moves laterally and often forward, and the door intrudes into the area with the car seat. When car seats for children became law, it was not intended for side crashes and it took 15 years to upgrade the initial law to cover crashes from the side. Installing car seats is only as effective as how well, and properly, they are installed. Often, parents have to seek help from certified professionals, usually a member of the local police department who has taken a 40-hour hands-on instruction course and is certified in how to use the various required anchors and tethers that automakers are required to provide to mount the seats.

According to KidsAndCars.org and NoHeatStroke.org statistics, an average of 38 children die of heat stroke each year after being left in a car. For children under 14, it's the second leading cause of accidental death. Recommendations to prevent these accidents include:

- Put a purse, briefcase, or another important item on the floor of the back seat in front of the car seat to ensure that the back door is opened to collect the belongings.
- Putting a large stuffed animal in the child seat when it's not in use and moving it to the front seat when the child is in the car. It serves as a reminder that a child seat is in use.
- Making it a habit of looking in the vehicle – front and back – before locking the door and walking away.

  Since the number of vehicular heatstroke victims was first recorded in 1990, more than 800 children have died in hot parked cars as a result of parental neglect. It is a sobering statistic in an automotive industry that offers technology to steer vehicles, alerts drivers that a car is passing in the next lane, or that the gas tank is open; but, there is nothing to tell drivers that a sleeping child was left in the back seat of the car. Congress has studied the issue, but all recommendations were met with opposition from auto industry manufacturers.
- One argument against legislating mandatory notification technology: it misses the targeted population because parents of young children may not be buying new cars.
- Warning devices would initially have a minimum impact because a majority of cars on the road are many years old, and the owners would not have any incentive to pay for a modern warning system.
- There is now technology in most vehicles to alert drivers to check the back seat whenever the rear doors are opened and closed prior to putting the car in gear.

9 TRAFFIC HAZARDS AND DRUGS

Though national statistics lag by some years, the Governors Highway Safety Association reports 43.6% of highway fatalities in 2016 involved drugs. Many states have some form of law mandating drug testing of drivers in accidents, but others require some evidence before such a search can be conducted.

Another problem concerns the difference in handling the legal considerations between taking a potentially addictive drug like cocaine and, say, a drug like Vicodin that is legally prescribed. Until more data becomes available and more states start testing for drugs, the legal implications and criminal penalties will remain in the gray area.

10 INFOTAINMENT

Since mid-2010, car manufacturers have begun integrating screens on dashboards. Using special software that utilizes the car's battery power, these screens display more than cell phone information, digital identity of the radio station, and the time of day. The dashboard panel became a fully programmed desktop computer terminal with a ten screen offering high-definition maps and additional websites, almost like the display on the panel in an aircraft cockpit. The driver is supposedly prevented from using some of the display functions while the car has its ignition on, but some functions are usable while the car is in motion, and the programs can be changed via buttons on the steering wheel, console or a touchscreen. This technology continues to advance in spite of all the research indicating that such technology distracts drivers and can lead to more vehicular accidents, in addition to those already associated with cell phones and texting.

Customers now ask auto dealers about a vehicle's technology package, cell phone connectivity, or about the latest collision avoidance systems more often than they ask about how quickly a car goes from zero to 60. Most of the questions revolve around smartphones and apps. For example, an iPhone app works with a cable connecting the phone to a vehicle's diagnostic port under the dash that mechanics use to determine why the engine light came on and why it is or is not a problem. Another highly popular program automatically tells the tow truck where the stalled vehicle is located and estimates when help will arrive. Some programs allow voice commands to control the GPS and lock or unlock the vehicle remotely – even if the owner is in New York and the car is in Los Angeles – a feature OnStar has been offering for General Motors cars for several years now.

11 FUTURE INNOVATIONS ARE AVAILABLE

Many car manufacturers now offer collision warning systems that use radar to detect an imminent crash which prompts the vehicle to alert the driver (loud noise or flashing notifications on the dashboard). Pedestrian detection uses a combination of radar and video camera sensors that can pick out people near the moving car. If a pedestrian does suddenly appear in the car's path – and the inattentive driver does nothing – the car sounds a series of warnings and then automatically applies the brakes to bring the car to a complete stop. It functions at car speeds of up to 22 miles per hour; at higher speeds, it decelerates the car but will not be able to prevent an impact. Self-driving cars have gone from a thing of the future to something from the present. While not yet in wide use or distribution, there are those who believe that human error causes most crashes. If vehicles could learn traffic patterns, sense objects and vehicles around them, and be able to speak with each other, many argue that our roads would be a lot safer – if markedly less fun, in that people wouldn't be driving them.

Cameras mounted on the rear end of vehicles have become standard equipment on many models since 2007 and manufacturers have expanded the system by adding a series of cameras and sonic sensors to determine if a parking space is big enough. The parking assist system directs drivers into the proper starting position and AUTOMATICALLY steers the car snugly into a parking space; all the driver has to do is operate the brakes and gas pedal. There are, of course, cars available now that will self-park into spaces as well.

12 FBI MOST WANTED–DIGITAL BILLBOARDS

TELEVISION ON-A-STICK, a term used to describe a new and possibly more dangerous distraction in the form of digital roadside billboards that continually change the advertisement displayed to deliberately attract the attention of drivers moving at high speeds along heavily trafficked highways. Digital billboards became a fast-growing segment of the outdoor advertising market since a federal rule against them was eased in 2007. As of 2024, the number of digital billboards rose to 16,600 out of the nation's 350,000 billboards. The billboards are called "Television on A Stick" and give drivers, many of them already busy calling and texting, another reason to take their eyes off the road. Some of these huge signs have the glow of a Las Vegas billboard and the images change every four to ten seconds, each complete with colorful displays that have proven to be highly dangerous. The signs are really weapons of mass distraction; a cell phone can be turned off, but a billboard attracts attention whether the driver is interested in the message or not.

A digital billboard costs almost $300,000 as compared to the $5,000–$50,000 for a traditional billboard. Besides the obvious visual attraction is the fact that the digital display can project six different advertisers at a single location. The monthly rent for the static sign or the digital billboard remains the same.

The Federal Highway Administration relaxed a rule against digital billboards in 2007 since it was determined that they didn't violate the 1965 Highway Beautification Act's ban on "intermittent," "flashing," or "moving" lights. The same administration researched the digital signs, using eye trackers inside volunteer vehicles to determine whether and for how long drivers look at billboards, and found that the billboards caused only a few seconds of distraction and did not pose a great danger to the driver.

Since their introduction to highways in 2007, the use of digital billboards has helped the FBI capture elusive criminals. The program has steadily been expanding; by 2010, the signs were seen by drivers in over 40 states and have led to the capture of about 40 criminals a year. The number of cases solved with the help of the digital billboards is probably higher, according to the FBI, because tipsters don't always mention where they saw information about a suspect. This number outpaces the Internet and rivals the TV show, *America's Most Wanted,* in leading to an arrest. They are only about 1% of the estimated 350,000 billboards in America with many of the signs in highly populated areas.

Section VIII

Gadgets in All Aspects of Security

Information overload is causing consumer brain fatigue as more and more electronic gadgets are being updated, modified, and offer new applications every six months or so. A willing mass of consumers seems to enjoy lining up at storefronts every year to purchase more or less the same product they purchased the prior year – only the newer device may be thinner.

Since 2014, there are more mobile phones in use throughout the world than toilets.

In 2008, when the iPhone was just a year old – with 10 pre-loaded apps, including YouTube, Google Maps, and the camera – Apple launched the Apple Store. In 2010, Apple trademarked the word "app" to describe software that provides information for every need on a smartphone simply by downloading the information from the Apple store. There were an astonishing 10 million downloads the weekend the store opened and since then apps have been downloaded over 370 billion times worldwide, which have generated combined over $320B for iOS developers as of 2022.

Younger generations, in particular, are rejecting desktop PCs and laptops for smartphones and iPads to play games, text, and email since these activities do not require large screens, though gaming enthusiasts can benefit from large screens. The public has moved from the general- purpose PC to task-specific devices.

The largest social network service provider on the Internet in terms of hours of use and the fastest-growing company in the nation's history is Facebook. It has grown into something much more than a fad and since its beginning in 2004, this network has become a way of life for its 1.6 billion active users worldwide (2016). Roughly 1.4 1 billion people spend an average of 0 minutes a day on Facebook.

Changes in the computer marketplace have shifted the focus to mobile devices, but it doesn't mean that the familiar PC is necessarily going the way of the dinosaur or pay phones. The sudden rise of special software and cloud computing has paralleled the waning dominance of the PC and contributed significantly to its fading prominence.

DOI: 10.4324/9781003091103-28

20 Innovation Fatigue

1 CHANGING TECHNOLOGY

Information overload is causing consumer brain fatigue as more and more electronic devices are updated, modified, and offer new applications every six months or so. A study conducted by product testing and certification organization Underwriters Laboratories found that half of the consumers interviewed in four countries believed that high-tech companies are introducing products faster than they feel is necessary. Companies like Apple and Microsoft push new products out the door every six months, often with only a few new features or simple design tweaks (practically the same as last year's device).

Companies in the physical security industry seem to take some pride in abusing the word "innovation" more than any comparable industry, especially in the digital communication age. Innovation can simply mean something new, but often it is used to refer to an idea, device, or technology. Something that is novel. Using the "I" word to convey monumental change is an exaggeration when the product being introduced is really ordinary. Like the once-popular buzzwords "synergy" and "optimization," "innovation" is in danger of becoming a cliché.

Not only is the word overused in describing products, but many companies established the position of "Chief Innovation Officer," and there are innovation consulting firms charging $300,000–$1 million for three types of professional services:

- Efficiency innovation: Producing the same product more cheaply, such as automatic credit checks.
- Sustaining innovation: Turning good products into better ones, such as the hybrid car or electric vehicles (though it could be argued that e-vehicles have been more disruptive than sustaining).
- Disruptive innovation: Transforming expensive, complex products into affordable, simple ones, such as the shift from mainframe to personal computers. These types of innovation may take the longest to develop but offer the biggest potential rewards since the other two types could just as well be called ordinary progress and normally don't create more jobs or business.

The next buzzword is expected to be "inventive," but it will take at least a generation before the ubiquitous "innovation" completely loses its meaning.

2 CELLULAR NETWORKS

2011 served as a technology tipping point: for the first time in history, smartphone sales outpaced personal computers in 2011.

Once a smartphone's capabilities are fully discovered and experienced, it is hard to go back to older model cell phones, which are limited to only making phone calls. The disappearing, older model cell phones (used for calls only) were generally used to replace a landline phone and had incorporated some added features like calendars, clocks, music, global positioning System GPS) maps, and cameras over time. The extra features were impressive until the introduction of smartphones, which, in effect, gave users a handheld microprocessor. These smartphones duplicated almost all the data and communication features found in a computer. Smartphones allow users to install compatible apps, creating a system unimaginable only a decade ago. According to current estimates, there are nearly 9 million apps, many of which can be downloaded for free; that number will continue to grow.

DOI: 10.4324/9781003091103-29

3 APPLE FIASCO

With the help of the public's overwhelming acceptance of the iPhone, Apple redefined the mobile phone business. The iPhone changed the way people work and socialize while reshaping a variety of industries from music to hotels. It also heralded a revolutionary change in the way people communicate with the world around them.

When the first iPhone was introduced in 2007, people stood in long lines in front of Apple stores for 24 hours or longer to purchase this top-end device that offered wireless Internet connections via AT&T. About a year after its introduction, Apple's competitor Verizon began distribution of its Storm unit, followed by Palm, Microsoft, and Google each rushing out their own catalog of mobile applications with significantly less fanfare. None of these competitors had much effect on the iPhone phenomenon and cellular providers joined AT&T in supporting Apple products. Long lines to purchase new iPhone models became an almost annual ritual.

Ten years after its introduction, the iPhone became one of the best-selling products in history, with about 1.3 billion sold and generating more than $800 billion in revenue. Those figures made Apple the world's most valuable publicly traded company at that time. The growing number of Apple retail stores has become a strategic contributor to its core business. As of 2025, there were 535 Apple stores in 25 countries around the globe, 273 of which are in the United States (Statista.com).

The BlackBerry enjoyed phenomenal success from its founding in 1984 and especially between 2007 and 2010. After that, it went from being a hot gadget with pent-up demand to a discounted product, eclipsed by the next generation of devices, which came to define the "smartphone" industry, the iPhone. Other companies in the wireless communication industry have similarly made marketing mistakes that often proved fatal in the fast-moving and highly competitive field.

Research In Motion, the BlackBerry that revolutionized the way workers communicated, did not competitively upgrade its technology like the other major providers in the smartphone industry and struggled to find its niche. The first BlackBerry was launched in 2003 and was initially marketed to business users. By 2008, the company became a leader in the wireless phone sector of the communication industry and was valued at over $80 billion. The introduction of the iPhone four years later, followed closely thereafter by the iPad and the Android, which offered more versatility and innovation, adversely affected the sales of BlackBerry devices. This resulted in a disastrous loss of popularity for a company that once enjoyed 50% of the smartphone market – significantly lowering its market value and threatening its future.

- T

Even Apple, with all of its phenomenal successes, had encountered some monumental marketing and design mistakes not mentioned in any of the eulogies for the company's chairman and co-founder, Steve Jobs, who died in 2011. Jobs was the heralded genius behind Apple's iPhone, iTunes, and iPad. He and Microsoft's Bill Gates were the two innovators who pioneered the computer industry by revolutionizing the way people interacted with digital technology. The public considered Jobs' statement that eventually everyone will have a computer of their own someday to be ridiculous. It was like saying that everyone will have a railroad in their homes by the turn of the century. Jobs, and his vision for Apple, proved to be prophetic, and the company kept reinventing itself by continuously redesigning and upgrading their product line and introducing new products to become the most successful corporation in American history.

There were serious marketing fails that were unmitigated but remain long-forgotten disasters that did little to stop the amazing growth of the Apple juggernaut:

- ***APPLE III*** was introduced in 1980 to the commercial market for the astonishingly low price of $8,000 each. It was introduced following the success of the Apple II personal computer, but critics called the Apple III the worst computer ever built. Steve Jobs, Apple's president, had decided that the Apple III did not need an internal fan, and, as a result, chips overheated and detached from the boards – a problem Jobs suggested could be fixed by picking up the device and dropping it.

- ***LISA, named*** after Jobs' daughter, was another business-oriented personal computer that came out in 1983 to almost universal scorn. Even though it was the first commercial computer with a mouse and graphics display, it flopped due to its $10,000 retail tag. Lisa was taken off the market almost immediately and replaced with the cheaper Macintosh in 1984.
- ***THE CUBE*** was a small desktop computer beautifully encased in a cube of clear plastic that won many design awards. It flopped with the public because it lacked a monitor and cost $1,800, an amount the public was not willing to spend.
- ***NEWTON MESSAGEPAD*** debuted in 1993 at a cost of $1,200 and never caught on with buyers. The device had been promoted as the first unit to use handwriting recognition software, but it was wildly unreliable. Its clunky size, eight inches long and five inches wide, was described as palm-sized by Apple, prompting techies to comment that the size comparison was true if one was Yao Ming – the seven-foot, six-inch NBA star!

4 IPHONE MODELS: SHORT LIFE CYCLE

There seemed to be a yearly ritual that involved the introduction of a new model iPhone to an eager population waiting anxiously to replace their year-old model with the new one. Since the debut of the iPhone in 2007, each new model has broken sales records. The annual process occurs even though improvements are sometimes minimal, cosmetic, or mostly unnoticeable.

By example, in 2012, the iPhone 5 came out with little change from the previous model iPhone 4S, which had changed only slightly from the iPhone 4, which was much like the iPhone 3, etc. The iPhone 5 increased the screen size to 4″ from 3.5″ and was 18% thinner and 20% lighter than the previous version. The only significant change between the iPhone 4 and iPhone 4S was that the latter had the Siri feature that verbally responded to the inquiries of its owner. The 4S model sold 4 million units in its first weekend, while the iPhone 5 sold out its entire stock of 10 million units in two days.

And so it continues with the current iPhone models of today.

> Regardless of the small improvements, millions of each succeeding model are sold to people who replace the "old" models they have owned for less than a year. Apple still faces stiff competition from other manufacturers; Samsung phones comprise just over 22% of the smartphones in use in 2025 with Apple having risen to roughly 25.5% share of the market.

5 WORLD WIDE APPS

Launched in 2008, Apple's App Store initially had ten pre-loaded apps, including YouTube, Google Maps, and the camera. There were an astonishing 10 million downloads the weekend the store launched and in 2021, 143.6 billion apps were downloaded worldwide. Five years after the first app appeared, there were almost a million different applications available, transforming the lives of people around the world, even those living in third-world countries. The apps vary from using the iPhone as a high-frequency bug repellent, flashlight, radio, GPS locator, translator, game machine, and hundreds of thousands of other functions. By 2016, the apps market had generated almost $73 billion in income for its developers and in 2025, there are nearly 21.96 million apps available for download.

The app era made computing more fun and functional, and it ushered in the decline of traditional desktop computers. PC sales stagnated, and Microsoft's long dominance in computing took a hit as "apps" became the new buzzword. The versatility of the iPhone and other smartphones has proven to be so unlimited that people from all walks of life have developed their own apps, which are either sold or provided for free over the Internet or from Apple's App Store or Google's Play Store. Apps created by both amateur and professional programmers have frequently made large sums from their efforts.

For developers to offer their wares in the App Store, Apple's standard commission is 30%, though it can be less under certain circumstances; Apple paid out $406 billion to developers in 2024 alone.

Stanford University and New Jersey Institute of Technology were the first to offer a three-credit "app" college course (2008) to teach students how to design and market their own smartphone apps. Many other colleges added similar courses to meet the demand for this new form of communication technology. Creating video games for the handheld communication network has also become a popular college program where students prepare to enter a growing industry offering many career opportunities. The expansive apps programs are an important marketing feature for any smartphone, as illustrated by the success of the iPhone.

The app phenomenon accentuated the trend of developing open-source software and "cloud" services, practically eliminating the slow process of taking years to go from an idea to the marketplace. It now takes a few days or weeks to accomplish the same. Getting a concept to the consumer has turned into a sprint since the age of Internet technology has taken control of the universe.

A perfect example of this change was that of an assignment given to a Stanford University class of 75 students broken up into teams of three. They were to create an app and get people to use it. When it was offered in 2007, it was the first such course in higher education, and it became known as the "Facebook Class." The 25 teams in that landmark class ended up getting 16 million users in just ten weeks. The free apps ran on Facebook and, as advertising rolled in, some of those students started making far more money than their professors, even though some of them were considered "C" students. That class helped to pioneer a new business model that overturned the tech establishment by getting ideas to market within a week. One of the teams in the class designed an app that, by the end of the term, netted $3,000 a day and morphed into a company that later sold for a six-figure sum. Some of the students dropped out of school to make fortunes based on their experience... like the founders of Microsoft and Facebook had done before them.

Four weeks into the Stanford class, one team created an app that would allow Facebook users to send "hugs" to one another. It took all of four hours to develop. Then the team moved on to apps for "kisses," "pillow fights," and other digital interactions – seven in all. The apps caught on with millions of people and were soon bringing in nearly $100,000 a month in ads. Each member of the team got an A for the course and may have been tempted to hire the professor after they graduated.

Consumers have spent billions of dollars in some years on smartphones and the different apps that cover everything down to finding the nearest pizza parlor, dentist, or bar. Rising prices for these exotic apps doesn't seem to discourage average smartphone users from downloading dozens of apps per year (roughly 95% can be downloaded free). Some app designers claim to have made hundreds of thousands of dollars in two weeks following the release of their programs, but for every one of those successes, there are hundreds of other entrepreneurs struggling to market their programs with little or no success.

6 DEMYSTIFYING THE MYSTIFYING MOBILE NETWORKS

At the beginning of the 21st century, a new and strange language created by smartphone manufacturers appeared that was meant to describe the superiority of their amazingly fast mobile devices. Each year additional and trickier communication terms like 3G, 4G, and now 5G are advertised as the fastest cellular connection ever developed to a public where the majority remain confused about even the most basic terminology in wireless cellular technology.

Some customers indulge in a new language form known as "smartphone-speak," which can only be deciphered by a select group of nerds posing as computer and phone salespeople. More confusing words are added every year to keep the general public baffled.

7 BANDWIDTH ADVANCES: FIRST GENERATION TO 5G

In the dark ages before handheld cell phones, people who really needed mobile communications capability, and could afford it, would install radiotelephones in their cars. In that era of the earliest wireless communication systems, there was only one central antenna tower per city, and perhaps 25 channels available on that single tower. This central antenna meant that the phone in the car needed a powerful transmitter – big enough to transmit 40 or 50 miles – and it meant that not many people could use radiotelephones – there just were not enough channels and it came at a very high cost. The great advantage of the newer cellular system was the division of cities into small cells. This allowed extensive frequency reuse across a city so that millions of people could use their cell phones simultaneously.

One way to understand the sophistication of cell phones is to compare them to a Citizensband (CB) radio or a walkie-talkie, and the duplex-type transmission signal.

- Both walkie-talkies and CB radios are half-duplex devices, meaning that two people communicating on a CB radio use the same *frequency.* Thus, only one person can talk at a time. Modern cell phones use a full-duplex device, which allows the use of one frequency for talking and a second, separate frequency for listening. Hence, both people on the call can talk at once.
- The important distinction between these communication devices is that a walkie-talkie typically has one channel, and a CB radio has 40 channels – and a typical cell phone can communicate on 1,664 channels or more.
- A walkie-talkie can transmit a distance of about one mile using a 0.25-watt transmitter. A CB radio, because it has much higher power, can transmit about five miles using a 5-watt transmitter. Cell phones operate within small cells, and they have the capability of switching cells as they move in space. Cells give mobile phones an incredible range. A person using a cell phone can drive hundreds of miles and maintain a conversation the entire time because the cellular approach transmits the communication from one tower to another tower as the caller continues to move.

The typical analog cell phone system in the United States allows the cell phone carrier to receive about 800 frequencies to use across the city. The carrier chops up the city into cells, with each cell typically sized at about ten square miles. Cells are normally thought of as hexagons on a big hexagonal grid and the individual phones and base stations use low-power transmitters. Each cell has a base station that consists of a tower and a small building containing the radio equipment.

In telecommunications, **5G** is the fifth-generation technology standard for broadband cellular networks. Cell phone companies began deploying 5G worldwide in 2019, and it will eventually replace existing 4G networks. Like its series of network predecessors, 5G networks are cellular networks, in which the service area is divided into small geographical areas called *cells.* All 5G wireless devices in a cell are connected to the Internet and telephone network by radio waves through a local antenna in the cell. The main advantage of the new networks is that they will have greater bandwidth, giving higher download speeds.

8 SIGNIFICANT ADVANTAGES OF 5G

When Apple finally entered the 5G era, following the introduction by AT&T and Verizon in 2020, it replaced the 4G network almost entirely with more reliable download speeds and all the applications and services that 4G was not able to provide and applications that it couldn't support. When 4G phones were launched around 2002, few users knew about ride-sharing services like Uber or Lyft or travel apps like Airbnb because they didn't exist and instead of Netflix streaming services, videos were mailed in a DVD format. Downloading a 4K video series from Netflix and other services became obtainable on the 5G network.

The increased 5G speed is achieved partly by using higher-frequency radio waves than older cellular networks. However, higher-frequency radio waves have a shorter range than the frequencies used by previous cell phone towers, requiring smaller cells. To ensure wide service, 5G networks operate on up to three frequency bands described as low, medium, and high, with each requiring different antennas and each type giving a different tradeoff of download speed vs. distance and service area. Low-band 5G uses a similar frequency range to older 4G cell phones, and their cell towers will have a range and coverage area similar to 4G towers. The mid-band 5G uses microwaves that allow higher speeds and each cell tower provides service up to several miles in radius.

5G can be confusing to new users in how it differs from 4G. With 5G, the signal between the cell phone and the call tower can use low power or mid-band and millimeter waves (mmWaves).

- The low-band 5G provides miles and miles of coverage and little limitations or interference factors, but this service is generally not any faster than existing 4G.
- Mid-range offers a compromise essentially 10–15 times faster than 4G and significantly better coverage than mmWave, though it still is not as broad as the low-band 5G.
- mmWave 5G can offer up to 50 times faster performance than the average 4G service, but its range is only measured in city blocks… while it may be superfast, it is also super hard to find. It is unable to pass through walls and windows, and can suffer from interference from trees and other objects.

These levels of service became available in most metropolitan areas following the iPhone 12 that Apple introduced in 2020. The initial units had trouble passing through some types of walls and windows and were deployed in dense urban environments and areas where crowds of people congregate, such as sports stadiums and convention centers. These limitations were corrected and speeds are expected to increase during the next few years …until 6G service is introduced.

9 APPLICATIONS

- **BANK TRANSFERS** Web-enabled cell phones have opened up new horizons for money transactions never imagined when first introduced in the early 1990s. Banks now allow customers to view account information, transfer funds, and pay bills by phone. As mobile banking became popular, cell phones became the target for thieves and users should exercise great caution when using this technology.
- **INCOME TAX** apps for smartphones allow taxpayers who use the abbreviated tax form 1040EZ using the standard tax deductions, and no dependents to file on their phone. Users of this app have been made aware that criminals have attempted to hijack information from any online banking customers by creating apps that resemble those offered by major banks and the best way to discourage hacking of this type is to download apps only from Apple and Android stores, verify the identity of the developer and always use passwords.

3G, 4G, 5G, WIMAX, and HSPA are mysterious and confusing terms in our digital communication age. Each describes an aspect of smartphone operation. The "G" stands for generation and the higher the number next to the G, the faster the network, but these designations are not created equal and different carriers use different levels of 4G. Included networks, from fastest to slowest, are, LTE, WIMAX, and HSPA+, then come EVD and EDGE (these last two are in the category of 3G).

Choosing the fastest network with good reception in the appropriate price range is a daunting task for the confused consumer. For example, not every city has 5G available from every carrier operating in the area and the user needs to select a mobile phone that receives the 5G network… not all do. Most importantly, the speed of a network doesn't matter if there is no signal from that particular carrier where the cell phone is usually located. It is best to choose a carrier that has the strongest signal where the cellular phone is used, and it is best to ask a neighbor about reception – not the cell phone salesmen.

CDMA vs. GSM refers to different wireless standards, like AM and FM on radios. Of the larger carriers, Verizon and Sprint are CDMA, while AT&T and T-Mobile are GSM. The choice of a standard usually matters to international travelers; GSM is the standard in most of the world. A major difference between these standards is that GSM phones keep their identity and data, like contact lists, on a small, removable SIM card. The SIM card makes it easy to switch to a new phone in a matter of seconds, simply by removing it from the old phone and inserting it into the slot in the new phone. This is in contrast to the CDMA standard, which holds the identity on the worldwide network, so the information can be sent over the network to a new phone, a procedure which is usually done in the provider's retail store almost as fast as removing and reinserting a SIM card.

OS stands for operating system (logically, and it is one of the few terms in the smartphone language that makes some sense). It represents the underlying software that runs the phone and determines, among other things, the features of the phone and which apps it will run. The top two cellular OSs are Google's Android and Apple's iOS.

LTE stands for Long Term Evolution, and it's a technology that wireless phone companies around the world have adopted for their high-speed networks. It provides faster Web browsing and downloads that depend on such factors as the quality of the cell phone coverage in a particular area.

10 IPHONE CAMERAS

With hundreds of millions of users in the United States, cell phones have become a sort of digital Swiss Army knife for the 21st century. Besides having the capability to be used for GPS, it can chart routes for boats or function as a book reader, etc. It has also become a primary camera for some. Introduced in 2011, the iPhone 4S offered a sharper image than previous models with its 5-megapixel sensor; the iPhone 5 debuted in 2012 with an 8-megapixel sensor that matched the imagery of 8-pixel Android phones. There has been a concerted effort by hardware application providers to upgrade all high-pixel iPhone cameras to rival professional photo equipment. The more popular photo-related support items include,

- Steadycam is an attachment that packages a camera mount and a wide-angle lens. It attaches to the phone and has hand grips on all sides of the iPhone that provide a stabilized image.
- Lenses that can be attached to capture 360-degree videos and pictures.
- Tripods are to be used when the phones have a telephoto lens attached.

As cell phone camera technology has advanced – by incorporating higher resolution, image-enhancing software, and high-clarity screens – it has become very easy to capture high-definition photographs and view them later. Many people have a cavalier attitude toward using cameras to obtain copyrighted material, which has led to a new wave of lawsuits brought by bookstores and publishers. This has also created a gray area involving copyright laws when a cell phone is used to take pictures of pages in a book instead of buying the book. The act of photographing pages in a book while in a library or in a Barnes & Noble bookstore is much like wheeling a copier machine into the store, a ludicrous thought since no bookstore would allow such an activity. Yet, the ubiquitous smartphone is now a sophisticated copier, raising serious legal considerations.

The core issue is that creating a copy of printed material rather than buying it may be considered, by many, morally incorrect but may not actually be a violation of the law. To be illegal under copyright laws would depend on how much of the book is copied and what is done with the captured images. This has caused a perplexing dilemma in the publishing field, but the problem is not entirely new. The introduction of Xerox photocopiers in 1959 raised the same issues when they quickly found their way into libraries and people began to photocopy books and journals in large numbers.

Many students discovered that it was cheaper to photocopy a friend's textbook – or a copy of his/her copy – than to buy the book.

According to articles in the NY Times, the publishing industry saw the copier as the enemy and by the 1970s, publishers began to file lawsuits, preceding by several decades the music industry's lawsuits against those who downloaded free music and motion pictures over the Internet. After many legal battles, the publishing industry succeeded in persuading Congress to pass the Copyright Act of 1976, which defined "fair use" of copyrighted material and sought to establish a set of rules for ownership of content. It allowed limited replication of snippets of music, video, and text, and it gave teachers the right to photocopy certain material without having to pay fees to publishers. The Internet brought new challenges to copyright laws when it encouraged file sharing, enabling new forms of duplication that set off copyright disputes, including software.

It could be years before the publishing industry feels the real economic impact caused by people who take pictures of books or magazines with their mobile phones. It is also quite possible that digital technology will make this particular problem obsolete, since by the time this issue becomes a serious challenge to the retail bookstore – there might not be any bookstores.

11 SPECIAL SERVICES

The federal government introduced a national emergency alert system in 2011 that sends text messages to mobile phone users, including alerts issued by the president, information about public safety threats, and Amber Alerts for missing children. Known as the Personal Localized Alerting Network, the system is free and the messages appear directly on cell phone screens. A Federal Emergency Management Agency (FEMA) report stated that the technology could become a lifeline for millions of Americans who could send messages to wireless companies, which then use their cell towers to forward the messages to subscribers in the affected area. A New York resident who is traveling in Chicago at the time of the emergency in New York would not receive a message; a Chicago resident who is a customer of the same phone company would see the text Alert while in New York City.

Most apps are of the novelty type that fades quickly, but some of the more ingenious examples include,

- Hardware apps have become more common, including a program incorporating a noise-measuring decibel meter, offered by a maker of earplugs. A similar app turns an iPhone into a carpenter's level for construction purposes and for hanging paintings.
- Retail store mobile shopping has created an innovative industry where several Internet services provide information to customers, allowing them to compare prices across a range of retailers. The comparison can be done by scanning the barcode on an item and using the app to show a list of other retailers that carry the same item and at what price. This is all accomplished while standing in the store aisle. Before taking books for purchase to the cashier, one can simply type book titles into a web browser or the Amazon app on a smartphone. Online retailers, such as Amazon, may offer the same books for 40% less and with free shipping.
- In 2010, iBooks was introduced to compete with Amazon's Kindle and other devices that can connect to the growing number of online bookstores. iBooks was replaced by Apple Books in 2018, with iBooks being discontinued in 2020.
- Several apps now incorporate GPS technology that not only identifies the best routes to avoid traffic but also alerts drivers to the location of red-light cameras … and even bridges where police cars are hiding near speed traps.

There are several iPhone software packages available that offer features to assist the visually impaired. Users can enlarge the phone's graphics and text and there is even a voice over (easy-to-remember *VoiceOver*), which speaks text aloud when the phone's screen is touched. The most practical apps help millions of visually impaired people in the United States who face major obstacles when paying for something in cash. Unlike other countries that print currency in different sizes, there are no differences in the size and shape of US currency. The free app developed by the Bureau of Engraving and Printing to assist visually impaired consumers identify currency, EyeNote, uses image recognition technology and the mobile device's integrated camera to recognize a bill and indicate the bill's denomination. Another app, the LookTel Money Reader, recognizes US currency and can read $1, $5, $10, $20, $50, and $100 bills out loud.

Under the Rehabilitation Act of 1973, the US Treasury Department must change the currency to make it more accessible to the visually impaired, but after more than 50 years since the Act was made into law, no effort has been made to enact this change.

12 TWO-WAY RADIO RETURNS

After exhausting every possible gadget for handheld wireless digital communication since the early 1990s, in 2012, a walkie-talkie app called Zello was released that works with iPhones, iPads, Android, and Windows phones. It's essentially an improved version of the two-way radio since all it requires is tapping a name on the screen, holding down a talk button, and speaking. A second after speaking, the user's voice bursts into life on the friend's phone, wherever it may be in the world, even if it's in silent mode. The person called presses his own talk button and the conversation is underway, without the need to dial or use any of the cellular minutes under the contract with the wireless provider.

The initial reaction to the walkie-talkie app was that enough communication networks already clutter the airways and there's no need for another app. But this app has some new and useful functions since it is far safer than texting while driving. There is no need to look at the smartphone since it connects to the Internet by touching a button. A voice is also far more expressive than text messages; people cannot say that they misunderstood the tone of the text message. It can be used at any Wi-Fi hotspot, even out of the country. It's much like having an intercom rather than a phone, and it has a global reach – like a very long-distance office intercom. It's very similar to the old Sprint Nextel phones that worked for short distances as a walkie-talkie.

13 TABLET COMPUTER REVOLUTION

The tablet PC is best described as an upgraded cell phone/computer combination that revolutionized the communication market, mainly due to the introduction of the iPad in 2010 and iPad 2 in 2011. Smartphones and tablets duplicate features that a short time ago were found only in 24-pound desktop computers, then expanded to the laptop, netbook, smartphone, and finally the amazing flat-screen tablet.

The proliferation of smartphones was a major factor in the overwhelming acceptance of the iPad and other similar tablets in early 2010. The "tablet effect" was credited with changing consumer behavior by starting a new social trend that represented a post-PC era and a social shift caused by many unique technology advancements. Computing has become more ubiquitous, casual, and intimate and is no longer just a desk activity found primarily in the living room, dining room, and bedroom; it easily travels in a bag or handheld case that can be shared with friends. The tablet effect has significantly lowered consumer demand for the older mobile PCs, including laptops and netbooks, but it hasn't completely replaced the conventional PC or laptop completely … at least, not yet.

Apple has controlled the tablet market since the introduction of the iPad, but it wasn't the first tablet on the market. In 2001, Microsoft produced a Microsoft tablet PC, but it wasn't very successful and disappeared within two years. iPads operate using a touchscreen instead of keyboards (keyboard accessories can be purchased), and for all practical purposes, these wireless devices operate like an external USB connection to the computer. Individuals want to be wireless while taking their personal and professional life with them as they move from one location to another. The biggest drawback to an iPad or other tablet is that they carry a lot of information that can become an immediate target for security attacks.

The Apple iPad Mini, introduced in 2012, was sold for $329 and had a basic manufacturing cost of about $188 for the Wi-Fi-only version with 16GB. A research firm used a “tear-down” analysis to evaluate the price of each component in the iPad Mini at the time; that compares with the cost of $599 for the Surface 32GB unit and $199 for the Kindle Fire. The component cost of the iPad Mini amounts to 57% of the retail price, involving a gross margin of about 43%. Assembly is provided by several different sources in China, a practice followed by many in the electronics industry. The multiple provider strategies allow a customer to play one supplier off another for lower prices and minimize disruption if a single factory runs into production problems. Amazon’s Kindle Fire is another tablet line that sells for much cheaper than comparable Apple models, with the cost of production for each allowing for little room for profit. Having said this, and unlike Apple, Amazon’s tablet strategy is to make money on content from its e-book and digital video stores as well as on sales of physical goods from its online retail stores.

14 APPLE SUES SAMSUNG

In ***Samsung Electronics Co. v. Apple Inc., a*** jury found that Samsung had infringed on a series of Apple’s mobile device patents and awarded Apple more than $1 billion in damages. The amount may appear enormous, but it was not a big financial blow to Samsung, one of the world’s largest smartphone makers. At the time, the 2016 verdict sent shock waves through the mobile phone and tablet industry since other manufacturers could also be sued by Apple in future years. The two sides had enlisted some of the top and most expensive intellectual property lawyers in the nation. The verdict of $1 billion could have been avoided if Samsung had accepted an offer from Apple to sell a smartphone and tablet patent license for $24 a unit. Based on the 21.3 million smartphones and 1.4 million tablets that Samsung had sold in the United States that infringed on the Apple patents, Samsung would have paid $545 million had it accepted the licensing offer.

SMARTWATCHES were introduced to the general marketplace in 2015 and quickly became the newest digital trend. There are dozens of brands of smartwatches that can sell for as little as $50 and as much as $1200, plus subscription service costs. In the first year of the watches’ introduction, Apple sold 4 million units; in 2024 alone, it is estimated that over 450 million smartwatches were sold worldwide.

21 Social Network Phenomenon

1 FACE-TO-FACE CONVERSATION – A VANISHING ENCOUNTER

By most measures and as of December 2024, the largest social network service provider on the Internet in terms of hours is Meta, formerly known as Facebook and rebranded in 2021. It has grown into something much more than a fad, and since its beginning in 2004, Facebook has become a way of life for its over 3 billion active users worldwide. It has also become the world's largest repository of photos, particularly if you include Instagram, which was acquired back in September 2012. Facebook also acquired the voice-over-IP and instant messaging service WhatsApp in February 2012 for $19 billion. As the seventh-largest company in the world with an estimated valuation of over $1.5 trillion US, all this was inconceivable when it was started by Mark Zuckerberg in a Harvard dormitory in 2006. Microsoft offered to pay the 23-year-old founder over $4 billion for his share of the company in 2007. He didn't even consider it and now currently sits as the third richest person in the world.

The initial Facebook concept was to link together classmates attending Harvard, where Zuckerberg was a student, with a sort of dating service. Within a year after it started, the site was opened up to students from other colleges, to high school students a year later, and finally to everyone by 2006. After the first five years, about two-thirds of its 140 million American members were older than 26. In country after country, it has become a central aspect of people's social lives, particularly with friends or relatives who are distant geographically.

It has been claimed that the Facebook founder stole the idea for the website from a group of upperclassmen at Harvard who were building an online social network called the Harvard Connection. Zuckerberg launched his own site, which he called *The Facebook*, before those other students could complete theirs. There was a well-publicized lawsuit filed against Zuckerberg, charging that he'd ripped off their ideas. The case was settled out of court, giving the students stock worth over $60 million. The facts surrounding the startup of Facebook were greatly influenced by services already in operation, including Friendster (launched in 2003). Moreover, social networks were appearing at colleges all over the country the same year that Facebook was started, including Yale, Columbia, and Baylor. A sophisticated service called Club Nexus had started its service as early as 2001 at Stanford.

Facebook started as an electronic collection of profiles of undergraduates who were willing to answer as many questions about themselves as they felt comfortable in sharing, name, contact information, relationship status, favorite music, and possibly a few photos. This intimate information allowed members to search for friends, past or present, attending the same college, etc. When people accepted an invitation to connect – they could view their own Facebook pages and also see the accepted "friend" description. Facebook is mostly used for the sharing of news, photos, and videos of people who might care; for finding long-lost friends (or old lovers); for joining groups that support various causes of interest; and for sending messages. It is somewhat more streamlined than e-mail and is sometimes used for playing games. Facebook has grown to become the largest social network in the world and is in virtually every country, with the prominent exception of China, which has banned Facebook (now Meta) since 2009.

Most members live outside the United States and Canada, and an estimated 90% and virtually all users – an estimated 98%, log in on mobile devices rather than personal computers, with nearly a quarter of the users aged between 25 and 34 years old. Most of them are from Brazil, India,

DOI: 10.4324/9781003091103-30

Indonesia, Mexico, and the United States. Interestingly, millions of current users are not actually people at all. Meta estimates that between 10 and 16% of users are either duplicate or false accounts, which would amount to hundreds of millions of accounts falling into this category.

The phenomenal growth of Meta cannot be matched by any other communication network in history, with over 70% of all Internet users in the United States using it regularly. Over 97% of Facebook's income has come from advertisers who place ads on the pages of targeted members and reached nearly $56 billion in the United States alone as of 2024. This unparalleled growth in the number of messages sent in the United States was promptly followed by other nations worldwide, collectively spending billions of hours online every day, checking in with friends, writing on members' walls, and clicking through photos.

- **FACEBOOK BY THE NUMBERS:** By 2014, the growth and worldwide acceptance of Facebook as the leading social communication network had surpassed all the expectations of its founders and subscribers. Ten years in, it boasted,

201.6 billion friend connections, 400 billion photos shared per day.

1.23 billion monthly active users who send 7.8 trillion messages/day and 945 million monthly active users who use Facebook mobile applications 6,337 employees worldwide.

7.8 trillion messages sent per day.

Obviously, these collective numbers have grown astronomically in the ten years since.

WhatsApp is a subscription instant messaging service founded in 2009 and now owned by Meta. As of 2024, it has over 2.7 billion active users globally a month, with roughly 100 million voice calls and 1 billion messages sent daily. Obviously, it is one of the most widespread social messaging apps in use in the world.

- **MYSPACE** was launched in 2003 and was initially used for much the same purpose as Friendster. By contrast, Facebook was conceived as a much broader communication tool. Two years after its initial introduction and success, the company was sold to News Corp. for $580 million. Myspace (originally spelled with capital S, MySpace) was not the world's most popular social networking site. Blindsided by Facebook's ascendance, which had more or less sought out the same subscriber base as Myspace, it was dumped by News Corp. and picked up by another media company for only $35 million in 2011.

HACKED. The once mighty social network, Myspace, is now a shadow of its former self and faced a new challenge in 2016 when all of its information was stolen, leaving the data vulnerable to potentially being sold online. Names and passwords from more than 360 million Myspace accounts were compromised. The stolen data consisted of names, passwords, and email addresses from the platform before the site's relaunch (with stronger security) in 2013.

There's still a website called "Myspace" but it exists only in a read-only mode since January 2023, with no new articles being published since 2022. Also, no existing media or songs are able to be played.

- **FRIENDSTER:** a pioneer in online social networking in its day, it was created by a couple of men who were focused on helping friends find dates. The site inspired a wave of copycats and garnered over 115 million registered users at its peak. While its popularity held strong in Asia, it could not withstand the competition in the United States and in 2011, rebranded itself as a gaming site. Now based in Kuala Lumpur, it has almost all its users on that side of the world.

Friendster and Myspace were never as technologically sophisticated or achieved the popularity of Facebook, nor did their leaders possess a fraction of Facebook's paranoia toward competition – which ultimately served Facebook well.

2 MORE SOCIAL MEDIA AND CRIME

Technological advances continue to revolutionize law enforcement. In addition to GPS tracking apps and mobile devices, as well as transponder information for auto-paying tolls, social network information can also be used as a potent weapon against crime. Some of the information shared on the Internet leads to arrests, sometimes from pure criminal stupidity. Cell phone pictures voluntarily displayed on the web by a criminal perpetrator, or other such information that can be gleaned publicly, or by being granted access to the suspect's accounts via the networking companies, have increasingly helped information cases.

As a corollary, with individuals sharing so much personal information online and through such sites, criminals can glean all kinds of information, using it to hack social media profiles in order to harass people, gain access to financial or other accounts, impersonate individuals, and commit various forms of fraud. People would be wise to minimize the information they share and ensure strong passwords and privacy settings to protect both their accounts and their identities.

3 AGING MEMBERSHIP

Among the millions of Facebook members, the social network attracts many senior citizens, a group that has the network's highest mortality rate, so removing the pages of people who have passed away from the network will become an almost insurmountable problem. Before removing a profile, Facebook must determine if the person is actually dead. Death, of course, is unavoidable, forcing Facebook to find a practical way to integrate the information into the social experience online. The task gets more complicated when members desire to commemorate the deceased.

Facebook has implemented software to address the "deceased" problem by scanning for repeated postings of phrases like "Rest in peace" or "I miss you" on a person's page and then dispatching a human to investigate the account. That scanning approach is only marginally effective since it could easily lead to pranks – or mistakes. A member had his name removed after someone sent Facebook an obituary with a similar name, and the live member was locked out of his own page. Since there was no easy procedure to get reborn again, the member wrongly considered deceased had to create a website and post it on Twitter to get the news of the mix-up to spread. They finally got Facebook to take notice, and their page was restored, with an apology.

4 INFORMATION SATURATION

Some Facebook members cannot add any more friends to their page, having reached the maximum number allowed by the company. Thousands of Facebook junkies believe that there's no such thing as too many friends. Technically, however, there is: the site caps the number at 5,000 per person, according to Facebook rules. Only a fraction of a percent of the site's global users have reached that point – making them the elite few who have achieved 5,000 pals. The restriction is not for technical reasons and users can create "fan" pages, which have no limit. The person with the most followers is Portuguese soccer player Cristiano Ronaldo, with over 170 million followers and, across all social media platforms, has over a billion followers.

Social networking requires vigilance, not only in what information is posted, but also in what friends post. Some of the admissions and hilarious pictures can conceivably come back to haunt members if they run for political office or are considered for promotion to a high administrative position. Users can run the risk of sharing details of everything they do during their daily lives, including pictures of being skimpily dressed and possibly intoxicated, all funny to the high school or college student. However, more and more Human Resource departments are using not only Google but also social networks in background checks; viewing pictures on Facebook from rowdy high school or college days could discourage an employer or college admission officer from contacting the person.

Potentially inappropriate or insecure data shared by users can include relationship status, birth dates, including the year, home addresses, broadcasting when they go on vacation and for how long, and publishing their mother's maiden name.

A series of suicides in recent years by teenagers and college students who had embarrassing pictures and descriptions appear on Facebook caused a public outcry and may result in some form of censorship on the information posted on any of the social networks. The prosecution of offenders and prison sentences may cause another debate on how much free speech protection such postings are granted under the First Amendment.

5 GROWING SOCIAL NETWORKS

Social networks like Instagram, YouTube, X (formerly Twitter), and especially Facebook have revolutionized the way people socialize and, at the same time, have altered the English language so that people now communicate in cryptic tongues, sometimes meaningless to anyone but the communicators. With more people revealing the details of their lives online, sites like Facebook and others are providing information that is being used in legal battles, mostly as prosecutorial evidence. One such event occurred in a criminal case in which the burglar used the victim's computer to log in to Facebook and forgot to log off. The police followed the digital trail to the thief, who was promptly arrested. Another reported incident involved a murder suspect who claimed that he was not the kind of man who could kill his girlfriend's child but remarks the suspect posted on Myspace were highly self-incriminating, which contributed to his conviction, and he was sentenced to life in prison without parole. In civil cases, online communications have helped strengthen evidence, especially in divorce cases, where they are often used as proof of infidelity.

LINKEDIN, acquired by Microsoft in June 2016, is a networking site directed more toward professionals interested in establishing a "who you know" (and who you *want* to know) network of current, former, and potentially future business colleagues. As of June 2013, it reported more than 259 million members globally who use this social network to find jobs, recruit candidates, and make useful contacts. LinkedIn doesn't try to compete with Facebook and its billion users or X and the niche it has carved out for those interested in broadcasting their thoughts a handful of characters at a time. LinkedIn has built a presence in social media and is growing in popularity for those interested in a career or business or who have ambitions to climb the corporate ladder. For the LinkedIn novice, the first step is to create a profile, which is much like putting together a résumé, listing education, professional experience, and skills, but allows for much more information than found on a printed résumé. That means listing not just major positions previously held but also minor internships and summer jobs (including any special skills and endorsements from colleagues).

It is also easy to look for jobs using the company's search tools, which can involve reviewing the information on the millions of pages of companies that may be interested in hiring someone who is a member of the service. Email notifications and looking on the site itself will reveal the names of people who were recently hired or who have left a company, as well as those who have changed positions. There is no need to ask immediate coworkers for referrals since a LinkedIn member can ask colleagues of colleagues, greatly expanding the reach of job seekers. A popular feature – called Answers – lets an individual ask business-related questions of people who might have advice on everything from résumé formatting to business software. Like every other social network, this one may carry more serious communication, but it is just as susceptible to illegitimate connections and can attract impostors and troublemakers.

- **FOURSQUARE** is a relative latecomer to the social networking scene and, in fact, as of July 2014, became a program solely meant for personalized, local searches. It used to combine the location services of cell phones with GPS with a database of other users of the app to produce crazy new possibilities. Foursquare knows where a person is located within a few feet once the app is opened on iPhone or Android phones. A picture on the screen will list restaurants, bars, and shops near where the phone is being held. It also shows where friends are located, making it easier to meet up with them. As the phone moves past businesses, the app notes what products they can offer. If a user of the service walks by an establishment like a restaurant, others on the network can leave pointers about the food ("avoid the halibut") that appear on screen as a person enters.

- **TUMBLR** is a blogging service that some consider a longer-worded version of Twitter. Tumblr's founder was a 20-year-old high school dropout who launched the site to allow short-form blogging in 2007. This service combined blogging and social media, enabling brief posts – often just a single photo or a sentence. Tumblr, at its start, was especially popular with the fashion community, owing to its relative ease of use and visual emphasis. The Tumblr name stems from tumblelog, which is a short-form mixed-media blog. Tumblr allows up to 4,096 characters of text per post and includes photos, videos, and excerpts from other users' posts (with a tag character limit of 140). Tumblr is owned by Automattic Inc., which operates WordPress.com; Tumblr had been owned previously by Verizon, and AOL before that. There are currently nearly 600 million blogs and over 135 million active users as of January 2025.
- **PINTEREST** is a "visual discovery tool" that lets people find, collect, and share images of items they like with others. They bookmark or "pin" visuals on online boards that are arranged thematically. Pinterest activity has become an important source of information for corporate marketers. According to the Internet activity measuring service comScore, it became the first site to break the 10 million unique visitors per month barrier so quickly; there are currently over 537 million active users worldwide as of September 2024.
- **ORKUT** was used mainly by teenagers in Brazil and India and owned by Google. Founded in 2004 as a way to meet and maintain friends, it announced in June 2014 that it would be shutting down three months later. Competing services and legal issues were too much for it.
- **REDDIT** was originally created as a forum to ask anyone anything. The name is a play on the words, "I read it." Users post content and then others vote the content up or down and can comment on the submissions. There is a feature called "Ask Me Anything" or AMA, where people can have online press conferences. President Barack Obama's AMA in 2012 was so popular it brought down parts of the site. Reddit averages about 1.2 billion unique visitors per month and is estimated to have more than 500 million accounts.
- **SNAPCHAT** lets people send photos and videos (and add text or drawings) that disappear entirely after they are viewed for a few seconds. Although originally used to send "funny" content to friends, it's also perfect for users who don't want risqué images, texts, or any other personal correspondence disseminated. Snap added the "Family Center" feature in August of 22, so that parents can monitor the activity of their children (aged 13–18) within the app. Snapchat declined previous acquisition offers, including $3 billion from Facebook and $4 billion from Google. It is estimated that over 4.75 billion Snaps are created each day, with Gen Z making up over 51% of Snapchat's monthly active users.

6 YIK YAK – ANONYMOUS AND LEADS TO COPYCATS

Special mention should be made of how an innocent-looking app can be turned into a very hateful type of communication network due to its complete anonymity. In 2013, two college students offered a mobile app, Yik Yak, to be used on iPhones that combined the technologies of GPS and instant messaging to send comments anonymously to other iPhone users who don't necessarily have an account. It differs significantly from other anonymous sharing apps in that it is intended for sharing primarily with people within a proximity of five miles, though originally it had been much more. Six months after its introduction, Yik Yak became the ninth most downloaded social media app in the United States.

Used mostly on college campuses to communicate messages that are often critical of other people, it has been criticized by parents and others as being too vicious and hateful since the messages can't be traced to their source. Due to controversies and the inherent ability to perpetrate cyberbullying, many schools and districts took to banning the app, which resulted in a rework of it and a relaunch in 2021.

A more intrusive and more disturbing app that was, fortunately, short-lived was Peeple (Peeple is not misspelled) that let anyone rate anyone they know on a scale of one to five for all the world to see. The creators believe it was a way for people to learn what the world thinks about them. Anyone with a cell phone number could be rated. A text message notified the person being reviewed that they are to be rated and there was no way to stop the action. Any rating with two stars or less went to the inbox of the person being reviewed. Only positive reviews showed up in the profiles of people who hadn't signed with the service. This created an obvious problem because to see what people have said about you, you would have had to sign up. In doing so, it would automatically allow negative reviews to go up, reviews that wouldn't be posted if you hadn't gone on the site – a sort of modern-day digital twist on *Catch-22*.

7 PROTECTIVE TEXT

Text messaging – also known as SMS (Short Message Service) – has already replaced many face-to-face conversations and is slowly replacing phone service, two seemingly fading means of communication. Sites like X, with postings originally of no more than 140 characters (now a maximum of 280 for free accounts), create and reinforce the habit of communicating in cryptic, sometimes unintelligible, microbursts. Many Facebook and X devotees enable settings that earmark each instantaneous contact. The growth of text messaging has been phenomenal, and it is estimated that 23 billion text messages around the world are sent each day, with roughly 6 billion in the United States alone – and 18–24-year-olds sending the most, averaging nearly 110 messages per day.

Texting is the college student's predominant method of communication since their goal is to pass along information in as little time as possible, with as little small talk as is acceptable. What students like most about their mobile devices is that they can reach large groups of people; what they like least is that other people can reach them. Texting seems to satisfy these conflicting feelings for some unknown psychological reason.

That feeling seems to be shared by the general public since texting has changed the way people interact and, while it may save time, it has almost eliminated phone conversations. While the average "textophile" thinks voicemail is a waste of time, talkers still like to talk and phone conversations will never be completely obsolete. Although work can be conducted and ideas exchanged over e-mail, many professionals in innumerable fields are still forced to do business on the phone, where the conversation requires more than 280 characters.

Of course, "verbal texting" has reached an all-time high. Reacting to an obviously funny story would not actually get a laugh; instead, there would be a response with the acronym "LOL." A long list of abbreviations that were meant to be written have become part of the spoken language and Internet slang – often used by a person's "BFF" (best friend forever) or by coworkers who preface everything with "FYI." First developed in the early 1990s to streamline conversation on chat platforms and popularized on AOL Instant Messenger and Gmail chat, terms like LOL (laugh out loud), OMG (oh my God), and BTW (by the way) now seem to be popping up IRL (in real life). There are many adults who reject these letter languages and insist on laughing out loud rather than sounding like an idiot by screaming ROFL (rolling on the floor laughing).

Whether this new practice will leave a permanent mark on the English language remains to be seen, but it would hardly be the first time an acronym has become so commonplace that people eventually forget that it wasn't an actual term. Americans know AWOL is a military term, but many may have forgotten that it stands for "absent without leave." Celebrities enjoy plenty of "swag" at Hollywood parties, probably not knowing that it means "stuff we all get." And the popular NASCAR race is hardly diminished if it is known to officially stand for the "National Association of Stock Car Auto Racing." There are probably hundreds of such abbreviations that are used in daily conversation and the number will grow exponentially with the widespread acceptance of texting language.

EMOJI are the shorthand symbols used on the social networks to express thoughts and feelings.

In 2016, Facebook offered new options to express reactions such as "like," "love," "haha," "wow," "sad," or "angry." The six emoji symbols are recorded and the use of each symbol will be published by Facebook to show trends and changes in members' emotions over a set period of time.

The emoji pictographs have become a commonly used expression in text messages but they differ slightly based on the software. Apple's iOS, Google's Android system, and Microsoft's Windows 10 platform all have their unique interpretations of the various emoji characters as do the other phone makers. Even social media providers like Facebook Messenger and Twitter have their own variations.

8 FREEDOM OF TWITTER SPEECH

When posting on X or texting becomes compulsive, some messages are so rapid-fire, off-the-cuff, and often satirical that they seem like spontaneous twaddle (British word) or gibberish (American) and they may not be protected as free speech. In fact, they can be considered a criminal act since the messages are not treated as a form of free speech. In the UK, in 2010, a man about to get on a plane to visit a girlfriend was thwarted by a snowstorm that grounded flights, and in a spout of frustration tweeted the person who was to meet him at the destination airport with an outburst: *"You've got a week to get your shit together otherwise I'm blowing the airport sky high!!"* The message would normally have gone unnoticed except that someone accidentally saw the message and reported it to authorities and within hours, police arrived at his place of work, arrested him, and interrogated him for eight hours. They also seized his phone and computer.

A judge convicted the frustrated tweeter of sending a menacing message over a public telecommunication network under the Communication Act of 2003. He was fired from his job, fined $4,500, and got another job and was again fired after that company found out he was a convicted felon. The entire episode became a high-profile case and inspired a band of angry tweeters and bloggers to start a crusade to have the charges dropped since the message was harmless and simply an expression of free speech. They backed up their cause by sending hundreds of messages threatening to blow up other random things, the White House, the courtroom, the town where the courtroom was located, the nearest airport, the NBC TV station, etc. None of these sympathizing copycat messages caused any alarm, and no one was arrested. Fortunately, upon a second High Court appeal and before a panel of three judges, the case was quashed and rendered invalid – but only after years of legal proceedings and appeals.

Twitter swept the country almost immediately after its introduction in 2007 and within two years had attracted over 12 million subscribers; most, it appeared, were taxi and car drivers in major cities. By the middle of 2013, there were 400 million tweets a day. With the purchase of Twitter by Elon Musk in October 2022 for $44 billion and the subsequent rebranding of it to X in July 2023, Twitter has had many changes in policy and various ups and downs, mainly due to the capriciousness of its head. Twitter currently boasts approximately 600 million active users around the world, give or take. But, with bots, harassment, partisan advertising, and toxic messaging cited as the most common complaints, many users have started to look for alternatives. As a result, Mastodon, Threads, and Bluesky, created by Twitter founder Jack Dorsey, have all entered the market in efforts (with mixed but increasing success) to siphon off X users to their services and platforms. What the future of these companies and their user base holds remains to be seen.

9 ONLINE ADDICTION AND COMPULSION

The field of therapy has developed a new niche that deals with mental health care for those suffering from an addiction known as "onlineaholic" or "video games addiction." Internet addiction disorders have proliferated with the explosive use of Facebook, Instagram, X, text messaging, and the most contagious mental disorder – the gaming software obsession. US Internet users have developed a dependency that can be as destructive as alcoholism and drug addiction. Some skeptics consider this particular addiction to be a fad illness, while mental health professionals feel that a majority of obsessive users are online to feed their addictions and have become much more dependent on this vice because they can excuse the time spent on the Internet as educational. The condition is not recognized in psychiatry as a disorder, and insurance companies do not reimburse for treatment, but there has been a growth in the number of mental health professionals who have a diagnosis: Internet addiction disorder.

10 PEDESTRIAN DISTRACTION

A growing number of communities and jurisdictions are attempting to stave off pedestrian accidents that can happen when people walking become so engrossed in staring at their cell phones and wearing headphones. Some cities have issued fines for careless walking; the Utah Transit Authority made walking around trains while distracted punishable with a $50 penalty. Delaware has taken a different approach, placing stickers with the words "LOOK UP" on sidewalks near crosswalks in several cities, urging pedestrians to pay less attention to their phones and more to what's going on around them.

22 Decoding email and Personality Traits

Digital forensic stylistics is a subfield of both digital forensics and forensic linguistics. The practice aims to examine the style of written or spoken text in legal contexts – using linguistic analysis – especially to determine authorship or authenticity in some capacity. The legal term "forensic stylistics," when applied to email or text messages, uses a cross-textual comparison of eleven different style markers, including variant forms of punctuation, spelling, and grammar. The reliability of applying techniques developed originally for the verification of handwriting samples to digital signatures is causing much controversy in the forensic linguistics community.

Very little attention had been focused on this unproven science until a plaintiff in a 2011 lawsuit against the founder of Facebook. The plaintiff claimed he had email messages from the founder during the early stages of Facebook's formation that proved that he was a co-founder, and he demanded 50% ownership. The credibility of the lawsuit rested on proving that messages kept for eight years by the plaintiff were actually written by the Facebook founder. For example, the questionable email messages had not capitalized the word "Internet," while a known sample of email messages sent by the founder had two capitalized instances of the word "Internet." This was one sample of a digital fingerprint that was inconclusive and did not help the case of the accuser. The problem of identifying the authorship of a document produced by a computer to which multiple users have access is called "keyboard dilemma."

Many linguists have challenged the notion that a "digital fingerprint" could be treated as a unique identifier since there is no recognizable linguistic DNA.

There are others conducting research on the subject because they are convinced that digital messages can be identified and will eventually be legally acceptable in court as forensic evidence. Some laboratories working in this new field have developed computer software that categorizes grammatical structures as "marked" and "unmarked": an unmarked noun phrase has its main noun at the end of a simple phrase ("our *marriage*," "a *divorce*"), while a marked one has the noun at the beginning of the phrase ("*anything* you ask") or in the middle ("the *rest* of our lives"). These aspects of a writer's syntax are relatively stable across different styles of writing and are less prone to technological intervention – such as spelling and punctuation, which are changed on the fly by spellcheck and autocorrect features built into computer software.

A team of such forensics experts developed a program that can look at an anonymous email message and predict who wrote it out of a pool of known authors with an accuracy of 80–90%. The sample data involved thousands of messages sent by employees at Enron, which were used as the sample "pool." The data collected included the position of greetings and farewells and the use of symbols ("$" or "%") instead of words ("dollars" or "percent"). Combining all these features allowed the researchers to determine what they call a person's "write-print." Without the defined pool of Enron users, identification of email features could not be accurately developed.

1 ARTIFICIAL INTELLIGENCE

The idea of creating a computer that can think, analyze problems, and act autonomously to DUPLICATE and exceed human capabilities is as old as computers themselves. Experts working in the various disciplines that make up the artificial intelligence (AI) field feel that someday, software may indeed mimic and rival human intelligence. Its development could prove to be the biggest threat facing the world – only a little less threatening than a giant asteroid strike or all-out nuclear war.

DOI: 10.4324/9781003091103-31

Having said this, there is also the potential for higher productivity, solutions to health and medical, mathematical, and scientific puzzles that humans have heretofore been unable to solve. AI has seen meteoric growth in its applications and hundreds of billions of dollars being invested into the technology in recent years.

Where and how far AI will develop in the coming years is anybody's guess but with programs like ChatGPT, Midjourney, DALL-E, Gemini, DeepSeek, and Veo 3 – and basically all major technology companies investing heavily in the area – there lies both great risk and great opportunity in the advancement of human knowledge.

2 MENACING X (FORMERLY TWITTER) POSTINGS

X (formerly Twitter) posts have been the basis of defamation suits in courts worldwide. One of the few ways to make activity on one of the many social networks a criminal offense is to invoke a rarely used law on cyberstalking. Twitter is an especially vexing new problem in the criminal justice system since it prompts ordinary people who use it to create public personas, and it can put celebrities in direct contact with a large following, including many who insist on using pseudonyms. The people being cyberstalked can either block them – or – make a federal case out of the posting. Arresting someone making 8,000 posts has exposed the grey areas in the law and how the legal system is wrestling with nasty behavior in new and rapidly changing modes of communication. Similar communication issues that have gone to court involving similar circumstances:

- A boy who hacked into the Facebook account of an acquaintance was charged with felony identity theft.
- Students who bombarded their professor with mean emails were accused of disturbing the peace.

In almost all the cases relating to Internet abuse, the defense team points out that people post things that may cause emotional distress to others, like an unkind review of a book on Amazon, and even an unvarnished assessment by a college student on RateMyProfessor.com. It has also been pointed out that while not all speech is protected by the First Amendment, the idea that the courts must police every inflammatory word spoken online challenges the very basic principles of freedom of speech and is unsupported by legal jurisprudence.

3 SEXTING UNDER ATTACK

Sexting is the term used in most states to describe those who send or receive sexually explicit photographs by cell phone or computer, an action for which someone can be charged with a child pornography felony, resulting in the person being placed on a sex offender registry for decades to come. The law is being applied to teenagers who send nude pictures of themselves over the Internet, and while not exactly the obscene act associated with a pedophile, sexting is being considered a crime by prosecutors in many jurisdictions. Many lawyers have criticized the application of child pornography laws to adolescent cyberculture in which all kinds of sexual pictures circulate on sites and messaging services. In accepting teenage facts of life rituals, many states have changed, or are considering changing, their laws to reduce penalties for teenagers who engage in such activities by treating sexting differently from adult pornographers and sexual predators.

Some state lawmakers around the country have struggled to decide if teenage "sexting" sending sexually suggestive photos by cell phone is a serious crime or a harmless juvenile coming of age. Generally, states have laws aimed at treating minors in a more compassionate fashion than if they were prosecuted under existing child pornography laws, which include the possibility of prison time and sex offender status. The states of Connecticut, Arizona, Illinois, and Louisiana have enacted laws that impose relatively modest penalties against minors who text, while maintaining harsher penalties for adult offenders. Although sexting may be high on many parents' worry lists, it's difficult to estimate how common the practice has become. Some law enforcement agencies believe it could be as high as 20% when the entire range of texting is taken into consideration. It may be the exchange of a revealing photo between two romantic partners or a rapid-fire humiliation campaign with a photo spreading throughout a school. A single photo could start out as the former and end up as the latter.

Some attorneys view teenage sexting as a comparatively tame activity compared to other youthful endeavors that can be far worse. Sexting is probably better handled by parents and teachers than by prosecutors, but there are many law enforcement experts who insist that the practice is a crime punishable by a jail term, since some of these so-called harmless messages and pictures can ruin reputations when the communication oversteps the bounds of common decency. A Louisiana law takes a graduated approach to penalize minors who forward lewd images of other minors by imposing a maximum sentence of 10 days for a first offense and 30 days in jail for a second offense. There are privacy advocates who believe that controlling what one does with information about oneself is a serious violation of the First Amendment and that by sentencing someone for any sort of sex offense, whether it be a juvenile or adult, a scarlet letter will be attached to their reputation forever.

Summary of legislation that deals with underage sexting

Louisiana, Arizona, Connecticut – Impose possible fines and incarceration
Illinois, Missouri, North Dakota – Treat certain violations as misdemeanors
Nebraska – Felony charges make allowances for certain teenagers
Colorado – Makes allowances for first-time sexting offenders
Pennsylvania – Requires teenagers to follow educational programs if caught sharing sexually suggestive material
Oklahoma – One set of penalties for "consensual sexting between two people ages 14 to 18, but possible jail terms for another type of sexting."

While sexting is a bad judgment, it is not what the US Supreme Court had in mind when it crafted the child pornography law. Prosecuting the pornographer and the victim doesn't seem reasonable when they are one and the same person. One state established the law that doesn't criminalize children under 18 who send out their own photo to a willing recipient who is at least 15. On the other hand, a teenager who passes the photograph on to friends could face a felony child pornography charge and five years in prison.

Sexting gained notoriety after school officials around the country increasingly learned of photographs of nude and partially clothed teenage girls on several students' cell phones. Male students had been trading the images over their phones, and within hours, the photos could be sent to hundreds of other cell phones. Other noted incidents:

- A teenager was placed on the sex offender registry because, at 18, he was convicted of distributing obscene material to a minor after he sent a picture of his genitals by cell phone to a 14-year-old female friend who had requested it.
- Another 18-year-old was charged with distributing child pornography and put on the sex offender registry after sending photographs of his nude girlfriend by email to dozens of people after he had a fight with the girl. One picture went to the girl's parents.
- A 14-year-old girl was arrested and charged with possession of and distribution of child pornography after posting dozens of sexually explicit photographs of herself on a prominent social media app.

A landmark decision in 2017 was when a County Juvenile Court judge found a female teenager guilty of involuntary manslaughter and sentenced her to five years in prison. The crime was that she literally killed her boyfriend using her words as the murder weapon. The two teenagers had built a virtual relationship through texting for two years and were constantly texting each other. When the boyfriend had his car retrofitted to feed carbon monoxide fumes into the car in his attempt to kill himself and had second thoughts and left the car when the girlfriend texted him: "The time is right and you're ready, you just need to do it." He committed suicide a few minutes later.

The judge found that the girlfriend had a duty to help the victim but instead ordered him to get back into the truck. Legal experts contend that the manslaughter charge was probably illegal since the girl on the phone was not at the scene and the boy acted alone. While it may be true that her words caused great harm, they didn't kill the young man. In terms of morality, what she did was despicable but that doesn't constitute manslaughter. The appeal to a higher court found the decision was unconstitutional since liability with regard to a suicide – for someone who failed to act and who wasn't present nor did she provide the instrument for the suicide – does not constitute involuntary manslaughter.

Preventing kids from exchanging inappropriate text or email messages with persons unknown cannot be left to the wireless provider due to privacy concerns that forbid monitoring of text messages and emails. A number of small technology companies have entered the market by offering software programs that allow parents to monitor the contents coming into their children's communication devices, including cell phones. The software is designed to alert parents when their children exchange lewd text messages, communicate with predatory adults, or taunt each other via social networks. These available programs vary in what they offer; some can alert parents when messages containing keywords such as "suicide" or "drugs" are sent to their child's phone, while other programs draw from a unique library of 6,000 phrases deemed inappropriate, including slang and online abbreviations.

4 MANDATORY PRISON SENTENCES

The mandatory minimum prison sentence faced by defendants in federal court charged with receiving child pornography is five years, even when the defendant is a juvenile. (State courts usually have maximum sentences of four years). People who amass a vast collection of child pornography, but never sell or take any of the pictures, still face the mandatory sentences in federal cases that often exceed the prison terms given for actually sexually abusing a child. Sentencing laws, regardless of the crime, are highly controversial, but more so in cases involving child pornography than in any other type of criminal activity. Usually, juries are given the standard instruction not to consider possible punishment during deliberation since most juries will assume that the sentence has been left to the judge's discretion and have no knowledge of any mandatory sentencing minimums.

Another more serious effect of social networking is posting messages, private messages, or posts that get referred to law enforcement. A teenager logged onto Facebook and posted a threat against the gym teacher using the phrase: "Ima kill em all." The sender was suspended the next day, permanently expelled a week later, and had to be homeschooled until they met graduation requirements. School officials in that district had suspended and expelled students in the past for infractions that involved social networks, text messaging, email, and other technologies. The issue confronting many school administrators is balancing students' free speech granted under the First Amendment and privacy rights online. Mouthing off to a friend about a teacher is almost a rite of passage for high school students, but technology now allows the user to do it anonymously, permanently, and publicly, possibly destroying a reputation for life. Not all insulting passages are done anonymously, which can lead to a whole set of problems that have challenged certain basic rights.

Internet crime transcends borders, as it does in child pornography cases, where the FBI often initiates cases against defendants in far-flung locations. If someone from New Mexico sends an image to Maryland, the case could be tried in Maryland even though the defendant had never been in Maryland. The government can get jurisdiction everywhere once an FBI agent uploads documents and is located far away from the defendant. Defense attorneys maintain that jurisdiction should be based on the locale of the alleged perpetrators to make it more convenient for them to go to trial, but most courts have held that inconvenience is not a valid defense.

5 CYBERBULLYING

A study by the National Institute of Health indicates bullying has moved beyond the schoolyard and onto social media pages, online chat groups, and text messages. The study, based on a 43-country survey of school children, revealed almost 14% had experienced electronic bullying – the latest incarnation of an eternal problem that leaves victims feeling hopeless and depressed. Traditional, in-person bullying differs from cyberbullying in that the online version might be anonymous. Still, in-person and online bullying are not necessarily distinct events as one often flows into the other. The cyberbully often makes their victim feel isolated, dehumanized, and helpless, more so with repeated attacks.

Victims of cyberbullying attacks, particularly high school students and teachers, experience a loss of privacy as peers on social networks like Facebook act as an unseen audience. A Florida teenager sued her former principal over a three-day, senior-year suspension meted out in retaliation for creating a Facebook page critical of a teacher. The suit sought to have the suspension expunged from her disciplinary record since comments made on Facebook are theoretically protected by the First Amendment.

While schools do not have the authority to punish students for speech outside of school, true threats are not protected by the First Amendment. School officials had probably overreacted by ordering the suspension, claiming the words were a form of cyberbullying. The situation was created when the Facebook page appeared with the title "Ms. [name of teacher] is the worse teacher I've ever had" and invited past and current students to post their own comments. The page appeared for only three days, but it was enough to receive the suspension. The case actually went to trial in 2010 and the decision forced the school to expunge the suspension from the record and pay legal fees. The student enrolled at a college and majored in pre-law, specializing in the First Amendment guarantees of free speech. Determining what is free speech and what constitutes a threat is vaguely defined, and the issue has not reached the US Supreme Court. School officials have had to make decisions that may appear to be in violation of the First Amendment, but their actions may be justified if they relate to maintaining a secure and safe environment for students.

There is a criminal element attached to cyberbullying, which came to the attention of the public in 2010 after a 15-year-old girl hanged herself. The girl had moved to Massachusetts from Ireland and was tormented verbally, on Facebook, and through text messaging. Prosecutors became involved since the situation entered the area of criminal activity, with six fellow students being charged and eventually given either probation or jail time. The case also raised questions about the actions of school officials who knew about the incidents of abuse but did nothing to intercede.

6 SELF-INCRIMINATION BY SOCIAL MEDIA

More and more videos are appearing on social media, proudly posted by some not-so-bright thieves during the course of an actual robbery. Analysts at one city's Real Time Crime Center regularly identify robbers on social media and use the video evidence to arrest them. With more "netizens" flaunting their thoughts and actions on social media, it has become an important mainstay in police work. Police departments and federal agencies are aggressively seeking information from social media companies during investigations, igniting heated debates about privacy and the limits of current laws that spell out how police can legally retrieve what is considered personal data.

Existing laws about evidence collected from the Internet are arcane and confusing, and there is a significant debate on what governs what. There is a learning curve that has yet to be completed before the laws are better defined. Meanwhile, an overwhelming majority of law enforcement agencies have used social media sites, Facebook being the most frequently used, and some of the evidence collected has been challenged in court. Fewer than half of all law enforcement agencies have no social media policy, according to the International Association of Chiefs of Police. Social media companies use standard legalese to spell out their policies for granting access to individual accounts. After a crime is committed, the Facebook page created by the suspect may be shut down; law enforcement advocates claim that the site may contain evidence and should not be immediately shut down.

When the Electronic Communications Privacy Act was passed in 1986, email was stored on a personal computer and the law required police to obtain a warrant to access some types of information. With social media sites, information is now stored in the cloud on remote, third-party servers; owners of the site have in effect traded control of the data for convenience. Legal standards for information stored in the cloud are still unclear.

7 NAVIGATING CELL PHONE CHOICES

With the iPhone, Android, and other smartphones becoming commonplace, the process of demystifying the features of each and appreciating the difference in available networks has become more difficult each year as new phones are upgraded and more apps are added on an almost daily basis.

The most important features to look for include:

Network – This choice is not as simple as just selecting the fastest network with the best signal and the lowest cost. A network can have a great signal on one block and a terrible one on the next one. Most networks will allow customers to return a phone it if it doesn't perform as hoped but they may require a re-stocking fee and other costs. The best method is to ask acquaintances who live or work in the neighborhood about their network reception.

Speed – All the networks try to improve the speed with 4G and 5G networks. As with past generation upgrades, there are improvements in speed and capability, but it takes a long time to fully migrate systems and networks, in addition to the phones and hardware. 4G will likely remain operational for some time before, invariably, it too will be phased out at some point.

Contract – Contracts can be complicated and the biggest mistake people make is buying more minutes and data than they need. A contract can be negotiated, especially at renewal and there are often deals frequently offered. As technology advances, consumers are often the ones who bear the cost of improved functionality and rising costs in phones and plans.

Phone – Crucial considerations are the phone's size and operating system. Smartphones vary in size and features, in addition to differences in thickness and heft.

8 SWITCHBOARD OPERATORS – BEFORE THE DIAL TONE

After its invention in 1876, the initial public use of the phenomenal new voice communication device called the *telephone* was limited as telephones were only rented in pairs for conversations to be conducted between two parties operating two separate instruments connected by a direct wire. About two years later, when a few hundred telephones were installed, a Boston company started hiring boys in the newly created occupation of "telephone operators," but these low-paid young men were found to be less attentive, lacked patience, and their behavior was generally unacceptable for live telephone contact. The company switched to hiring only women operators and as the industry grew, it became an all-female occupation. Some of those pioneering operators in the early years, particularly in small towns, often had the small switchboard installed in their homes to monitor a few dozen subscribers in their neighborhood 24 hours a day. It was the beginning of the "Central Exchange" which revolutionized the telecommunication industry until the introduction of the smartphone.

Switchboards in large cities were mounted floor to ceiling to allow operators to monitor all their lines in the exchange. Ladders were used to access the higher jacks until the lower-constructed Divided Multiple Switchboard was created and quickly became a familiar sight (Figure 22.1). The

FIGURE 22.1 From 1921, workers operating the switchboard.

telecommunication system in the early 1900s relied on thousands of trained operators to manually connect incoming calls. The leading telephone provider at the time, the Bell System, relied on its operators to connect lines at large switchboards many years before the dial tone was introduced (men began joining the operator field again in the early 1970s).

Many local manual branch exchanges remained operational well into the second half of the 20th century and didn't disappear completely until the 1960s, when the automatic dial system took over. Prior to the introduction of the tone system, customers came to know their operators by name. Callers had to inform the local operator of the destination of the call and the operator became familiar with the customers who lived nearby who usually called the same group of friends and family members.

Eavesdropping was always a temptation to operators, especially when there were long periods when the calls were few and boredom set in, particularly in remote areas that often installed switchboards in small offices that employed a single operator for the night shift. In many situations, the operators would become very familiar with frequent callers and hold friendly conversations with subscribers during the slow hours of the day.

9 WALL PHONES AND LANDLINES: MUSEUM ARTIFACTS

Over 90% of American cell phone users program their contact phone numbers into the phone and dial by name – not numbers – on the screen. No one seems to memorize important numbers any longer since they are stored in their pockets in the cell phone directory. With a complete reliance on the convenience of dialing shortcuts, it may be that the only phone number people remember is their own number and 911. In the days of the recent past, when there were such things as rotary phones and corded wall phones, a call provided an active rehearsal of the information, so it was never forgotten.

Before the advent of digital address books, dozens of frequently used phone numbers were memorized, but digital technology has rendered those numbers obsolete relics of the dark ages. None of the social media communication exercises requires a specific number to be memorized and, in fact, there may be no room left in a person's long-term memory bank for the phone numbers of even their loved ones and business associates because phone numbers have been replaced by the multitude of passwords that much of modern technology now demands.

There is no longer any justification for the copper wire phone connection, yet some communities still fight for this obsolete method of communication. Those small rural communities that still cling to analog appliances are inconvenienced by the universal switchover to all digital equipment. Faxes, credit card readers, and dial-up modems have become museum relics much like horseshoes, typewriters, and telephone party lines.

Defenders of the status quo contend that the copper phone network has its own power source, so landlines continue to work when the electric grid goes down – but of course, the same winds or flooding may also take out the phone lines. During Hurricane Sandy in the fall of 2012, it was the landlines that failed while the cell towers kept working. Another excuse to subsidize the copper network is that landline numbers are confined to a fixed location, so 911 knows where the call is originating from – even if the caller is unable to speak. Triangulation on a cell phone signal is still a business of approximation and that small advantage to a hard-wired phone line is no justification to waste millions of dollars to keep copper lines intact for a relatively small number of homes in the middle of nowhere. The Federal Communication Commission has determined that plain old telephone service must disappear since nobody is willing to pay for its high upkeep.

DIAL TONE – In 1919, an article appeared in a major city newspaper directed to the few readers who had mounted a new telephone on their kitchen wall. The article offered the following instructions:

> Remove the receiver from the hook and listen for a steady humming sound and if you hear the dial tone, place your finger in the hole containing the first figure of the number to be called and pull the dial as far as it will go…

More than a century later, smartphone instructions are far more complicated!

> The next major milestone took place in the 1950s and may be unrecognizable to people under the age of 50. Phone exchanges used to use names instead of a string of nine numbers that were easy to remember. The names preceding numbers like Pennsylvania 9 200 or Butterworth 4 677, etc., were once famous but are now long forgotten. The round dial posed a challenge before it was decided to place the numbers outside the round dial instead of in the hole. This slowed the time it took to dial all the numbers – but, by adding a dot in each hole to help aim the finger, the problem was solved. Another solution that evolved from Human Factor Engineering studies was the length of the phone cord which had been introduced at three feet in length. In the post-war years, the copper used inside the cords remained scarce and the engineers thought of an unscientific way to determine the minimum length that would be comfortable. The method was to sneak into the executive's offices every three days and shorten the phone cords an inch at a time. No one noticed until the cord lost an entire foot. From then on, all cords came with the shorter cords.

In 1963, the round rotary dial was replaced with all-digit buttons that were developed at the famous Bell Laboratories where the first Human Factor Engineering Department was established and where the design and shape of the keyboard were selected to be rectangular instead of round and the shape of the buttons was determined. The position of the numbers was chosen to have "1-2-3" on the top row of the buttons instead of on the bottom as on calculators. The shape of the keypad design has now become the international standard on objects as diverse as ATMs, gas pumps, door locks, vending machines, and medical equipment.

9.1 Internet Hunting

Aerial mapping is a phenomenon of the Internet age where services like Google Earth, Microsoft's Live Search Maps, and Zillow.com can pull up views of buildings anywhere in the world. Small businesses use this mostly free service to find new clients and inspect properties, saving time that was spent driving to sites. These services provide access to aerial images of buildings taken by airplanes and satellites and are so accurate that cost estimates can be made without leaving the computer screen. In the past, businesses would buy aerial photos, but that was an expense only a few firms could afford. Roofers, landscapers, pool maintenance companies, real estate agents, and insurance companies use these services. Not only are Americans under surveillance while in the workplace but they are also being watched while in their homes.

While this new use of established technology has many positive applications, it can also be used for malevolent purposes. The technology is ideally suited for would-be terrorists, who no longer have to leave their hidden locations to discover American military sites. The enemy can now use Google Earth to pinpoint military targets and then locate the exact coordinates with the use of GPS devices. Information technology often finds unanticipated applications. This "dual-use" of hardware forces our Defense Department to weigh serious national security issues against commercial ones.

23 Low Battery Anxiety

1 BATTERY CHARGING MISCONCEPTIONS

The standard battery is not very complicated since it operates by placing two different metals in a conducting liquid, called an "electrolyte," causing ions with extra electrons to accumulate on one "electrode" while ions short of electrons gravitate toward the other. This definition seems quite simple, but it represents a long chain of discovery stretching back to the 18th century, when the term "battery" was first coined by physicists who captured and stored electricity in Leyden jars. In 1800, Alessandro Volta captured electric impulses by stacking thin layers of zinc and copper, each separated by cardboard soaked in brine, creating the first electric battery.

Despite the battery's seemingly endless potential, it remained little more than a novelty and a tool for scientific experiment until Samuel Morse realized that it could be used to transmit messages. In the mid-1840s, Congress granted Morse $30,000 to string a wire from the US Supreme Court building in Washington to Baltimore, carrying the world's first telegraph message, "What Hath God Wrought?" By October 1861, the telegraph had stretched 12,000 miles of telegraph lines across the nation, and California's chief justice wired a message from San Francisco to President Lincoln in Washington.

As with today's rapidly changing technology, many innovative concepts in the past were suddenly replaced by newer and better technology – in 1860, a wire strung coast-to-coast was so effective as a communication method that it completely replaced the Pony Express, which was replaced by the railroad, then cars, then planes.

Inventors throughout history placed too much value on the battery. One such man was Thomas Edison, who tried installing batteries in people's homes, but the power simply wasn't there. The future belonged to central generating stations, which converted coal or falling water into electricity. Batteries, meanwhile, were relegated to powering small appliances such as flashlights and ham radios. It remained that way until the 1990s, when wafer-thin lithium-ion batteries were marketed that possessed three times the capacity of their alkaline and nickel-cadmium predecessors. Lithium-ion batteries were the guiding force that helped the industry to go from AA battery-powered Walkmans to laptops, cell phones, iPods, and personal digital assistants (PDAs). Batteries have come a long way in 200 years, but before we can power cars without the need to charge them overnight, another revolutionary breakthrough in battery technology will have to become a reality. Although the newest batteries generally don't last beyond three years, neither do most of the products they power.

1.1 Proper Charging Procedure

Battery operation is a subject that has confused the public since portable radios and wireless electronic devices became so popular. Some wireless device salesmen claim that the rechargeable batteries used in almost all handheld communication devices can be damaged by overcharging and that the batteries should be drained completely before charging to increase the battery life and attain a full charge. This practice of depleting completely before recharging may still be valid in digital cameras, flashlights, and power tools, all of which are powered by nickel-cadmium cells. Older generations of nickel-cadmium batteries can lose capacity if they are not discharged completely before being fully recharged; this "memory effect" causes the battery to hold a progressively smaller amount of power.

Selecting batteries from those huge display racks at stores can be challenging and very confusing. Not only are batteries different in size and material, but there is a significant cost difference

DOI: 10.4324/9781003091103-32

between the national brand names like Duracell and Energizer and the less expensive store brands. Other mysterious areas involve the advantages of rechargeable batteries and the difference between titanium and top-of-the-line lithium. Regardless of the battery selected, and contrary to popular belief, it's not a good idea to store batteries in a refrigerator. It does not prolong shelf life and the condensation that occurs when batteries are taken out of the refrigerator can actually damage them.

With the many choices and the scarcity of accurate information, there is far too much unsubstantiated hype associated with the subject of batteries used in the vast array of communication devices. This has led to some confusion, and the following description may offer some clarity on a subject that is growing in importance since everyone in the world now seems to be carrying at least one battery-operated device.

- Alkaline batteries are the backbone of the battery industry and are ideally suited for items that do not use much energy or are considered "low drain." Some samples are smoke alarms, children's toys, kitchen clocks, etc., where there is no practical advantage to moving up from the standard alkaline battery.
- Brand names are really not that important when buying AA or AAA batteries, but the public seems to be unaware that carbon zinc batteries (which generally use the misnomer "heavy-duty") do not last as long as basic alkaline batteries.
- Brand names will probably last longer than off-brand names in the C, D, and 9-volt sizes.
- Titanium is best for high-drain devices such as digital cameras. They may be more expensive but will last 50% longer. Titanium compounds are added to the battery to improve performance by lowering resistance so that less charge is lost to heat.

Most round rechargeable batteries (AA, AAA, etc.), as opposed to those in standard cell phones or laptops, are made of nickel metal hydride, which is largely replacing nickel cadmium. These newer models claim to have a shelf life of five years. Rechargeable batteries can be a better value over the long term if used to power products that require short bursts of power, like digital cameras, handheld games, and some toys. A good rule of thumb to follow in deciding when to replace a battery is whenever the user has forgotten the last time the battery-powered device was used, which is probably a good time to keep using rechargeable units.

2 BATTERY LIFE

There are several factors that affect the battery life in laptops, including screen brightness and wireless features turned on, but there is no industry-accepted standard. More reliable statistics are better defined in other industries: cell phone batteries die a lot faster when making calls than when they silently rest in a jacket pocket and battery times are listed as "4 hours talk time/300 hours standby," much like the fact that cars achieve better mileage on highways than in the city. iPods enjoy much better battery life when playing music rather than video, etc. There are similar descriptive and reliable data that can be applied to the expanding laptop industry, but battery life claims remain misleading and highly exaggerated. With the introduction of smartphones like iPhone, BlackBerry, Android, etc., owners soon discovered that the more information the device furnishes, the shorter the battery life. Standard cell phones may sometimes hold a charge for three days or more, but generally, most smartphones require charging at least every 24 hours. A few simple practices can lengthen the time between smartphone charges; for example, when in a dimly lit room, the LCD screen's brightness should be turned down since the brighter the screen, the more juice is being used. Since smartphones are continually looking for a wireless signal, the unit will work harder in areas where the signal is weak and where there is no signal. In that situation, the unit's wireless capabilities should be turned off. Another way to cut power to a minimum is to put the smartphone into "airplane mode," which may extend the battery life up to a week or more.

3 LITHIUM-ION BATTERIES

Lithium-ion batteries cannot be overcharged; a device's circuitry cuts off the power when they are full. However, manufacturers still recommend that a laptop not be continually connected to power once the battery is at its capacity. There is no problem charging overnight, and the practice of draining the battery completely before recharging is another lingering myth from an older generation of batteries. It is best to recharge the phone when the battery is only partly discharged; in fact, the best scenario for extending battery life is to get it below half charge and then recharge it. If a laptop won't be used for several months, it should be stored with the battery in a 50% charge state. All batteries can be fully charged and discharged for a fixed number of cycles; lithium-ion batteries typically last between 300 and 500 cycles. Some iPhones have to be sent back to the factory for battery changes, but those devices are an exception to the rule. Returning the unit every two years is a big inconvenience, and users will replace the entire unit every two years rather than deal with the downtime incurred to send it back to the factory for battery replacement; a sort of built-in obsolescence.

Digital camera battery life claims are far more reliable since they are based on industry-wide standards, a fact that is not shared by laptop computer manufacturers. Until 2003, the digital camera industry faced the same challenges that laptops face. Until that year, cameras advertised battery life in overblown terms, complicated by the fact that each manufacturer used its own testing protocol (none of which was representative of real-life experiences). Eventually, the Camera and Imaging Product Association, an industry trade group, developed a standardized battery life test. The test was based on the following way people tend to use a camera in the real world.

- Snap one photo every 30 seconds after turning the camera "on."
- Half the pictures with the flash on, half with the flash off. The screen is left on all the time between shots.
- After every ten shots, the camera is turned off, etc.

Research on using radio waves as a low-power source is being conducted at several university laboratories. One day, power from batteries on low-consuming devices may be replaced by power from Wi-Fi transmitters, cell phone antennas, TV towers, and radio stations.

A dying iPhone battery can be extended until a charging outlet can be found by turning off the cellular data, even though the phone will be cut off from wireless Internet service – meaning the device will be without access to email. It will still be able to make calls and send and receive regular texts. Also, placing the phone into airplane mode will disable all communications while dramatically cutting down battery usage.

Over 5.5 billion AA, AAA, C, D, and 9V batteries are sold each year in the United States. Most of the alkaline and lithium replaceable battery market is shared by the two main providers, Energizer (33% of the market) and Duracell (36%), but that market has stagnated since the introduction of mobile devices that come with proprietary rechargeable batteries, which aren't powered using the AA and the more powerful 9-volt throw-away batteries.

CHARGING STATIONS can be found at airports where waiting passengers can power their cell phones and laptops. The lines at charging stations can be avoided by using an external charger that doubles the battery life of most handheld devices. There is more of a need to charge smartphones and tablets than to replace standard AA and AAA batteries. The trend has forced manufacturers like Duracell and Energizer to produce charging devices for the newest mobile units. This evolving smart charging market is unlimited, mainly because battery drain is higher than ever on mobile devices. The battery units that charge mobile devices may use AA batteries, but they may also have a proprietary battery like those found in tablets or smartphones. Two of the more popular chargeable devices:

- The PocketCell unit plugs into the computer for an hour to get juiced up, then it can be used to revive a dead smartphone or tablet.

- Mophie battery fits tightly over the iPhone and is recharged simultaneously when the phone battery is being charged – doubling the operating hours of the unit.

Battery chargers are not necessarily interchangeable, and most indicate on their packaging whether they are for use with smartphones or tablets. If the packaging doesn't specify a device, then the power output can be used to determine if it is for a phone or tablet:

- One amp for phones.
- Two amps for tablets.

Attempting to charge a tablet with a smartphone accessory will result in slow charge cycles or none at all. Charging a phone with a two-amp device won't damage the phone, but the phone will use only one amp of the two available. Furthermore, external batteries are rated in terms of mAh, or milliampere hours. The rating is analogous to the capacity of a car's gas tank; the higher the number, the bigger the battery capacity and the more devices that can be charged.

There are many external charging devices available that store power from a convenient outlet for future use. There is also a charger that can be used if the phone is in the wilderness – miles from a power source. The "PowerPot" is a cooking pot that, when heated, will cook food over a campfire and power up the smartphone as well via the pot's attached USB cable.

But lithium-ion batteries have caused concern because, when improperly manufactured, they continue to overheat, combust, and explode – a problem that has led the airline industry to establish guidelines specifically for these batteries when carried on a plane inside the digital communication devices.

Under the International Civil Aviation Organization's dangerous goods regulations, lithium batteries with power exceeding 100 Wh are not allowed on passenger planes without advance approval from the airline. Batteries with a power greater than 160 Wh are not allowed in passenger carry-on or checked baggage. On a wide-body jet, that could easily add up to more than a thousand batteries in the passenger cabin and sometimes they catch fire (especially on hoverboards and certain cell phone models).

Beginning in about 2016, some energy storage facilities were constructed to hold thousands of oversized versions of the lithium-ion batteries much like those used in smartphones and other digital devices. The largest facility of its kind in the world is located near San Diego, which contained 19,000 battery modules the size of a kitchen drawer, wired together in racks and made by Samsung. The batteries are designed to absorb low-cost energy particularly solar power, during the day and feed it back to the array of batteries after dark. These battery storage facilities will back up conventional electric grids in case of a disaster and are in the pilot stage but seem to be more acceptable than other power sources like wind farms.

4 LITHIUM BATTERIES

These powerful lithium batteries are considered top of the line but cost more than alkaline or super alkaline units. Some tests have indicated that lithium batteries can last for seven times as long as alkaline batteries, justifying their higher cost. Another factor is that lithium batteries hold their power after being unused for up to 15 years, compared to alkaline units, which have a shelf life of about seven years.

5 BATTERY RATING

Battery rating is generally defined as *watt-hours*, a figure that measures how much energy the battery can hold as well as how much energy it can deliver over time. Given the potential for lithium batteries to catch fire and the toxic fumes the battery emits, airlines established a 100 Wh limit, which is typically written as 100 Wh.

The federal Department of Transportation considers watt-hour rating as a function of the amount of lithium in the battery. Eight grams of Equivalent Lithium Content are about 100 Wh. A battery in a smartphone is only 12.4 Wh but professional broadcast cameras have batteries that exceed 100 Wh. These are allowed in the cabin of airliners up to a maximum of 160 Wh and only under restricted conditions. Portable chargers are considered spare batteries and can travel only in carry-ons – not in checked luggage, where a fire can go undetected.

6 BATTERY DISPOSAL

Battery disposal has become a serious problem for waste management agencies due to the proliferation of electronic devices and the millions of batteries used and discarded every year. The manner in which they are disposed of can become a potential danger to the environment. Various Department of Environmental Protection agencies operating at the federal, state, and municipal levels have mandated disposal procedures in an attempt to eliminate the dangerous toxic waste from rusting electronic devices discarded in in-ground burial pits. The problem begins when decaying components bleed into groundwater and the federal government is forcing manufacturers to make provisions to accept returned old equipment and dispose of the parts without polluting the ground. It is hoped that such federal involvement will force manufacturers to design equipment that produces less toxic residue.

The enormous number of electronic gadgets found in American homes has proliferated beyond all expectations, and most of the items use batteries that can fall into the hands of children. Their small size makes batteries a target for children and can cause severe injury if a child attempts to ingest them and they get lodged in the esophagus. A study published in the European Journal of *Pediatrics* in 2018 found that emergency room visits by those under age 18 involving ingestion of these small batteries were almost 90% resolved without surgery, though the symptoms – vomiting, diarrhea, and signs of poisoning are ominous. (Eur J Pediatrics. 2018; 177(7): 1063–1070. Published online 2018 May 2. doi: 10.1007/s00431-018-3154–6).

Batteries in electronic toys weren't the only culprits, since only 29% of batteries swallowed by children came from their games and playthings. Other devices that use flat, round batteries are hearing aids, calculators, flashlights, and remote controls. The data available on infant injuries is probably very conservative because it reflects only emergency room visits and doesn't include deaths because those wouldn't generally occur in the ER. Research on the subject of children and batteries has highlighted the dangers of button batteries, particularly a commonly found type called 3-V 20-mm lithium button batteries. While most batteries will pass through the gastrointestinal tract without injury, the 3-V battery can become lodged in a child's esophagus and give off electricity, possibly killing nearby tissue within two hours. The battery can also potentially burn a hole in the esophagus with no initial symptoms to alert the parent that the child's life is in danger.

7 TEMPERATURE EFFECTS

Older iPhones, smartphones, and laptops had been affected by the rise or fall in temperature, which can impact battery capacity and performance. Smartphones around the world have two types of batteries: lithium-ion and lithium-polymer batteries. Of these two, lithium-ion batteries do not take well to rises in temperature and can adversely harm your phone or the battery itself in case of high temperature. Whereas lithium-polymer batteries don't have such issues, other than the fact that they cost more, which serves to justify the use of lithium-ion batteries by most smartphone companies. These less expensive batteries have a problem known as "thermal runaway," whereby excessive heat promotes more heat and ultimately leads to a damaging overheating issue. It is always advisable to avoid excessive exposure to heat sources such as direct sunlight, leaving the phone in a hot car, using it near an area where there's a lot of heat (while cooking or near a fire), or even using your phone near machines emitting a lot of heat.

Battery performance is usually specified for temperatures somewhere in the + 20°C to + 30°C range; however, the actual performance can be substantially affected by this if the smartphone is operated at higher temperatures. If the temperature rise is excessive, the smartphone's battery will display certain characteristics:

- Active chemicals expand, causing the smartphone's battery to swell.
- The mechanical distortion in the cell components of your smartphone's battery may result in a short circuit.
- Prolonged operation at high heat/temperature can even cause cracking in the plastic parts of the smartphone's body.
- Pressure builds up inside the cell.
- The battery might eventually explode.
- Toxic/inflammable chemicals may be released from the battery.

At an average temperature of 0°C, a lithium-ion smartphone battery could lose up to 6% of its maximum capacity per year. At 25°C, that number jumps to 20%; at 40°C, it's a simply exceeds a whole lot of 35% of its maximum capacity every year. So, it's clear by the numbers that you should avoid exposing your smartphone to excessive heat.

A drop in temperature can also affect a smartphone's battery. Some smartphones are known to be able to withstand cold temperatures much better than hot temperatures. However, extremely cold temperatures can also have adverse effects. Recent studies have shown that many Android smartphones perform better in the cold than most Apple devices. Though it may feel like it's –40 degrees outside, recent temperatures are actually low enough to produce negative effects on a smartphone's battery.

Excessive battery drain is one of the most instant effects of extremely cold temperatures. Phones may eventually shut down with a dead battery in extreme cold and, over time, persistent exposure to cold temperatures can kill a smartphone battery altogether.

8 LOW BATTERY "NOMOPHOBIA"

Smartphone research has covered many topics, including an English study from 2008 involving 13 million subjects, which concluded that almost all of them thought losing their phone or having it run out of power to be "most stressful." This fear is termed "nomophobia," a word that entered the Merriam-Webster Dictionary in 2016. In 2014, public health scientists began lobbying to have nomophobia designated an official illness and recognized in the Diagnostic and Statistical Manual of Mental Disorders (DSM-5). As of 2022, it has not yet been recognized.

9 WATER-DAMAGED COMPUTER TERMINAL

With the growing use of laptops in coffee shops, there is a very likely chance of spilling a $7 latte all over the keyboard. This could result in an unrepairable tragedy since liquids can destroy the electronics. It is possible to salvage the computer if action is taken immediately. The power must be turned off completely and the power cable unplugged. If the power is left on, even for a few moments, it may cause a serious short circuit that can't be undone.

Ideally, the battery should be removed, and on older models, this is easily accomplished by simply undoing the easy-to-release latch cover. Some of the newer models require unscrewing the entire backplate. Any USB drives/devices must be disconnected to ensure nothing is trying to draw even a trickle of power from the terminal. Then proceed with the cleanup:

- Use a dry cloth to wipe away any visible liquid from the surface.
- Open the lid and hold upside down allowing the unit to drain.
- Allow it to dry for at least 4 hours, though 24 hours would be ideal.

When plugging the laptop back in, if it lights up that doesn't necessarily mean the problem is solved. Once fluid intrudes, even after it has been allowed to dry, it can leave behind minerals that carry electricity, causing corrosion over time, especially if the liquid is soda or coffee.

10 SOLAR POWER

Alternate fuel sources have gained in popularity since the obsession with energy consumption and the practical uses of wind, hydro, wave, tidal, geothermal, and solar. This last source of energy is free, unlimited, clean, reliable, and readily available anywhere on earth. The now familiar solar panels first caught the attention of the public when used to power satellites but by the second decade of this century, they have appeared everywhere from walkways to illuminate paths to rooftops to supply power to entire houses. The components of a typical solar system include a photovoltaic panel along with the charge controller, batteries, power inverter, a monitor (meter), and the electrical wiring to interconnect the components.

SOLAR PANEL. Most solar panels are made up of many individual silicon cells built into a single frame. When sunlight strikes the surface of the cell, a small electrical current produces about 0.5 volts. A typical 12-volt panel will contain 36 cells wired in series to produce 17 volts. The panel array is wired in parallel to increase the current capacity and the more panels, the more appliances can be used simultaneously. When wiring in series, it will increase the voltage to 24, 48, or even higher voltage levels. The standard panel is mostly about 1 to 2 inches thick and can weigh up to 30 pounds.

CHARGE CONTROLLER. This component is necessary to maintain the proper charging voltage and prevent overcharging the batteries. The more direct the sunlight becomes, the more voltage the solar cells produce, and excessive voltage could easily damage the batteries. Once a battery reaches a state of full charge, it can no longer store any additional incoming energy.

BATTERIES. Recreational vehicle (RV) or marine-type batteries are not suitable for solar systems since they do not have the capacity for continuous service with many charging cycles. Car batteries are the worst because they cannot be discharged much more than 10% without internal damage. The deeper the battery is discharged and the more frequently the battery is discharged, the shorter the lifetime of the battery.

INVERTER. Unless DC power is used on all appliances, there will be a need to convert (or invert) the low DC voltage to the 120 volts AC that is used by most appliances and devices. This inverter can also be used to charge the batteries and handle a wide range of loads and can regulate the power output within a very tight parameter with a minimum of power loss.

MONITOR METER. Provides information needed to protect the batteries and indicates when the batteries are approaching a full charge.

Section IX

Network Insecurity

Antivirus protection is among the fastest-growing industry security sector since 9/11. Safeguarding computer data and network is top priority for governments and private sector corporations. An advanced cyberattack could be the next Pearl Harbor. Cyberweapons are the most dangerous innovation of the 21st century since they could easily be used to disrupt power grids and financial systems as well as play havoc with military defenses.

Superpower treaty negotiations to ban or at least limit computer warfare haven't slowed proliferation of a very real threat. Online weapons are thousands of times cheaper than conventional armaments. Common antivirus software can detect and neutralize most common variety threats, but military and spy agency-level hardware and hacking skills present such a wide range of exploits that nothing can totally protect from a disciplined and skilled attacker. It is unlikely that a ban similar to those on poison gas, expanding bullets, or nuclear weapons will be effective.

After a series of cybercrimes in the first decade of this century, there was a huge infusion of investment in security against cybercrime. The growth in this industry is similar to the wild-eyed dice-rolling dot-com boom of the late 1990s, and the big bets placed on social media and cloud computing start-ups in the mid-2000s to the present.

DOI: 10.4324/9781003091103-33

24 Domain Names

1 UNIQUE WEBSITES

Collecting dozens, or even hundreds, of domain names has long been a full-time and highly profitable occupation for enterprising entrepreneurs. Corporations and institutions will pay large fees to gain access to popular domain names, and they have been known to bid hundreds of thousands of dollars for unique names that identify their products or category of products. Firms or individuals that specialize in accumulating domain names have become part of a growing industry; the wide-ranging listings can surpass 80,000 names since every name in the dictionary has a potential market value.

The most desirable names are sold at auction for large amounts of money; even common names like canteloupe.com had an asking price of $25,000. One domain name specialist leases names to relevant companies that test their marketability for six months to a year before making an outright purchase. It's an unpredictable business, but one that could prove to be very profitable if a registered name becomes trendy and catches the imagination of the public. The domain name business is estimated to be worth in excess of $3 billion annually and is growing in value.

Reselling domain names directly to the end user – avoiding the auction process – has grown into another major industry with over 35 million names registered and 25,000 domain names dropped (deleted) every day. Many of these registered domain names are sold for millions of dollars in the resale market. Some examples of the more expensive domain name sales *slots.com* that sold for $5.5 million, *VacationRenta.com* for $35 million, and *CarInsurance.com* for $49.7 million.

Selling domain names was once part of a small, insular, and sometimes underground world of so-called *domainers*. They shared a negative reputation, and the field was dominated by offshore practitioners who often used secretive, if not illegal, tactics. As the Internet grew in importance, domain names became valuable, cash-generating assets, just like stocks, bonds, and property, and the sale of names evolved into a legitimate business. Firms buy domain names at public auctions where enormous amounts of money are generated if the name becomes a catchphrase or reflects new concepts. Confusing names are sometimes deliberately selected to create undeserved interest, while many names accidentally have more than one meaning, depending on how they are read.

"Name-jacked" is an apt description of how some domain name owners (usually their real name) find that their domain name is now in the public domain and they no longer own it. Adding insult to injury is that the name is placed back into circulation and offered to the highest bidder. What often causes this "theft of domain name" is that oftentimes people forget to renew the domain for various reasons. When the victim of the lost name tries to buy it back, it can be too late because someone else has already purchased it. The best course of action in retrieving the personal domain name is to contact GoDaddy, an Internet company with a domain buying service that gets hijacked names back, often for under $100. That cost could be much higher if the original owner were desperate enough to pay thousands of dollars for the return of the name. A more complicated alternative would be to file a complaint with the World Intellectual Property Organization and sue the hijacker by claiming that the name is unique, in effect a trademark, and is being used in bad faith by being stored in what is called a "parking site." The cost of suing has averaged $3,500 but can be higher if the lawsuit has to be translated into another language in cases where the hijacker lives in a different country.

DOMAIN PROTECTION: Domain name abuse and hijacking are often initiated by so-called "cyberpunks" who create embarrassing sites to tarnish the image of the primary site owner. For example, the former mayor of New York City financed the purchase of domain names that sounded much like his name, e.g., MikeBloomberg.nyc. Altogether, over 400 domain names were re-secured

DOI: 10.4324/9781003091103-34

that were deemed inappropriate. The names cost between $25 and $40 per year, making Bloomberg's annual payment $10,000–$16,000 for securing the sites.

2 DOMAINERS AND PROPOSITION 19

California was determined to make marijuana legal in spite of the federal law against its universal use. In 1996, they passed a law that was copied by 14 other states, making marijuana legal for medical purposes only. The state shocked the nation in 2010, when California placed Proposition 19 on the ballot to legalize the sale of up to one ounce of marijuana for recreational use, with a medical justification, throughout the state. That event was treated as a mini gold rush by *domainers,* some registering more than 1,000 marijuana-related domain names that they hoped to sell as they thought other states would gradually begin to legalize the use and growing of marijuana. The domain names that suddenly appeared included such oddities like icecreammarijuana.com, marijuanapastry.com, and even potfordogs.com and marijuanadogbone.com. Proposition 19 was defeated but was followed up by the Adult Use of Marijuana Act in 2016, which successfully passed.

ICANN approved the xxx top-level domain in 2011 to clearly establish an indicator for adult entertainment on the Web. The xxx designation was created to allow people who wanted to see adult entertainment to easily find it and to identify it for those who wished to avoid it. The xxx surprisingly received an overwhelming response from some of the best-known colleges and universities in the nation. They purchased, for about $200 each, xxx domain names that included the college name in order to protect the integrity of their institutional name and to ensure that the domain name could not be used for adult content.

3 NAME REGISTRATION

Self-employed and creative professionals have discovered that the most effective way to attract new clients is to have a dedicated website. If properly constructed and designed, a website is all certain businesses require to succeed, and it can replace the need to advertise in print media. The task of obtaining a respectful and meaningful name is easier than creating a PowerPoint presentation.

- There are several sites that sell domain names, such as Gandi.net, Network Solutions, and Hover, but the big daddy of registrars is still GoDaddy.
- Registrars make their money on the services they try to tack on, including private registration, email boxes, and hosting. A cheap hosting plan is like parking your car in a ten-story garage, whereas a deluxe package is like putting it in a private driveway. A garage at roughly $10–$15 a month is probably good enough unless the owner of the site expects a million hits a month.
- After buying a domain name, a hosting company is then selected (e.g., InMotion, Rackspace, and WordPress).
- In designing the site, some unique personal features should be included and that may mean the services of a website designer or the use of a standard template offered by most hosts. The site should tell a story the way the owner feels is best suited for the services offered.
- Once the site is built, it may be advisable to promote the site by spending $100 to peddle the message or product through Facebook, Google, or other online ad services. The entire process to find the proper digital landscape is easier to navigate than most people realize … and if the task proves too difficult, there are many 20-year-old graphic designers who can help, and most of them can be found on their websites.

A typical scenario for obtaining exclusive use of a domain name was the experience of a small alarm manufacturing company. They developed a small, self-contained camera package that was motion activated and powered by a solar panel; it also had a built-in voice synthesizer to alert intruders that they were being captured on camera and should leave the premises. Meanwhile, the image was immediately transmitted to a centrally located control center via the Internet for viewing and dispatching of the police, if necessary. It was specifically designed for use at remote sites with no infrastructure, like school playgrounds at night, public parks while closed, and even commercial sites. In order to separate the product from similar units and to ensure that potential customers could find the product, the company went to GoDaddy and found that they had a legitimate and unique domain name. The domain name was purchased for $300.

4 GENERIC DOMAIN ADDRESSES

A powerful non-profit organization that few know of, ICANN, has managed several Internet-related tasks since it was given the job in 1998 as a substitute for government control. It is responsible for establishing the suffixes to the right of the dot in the domain name, such as.com,.org, and.gov. These are called top-level domains, and until 2012, there were only 22 such names. ICANN started the process of expanding that number to as many as 1,000 as a way to promote innovation and relieve dot-com crowding. Those seeking to buy the right to operate the new generation of domain names paid $180,000 to apply.

Banks, for example, may adopt .bank or .financial in their domain names, attorneys might capture .law and restaurants .food. The added names provide brands with an opportunity to carve a slice of the Internet market unavailable before 2012. The compliance office at ICANN had trouble keeping up with problems involving the current names and operators of the new top-level domain names that were required to complete a rigorous application process, involving 300 pages of rules established by ICANN to ensure that the Internet will be as secure as possible. The few critics of the expansion program exposed new domain spaces to the possibility of online malfeasance, which is like building more homes to increase the possibility of home burglaries. The experts working with ICANN spent six years developing the expansion of the domain-name program, which they believed to be more secure than what predated it.

5 CYBER CRIMINALS DIVERTING DOMAIN NAMES

Cybercriminals steal more than customer passwords and credit card numbers, sometimes misappropriating the domain names of the businesses they attack. Cyber thieves can hijack company domain names, often to transfer the business to a foreign-based interloper. Financial disaster ensues for the true owner, who still bears all the marketing costs.

ICANN has received a record number of complaints since 2014 about domain name thefts. Small firms with memorable domain names and entrepreneurs who collect domain names for resale are especially vulnerable. Most have unsophisticated Internet security controls. In practice, ICANN is not a regulatory authority or law enforcement agency, has no power to stop domain name abuses, and relies on local courts and law enforcement. Savvy international cybercriminals face spotty prosecution and few penalties.

Domain names can be hijacked by scraping public directory listings for owner contact data. Next, phishing emails surreptitiously capture a domain owner's keystrokes or passwords, among other intrusion tactics. Using that data, the hackers take control of the domain name and transfer it to another registrar under accounts they control. A phishing message might claim that "your GoDaddy service shown below has been suspended because some of the purchases on your account remain unpaid." The message instructs the customer "to click on a link to make payment and reinstate your service." Once clicked, the domain name will operate out of China, selling much of the same product.

25 Open Networks

There are roughly 1 billion PCs in use worldwide. Since there are currently just over 8 billion people on the planet, this works out to one personal computer for every four people. Not to mention the estimated 20 billion wireless devices. Given this, computer hackers are continuously trying to gain access to all of them.

1 ENCRYPTION

Encryption has been used by military and government agencies as a secure wireless network by incorporating codes, or code words, to prevent an unauthorized interloper from entering a computer network. Unwanted phone calls can be eliminated by keeping a phone number unlisted, only giving the number out to family or friends. In a wireless network, protecting privacy is much more complicated, but the application of encryption is extremely effective. Without adding encryption, the network is open to anyone, as if the phone number, name of subscriber, and address were listed in an advert in the newspaper.

Much of the effort to improve the security of wireless networks and computer terminals has been focused on developing an entirely new concept that would replace current technology. The Internet's original designers never foresaw that the academic and military research network that they created would one day bear the burden of carrying all the world's communications and commerce. There was not one central control point and its designers wanted to make it possible for every network to exchange data with every other network. Little attention was given to security at the time. Since then, there has been an immense effort by both the private sector and government agencies to improve wireless security, but with mixed results. Breaches of government agencies, corporations, and hacks of major network providers and software companies occur all the time. Far too much of the research and effort has been devoted to patching the current weaknesses, rather than investing in the redesign of the infrastructure.

At the same university where the software protocols for the Internet were designed, researchers are creating *(e.g., InMotion, Rackspace, and WordPress)* a system to make it possible to slide a more advanced network underneath the current operation of data transfer. The goal of all these uncoordinated efforts is to build a new Internet with improved security and the capabilities to support a new generation of not-yet-invented operating systems, as well as eliminate the known shortcomings of the current Internet. A more secure network is one that would offer less anonymity, which privacy advocates are not happy about. One method seriously being considered would be to require the equivalent of driver's licenses to permit someone to connect to a public computer network. That will not easily be accomplished in a world where taking over someone's computer from half a world away – and operating it as your own – can be easily accomplished. As long as this remains true, building a completely trustable system will remain virtually impossible.

Using any form of encryption poses a dilemma with the always-present threat of hacking by terrorists who use encryption to prevent governments from determining what communication is going on between the various groups that are determined to kill.

2 SECURING THE INTERNET

Virus – a self-replicating program that spreads by inserting copies of itself into other programs.

Vulnerability Scan – a process where a computer or network is checked for security issues, missing patches, or misconfiguration. The scan results are typically compiled into a report, identifying the vulnerability along with remediation steps to correct the issue.

DOI: 10.4324/9781003091103-35

Worm – a special type of virus that can spread automatically via email, Internet Relay Chat, or other network transport mechanisms.

Spam is the short word for unwanted junk email that has become big business, frequently backed by illegal enterprises and organized crime. Most of the messages originate from networks of compromised computers, called zombies or botnets, that push ads for subprime loans, herbal remedies, and get-rich schemes. It includes phishing emails that lure recipients into typing sensitive data onto bogus websites. There is also "pharm" which is a form of spam pitching fake pharmaceutical drugs. Spam email messages may exceed 1605 billion per day (or over 45% of all e-mail traffic) and that total grows every year. These unwanted communications represent a potent weapon that can overload a secured system server and cause it to crash. Knowledgeable spammers can use a botnet to send out millions or even billions of messages simultaneously. Some active spammers can collect about $500 to send an email to 20 million addresses for 15 days or sell about 80,000 email addresses for the same amount. The spam industry has prompted the creation of an anti-spam industry where a growing number of companies are dedicated to blocking spam messages.

Backscatter Spam is considered one of the most annoying of the various types of spam, clogging email accounts and slowing victims' inboxes to a crawl. Spammers collect real email addresses through computer viruses that steal the addresses from corporate databases. They then fake or "spoof" those addresses to send spam that appears to come from an individual. The problem occurs when the stolen email is aimed at the email that doesn't actually exist. The bounced email is returned to the email address of the victimized user. Backscatter primarily afflicts those who had the same email address for a long time and therefore have it floating in cyberspace.

Botnets are the systems blamed for huge spikes in spam and have helped feed a surge in various forms of online fraud and increased data theft. The war to make the Internet safe was lost long ago and computer security experts are still trying to neutralize the impact caused by programmers who secretly install their code on millions of personal computers and band these computers together into an army of zombies, using the collective power of the network to commit Internet crimes. Some experts estimate that botnet (Robot Network) programs are present on more than 650 million computers connected to the Internet.

In recent years, attacks have become endemic, forcing increasingly stringent security responses. Botnets have predominantly infected Windows-based computers, but there are reported attacks on computers running Linux and Macintosh operating systems. The public will always have trouble eliminating viruses and unwanted spam, but there are certain practices that may slow the infection:

1. Don't share a computer on which bills are paid with children who are prone to download games or other suspect content.
2. Don't use the same password on more than one financial site.
3. Don't let the browser store the passwords for these sites.
4. Don't buy anything offered by a spammer.
5. Most important: Don't click if someone offers something too good to be true. It is.
6. When asked for a social security number – respond only after determining:
   a) How will it be used?
   b) How will it be protected?
   c) Is it really necessary?
   d) What will happen if the person called refuses to provide the number?

Bot-herders collect a network of several thousand compromised PCs and sell the batch (sometimes listed online). They are used by criminals to take down businesses unwilling to pay tens of thousands in protection money. Much of this crime goes unreported because it often targets sketchy and illegal enterprises like unlicensed gambling sites. These victims pay to stop the attacks since they are losing money and unlikely to involve law enforcement agencies.

Malware is closely associated with botnets and refers to "malicious software." Attackers send malware-laden email that infiltrates specific machines and then download data for targeted information.

Scareware has been a prominent part of the Internet since its early days. It quickly became the scourge of Internet users by innocently appearing as descriptive promotions crafted to panic the unsuspecting user into spending from $50 to $500 for worthless antivirus protection. The scam starts when a dummy agency places innocuous-looking ads for big-name corporations, without their permission, on popular websites.

A scanner window appears as a red-letter warning and lists strange-sounding viruses purported to be infesting the hard drive. After being lured into a fake "free" scan of the PC, the bogus scan will claim to find virus infestations that do not exist. A series of dialog boxes then give choices all leading to the same screen – a sales pitch. Once the victim triggers the program, they get caught in an unending loop, impossible to abort. The scareware purveyors take advantage of merchant banking rules that enable them to accept credit card payments from victims across the globe. Criminals can be confident that there likely will be no repercussions since the process is practically crime-proof.

Ransomware encrypts files of a targeted business and brings commerce to a grinding halt. The perpetrators then demand a ransom for the unlock code. The FBI has, of late, advised corporations and even major cities to pony up to these criminals, especially if they need to quickly free their data. Many targeted organizations face immediate and devastating costs without access to their computer systems. For instance, an online retailer might lose hundreds of thousands in sales each day its systems are locked, a hospital is crippled without its patient histories and records. While many companies plan recovery strategies for these attacks and harden their database defenses, many others are easy target – especially for multistory buildings filled with seasoned and, in some cases, trained hackers.

3 ONLINE CRIMINALS

When Internet users check the box marked "I Agree" to the terms of use, they often never read the agreement that follows and may be committing a felony. If a person chooses to read a book on an e-reader, Amazon advises that if the person is not willing to sign a 2,500-word agreement, they cannot use the Kindle, any reading application, any digital content, or the service. Should the same person want to acquire music, books, or apps digitally from Apple, they must sign off on lengthy contractual language – more than 17,000 words of boilerplate. The public today enters into more legal agreements in a year than their grandparents made in a lifetime – and mostly without reading beyond the contract title.

4 VIRUS ATTACKS

The different terms like malware, worms, and Trojan horses are some samples of computer viruses and have been around as long as the networked computer. The 40th anniversary of the first known computer virus was in 2011; it was a laboratory experiment that didn't cause damage but proved to be a harbinger of the risks that plagued the digital community for the next four decades. A listing of the most significant milestones involving major viruses follows:

1971 – Creeper was the first virus-like program and was developed by a company that built the predecessor of the Internet. It was a lab test to see if it was possible to create a self-replicating bit of software.

1982 – Elk **Cloner** was a self-propagating program written by a high school junior. The program spread via floppy disks on old Apple II computers and displayed a poem-like message on infected machines.

1983 – **Virus** was the word first coined by a researcher at a university to describe self-replicating programs. Antivirus software is the basic tool to protect data and computer hard drives from hacking. It is designed to capture any potential threat before it reaches the system. Whichever antivirus program is chosen, virus scans should be done at least once a week.

1987 – **Jerusalem** was named for the place it was first detected and represented the first destructive virus that spread around the world. It affected computers running the DOS operating system by deleting some programs in an infected machine on Friday the 13th.

1992 – **Michelangelo** was designed to activate on March 6, the painter's birthday, and its discovery caused a minor frenzy among security experts who warned that millions of computers would crash on that date, much like the warning given a few years later that Y2K would cause planes to crash when the calendar changed on January 1, 2000. Both warnings proved to be false.

1990s – Phishing was used for the first time by hackers to collect confidential information by pretending to be a trustworthy source. The tactic was groundbreaking because it didn't rely on malware; instead, it took advantage of computer users' gullibility to collect information like credit card numbers and passwords.

1999 – The **Melissa** virus was spread by tricking millions of users into opening an email file that contained a program that sent messages to people in the infected computers' contact lists.

2001 – **Code Red** was a virus designed to overload the White House website with a flood of messages from infected servers.

2004 – **Sasser** was the first Internet worm whose effect was to travel over the Internet instead of via email. It affected more than a million computer systems and was blamed for airline flight cancellations and interruptions in satellite communications.

2005 – **MyTob** was a worm that was distributed through mass email, creating a "botnet" network on controlled computers that could be used to distribute spam or launch phishing attacks.

2010 – **Stuxnet** was a highly sophisticated computer worm and an example of malware as a cyberweapon. It could be spread through a USB device and this virus was designed to affect the operation of the specialized industrial control system. Some security researchers strongly suggest that Stuxnet's attack was aimed at centrifuges used to enrich uranium for Iran's well- publicized nuclear program.

5 CYBERTHEFT

A cyberthief is a hacker who operates in a secret often illegal underworld organization. Cyberthieves often exploit computer software flaws to break into websites. Aggressive hacker communities appear to be concentrated in China, parts of Eastern Europe, and Russia. Technology experts mostly agree that it is frequently difficult to trace hackers behind any specific attack. China stiffened its law that makes hacking a crime with a punishment of up to seven years in prison, but it is not strictly enforced. And, it should be added, that the Chinese military uses the same techniques to "legally" conduct cyberwarfare across the globe.

Computer hacking has become something of a national sport, and a lucrative one, with regularly scheduled hacker conferences, hacker training academies, hackathons, and a growing number of magazines dedicated to the topic. The articles offer tips on how to break into computers and build a virus like a Trojan horse step-by-step.

More headlines are given to hacking attacks against easily recognizable major corporations or government agencies, but there has been a slow but significant trend by hackers to expand their attacks to include any business that stores data in electronic form. Since 2009, small companies have become the main targets of hackers' and a surprising majority of reported attacks involve companies with fewer than 100 employees, some involving firms with as few as two employees. With limited budgets and no technical experts on staff, small businesses generally have weak security – or none at all.

In the time it takes a hacker to break into a major company like Citigroup, that same hacker could steal data from dozens of small businesses and go undetected, according to an online message board for hackers called Digital Gangsters. Even newsstands and pizza places have addresses, names, and credit card information. There are so many types of security threats that it is difficult for small firms to protect themselves. The costs to the owner of the establishment where the attack takes place can put a small company out of business. Financially motivated attacks typically rely on computer code that hackers plant in the victim's computer, often as attachments or links in emails sent to employees. While these malicious programs are well known to security experts, hackers tweak them frequently enough to render them undetectable to antivirus software.

Payment Card Industry (PCI) standards are required by businesses that accept credit cards, but there are few measures to enforce them on small businesses. Thousands of small enterprises have moved away from their old "point of sale" (POS) system that wasn't connected to the Internet but instead required a telephone modem connection to a bank to process credit card payments. The POS has all but disappeared, having been replaced with a wireless Internet connection. It's a much faster system but easier to compromise. Programs like Kameo – out of Russia – can capture every transaction processed on a computer system (even before it reaches the credit card bank), providing the hacker with credit card numbers, customer names, and other personal details. The small business hacking issue is out of control. The few law enforcement agencies focused on such attacks suffer strained resources from the volume of small business cybertheft cases.

Cybercrime, in its various forms, often targets federal agencies. More often though, these cyberattacks seek softer targets in private industries. The end result, at a minimum, strains – if not threatens – markets, financial exchanges, bank accounts, trade secrets, and intellectual property. Personal privacy is also at stake. The problem is exacerbated by private industry's lack of awareness of the problem. Many companies have no plan to deal with cyber intrusions and the resulting theft, not even knowing where to start or who to contact in law enforcement.

A common attack on small businesses targets easily fungible credit card information that hackers can sell or use the proceeds to make fraudulent purchases. In 2006, major credit card companies formed an industry-wide group called the Payment Card Industry Security Standards Council (PCI SSC), which established minimum technical protections for businesses that accept credit cards.

In 2013, a record-setting cybercrime took place involving stolen prepaid debit card numbers in a scheme known in the cyber world as an "unlimited operation." One of the attacks resulted in the theft of $45 million and involved 36,000 ATM withdrawals in more than 20 countries over a ten-hour period. In one instance, the criminals used 12 stolen prepaid debit account numbers to withdraw about $42 million of the $45 million. The other attack took place a few months prior in New York City; the criminal leader walked from ATM to ATM with his backpack getting heavier and heavier as he filled it with $20 bills dispensed from the machines. He and his colleagues siphoned off about $2.4 million in a 24-hour period, making it the second-largest cash heist in New York City history.

No individual bank accounts were compromised since the cybercrime group simply hacked into the computer networks of prepaid debit card processors. The thieves increased the available balances and eliminated withdrawal limits on the cards. When the time came to cash out, criminals around the world received texts or emails with PINs for the stolen cards and executed ATM transactions at the same time. In order to conceal the proceeds, the money was often used to purchase luxury goods, such as automobiles and Rolex watches, that were later resold to others or on eBay.

6 HACKERS POSE AS BANK EXECUTIVES

Hackers discovered an easy way to steal $1.2 billion from 7,000 businesses in the two years between 2013 and 2015 simply by sending phishing emails to the CEO and other top executives to gain access to their accounts. They then sent emails from the executive's accounts and created dummy email addresses to fool the finance department into thinking the messages were coming from the unsuspecting CEO executives. The recipients of the phony emails hit the "reply" key quickly without paying attention and a reply was sent to the dummy email address. After securing the necessary information about the company's accounts, the hackers wired money out of the company into their own accounts.

- Similar scams, while less sophisticated, are widespread and have been reported in all 50 states and in virtually every country.

7 HIRING A HACKER

There is a growing underground industry that advertises itself in law journals as so-called "hackers for hire" who help lawyers and investigators for a fee. After all, not all hackers are hidden in a remote room in some undisclosed location and there are people with the same skills as those working for theNational Security Agency (NSA) or Central Intelligence Agency (CIA). In recent years, federal authorities have focused on private investigation services that work with such "white hat" hackers for hire to illegally gain access to email and social media accounts to gather information for ongoing lawsuits where such information could be helpful.

While there may be a good reason to find out what is in an employee's email, particularly in a thriving business, it is not allowed by law. Many hackers have websites where they openly claim to be able to perform tasks like Internet profiling, email tracing, and computer monitoring, and some advertise that they specialize in insurance claim investigations. Hackers-for-hire get jail sentences of about four months for each intrusion – not much of a deterrent given the potential rewards.

8 CYBERTERRORISM

The growing risk of computer-launched foreign assaults on US critical infrastructure poses a bigger security threat to national security than al Qaida threats against the power grid, transportation hubs, and financial networks. In recent years, cyberterrorism has consistently been ranked higher in the US intelligence community's annual review of worldwide threats than worries about transnational crime and the proliferation of mass destruction. A cyberterrorist may hack into US critical computer networks in order to cause grave harm such as the loss of life or significant economic damage. Such attacks are aimed at wreaking havoc on information technology systems that are an integral part of public safety, traffic control, medical and emergency services, and public works.

The United States addresses cyberterrorism under the Patriot Act of 2001 which prohibits and provides penalties for cyberterrorism. Individuals violating this Act – and causing physical injury, loss of life, or in any way threatening public health or safety as a result of gaining unauthorized access to a protected computer – can be sentenced to long prison terms. The purpose of a cyberattack is to intimidate or coerce a government for ideological, religious, or political reasons. These attacks are specifically addressed to infrastructure computer systems to:

- Disrupt air traffic control and emergency systems.
- Shut down electric utilities and damage gas utilities.
- Contaminate water, food, or medical supplies.
- Block communication systems.

9 LAPTOP HARDWARE SECURITY

Cyberterrorism is not limited to government mass network terminals; it can also include the theft of laptops containing sensitive information. Physically securing computer terminals was once a major industry, where metal brackets and cable arrangements were used to lock down desktop terminals that were rarely moved from their initial location. With the advent of laptops and iPads, a new approach to securing portable devices had to be devised, since relying on wire cables and padlocks, much like securing a bicycle to a fence, was no longer practical. The expense of a few hundred dollars for securing a laptop was justifiable when laptops first appeared for $2,000+, but as laptops became less expensive, the cost of securing the devices became unjustified. Stolen laptops could be replaced with less expensive models, locking devices were discarded and the "computer-lock hardware" industry nearly vanished, replaced by digital programs that had built-in tracking software, passwords, and anti-hacker systems.

As laptops replaced desktop computers, they brought with them a new vulnerability that changed the direction of computer security from securing the hardware (computer terminals) to protecting the sensitive data stored within. This concern with securing software created an industry that, among other security controls, required users to type in a password to activate the laptop. Other innovative programs were developed that replicated the function of the LoJack automobile tracking system. The pre-programmed software resides within the stored circuit of the laptop and the moment it is connected to a network, a signal is automatically sent to a centrally located server, reporting the missing computer. This allows the laptop owner to track its whereabouts and even take a photo of the thief if the computer has a built-in camera.

An estimated 1,800 laptop computers are stolen daily in the United States. Laptops, iPads, e-readers, and smartphones are lightweight and portable, and they are easy to steal or misplaced, with most people assuming they are gone for good. To protect sensitive information stored on laptops and hopefully retrieve those that were lost or stolen, a software industry has evolved that provides the backing of an Internet theft recovery team that tracks the stolen device, reports its location to the police, and can be programmed to remotely delete any personal data.

Popular Anti-Theft Systems Available:

- Software that can display an on-screen message letting the user (who may not be the thief) know the device has been stolen and where it came from.
- Takes a picture of the user with the computer's webcam, either every 30 seconds or whenever the computer is awakened from sleep. Pictures, along with location data gleaned from a program that triangulates cell phone tower locations along with GPS data, are sent to the user's Flickr (or other such) account, allowing the creation of a map that pinpoints the laptop location.
- Sends pictures of the laptop's user every time the device is turned on. Once the machine is connected to the Internet, it uploads those photos, as screenshots of the user's activities are taken every six minutes.
- Laptop owners can subscribe to an annual service that gives customers the ability to selectively lock out, delete, and retrieve data from the hard drive regardless of its location. Even if the laptop is not recoverable, the data is.
- A program that allows users to send messages that appear on the screen of the missing laptop, such as "Give me my computer back, you crook, or you're going to jail." (This is the last step that a victim of a theft should take since the warning would most likely encourage the thief to stay off the Internet to avoid being found)
- LoJack, the best-selling anti-theft system for vehicles with 1.2 million users in 51 countries, has a version for laptops that monitors the computer's activities if it is reported stolen. When the missing computer is located, the service provider alerts the local police of its status and location.
- An app is available that can track the stolen computer through GPS, but the computer must be on.

Regardless of the laptop retrieval program selected, the police are not likely to make retrieving a smartphone or laptop as high a priority as they do stolen vehicles. The less expensive – and probably the most cost-effective method of securing stored data, even for a stolen laptop – is to install a password protection program and set up an account without password protection at the same time. This will hopefully safeguard important personal data while thieves unwittingly continue to keep using the computer for their own purposes while allowing the computer to be tracked.

Signs a computer has been compromised

Abnormal response time or unresponsiveness
Unexplained lockouts
Password not working
Running unexpected programs
Bounced back emails
Inability to connect to the network
Constant or increasing crashes
Abnormal hard drive activity
Connecting to unfamiliar sites
Browser settings changed
Extra toolbars that cannot be deleted

Recovery

Remove vulnerabilities and update firewalls to prevent future access
Reinstall clean versions of the operating system
Change passwords
Reconnect to network
Monitor system closely

10 CAPTCHA

In 1996, an opinion poll undertaken by a presidential candidate that used a website to collect data from potential voters uncovered a problem: hackers who wanted to neutralize the technology would use bots to automatically fill the site with useless spam, compromising the data. The company that created the site attempted to stop hackers by placing an image of the US flag at a random spot on the screen which users had to click before they were able to register their vote, essentially proving that the user was human. Some hackers developed simple programs to defeat this method, but the idea was eventually copied and expanded with the goal of finding a way for computers to respond to a human and not a bot.

Instead of a flag, an idea was developed to add a string of random letters and numbers that were distorted on the screen and users were asked to type the characters as they appear. It was, for all practical purposes, impossible for a machine to read the same random set of *captcha rds*. Cryptographers devised the acronym that describes the task as a **C**ompletely **A**utomated **P**ublic **T**uring test for telling **C**omputers and **H**umans **A**part.

Thus, the current and common added security feature of CAPTCHA was born.

11 SOCIAL MEDIA BOTS

It used to be difficult to produce nonexistent social media bots, but that has changed and someone with minimal programming skills can create a few thousand fake "friends" for about $5. As a result, a giant pyramid scheme emerged where fake friends now command real money. It is done by downloading software called Twitter Supremacy, which charges $50 for a six-month license that allows the purchaser to fabricate an unlimited number of friends. There are other similar services that do this for websites like Instagram, Twitter, Tumblr, YouTube, and Facebook. Reliable reports have found that celebrities, politicians, and companies often buy fake followers to enhance their perceived importance online.

Once an army of fake friends is assembled and given names, the person who enlisted the fake army charges others who want to increase their follower base on social media. There are entrepreneurs who manage hundreds of thousands of bots and make a profit by pushing posts to the app's popular page. The client base for the fake bots includes well-known celebrities and fast-food brands, as well as everyday people who want an ego boost.

Celebrities and other users often have no control over the number of fake followers they have. Bot farms will follow big-name celebrities regardless of whether a celebrity is a client or not. All major social media networks are subject to such schemes. The practice is so widespread that some social media consulting companies now offer a web program that can identify how many fake followers a person truly has.

12 IP THEFT

Many tech-savvy thieves get a free ride through IP theft, stealing another person's Internet access by tapping into their home router or cable modem. This method of piggybacking on an unsuspecting neighbor's Internet connection for illegal activity could leave the owner of the Internet service an unwitting target of police investigations. The Federal Communications Commission considers IP hijacking, particularly of large amounts of Web traffic, one of the top three biggest cybersecurity threats associated with the Internet. The others are botnets and domain hacking, which allow criminals to redirect traffic to fake websites. Though thousands of cases exist, only a small percentage of IP thefts are ever investigated due to a lack of resources and privacy laws under the Electronic Communications Privacy Act.

13 PRINTER DATA VULNERABILITY

Printer security is almost always overlooked when conducting a risk analysis, since the primary focus is often on securing computer terminals, hard drives, and software. Yet, the printer is a necessary computer companion and may contain sensitive proprietary information that resides inconspicuously within the unit long after the data was reproduced, removed, or scanned. Treating the printer as just another ancillary device, like a mouse or a memory stick, is no longer an acceptable practice. One of the basic security weaknesses in any computer network is the practice of never changing the printer password, reducing the level of security against hackers. Most organizations in this day and age have come to the conclusion that leaving sensitive data to reside endlessly on printers could lead to a serious breach in security.

Most major printer manufacturers have incorporated unique technology features that leave microscopic yellow dots on each printed page that identify the serial number of the printer. The dots, invisible to the naked eye, can be seen using a blue LED and are used by the US Secret Service to investigate counterfeit bills made with laser printers, which is a federal offense. This faint, hard-to-notice, and unique mark has led a growing number of privacy advocates to warn that the "dots" could be used to ensnare political dissidents, whistleblowers, or anyone who prints material that includes the tracking feature. These inconspicuous dots are produced only on laser printers, not on inkjet printers which are more commonly used home printers. These controversial dots provide not only the serial number of the printer but also, in some cases, the time and date a document was printed. Supposedly, the Secret Service is the only US agency that has the ability to decode the information, while other agencies have shown interest in the same.

Governments and banks feared an explosion of counterfeiting, but while that fear was justified, it was not a serious or lasting activity as first feared. Laser printers have become increasingly popular as their prices have dropped enough to make them an affordable alternative to inkjet units. Meanwhile, the desktop printer seems frozen in mediocrity while PCs, smartphones, and tablets have evolved beyond most people's wildest dreams.

A printer is relatively inexpensive to buy but not inexpensive to own. The initial purchase price is only a small fraction of the cost of ownership since ink needs to be replaced and paper needs to be refilled, so the lifetime cost of a printer can be ten times the initial purchase price. Fighting the overall cost of the ink cartridge can be a daunting task, but there are a few practices that may neutralize the seemingly unending commitment.

- Consider establishing a printer-free home by instead using an office printer or by storing documents in the cloud or on a thumb drive and visiting a print store when needed. A printer-free environment could add years to the life of anyone who had become frustrated by the stress associated with printer maintenance. Of course, the information remains in the memory of the printer regardless of its location.

One of the primary reasons for owning a printer is that many documents require an original signature on paper. For the technically uninitiated, getting a person's name on the bottom of a piece of paper in the digital world can be a time-consuming and infuriating process: first, the document must be printed, signed where indicated, then it must be either mailed, scanned, and emailed or faxed back to the sender. This process has since been streamlined by several online e-signature services such as DocuSign. Also, software packages such as Adobe and NitroPDF – among others – allow you to store a digital signature, which can then be positioned and resized onto any document uploaded to the service.

Realizing that printer manufacturers want users to purchase ink refills and cartridges from their brand, purchasing off-brand ink can save hundreds of dollars over the life of the printer. Manufacturers also configure their printers to alert users to when the ink is low. These alerts are more often than not inaccurate. It is recommended that when the alert first appears, remove the cartridge, shake it once or twice, then replace it, and forget about it until printed documents appear to actually be running out of ink (evidenced by lighter print on the page).

Hospital practitioners use the internal IP network to print a prescription from a doctor in one department to a printer which has the actual prescription paper in another department – usually in the pharmacy. By 2012, at least three states established regulations mandating that hospitals lock the paper trays that hold prescription forms, for obvious reasons. Staff, visitors, and others understand the unique value of these blank forms, since a simple forgery can result in obtaining vast amounts of controlled prescription drugs from the nearest retail drugstore. Some printer manufacturers have models available with a cam lock on the paper drawer.

One security lock manufacturer developed a key-cylinder access control system that stores data in the memory of the key as well as in the cam cylinder and identifies the key user and the time of day and date. The key with its event record seems to be the only device that complies with the state law.

INK CARTRIDGE. Printers have steadily improved in resolution and speed and the costs have decreased substantially. Printers can now be connected to Wi-Fi networks and most work with every type of mobile device. However, the one item that has remained static is the ink cartridge and the need to continuously replace it. Many reliable printers are available from manufacturers such as Epson, Hewlett-Packard, Canon, and Brother, each for about $60. Manufacturers can afford to lose a little money on the hardware and recoup any losses with the ink cartridges which can cost $40 per refill. A growing number of retailers sell off-brand ink refills at much lower prices. To solve the recurring ink problem, one supplier offers a long-term ink reservoir that stores enough ink to last for several years under normal usage.

14 STEALING THE SIGNAL

Wireless IP and Wi-Fi systems offer a low-powered radio transmission format operating in the 2.4 GHz band, making it ideal for short-range, high-speed over-the-air data connections – about a 100–150-foot radius – in ideal conditions. Using Wi-Fi in a private network is always accompanied by concerns of reliability and security which limit its use in more sensitive locations.

The swiping of high-speed Wi-Fi access signals is widespread and protecting the bandwidth is not being done very effectively. There are several mobile devices that allow laptop users to make free or low-cost phone calls over the Internet even when sitting in coffee shops, parks, and/or in homes (a service offered for a monthly fee). There are just as many people sitting in the same room that pay the fee but they are successfully tapped into some nearby phone service Wi-Fi network without permission. This raises some tricky ethical and legal concerns, but the swiping of wireless connections hasn't, as yet, been tested in court. The Wi-Fi radio frequency spectrum is still unlicensed and not maintained by any one company. Once these signals spread throughout a city area, they could threaten the dominance of traditional telecommunications networks by giving the user an alternate method for voice and data transmission. The proliferation of Wi-Fi-enabled laptops and the "stealing" of signals could be considered breaking the law when using the bandwidth is done without the knowledge of the wireless connection owner. There are conflicting attitudes between those that consider using the Wi-Fi connection without permission to be the same as unauthorized access to computers, which is a crime, and others who feel that using Wi-Fi with or without permission is not the same as trespassing since the signals travel beyond property limits. (It's the same as sitting outside someone's yard and listening to the radio). If the owner of the wireless Wi-Fi connection wishes to protect the bandwidth, then a password should be added; otherwise, the public can steal/use the signal whenever it becomes strong enough to operate the laptop.

The use of a "mesh" network was introduced in 2006 and its flexibility and quick installation made it the choice for video surveillance in such facilities as schools, gas stations, and strip malls.

WiMax (Worldwide Interoperability for Microwave Access) represents a wireless technology that works like Wi-Fi but operates over a much larger area and has turned entire cities into giant hot spots. Eventually, WiMax coverage will be used by 100 million people throughout the nation. The first city to employ this wide area coverage was Baltimore in 2009. It is designed to cater to the needs of laptop, not cell phone, users for a monthly fee of $30.

15 CYBERTHEFT AND ICRIME WAVE

The theft of handheld communication devices has increased over the years, with used iPads and iPhones selling in lucrative second-hand markets. One way to attempt to control the crime would be to convince the wireless industry to adopt a registry system that would deny service to stolen devices. When a smartphone is reported stolen, its ID number can go into a nationwide database. When another user tries to make a call or download data, the device can pop up on a blacklist and the carrier can deny service. Two of the four major US carriers – Sprint and Verizon – have had blacklists since 2011 and identify their customer devices by serial number, so stolen devices can't be reactivated. AT&T and T-Mobile haven't created databases because their networks identify devices using SIM cards. While SIM cards can be disabled, a thief can easily install a new one.

Similar identity systems are in place in parts of Europe, Latin America, and Australia. In London, where a blacklist is in use, mobile device robbery incidents declined significantly. Expanding the blacklist in the United States has some inherent loopholes for tablets. Most iPads rely on Wi-Fi for connections, rather than a carrier network (4G, 5G, etc.), so the signals can't be blocked. Another problem is that devices blocked in the United States will have full functionality in many other countries. Thieves in the United Kingdom have capitalized on this weakness and have been exporting more stolen phones to places like Africa and India in recent years.

16 PHISHING AND OTHER PHRENZY THEFT

Data theft is a growing problem at an all-time high and this includes the incidents of personal identity theft, which have resulted in Social Security numbers and other critical data being stolen. This can lead to unauthorized credit card use, fraud, and even false tax filings with the Internal Revenue Service to defraud.

The word "phishing" may at first appear to be a misspelling but it represents a very serious threat that affects Internet users of all kinds. Phishing expeditions are a form of identity theft using phony emails in an attempt to gain Social Security, credit card, and bank account information. In recent years, the number of phishing successes exceeded 800,000, costing the public over $10 billion in losses.

The concept has been around for years and the term phishing probably comes from "phreaking," which means hacking or cheating a phone system to make long-distant calls. A more modern interpretation is that phishing is a variation of "fishing" as that's what cyber attackers are doing… fishing for data. It is defined as a scam or trick of sending fraudulent email to an organization's customers or employers in an attempt to trick Internet users into divulging sensitive information. The term "**whaling**" is a play on phishing, an approach that usually involves tricking big fish into divulging personal information like credit card numbers. Phishing attacks that are directed at a particular person, rather than blasted out to millions, are also known as **spear phishing**.

Most online criminals believe that institutions do not possess the knowledge, experience, or resources to protect themselves or respond to a phishing or hacking scam. The F.B.I. has been swamped with computer crime waves. Agents investigate more phishing incidents as a result of bank customers and brokers falling victim to hackers and inadvertently revealing information. Losses are in the hundreds of millions of dollars collectively, can be devastating for individuals, and many of the fraud attempts are made by organized groups.

Larger institutions (banks, retailers, etc.) have implemented countermeasures but only a few other facilities have started programs to stop illegal attacks on their networks. Some organizations have the misconception that if they maintain an in-house IT security staff, they are protected from phishing attempts, but this is not the case since phishers do not require access to the facility network to implement a scam. Phishers can simply browse websites, take screenshots of the log-in page, and then send email messages that appear to be from a legitimate source.

In recent years, the most popular method used is when someone creates a fake email that appears to come from a legitimate institution, like a bank or other online financial company such as PayPal, and asks the recipient to confirm their account number, sometimes suggesting that there may be a problem with the account. When an email appears to be phishing, it should be reported to federal authorities at spam@uce.gov. If the information is already being used improperly, a complaint can be filed at identitytheft.gov.

Thieves are using new lures, impersonating banks less often but attracting victims with phony sweepstakes or retail offers. They're also going "spear phishing" – targeting consumers with incomes of $100,000 or more. To avoid falling prey to these Internet thieves, don't open email from anyone unknown or even from "recognized" businesses that do not normally send messages and check with the bank or credit card company first when an email is received requesting an update of account information online.

E-fencing is another word for cyber money laundering, which scammers have perfected as they convert stolen data into cash while sitting at a computer. Thieves use stolen credit card numbers to buy gift cards on online auction sites or for a set discount at a gift-card exchange website. This practice of purchasing gift cards with stolen credit card information and selling them online before the victim has a chance to deactivate the account is easily accomplished and very lucrative. Gift cards have become a multibillion-dollar enterprise. The fraudulent cards are commonly advertised on eBay, Craigslist, and on card-exchange sites. There are some legitimate sellers who are reselling cards they received as gifts which they know will never be used, but when hundreds of gift cards selling at a 50% discount suddenly appear, the buyer should be suspicious. Some sites require any card issued by a major retailer to be ten days old before it can be posted for resale, but the laundering scheme continues to grow almost unabated.

Smishing is another popular identity theft scam, only it is perpetrated through messages sent to a mobile phone, not a laptop as is targeted by phishing scammers. An automated, prerecorded, bogus voicemail message is left that may say that the victim's bank account had been compromised, along with a call-back number. Those who return the call are then asked for account information, Social Security numbers, and PINs over the phone. Phishing and smishing are more or less the same terms for a brand of theft, but a message popping up on the mobile phone, as opposed to a computer, holds more urgency. Anyone recognizing the fake call should notify their bank.

Spyware can secretly gather information about someone's Internet activities such as web surfing and viewed sites. Though once reserved for hackers, this flavor of malware can be automatically inserted just by visiting a tainted website.

17 RANSOMWARE

Ransomware often starts with an unlisted phone call from a scam artist claiming to be a representative of a fake computer firm followed by a warning that the computer has been infected with malware. The scammer explains that malware will corrupt the data stored on the hard drive, tricking the victim into giving the imposter full access to the files which then allows the scammer to infect the computer, then charge a fee to restore the files to the original space. In the meantime, the scammer looks around the computer files to steal the user's identity. The best defense against this attack is to never click on anything suspicious online – antivirus software offers no defense.

Ransomware hacking practices have grown into a lucrative form of cybercrime that threatens to not only wipe a computer, but it can also capture private data and threaten to release information or – in many cases – lock cell phones or freeze computers by hijacking files or encrypting them and then holding them hostage until the owner pays a ransom to unlock them. In the days before the Internet and Facebook, kidnappers used to make ransom demands with letters cut out of magazines to form the ransom note. Now, the more knowledgeable criminals, including organized groups perpetrating such crimes, use modern technology to simply pop up on the computer screen and the hostage is the computer. At one time the encryption was easy to crack, but ransomware is getting harder to overcome since the availability of more powerful and sophisticated algorithms.

Attacks usually involve infecting a user's computer with a virus that locks the screen with a message to call a phone number that is answered by someone explaining the terms and cost to remove the virus. Experts discourage anyone from paying ransoms and strongly recommend that files be backed up on an external drive or in the Cloud. All too often, the only method to get rid of the virus is to wipe out the hard drive (a disaster if the files are not backed up regularly) and reinstall backup files and software if available.

CryptoWall A dreaded virus that appears as an announcement on the locked computer screen:

> Your files are encrypted. To get the key to decrypt files you have to pay 500USD. Failure to pay within one week, the price goes up to $1,000. After that, the decryption key will be destroyed and any chance of accessing the 4,500 files on your PC will be destroyed.

This immuno-resistant strain of computer virus is the definition of the term *ransomware* and is thought to infiltrate a computer when a legitimate-looking email attachment is clicked on or through existing malware lurking on the hard drive. Experts believe that it would be technically impossible for anyone to decrypt CryptoWall once it has locked a computer. According to the Internet Crime Complaint Center, a complaint should be filed, but of course, that doesn't get the files back.

- A Los Angeles hospital paid a ransom of $18,000 to hackers to restore its computer network and all of its digital medical files.
- A law firm's files were hacked and encrypted. The cybercriminal demanded and received a $50,000 ransom and the case was never reported to authorities since it would have compromised the integrity and reputation of the firm.
- Even police departments have been hacked and had to pay a ransom to restore the criminal or personnel records stored in the system.

- In one such example, the Sheriff's Office in one Tennessee town paid a CryptoWall ransom to unlock their 72,000 autopsy reports, witness statements, crime scene photographs, and other documents. Their ransom payment, like most ransomware money, was paid in Bitcoin.

18 SEXTORTION (NOT AN ILLEGAL CRIMINAL ACTIVITY)

Using nude photographs of someone to demand more photos that contain sexually explicit content has become increasingly common, especially on college campuses and military bases. It usually involves hacking into a computer to remove sexual images or even commandeering a webcam, then using the images to extort money or something else of value from the victims with threats to place the images on the Internet. It can be perpetrated by strangers but has increasingly been used by individuals who are in, or have previously been in, intimate relationships – often used as a means of leverage over the other individual. Law enforcement agencies recognize the growing use and danger of sextortion but have been hampered by a lack of laws on the books to enforce in order to better protect people. Increasingly, though slowly, laws are being put in place to prevent and retroactively seek damages for victims.

19 PRETEXTING

Data breaches once used to be relegated to only laptop and hardware theft. Now mass data breaches are all too commonplace. It is estimated that up to 30% of Americans have suffered identity theft, and according to Pew Research, as many as 64% of US adults have been impacted by at least one type of data theft. Listed here are some examples of computer hardware and software theft, representing a small sample of the enormity of the problem.

- An Ernst & Young employee's laptop was stolen from a car. The laptop contained customers' personal information as well as social security information for 38,000 employees.
- In 2007, British officials disclosed that the government lost two computer disks containing detailed personal information on 25 million Britons. Though the data was protected by a password, the disks were not encrypted and contained detailed personal information on 40% of the British population, including bank account numbers and the British equivalent of Social Security numbers.
- The L.A. Department of Social Services lost file boxes containing names, dependents' social security numbers, telephone numbers, medical information, and W-2 forms for 2,000,000 people.
- Hackers accessed the computer system of Ohio University's alumni relations department, gaining access to the social security numbers of 137,000 alumni.
- The US Department of Veterans Affairs had a laptop stolen that contained data on all American veterans who were discharged since 1975, including social security numbers, birth dates, and addresses of 28 million people.
- UCLA had entered a restricted database and retrieved the names, addresses, social security number (SSN), and other private information of current and former students and faculty for over a year before the breach was discovered. 800,000 records were stolen.
- Boeing, the aerospace giant, had a laptop stolen with 390,000 current and former employees' personal data.

Some experts in the field believe the threat of identity theft from data losses is greatly exaggerated, leading the population to fall into the trap of equating data loss with identity theft. With millions of names on the list, it doesn't mean that everyone suddenly had their identities stolen and their credit cards being illegally used. There is simply too much data available and criminals have not figured out a way to commit crimes against a million individuals all at once. However, at some time in the near future, organized crime will get truly organized and figure out what to do with the millions of identities and user accounts.

20 PASSWORD UNORIGINALITY

Before the introduction of passwords, the public only had to memorize seven- and then ten-digit phone numbers and five-digit zip codes. That all changed in the 1990s when the public was forced to place in memory a series of passwords to activate phone cards, credit cards, websites, and ATMs … and then words for iPhone apps … and on and on infinitum. The explosion in codes, passwords, and PINs was meant to protect users from unauthorized theft of anything of value from their personal devices. Most people's choice of passwords is easily predicted since many select numbers and words from a relatively small grouping of commonly used letters and numerals or a combination thereof that is easy to memorize. At the beginning of the password era, the most popular numbers were "12345," but as hackers and other types of computer wizards became more aggressive, the public reacted accordingly, and by 2012, the popular numbers advanced a digit to "123456" and "password" remained the most used word.

According to the most reliable data used for analysis, one out of five Web users still decides to leave the digital equivalent of a key under the doormat, choosing a simple, easily guessed password like "abc123" or "iloveyou" to protect their personal data. Most sensitive computer networks require four to ten digits, or characters, as a password in order to maintain the privacy of the data, especially in this age where the threat of terrorism has forced the digital communication community into becoming more network security conscious. It's important to avoid using passwords that identify a particular facility so that it is difficult for an unauthorized individual to recognize a specific network. For many years, hotels had attached a room key to a key chain with the hotel's address and instructions to drop it in a mailbox if their guest checked out without returning the key. That is no longer the case and metal keys have been replaced with thin plastic cards with no room number on them.

A research company examined a list of 32 million passwords, intending to design software for social networking sites. The list was stolen in 2010 by an unknown hacker and posted on the Web. Security researchers (along with hackers) then downloaded the information to study computer users' password habits. The results confirmed what experts have suspected since the dawn of passwords in mass communication. A significant percentage of the nearly 32 million people in the study had used "123456" as a password, and if they were required to use at least one word, they chose "password." Another, more unexpected finding was that about 20% of people on the list picked the same password for all their devices. Table 25.1 illustrates some of the most common passwords or password patterns.

Realizing the relatively limited number of passwords in use, hackers can easily break into many supposedly secure accounts just by trying the most common passwords. With the prevalence of fast computers, speedier networks, and special software programs, hackers have developed programs that fire off tens of thousands of passwords per minute in their attempt to find the right one without ever touching the keyboard. The previous common thinking was that password guessing is a very

TABLE 25.1
Some not-so-inventive password choices!

Same Digits	0000, 1111, 2222, …etc.	7.4%	"1234" most used four-digit	11%
Years	from 1900–2012	7.3%	"1111" second most popular	6%
2580 or 85454	center of keyboard	3.0%	"0000" third	2%
Qwerty	top of keyboard	3.0%	Racecar same backwards	.02%

time-consuming method of hacking since there are so many combinations of name-and-password combinations but that is no longer a problem. Websites have tried to thwart attackers by freezing accounts for a certain period of time if too many incorrect passwords are typed. It didn't take the hackers too much time to overcome this practice by making guesses at an acceptable and unsuspicious rate.

The reason why 30% of people continue to use the same password in all situations is thought to be because they are overloaded by the sheer number of passwords that they have to remember in the digital age. It is estimated that the public is now expected to keep ten times as many passwords in their head as they did less than ten years ago … including ATM PINs for each bank account or credit card and different passwords for eBay and Amazon. Using a different password for each website visited and memorizing all the passwords is almost impossible. In reality, it is suggested by most experts that it is ideal to have unique passwords, use capital letters along with symbols and numbers, and – when possible – use 2-step authentication, where a pin code is immediately sent to the user in order to confirm their identity.

Social networking and entertainment websites attempt to make life simple for users and are reluctant to place too many controls, with some exceptions:

- A few sites prevent people from choosing common passwords.
- Commercial sites like eBay have to weigh the consequences of freezing accounts, since a hacker could try to win an auction by freezing the account of other bidders.

It may be difficult to accept, but the password "MylasydogRex" is much harder to hack than "H7%doss" because a program that cycles through every possible combination of characters could crack the eight-character string of gibberish in less than four hours. The 12-character plain English phrase would take 317 years to crack. A single high-end computer running a cracker program can process 17 billion combinations in an hour. Some basic rules in selecting a password:

- The longer the password, the harder it is to crack.
- Use passphrases rather than passwords.
- Length is more important than complexity.
- Created different passwords for different sites (rarely done).
- Change passwords every few months (also rarely done).
- Use a random selection of words with intentional letter swap-outs with numbers, i.e., a "3" instead of a capital "E."

If you're curious, feel free to visit NordPass to view 200 of the most common passwords – i.e., passwords to *not* use – at: https://nordpass.com/most-common-passwords-list/.

21 DEARLY DEPARTED PASSWORDS

A significant portion of everyday life is now lived online – bank statements, stock trades, and email archives are all available with a few keystrokes, some requiring complicated passwords to access the information. The Internet has made life less complicated in some ways, but when people die, their passwords often go with them, along with access to their digital assets.

There are paid services that can provide "digital" estate planning. Some estate planners criticize the concept of Internet password storage since they can be compromised one way or another and are no more secure than writing the passwords down on paper and informing the executors or family members where the list is located. Having said that, there are now several apps that can securely store account passwords and automatically retrieve them whenever the user opens a website requiring a password.

When someone dies before leaving their heirs a list of usernames and passwords for their online accounts, the procedure to retrieve the information varies widely. For example, Yahoo will make email records accessible to legal heirs if contacted by the estate planner, who must go through a complex procedure. The policy of this particular service is to get users to agree to a non-transferability clause and if a representative of a deceased user contacts Yahoo, it may close the account. Each Internet-based service treats the passing of a subscriber differently. Facebook memorializes accounts when users pass away, making the account private so that only confirmed friends can see the profile and leaving a space for friends to post remembrances and preventing anyone from logging in. Google sometimes allows an authorized representative user to access email but first, it requires a death certificate and other information; if it gives the go-ahead, the representative would need to get a court order or submit more materials. Some people have split their assets into three categories: the house, the bank account, and the list of passwords.

26 Network Oddities

1 COMPUTER MISAPPLICATIONS

Power consumption is rarely discussed, but the fastest way to lower a PC's power consumption is to, obviously, shut the unit off after an eight-hour workday, providing 16 hours of savings each day. In addition, the terminal/monitor should be set to go to sleep after periods of inactivity, and different parts of the system can be set to sleep at different times. Setting the energy-saving preference to put the hard drive to sleep after 15 minutes of inactivity is an acceptable benchmark; the entire computer should be set to go to sleep after 30 minutes. The laptop alone, turned off but plugged into the wall, will cost about $9 a year. Cell phone chargers that aren't connected to a cell phone cost 14 cents a year. With hundreds of millions of chargers in circulation, that seemingly low figure can add up to huge losses.

PHANTOM POWER is an industry term that refers to the fact that even when a desktop PC or laptop is shut down, that doesn't automatically mean that the units are not using energy. Many electronics use power even when they are turned off. One method to stop this power loss is to plug the units into a power strip that senses which device has been turned off and then cuts power to it. It is a procedure that should be used to replace screen savers, which are throwbacks to the days of really old-school cathode ray tube monitors.

Piggybacking is a practice that started innocently enough with individuals adding their spouses or children to their credit card accounts as authorized users. This practice has been embellished by companies offering to add a person to four separate $20,000 credit lines with ten years of a "perfect payment" record for a fee of $4,000. This would add over 200 points to a person's credit score in 90 days, making the difference between qualifying for a lower home mortgage loan or not. Another ingenious concept is one that ostensibly hires someone as an independent contractor, providing a paycheck stub showing an "advance" with the corporate name and address (for $55). Another $25 will assure telephone verification of employment if a lender calls to check. When confronted by authorities, these websites claim that their products are used for "entertainment purposes only."

2 CHEATERS, ADULTERERS, AND PHILANDERERS

Alibis can be easily obtained through a variety of services offered on the Internet. These services can provide foolproof alibis for any indiscretion, for a fee, of course. With the proliferation of surveillance cameras, EZ Pass, FBI Data Banks, phone and global positioning system (GPS) tracking programs, and bus pass cards, it has become difficult to hide surreptitious behavior from prying eyes.

If someone needs to prove he or she attended an out-of-town seminar, the "network" can provide records of hotel stays, seminar programs, and souvenir pens, hats, and tee-shirts sometimes offered at these events. Also, appropriate charges for restaurants (actual and fabricated), sports event tickets for any date requested, and even a virtual hotel number that will be answered by a virtual front desk clerk, greeting the caller with the appropriate hotel greeting message and handling the call according to specific instructions. A virtual travel agent will confirm the travel arrangements (that never occurred) on a cruise ship that never sailed, etc. These fictional businesses claim to be legitimate because the fake bills represent invented companies, hotels, and restaurants and do not require an actual doctor or the need to forge real documents. The company can buy illicit gifts for the client so that suspicious purchases of flowers, perfume, or chocolate shops don't appear on the client's bank statement or credit card invoices.

DOI: 10.4324/9781003091103-36

The service can also create jobs for the unemployed, including a virtual office with realistic-sounding office assistance and impressive-looking business cards. It sends emails from anywhere in the world from an individual's own email account and receives phone calls from anywhere, routing them to anyone in the world, with any number showing up on the caller ID of the receiving party.

3 INFIDELITY NETWORK

The desire to expose cheaters has created yet another cottage industry of online entrepreneurs who attempt to overcome alibis by offering support services more aptly described as a vindictive outlet for victims of cheating lovers they meet on or off the Internet. Such sites have proliferated in the hope of stopping cheaters by exposing them. The sites provide a gaggle of resources to identify con artists using a safe and anonymous space to commiserate with other potential victims. The websites post photographs and descriptions like "charming, pathological liar, an electrician by trade, lots of promises before taking your assets." There may be some legal exposure for listing these details, but no such site has as of yet been forced to defend against privacy violations. Dating services have no incentive to police their members. The Communications Decency Act absolves Internet service providers of liability because the sites are not considered the publishers of the information on their pages.

Text messages have replaced the lipstick on the collar and the mislaid credit card bill as the new telltale sign of adultery. Instantaneous and seemingly casual, text messages can be confirmation of a clandestine affair, a record of those who sometimes forget that everything digital leaves a footprint. In 2009, a Detroit mayor, a Nevada senator, and even Tiger Woods were all exposed by text messages that were never erased or destroyed. Unlike earlier eras when a dalliance might be suspected but not confirmed, the always-present text messages provide proof. Most don't realize that text messages remain somewhere in the network even if they were deleted. Communication companies store messages anywhere from days to a few weeks.

4 CONVENTIONAL DATING SERVICES

One of the fastest-growing sectors on the Internet since the turn of this century has been conventional dating services, and there is no indication that the growth of these virtual meeting rooms will slow down in the near or distant future. Online dating has become the modern-day matchmaking process that older generations sought out when they became old enough to marry. Even those who at one time would never have dreamed of pursuing romance online have entered the arena. Finding a soulmate is often called a numbers game using statistics supplied by such sites as Match.com, Chemistry.com, eHarmony, JDate, OKCupid, and hundreds of others. Today, there are probably 1,500 services arranging romantic meetings for people from all over the world. Estimates are that 350–450 million people worldwide visited Internet dating sites in a given year.

Dating sites have changed traditional dating and courtship practices, and online dating is now the most popular method of matchmaking. Internet dating services have become a multibillion-dollar industry.

The dating service True.com was the first major online dating service to screen members to determine if they were married, felons, or sexual offenders (about 2% are rejected for membership due to the check). While this process sounds simple, most online dating companies question whether such checks can be effective. The checks rely on data provided by state and county databases, which are often incomplete and the checks, therefore, give daters a false sense of security. Some counties do not keep digital records, and many others do not provide data about sex offenders.

Mobile dating has added a twist to online dating. Dating apps have taken dating to a whole new level and have refined an industry. Just as other industries have gone from laptops to mobile devices, so has online dating.

One of the biggest dating sites claims that on average, 550 Americans get married each day after meeting on its site, representing 5% of all new marriages in the United States. Many experts believe that the online and mobile dating industry is more than just a business, and outside of the medical industry, it will have more impact on society than any other technology developed during the digital era.

Tinder was launched in 2012, and instead of attracting members interested in marriage, it grew into a "hookup" site. From its inception, this app has been downloaded more than 630 million times. This hugely popular program requires users to create a profile that includes six photos and a bio. A user swipes on the left of the profile for "Nope" or on the right for "Like." If two users mutually "Like" each other, they are alerted to their "Match." Dozens of dating apps have since followed Tinder's format.

5 AVOID SLEEPING WITH COUSINS – AN APP

In the small island of Iceland, with its homogenous population of nearly 400,000 people, citizens share descent from a group of 9th-century Viking settlers. There is an online database of genealogical details for nearly the entire population, including family trees stretching back 1,200 years. Kissing cousins is a real risk and was the motivation for three engineering students at the University of Iceland to develop an app called Íslendinga-App that helps Icelanders determine whether a potential dating partner is too closely related or not before entering into a relationship. It allows any two Icelanders to see how closely related they are by simply touching iPhones. If the "Incest Prevention Alarm" is activated – the two are close cousins and should end their conversation. It has proven to be a very successful program that solves a uniquely Icelandic form of embarrassment!

6 WIKILEAKS

Exposing government and business secrets on the Internet is a violation of some law, according to a federal judge who ordered the disabling of a website devoted to disclosing confidential information. The site wikileaks.org, founded in 2026 by Julian Assange, gained global prominence – and in government circles condemnation – for publishing classified and sensitive documents from anonymous sources. It invites people to post leaked materials with the goal of discouraging "unethical behavior" by corporations and governments. It has previously posted documents that showed the movements of American troops in Iraq and other evidence of what has been called corporate waste and government wrongdoing. The site posted top-secret communication cables from different federal agencies concerning relationships with foreign countries and their leaders in 2010 that embarrassed and compromised relations with allies and enemy nations.

While it is most famous for its 2010 release of US military and diplomatic cables, it also played a supporting role in the 2013 Edward Snowden revelations. While Snowden primarily worked through journalists, WikiLeaks did facilitate his asylum process in Russia and later published additional documents from his cache. These disclosures exposed widespread global surveillance programs, sparking debates about privacy, government transparency, and national security.

Currently, Julian Assange remains embroiled in legal battles. With a decade of asylum in Ecuador's London embassy (2012–19) and then subsequent imprisonment in the United Kingdom, he faces extradition to the United States on charges under the Espionage Act for WikiLeaks' 2010 publications. He remains a polarizing figure, with supporters viewing him as a free-speech martyr and critics accusing him of endangering lives. WikiLeaks continues to operate but with reduced activity amid financial and legal pressures.

7 DEBIT CARD FRAUD

A series of cases surfaced in recent years where bank accounts were looted after the victims had used their bank cards at a chain store retailer. Thieves used electronic devices called skimmers that allowed them to record information from shoppers' debit cards and steal their personal identification numbers or PINs. Once the ID was stolen, the thieves were able to use the data to create duplicate debit cards and use them at automatic teller machines to steal money directly from victims' bank accounts, primarily in withdrawal amounts of $500 from ATMs on the other side of the country.

Fraud involving debit cards, PIN numbers, and card processing equipment has increased greatly in recent years, according to retail theft prevention organizations. A type of credit card embedded with a microchip instead of a magnetic strip is considered far more secure. Long standard in Europe, major US retailers have largely made the shift to chip reader cards, with the mag strip versions being gradually phased out.

Losses from US card accounts – both bank and credit – totaled approximately $12.5 billion in 2024, with over 70% as card-not-present transactions. The volume of fraud attempts denotes a more than 45% year-over-year increase.

8 ID THEFT DESTROYS CREDIT RATING

Unsuspecting credit card holders have lost millions of dollars all while having their credit rating destroyed when attacked by ever-resourceful ID thieves. This criminal activity has also created new and profitable opportunities for financial institutions that offer some form of credit card loss protection service. The credit card companies make billions of dollars from the 700 million credit cards in circulation by charging penalty fees for late payment which, in some cases, may have been caused by ID theft. The major credit bureaus such as Equifax, Experian, and TransUnion, along with direct marketers and banks, reap enormous profits from the millions of Americans who subscribe to the monitoring services. Identity theft has become a big business and many analysts estimate credit monitoring alone to be an estimated $5+ billion industry, which is remarkable for a business that did not exist until 1995.

Bank robberies have steadily declined, perhaps because would-be bank robbers became more focused on the ease and profitability, of stealing someone's identity.

It should be noted that purchasing credit protection from a monitoring service is no guarantee that a violation will be detected. An experienced fraud artist is capable of using someone else's personal identification – like a Social Security number – to take out a loan in their own name. This data mismatch can very well cause the bureau's computer system to route a loan request to a separate file so that the credit monitoring service never picks it up. Privacy advocates are in favor of providing more fraud prevention tools to consumers by allowing them to freeze access to credit records as a precautionary measure.

9 STOLEN CREDIT CARD NUMBERS SOLD FOR $1

Stolen credit card information eventually floods into underground hacker forums where customer names and addresses are sold for as little as $1 each.

The price of a stolen card is based on how recently the card was obtained and is highest before the victim knows they've been breached and cancels the card. The most sophisticated criminal hackers hide their identity on the "Dark Web," using a router known as The Onion Router ("TOR") to conceal the location of the computer servers hosting the websites. TOR ensures secrecy by randomly routing computer messages through several places on the Internet so no single point can link the source to the final destination. This makes the sites nearly impossible to trace. Criminals can break into company databases with malicious software purchased online from computer hackers.

The software is then used to infiltrate a database, spread its code like a virus, and remain undetected for months.

> Once the information is collected, the hackers test and sort it into bundles that are priced before being sold on the underground sites. Bundles range from 500 cards to 10,000 cards, and to ensure that the cards work, thieves use a system that automatically charges a small amount to 10,000 cards at a time. The tests determine the card's validity and credit limit. Cards with the highest credit limit, such as an American Express Platinum card, sell for the most money. A card with a low credit limit might sell for $1 or $2 as compared to $25 for the Platinum card. Regardless of the limit, the purchasers of stolen cards must move quickly before the victims notice fraudulent charges and call their banks to cancel the cards.

10 CELL PHONE DATA AS WITNESS

There was a surge in law enforcement's use of cellular tracking techniques to keep tabs on suspects before they are arrested to build criminal cases against them by mapping their past movements. Since the beginning of this century, investigators have been able to match an antenna tower with a cell phone signal to track a phone's location to within a radius of 200 yards in urban areas and up to 20 miles in rural areas. Global positioning technology makes it possible to pinpoint a user's position with much greater precision… down to a few dozen yards.

In the past, the only way to track a suspect's phone was to use company records that showed a phone's approximate location at the beginning and at the end of a call. To be able to track the same suspect in real time, law enforcement officials had to ask the phone company to "ping," or send a signal to, a phone; for the effort to succeed, the phone had to be turned on, though it did not have to be in use.

11 CIVIL LIBERTIES

Courts often allow federal prosecutors to track the location of cell phone users without a search warrant and in some jurisdictions; the data has been used in criminal investigations. Civil liberties groups obtained data from the Justice Department (as the result of a lawsuit) and complained that the information was an infringement of the privacy rights of citizens. Existing laws do not provide clear or uniform guidelines. Federal wiretap laws, outpaced by technological advances, do not explicitly cover the use of cell phone data, and local courts seem more confused than the public in deciding how to handle the subject.

Internet services, like Google, require a search warrant before releasing GPS data pinpointing the movements of customers who use its mapping applications. Investigators seeking warrants must provide probable cause that a crime has been committed, but that requirement is often ignored. The practice of many investigators is to obtain cell-tracking records under lower standards, which are granted routinely based on arguments that the information requested would be relevant to an ongoing investigation.

Organizations dealing in civil rights protection do not oppose using cell phone surveillance to solve crimes or save people in emergencies, but they worry that the legal gray area is enabling it to happen without much scrutiny or discussion. The five largest carriers receive thousands of requests a month from law enforcement. A Federal District Court ruled that a search warrant was required even for historic phone location records, which the government had requested to track drug suspects. The ruling established the fact that people have a reasonable expectation of privacy regarding their physical location.

The American public would be surprised to learn that the government can access their phone conversations without showing probable cause. There are a growing number of private advocacy groups supporting the ruling and laws that are needed "to keep Big Brother out of individuals' pockets."

Section X

Built-in Obsolescence

Popular devices that were once essential parts of everyday life, but are no longer in use or about to disappear, include:

- Wall-mounted, cord-connected home telephones replaced by cell phones and VoIP
- Teletypewriters and calculators
- Hand-delivered mail by bicycle messengers; replaced with FAX machines
- Mimeograph machine replaced, by FAX machines, replaced by email
- Single-use roll film cameras and digital throw-away cameras replaced with iPhones
- Record turntables turned into iPods (though this one is one of the few trends that has rolled back a bit out of nostalgia!)
- Rear projection TV became flat plasma and LCD screens
- Pagers into smartphones
- Incandescent bulbs replaced with compact fluorescent and LEDs
- Computer keyboards into touch screens
- Physical roadmaps into GPS
- Human contact was replaced by social media (Twitter, TikTok, Instagram, Snapchat, and Facebook)

Americans purchase millions of electronic devices each year, about a third of which are cell phones that are discarded and replaced within 13 months.

While the IBM typewriter is a long-forgotten relic of ancient times, like the rotary phone and mimeograph machine, IBM seems to adapt to changing technology and has made up for the loss in the typewriter business by developing the following devices:

1945 First calculator	1971 Floppy Disc
1946 First artificial arm	1972 ATM
1961 Magnetic stripes on credit cards	1974 First UPC bar code stripe – Juicy fruit gum
1961 Selectric typewriter	1980 Laser for LASIK eye surgery

DOI: 10.4324/9781003091103-37

Other Relics

Mimeograph copy machines	Beepers
Sony Walkman	Palm pilot
Commodore 64 computer	Atari
Reynolds lifetime ballpoint pen	Corded phones – wall mounted
Rabbit ear antennas	Yellow pages
White Pages	

27 Limited Life Cycle

1 ELECTRONIC GADGETRY: RADIO (HI-FI) TO WIRELESS (WI-FI)

"Neophiliacs" can be used to describe those who line up outside Apple stores days before the newest iPhone is released. The actual definition is the "love of novelty" and is more than 100 years old – the first citation appeared in the Oxford English Dictionary in 1899. The opposite meaning word is "neophobia," a fear of the new and three years older than "neophilia." This affliction forces people to chase the new at all costs.

Accumulating so-called gadgets has become an American obsession; consumers feel compelled to have the latest and greatest devices with the most buzz as soon as they're introduced and re-introduced with a new model every year (iPhone 16, iPhone 16e, iPhone 16 Pro, iPhone 16 Pro Max, etc.). The result is a growing pile of gizmos in need of disposal, including computers, TVs, cell phones, and other gadgets that make up the fastest-growing municipal waste stream in US history, according to the US Environmental Protection Agency (EPA). As much as 80% of electronic waste goes out with the trash, while only about 20% is properly recycled. Recycling consumer electronics has created a $350 billion revenue-producing industry, according to a US Recycling Economic Information Study commissioned by the EPA.

Most smartphones are retired because of psychological, not necessarily technological, obsolescence, simply because everyone wants the latest gadget. It is estimated that 60% of phones are replaced because people get tired of the design which translates to close to 150 million cell phones being discarded every year. People are trading in their phones in less than two years on average, but phones have a useful life of four or five years; the device is, after all, a minicomputer.

Consumers want the new apps – even if there's no real need for them and some are too complicated for the average user. Selecting the most effective phone can be daunting since there are roughly 270 cell phone models for sale in the United States, with about 16 new ones being introduced each month. According to the US Geological Society, of the estimated 150 million cell phones taken out of service each year, there are now more than half a billion old phones sitting in American drawers and closets. These and other abandoned electronic devices are not being properly recycled. The EPA awarded a five-year, $2.5 million grant to the United Nations University (an international community of scholars) to develop a more effective way to measure e-waste, address concerns about illegal e-waste shipping, and assess the routes through which used electronics exit the country. There are standalone kiosks that look like ATMs that automatically buy back old electronics. The cell phone is placed in the large stall and within a few seconds, the kiosk recognizes the model from a constantly updated database and determines a value based on the condition. The software determines the condition by testing the screen, buttons, and other functions once the unit is connected to the kiosk via a cable. It will then spit out the cash just like an ATM, but only after a driver's license is inserted to authenticate that the phone wasn't stolen and that the person is over 18.

2 PILES OF OLD ELECTRONIC GEAR

E-waste has become a growing problem that is not easy to solve and will continue to get worse as consumers buy new devices and abandon older devices that are considered obsolete after only a few years of use. Many Americans accumulate outdated technology in their homes, especially after Christmas when the average consumer spends hundreds of dollars on electronics gifts on average, according to the Consumer Electronics Association. That trade group claims that the public still has as many as 100 million analog TVs to dispose of and the average household has approximately

DOI: 10.4324/9781003091103-38

25 electronic devices that have a life cycle of six to eight years before being discarded. Americans own 3 billion electronic products with a turnover rate of 400 million units a year that could cause an environmental hazard. Cell phones contain $300 million worth of dangerous substances that could easily leach into the groundwater if they were to end up in landfills. Among other toxic materials found in consumer electronic devices are the lead in older TV tube glass and soldering and the mercury in older LCD displays.

In 2005, Panasonic, among other manufacturers, stopped using lead and other hazardous materials in the production of their TVs after about two dozen states passed laws mandating the recycling of electronic waste, according to the Electronic Takeback Coalition, which promotes responsible recycling. All of the laws require manufacturers to pay for the recycling of their products and will eventually have to make provisions to accept returns and destroy the devices without polluting the ground. Meanwhile, cell phone service providers accept used phones, and many retailers like Target and Apple offer device trade-in programs.

A small percentage of old cell phones still work and are donated to charities or distributed to the poor, but a growing number of companies have tried to fill the void as to how to discard the units without polluting. This pollution problem has created a cottage industry of firms salvaging old phones, creating a multimillion-dollar business dealing specifically with mining these phones for their recoverable materials, not for their communication capability. One firm, Gazelle.com, buys, refurbishes, and resells older electronic devices and disposes of those devices that are worthless. Companies such as LG, Panasonic, Samsung, and Sony have joined the Consumer Electronics Association's eCycling Leadership Initiative, with the goal of recycling billions of pounds of electronics annually.

Most states have made it illegal to simply throw away computers, printers, and TVs, with 17 states banning electronic waste from landfills. The laws in these states require electronics to be recycled so that their toxic materials don't leach into groundwater. The trouble with environmental laws relating to toxic appliances is that the onus is on the users/owners/consumers, and it's on the honor system; there are no green police officers to search a household's garbage for the stuff that doesn't belong in the trash. Electronic gadgets contain potentially dangerous chemicals like lead, mercury, and cadmium, and currently only a fraction, between a quarter and a third by some estimates, of these discarded devices get recycled. The problem will persist until the public is educated about the dangers in discarded electronics, and until then, banning the burying of the devices will not be effective.

3 COLLECTORS' ITEMS

MIMEOGRAPH COPY MACHINE: Using carbon paper to rapidly print out announcements, mimeograph copy machines were a mainstay in all school principals' offices but have all but disappeared. The word "mimeograph" was first used in a license granted to a manufacturer by the inventor of the device, Thomas Edison, in 1887. This once ubiquitous copy machine, which gave the user blue fingers from the ink coming off the carbon paper, was also known as a stencil duplicator, a low-cost printing press that worked by placing a stencil in a typewriter after the typewriter ribbon had been disabled in order to "cut the stencil" by using the exposed sharp type element to strike the stencil directly. The impact of the element displaced the wax, made the tissue paper permeable to the oil-based ink. Once prepared, the stencil was wrapped around the ink-filled drum of the rotary machine.

When a blank sheet of paper was drawn between the rotating drum and the pressure roller, the ink was forced through the holes in the stencil onto the paper. It worked something like a squeegee and it often smeared. Mistakes could be corrected by brushing them out with a specially formulated correction fluid and retyping once the fluid had dried (a forerunner of "White Out"). The carbon copy machine that our forefathers were familiar with has since joined other relics of a bygone era.

TYPEWRITERS were once found in every office and in many homes; a few are still being made today but only in very small quantities. People under 25 years of age may never have seen one. The first typewriter that would become commercially successful appeared on the market in 1874 in very limited quantities and was difficult to maintain. The next two decades were marked by intense experimentation, with hundreds of devices coming on the market. In 1894, one model offered a unit with no keys at all, requiring the user to slide a lever along a rail to select the type wheel, which proved to be too complicated. By the early 1900s, typewriters settled into a more standardized design and by the end of World War II (WWII), entire sections of the telephone yellow pages listed typewriter repair services – none exist today – even the yellow pages have practically disappeared. It is (was) called a typewriter because it used a keyboard, like a modern-day computer, but was manually operated until the late 1950s, when typewriters started to become electrically powered, peaking with IBM's most successful model, which the company called the "Selectric." It became the leading model in the industry, replacing manual units and allowing a soft touch of each key to print out the letters instead of requiring the keys to be hit hard to move the letter through the inked ribbon to the roller holding the paper. It made manual typewriters obsolete and was the forerunner of the computerized word processor that wasn't introduced until two decades later.

The typewriter was once used to address envelopes during times when people *actually* wrote letters by hand. The once-popular typewriter is a device now being sold at vintage auctions for high prices, especially units that were used by authors and celebrities who insist on using this device to type manuscripts and novels. Yet the reported death of the typewriter may have been a little premature since there still exists a small boutique-type market for the typewriter in places as diverse as police station houses and since 2011, in prisons. In the near nonexistent typewriter business, one American firm enjoys an almost monopolistic position in the field. A New Jersey company called Swintec still sells several thousand machines a year. The company developed a new style typewriter, one that is perfect for the prison market – it's transparent and prevents smuggling of contraband.

WORD PROCESSORS eventually killed typewriter production throughout the world and were the short-term, short-lived bridge between the typewriter and personal computers. Word processors did allow for the cutting and pasting of text and other features not able to be performed on typewriters, but that is understood by computer users around the world. The limited speed of printing and their single-function utility (document editing and production) gave way to the ability of computers to run various programs – not simply word processing, but games, art editing software, Internet browsing, and more.

4 INNOVATIVE BELL LAB

Some of the accomplishments of one of the most innovative scientific organizations in the world were the result of ideas developed by the scientists and engineers at the famous Bell Laboratories in New Jersey, a division of AT&T. That organization was best described as an ivory tower with a factory downstairs and much of the research conducted at that lab over a half-century ago benefitted society well into the 21st century. The following list of achievements is by no means complete, but it does indicate how one organization can produce ideas that influence the way people live and how technology can change how the world communicates:

4.1 1920s

1925 First high-fidelity sound recording

1925 Fax service successful telephotography system in the United States – and early data network 1927 Negative feedback amplifier – cut distortion in long-distance telephony

1927 Long-distance television transmission – live television images of Herbert Hoover sent via phone

4.2 1930s

1932 Radio astronomy – easiest detection of radio noise from the Milky Way galaxy

1932 Wide-range audio reproduction using three bands for bass woofers, midrange horns, and tweeters

1933 First Transmission of stereo sound – symphony concert broadcast live over telephone lines

1937 First electronic speech synthesizer – recreates human speech

1939 First binary digital computer – using on-off relay to determine quotient of two eight – digit numbers

4.3 1940s

1940 First long-distance computing – computer operation in NYC via teletypewriter in a remote location

1946 First commercial mobile phone service – three subscribers per city could make calls at one time

1947 The transistor – a landmark invention that replaced vacuum tubes and transformed electronics

1948 Information theory – enabled data compression and cryptography

4.4 1950s

1951 Direct 0 distance dialing – no operator necessary for long-distance calls

1954 Solar cells – first use of the Sun's energy to create a practical level of electricity 1956 First transatlantic telephone cable – could carry up to 36 simultaneous calls 1957 Digitized music – first computer-synthesized music

1958 The laser – crucial for communications and DVD technologies

4.5 1960s

1962 Digital transmission switching – first transmission of voice signals

1962 First communication satellites – Echo reflected voice signals from coast to coast

1962 Paging system – Bellboy pager introduced

1963 Touch-tone telephone – enables voice mail and call service

1965 Evidence of Big Bang – discovery of radiation beyond the Milky Way

1969 Charge-coupled device – solid-state chip transformed patterns of light into information

4.6 1970s

1970 UNIX operating system – makes large-scale Internet networking practical

1976 First fiber-optic network – first test of information to be carried by pulses of light

1978 First commercial cellular network – installed by Bell Labs in Chicago

1979 Digital signal processor – an essential component of cell phones, modems, PCs, and video game systems

4.7 1980s

1980 Digital cellular phone – better sound quality, greater channel capacity, lower cost

1982 Fractional quantum hall effect – discovery of the new state of subatomic matter – wins Nobel Prize

5 IBM (OR BIG BLUE)

There will always be people who resist adopting new technology and devices like the Kindle, iPad, iPhone, etc. Some of these people are collectors, or antique dealers, who retain a fondness for old equipment and devices. One entrepreneur attempted to profit from this attitude and integrated the latest electronics wrapped in a retro design. Retro designs seem to appeal to people who want to retain a sense of permanence to disposable items and some of these reproductions are considered art.

Another example of this loyalty to old technology is manual typewriters reconfigured to work with a modified keyboard connected to a computer USB port key that sells for about $800. The device involves a refurbished typewriter wired with a sensor board that recognizes when a key is pressed. A USB plug makes the standard typewriter work like any computer keyboard. Even if the type bar doesn't hit the platen, a computer recognizes the input, but if the keys are banged hard enough, it will make an old-school hard copy on paper while a computer also records the keystrokes.

Technology trends are so fast-moving that many companies rise quickly only to flame out – outpaced by rivals and often winding up in bankruptcy court or being absorbed by another company (Holmes Security, Mosler, Pinkerton, to name a few). While IBM has followed many winning patterns, it hasn't been infallible. It had troubles with its computer product and eventually, IBM sold the business to the Chinese firm Lenovo. Staying on top of any industry is a feat and IBM is one of the few major US firms in the computer and related technology fields to last 100 years.

The general trend in technology is that each new device or system introduced is better (or at least should be better) and less expensive than its predecessor – similar to the marketing approach successfully used in the automobile and appliance industries for over a century. Beepers became essential in most businesses during the 1980s and 1990s and kept doctors tethered to patients and government officials connected to underlings; at the peak of their popularity, everyone carried a beeper. These once ubiquitous devices have become technological fossils and faced eventual extinction as the world rushed to cell phones.

6 FALL OF AMERICAN COMMUNICATION ICONS

In the 1930s and 1940s, before television became a household item, the only entertainment available to the average household was enormous, four-foot-high radios. These radios served as entertainment for any household that had one. The radios were so heavy that any service calls required a technician to make an in-person house call to repair them.

Beepers offer an example of an intermediate product, a rarely used term describing devices from older times, like the floor radio pictured on the left, the first portable telephones that were the size of a brick with a storage battery that lasted minutes, not hours. Just as people have been obsessed with antique cars and typewriters, pagers may continue to be used by some collectors until the end of time, but at a certain point, there will be no one around capable of repairing them. There were about 45 million pagers in use nationwide as recently as 2000; the total in 2023 is down to about 2 million.

BEEPERS have become more of a business device rarely used in the marketplace. They are often used as pagers in hospitals because, unlike cell phones, pager signals reach into buildings without concerns about interference with medical equipment.

PAGERS were introduced in 1995 and are more or less the same as beepers. Pagers were initially attached to those who worked in fields where instant communication was essential which usually meant doctors, though the group expanded in the late 1980s to include other trades. The early beepers displayed only numbers, giving rise to a numerical lexicon that included codes like 911 (call me back immediately) and other recognizable symbols. The pager's fall from popularity was attributable to the mobile phone and in particular the introduction of the BlackBerry, which once commanded more than half of the American smartphone market.

SONY WALKMAN: Sony created "personal audio" in 1979 by popularizing transistor radios, the compact disc, and the famous Walkman. These innovative ideas pioneered an entire category of devices that helped the company become a technology innovator in the 1980s and well into this century. Sony acquired record labels and even movie studios to enhance its image as a technology innovator. For many years, Sony defined Japan's technological prowess and seemed invulnerable by impressing the world with not only the Walkman, but also the Trinitron TV and overwhelmed Hollywood with bold acquisitions such as Columbia Pictures. However, by the beginning of this century, the high-flying company started to fade dramatically because it ceased to create new innovative products. Sony's gravest mistake was that it failed to adapt to technological advances, the shift toward software, and the growing importance of the Internet. The company also lost its dominance in two crucial categories, television and portable devices, and was late to the market with digital music players and flat screen TV.

When the digital revolution hit, Sony was too focused on its proprietary formats and fell far behind the engineering genius that brought the world the Apple iPod and iPhone. Other past leaders in the communication industry such as Polaroid, Kodak, Commodore, and Atari made the same mistakes, allowing upstarts like Microsoft and Apple to revolutionize the marketplace.

7 POLAROID INSTANT CAMERAS

In the late 1940s, instant cameras were born by placing a darkroom inside a handheld camera, giving Polaroid a distinct advantage over traditional film cameras.

Edwin Land invented the instant camera after dropping out of Harvard. After starting his company, he won government contracts during WWII for such products as infrared lenses and night-vision glasses. Then in 1948, the Polaroid company introduced a four-pound camera, calling it "instant" even though its sepia-toned photographs took more than a minute to develop. The camera sold for $89.50, the equivalent of almost $1200 today, and within eight years over 8 million cameras were sold. The leading conventional camera manufacturer, Kodak, did not view the instant camera as a serious competitor and treated it as a gimmick. In fact, Kodak was, for two decades, the manufacturer of Polaroid's film.

In 1965, Polaroid made a great leap into the camera mass market for still picture cameras by introducing the Swinger model which weighed only 21 ounces and produced a B&W picture in 10–15 seconds. Once the picture appeared, it was to be coated with a chemical to keep the image from fading. That camera was followed by a color, single-lens reflex camera in 1972. By 1980, Polaroid was selling 7.8 million cameras a year in the United States – more than half of all the 15 million cameras, instant and traditional, sold that year. (Figure 27.1) This sudden growth finally convinced Kodak to develop an instant camera and Polaroid sued Kodak for patent infringement. In 1985, Polaroid won a major patent-infringement suit, forcing Kodak to abandon its own instant-camera efforts. Kodak had to pay Polaroid $900 million and left the instant camera business.

The victory was short-lived, however, because neither Polaroid nor Kodak was able to recognize the rise of digital cameras. While consumers were rushing out to purchase video cameras, Land spent millions of dollars to perfect a system for silent 8-mm instant color movie cameras called Polavision. By 1980, digital cameras began appearing on cell phones, placing cameras in millions of pockets. Polaroid declared bankruptcy in 2001. After that time, Polaroid's brand and assets were sold off; that eventually led to a successor company. After changing hands several times, the assets were purchased in 2017 by Wiaczesław Smołokowski, a Polish billionaire. Since then the name has been licensed out to various makers of LCD televisions and DVD players and Polaroid instant cameras have made somewhat of a comeback over the past few years.

Kodak declared bankruptcy in 2012 and stopped making digital cameras, pocket video cameras, and digital picture frames… marking the end of an era for the company that brought photography to the masses for over a century.

FIGURE 27.1 The Polaroid SX-70.

Source: Wikipedia, Public Domain.

With the products being phased out, Kodak has focused on photo printing and desktop inkjet printers but will never achieve the status it once enjoyed before digital technology passed them by.

COMMODORE 64 rose to prominence in the 1970s and 1980s, producing an inexpensive, immensely popular series of computers credited with igniting the personal computer industry in much the same way that Henry Ford's Model T kickstarted the mass production of black automobiles. The Commodore company introduced the first PC in 1977 that sold a million units (1980), which led to the design of the second-generation model, the famous Commodore 64 that in 1982 offered 64 kb of memory. It became the industry leader and sold for half as much as the Apple TRS-80 unit. It was also the first computer with color graphics and an audio synthesizer chip. It sold 20 million units – four times the sales of the Apple II. That 64 model is said to have established the personal computer market. The total sales of all the Commodore models exceeded $1 billion by 1984 and Jack Tramiel, the owner of Commodore, was called the "anti-Steve Jobs" for focusing on the residential home market instead of the commercial market. The goal of the company was "to serve the masses, not classes."

A share of Commodore stock sold in 1977 for $2 and by 1984, it was worth more than $70,000, but the company no longer exists as digital technology and Apple passed them by.

PALM PILOTS (1997–2007) were dazzling when they first appeared: all your contacts, calendar appointments, and notes in one slim, pocket-sized device. A touchscreen, which required a stylus, made navigation easy and the software could be purchased through an online store. In later years, Palm added a telephone feature, creating an all-in-one gadget. Palm's reputation as a source of innovative hardware and software endured until January 9, 2007, when Apple introduced the iPhone.

ATARI 2600 (1977–84) brought video games into homes and popular culture. Over its lifespan, more than 30 million were sold… Then came the PC, which could not only play games but much, much more. Atari rushed to release games, assuming that its customers would play whatever it released; however, the public had other ideas and millions of unsold games and consoles were buried in a New Mexico landfill in 1983. Warner Communications bought Atari for $28 million in its glory days and sold it in 1984… for no cash.

RABBIT EARS describes the shape of a popular TV antenna that was first used in the early 1950s and became as widespread as the proverbial kitchen sink. They were gradually replaced by cable, Wi-Fi, Voice over Internet Protocol (VoIP) and eventually disappeared with the advent of satellite antennas and cable. On February 17, 2009, these rabbit ears (along with rooftop antennas found on the roofs of apartment houses) that guided television signals into 20 million American homes became useless – unless they were tethered to a device called a converter which captured the more modern digital spectrum. This change was set by Congress and the Federal Communications Commission in the mid-1990s to give viewers better TV pictures and made it easier for stations to broadcast their signals in high definition. The digital signal provides a far superior on-screen image than the older analog signal. The government set aside $1.5 billion to help viewers pay for the converters at $40 per rabbit ear. The rabbit ear manufacturers have slowly faded away as have eight-track tape recorders, Betamax, VHS units, and the many other recording and storage devices.

8 FAX TRANSMISSION IN JAPAN

The once-popular fax machine remains in the gray area between computer technology and the digital communication revolution. Every printer that connected to the computer terminal had fax capability and it replaced the need to use bicycle messenger services. Gradual advancements in the Internet Protocol address format made fax messaging unnecessary because of the easier-to-use email message. Technicians have noted that in almost every developed nation, the fax machine has joined answering machines, eight-tracks, and cassette tapes in the dustbin of obsolete technologies. Against this trend, Japan, renowned for its robots and bullet trains and some of the world's fastest broadband networks, has not yet rejected the fax machine. The fax concept was initially developed in Japan in the 1970s and while it ran its course all over the world, the fax machine can still be found in numerous Japanese government offices, business offices, and even private homes.

Japan seems to feel the fax is safer from being compromised by hackers and viruses through messages sent over the Internet and its love affair with the fax can be traced to the nation's economic heyday in the 1980s, when the machines became a household appliance more common than automatic dishwashers. Japan quickly dominated global fax production, making 90% of the tens of millions of machines built worldwide.

9 DISAPPEARING CORD PHONES

The traditional hard-wired house phone has rapidly become part of the American telecommunication industry's past. The cellular phone revolution and the ease with which computers connect to the network through Wi-Fi have eliminated any need for a hard-wired connection via the conventional phone line. Wired networks (cables) and wireless computer networks operate in a similar manner, with the obvious difference being that wireless networks use radio waves to transmit the signal instead of sending the signal through a direct connection that wire cables provide. Currently, nearly 75% of adults in 2024 live in cell-phone-only households, up from 10.5% in 2006.

Mobile cellular phones were not easily interchangeable when first introduced in the late 1990s and most major phone service carriers protected their domain. It was entirely at the carrier's discretion to permit or disable the factory-loaded features. The wireless service firms maintained complete control of their customers' phones and decided which software could be installed and how it might be used. In the United States, wireless carriers have deliberately hobbled their phone contracts to make the flight to a competitor costly, very costly. In most European and Asian countries, a customer can switch carriers in a few seconds by removing the SIM card from their cell phone and inserting a different one from a new service provider.

In the days when cell phones were inexpensive and could perform only one or two functions, they could be considered disposable, but with the introduction of smartphones, the cost has increased and now represents an investment, making people heavily reliant on their phones.

It should be remembered that in the 1950s, the landline industry resisted the opening of its networks with all kinds of threats and excuses. AT&T was fighting to exclude competition for their landline customers by arguing that it would be extremely difficult to furnish "good" telephone service if the user were free to attach the equipment or use with it all the numerous kinds of foreign attachments marketed by firms who had no responsibility for the quality of the telephone.

LIGHTWEIGHT WHITE PAGES The communication revolution caused the steady demise of the large 20 lb phonebook along with the wall-mounted cord phone. That ubiquitous book was, for over 50 years, found in every urban American household and apartment lobby. Musclemen have ripped them in half as a show of strength but now even a 90 lb weakling can rip up that same publication as it has lost most of its girth. As landlines are given up for cell phones, names disappear from the phone book.

- There are no "white pages" for cell phone listings. Americans are not eager to list their cell numbers in phone books. Privacy advocates balked at the idea in recent years when most of the big wireless carriers expressed an interest in compiling a nationwide directory. It seems that average Americans are very guarded about whom they want to call them. People may be gaining more privacy, but they are losing a sense of community.
- At one time, when people met someone, they would want to know where they lived and could look up their name in the phone book. There was also a certain ritual aspect to receiving the new phone book which took on the form of social visibility and social involvement, an experience that has almost disappeared.
- Some cities have banned the delivery of these heavy, thick, bound white pages (which appear in five-foot-high piles in apartment house elevator lobbies) unless the resident or business requests copies.

10 JAMMING CELL PHONES

The critical need to jam signals during a terrorist attack, to divert or cancel the cell phone communication network used by possible foreign belligerents, has become a major challenge to law enforcement agencies. The growing possibility of a crisis caused by scattered groups of terrorists has motivated law enforcement agencies to investigate techniques to disrupt communication between them.

There are many technological hurdles to shutting down cellular service in narrow regional locations, such as hotels or within a limited number of city blocks. Care must be taken in pursuing jamming signal systems since widespread shutdowns could hamper emergency personnel or prevent civilians from making emergency calls. There is also the consideration that there may be advantages to monitoring conversations between terrorists which may prove more valuable than cutting them off entirely. Electronic jamming of cell phones or of global positioning systems is complicated but possible (and may be in use by foreign military agencies). Cellular service in a large region, like a city center, could be simply shut down but more compact sites, such as an official motorcade, could be jammed by devices in the cars. It is far more difficult to accurately jam a wireless signal around a small defined area. It is even more difficult to jam a satellite phone than a cellular phone since the antenna is pointed at the sky and not to the nearest tower.

11 KILL SWITCH FOR STOLEN IPHONES

Tracking systems for lost or stolen iPhones have been available since 2009. Starting in about 2013, Apple had added the so-called kill switch that renders devices unusable after they have been reported stolen to the cellular service provider and by 2015, other major providers, Windows and Android, did the same. The procedure is similar to canceling a credit card by calling the bank. Data issued by various police departments worldwide reported that in the months before and after this anti-theft feature was installed, theft dropped 40% or more. Older models in circulation that did not have this feature experienced a 40% rise in thefts. Many state governments have either passed laws or are considering new legislation to make the kill switch mandatory within their jurisdiction.

Smartphone thefts are on the rise, particularly in large cities, where as many as one in three robberies involve a mobile device, according to the Federal Communications Commission. It's estimated that roughly 4 million mobile devices are stolen every year.

A kill switch is a term used to describe software that allows a smartphone owner to lock down the device, only to be reactivated with the correct password or personal identification number. The shutdown, in effect, wipes out everything… data, memory, contact numbers, etc., making it difficult to sell on the black market. The device becomes more like a paperweight than a communication device and has no resale value. Smartphone manufacturers have accepted the need for the kill switch and encourage jurisdictions to make the system mandatory, very much like seatbelts, the safety feature eventually legally required inside cars. California and Minnesota were the first two states to make the kill switch mandatory.

While the requirement for a kill switch is necessary to combat smartphone thefts, this policy has alarmed privacy advocates who claim law enforcement agencies could take advantage of this requirement to remotely shut down phones for public safety reasons.

12 EVERYONE KNOWS WHERE YOU ARE – ALL THE TIME

Dating back to the mid-2000s, it can be strategically assumed that every adult and child over the age of 14 carries a smartphone of some type and that fact allows anyone with a GPS reader to accurately identify the user and the location of the phone. Police have been following suspects in cars for many years and since 2010, Uber has been pinpointing the location of those seeking a car service and parents have been tracking the movement of their children as long as they carry a cell phone. These omnipresent devices are even being used by dog owners to record the movement of their dog walkers.

When someone orders pizza to be delivered, they can see the movement of their order in real time on a map on their phone. UPS, the world's largest shipping company, shows customers exactly where their packages are, and even cable and repair companies offer real-time updates on where the technicians or repairmen are. You can track where your Uber or Lyft drivers are en route to pick you up.

GPS has become a universally accepted and easy-to-use system, but it can also be used to record a person's movement that can be retrieved for future tracking purposes, which some consider to be a violation of privacy. Automobile GPS guidance programs such as Waze provide travel routes that avoid heavy traffic congestion by monitoring the movement of the car. These programs identify situations such as traffic jams, extended delays, and complete stoppages and offer alternative routes. In other words, the driver is not only following the best routes, but the system is also following the driver whether or not he or she is following the suggested route.

28 Perimeter Protection

1 FIRST LINE OF DEFENSE IS DE-FENCE

In the not-too-distant past, the first line of security used to protect sensitive areas bordering building perimeters was an eight-foot wire mesh fence with threatening barbed wire affixed to the top. Before CCTV systems became readily available, this ominous fence was a common sight up until the mid-1980s. The federal government issued numerous guidebooks and technical manuals describing perimeter site protection design features for foreign embassies and remote atomic energy plants, but the guidelines proved ineffective and only served to provide a false sense of security. Climbing over or under a fence was a serious problem until the fence was protected with a wide array of sophisticated motion sensors and detectors that emitted an alarm when an intruder attempted to circumvent the aboveground fence or the underground cable that surrounded the site.

Fence sensors vary from highly sophisticated electromechanical vibration-type sensors to natural barriers like rose bushes with their annoying thorns. Bushes are often used at gated communities and private homes that are reluctant to allow the installation of unsightly looking wire mesh or razor wire fences. Fences installed in affluent communities are usually more decorative (aesthetically pleasing) than the wire versions described. Newer barriers are designed to be unobtrusive and blend in with the surrounding terrain. Hollywood celebrities still demand huge gates and fences along with a security detail outside their fortresses. As seen in many movies, enclaves built to house Mafia Godfathers probably had the most effective perimeter protection barriers, since the more menacing the fence, the better suited to the crime bosses. They were not into aesthetics like movie stars.

The rich and famous now have gates attached to decorative chain link fences to block off the property with a guardhouse located inside the property. Properties can be furnished with state-of-the-art technology, including face recognition devices, Internet-connected digital cameras and remote iPad controls that can easily drive up the cost up to a million dollars or more. A decorated automatic driveway gate is the ultimate sign of affluence, but those same gates are also the trademark sign of the top 1% of the nation's millionaires, which attracts robbers. This group of high-wage earners can afford to hire a round-the-clock guard patrol which, for a full year, can cost $130,000–$500,000 for an armed security force. Homeowners can also route camera images to off-site monitoring centers relatively inexpensively these days by comparison.

Interest in perimeter protection has decreased dramatically in the last two decades, having been replaced by CCTV surveillance systems, though the metal fence is still used at prisons, military camps, nuclear plants, outdoor parking areas, and government facilities (Figure 28.1). There is no single factor behind the drive to fortify the perimeters of facilities since crime rates have largely fallen over the last few decades. The motivating factor for gate and perimeter protection can be traced to the proliferation of personal information on the Internet which has increased some owners' feelings of vulnerability and a strong desire for privacy. Others believe that some of the insecurity may be the result of the recent focus on income disparity between America's rich and poor.

Perimeter devices continue to grow cheaper and less intrusive, which makes it easier to design a modest-looking gate that's invisibly fortified. Cameras and software can determine whether movement outside a gate is generated by a person, animal, truck, or plane, or based on temperature, even in harsh climates. The most notable technology allows homeowners to remotely view the perimeter of their property using cameras connected to an iPad or another mobile device. Prices for this type of surveillance system depend on the number of cameras. Extra barriers can be installed in the form of vertical posts that rise out of the ground when a car pulls up to or exits the property.

There is no such a thing as a perfect perimeter sensor system or a fence alarm system with no false alarms.

DOI: 10.4324/9781003091103-39

FIGURE 28.1 Old habits die hard. There are still one of many types of facilities – like this prison – that rely on towers, fencing, and barbed- or razor wire as a security measure. While traditional fencing can be an eyesore, there have been new vendors who still provide perimeter protection, for various applications, that are still effective while being slightly less obtrusive.

Source: Shutterstock.com.

2 EXTERIOR INTRUSION FENCE SENSORS

Protection systems designed specifically for perimeter security rely on several categories of sophisticated technology. Some require sensors to be mounted directly on the wire mesh fence surface while others are buried under the ground, parallel to the fence profile. Each device or system has a history of problems, whether mounted on the fence barrier outside a nuclear plant or along the 700-mile US border with Mexico. In fact, after spending almost $1/2 billion on the Mexico fence project, the Department of Homeland Security abandoned the project in 2012 because of false alarms, the difficulty in maintaining the integrity of the perimeter protection system, and the remoteness of the site. Likewise, and despite funding and vocal outcry – led mostly by the candidate himself – for a wall to be built during the Trump administration, very little was done apart from limited fencing that was installed along the United States/Mexico Border.

The objective of sensors is to detect a change in the electromechanical field or the vibrations caused by movement over, under, or through the barriers. The properties are usually far from observation and almost impossible to respond to in a timely manner, which compromises any sensor regardless of its sophistication.

During the 1970s and 1980s, perimeter protection systems constituted an important segment of the growing security industry and offered a practical method to protect sensitive and remote properties. It wasn't until the late 1990s that such electromagnetic-type sensors were replaced by advances in CCTV surveillance systems. The perimeter protection and detection industry is less than half as large as it was just three decades ago. Barrier sensors and protection devices for fences include the following most commonly used systems:

2.1 Most Popular Fence Sensors

- Electrostatic Field Disturbance Sensors
- Taut Wire Systems
- Electrified Barriers
- Vibration Sensors
- Fiber Optics Cable
- Strain Cables
- Capacitance Sensor-buried Geophone
- Razor/Barbed Wire

Taut Wire is an electromechanical sensor consisting of multiple horizontal wires along a segment of the fence (about one sensor every 200 feet) which transmit a signal to a reporting unit whenever the wire is cut or spread by a climber using a ladder to get over the fence. This tightly strung cable detects any physical tampering of the wire or deflection of the wire. If force is used against the wire, a signal is activated unless the force is less than about 30 lbs. Anything lower than that pressure is ignored, eliminating false signals due to small animals bumping into the fence.

Geese Flocks are an efficient alternative for wide areas that don't require high security and are distant from the main building control center. They are especially effective in guarding the fenced-off terrain for extended lengths of time. A perfect example of a geese flock guard service is when they are used to guard vats of aged whiskey that must be stored for up to 25 years before bottling is scheduled. The geese are self-sufficient, feeding themselves on the grassy terrain and they have the natural ability to become familiar with caretakers who provide extra food and care. When the geese see or sense a stranger, they squawk loudly and the noise is picked up on speakers mounted on the fence. Geese do not make sounds when they sense or smell friendly people in their midst. The speakers are connected to the main house on the property and when the noise level reaches a set limit, an alarm is activated, alerting the person to the unwanted intrusion.

The government issued guidelines for designing a perimeter protection system under the title "Crime Prevention Through Environmental Design" (CPTED). CPTED was met with high hopes when it was published since it was based on a fundamental hypothesis that proper design and effective use of the physical environment would encourage behavior that reduces incidences of vandalism, thus lowering the fear of crime and thereby improving the quality of life. CPTED has some common-sense principles and recommendations based around signage, natural barriers, and lighting that can be used in external design and layouts of corporate, university, or government campuses and buildings. However, much of the reliance on current security technology and systems has somewhat marginalized its usage and implementation; it has not marginalized its effectiveness. CPTED should be among the suite of tools used for risk analysis and physical security needs-based assessments.

29 Bollards and Other Portable Barriers

1 USED MAINLY TO DISRUPT TRAFFIC

Ugly and obstructive are words that describe 95% of all street barriers, especially those four-foot-high concrete blocks of cement found along the nation's millions of miles of highway that have separated the traffic flow in front of buildings and construction sites since 9/11. Barriers of all types (both moveable and permanent) have inundated most inner-city commercial districts and the perimeter of sensitive military and government facilities. The most recognizable protective bollards are still the ugly and inescapable "Jersey barriers" that have gradually moved from highways into cities and are now blocking pedestrians from entering buildings and rerouting vehicles away from major intersections. Originally intended to deter terrorist vehicle attacks post-9/11, they have since become all-purpose/multi-use barriers for various reasons.

Once building managers and security professionals expressed an overwhelming interest in bollards, officials from several municipal agencies in the nation's largest 100 cities tried to create some uniform models that were limited to either metal or concrete, no greater than 48 inches high and 18 inches in diameter that could be anchored soundly to the ground. In most municipalities, facilities must receive authorization from a municipal agency (usually a Department of Buildings or Department of Transportation) to install a post, pole, or other form of a bollard. The standard requirement is that stanchions be at least four feet apart and be joined with horizontal members. A quick informal survey of most eastern seaboard cities indicates that this formula was modified to the specific needs of each city.

Effectiveness. Most waist-high barriers come in a variety of forms, including clunky planters, globe shapes, and the ever-present but aesthetically offensive Jersey barrier. The sole intended effect of these monstrosities was to limit the movement of people or obstruct the ability of vehicles from entering.

Typical installation of bollards finds these concrete-filled steel pipes placed contiguously or every few feet along a curb or sidewalk to prevent vehicles from getting close to the premises. Bollards are generally embedded with a concrete foundation of several feet below street level. The height of the bollards has to be higher than the bumper of a truck, and spacing would need to be three to five feet to be effective. The Department of Homeland Security does not have an official policy on the size or construction of these barriers and leaves the decision to state and local officials. The State Department has issued guidelines since they are concerned with US property overseas but, in general, the size, shape, and location of these barriers remain under the regulations of the municipality. In NYC, the Department of Transportation oversees the approval process to install street barriers and often has the New York Police Department (NYPD)'s counter-terrorism task force assist in determining a site's vulnerability to threats.

Counter-terrorism experts now feel that some of these barriers did more harm than good and a poorly anchored planter, struck hard enough by an explosive force, can become a "force multiplier" because it shatters into deadly flying shrapnel. More discouraging to many city officials was the realization that many building owners had spent upwards of $100,000 for a series of connected 4-ft-high barriers but were not really at risk. Many inner-city buildings are bunched so closely together along very narrow streets that a car bomber could cause extensive damage without breaching the barriers.

Information gathered by counter-terrorism government agencies appears to reinforce the belief that street barriers are not the first line of defense. One intercepted terrorist message indicated that

DOI: 10.4324/9781003091103-40

the best way to get close to a building, particularly in the financial district of any major city, is to commandeer a big black limousine and pack it with explosives, then park it in the underground VIP car garage (like what was done at the World Trade Center in 1993) or on the street next to the building in Oklahoma City (1995).

The Oklahoma City truck bomb was parked across the street, almost 50 yards away from the federal building, and the impact destroyed the structure with a high loss of life. In this author's opinion, barriers are more often cosmetic security devices, just as dummy cameras were in the 1970s and card-reading turnstiles in office building lobbies today. Immediately after 9/11, entirely new industries emerged with the fabrication of lightweight plastic barriers.

Even military facilities had thousands of concrete barriers installed following the event of 9/11 as a knee-jerk reaction. Apparently, officials from all levels of government had no clue as to the effectiveness of such a barricade, but it looked ominous enough to scare a timid terrorist. When the threat wore off, a new problem arose that was not planned for – how to remove and dispose of these heavy, cumbersome monstrosities.

2 THE UBIQUITOUS CONCRETE JERSEY BARRIER

These concrete monstrous eyesores first appeared along American streets and highways (and in front of buildings) in the early 1950s and have since become the most popular and possibly the ugliest road barrier in use throughout the world. Credit for its appearance can be traced to the engineers at a New Jersey college under contract to the New Jersey Department of Transportation. After a few design changes, an extended lip effect was added to the base of the basic model and it has become known throughout the world as the "Jersey Barrier" (Figure 29.1). There was a patent issued around 1951 to the NJ DOT for the parabolic curved structure that is officially called "White Concrete Barrier Curb." It has been copied throughout the world after it was first installed as a road divider on a New Jersey highway in the 1950s.

The cost to remove a Jersey barrier was often as much as the cost to purchase and install the unit, creating a dilemma for most municipalities since there was no readily available place to discard them. Previous barrier designs in the 1930s and 1940s used to separate two-way traffic almost always included islands of grass strips with a narrow sloping concrete curb on either side. As the traffic volume increased after WWII, these narrow strips posed a hazardous condition for crossover vehicle movement and became obsolete. The grass was slowly replaced with rigid concrete median barriers that virtually eliminated the danger of head-on collisions. The original barrier was high enough to stop a car from riding over it and the two parabolic curved surfaces prevented the vehicle body from hitting the barrier at a flat side impact angle if it weren't for the added "lip." In 1955, the height of the barrier curb was raised to 30 inches to match the raised height of newer cars and, in the mid-1960s, the final design selected by the NJ Department of Transportation was 24 inches wide and either 32 inches or 42 inches high with a reinforced steel pipe.

Plastic-shaped Jersey Barriers became commercially available around 1985, are lightweight and hollow, and are used mostly for temporary street closures at construction sites. The barriers weigh less than a hundred pounds when empty and a hundred of them can be transported on a single truck, whereas only around 24 original concrete barriers can be transported per truck. Once filled with water, they can be an effective alternative to the concrete type since they have been successfully tested to withstand the force of a truck going 50 miles an hour. They may not be quite as strong as the concrete barrier but are a more practical alternative. The federal government has started to investigate the use of plastic barriers for some military installations but, at present, there is a strong effort to make the concrete units the barrier of choice for the military.

FIGURE 29.1 While prevalent and effective, Jersey Barriers are not the most attractive physical security measure.

Source: Shutterstock.com.

3 BUNKER MENTALITY

Other bollards have entered the barrier marketplace with some building owners actually hiring architects to design fanciful concrete globes and ellipses to blend in with the decor of some of the more picturesque and older buildings. A telescopic or retractable bollard is one that can be lifted out of the street to form an insurmountable barrier and when lowered, disappears into a hole in the street. Another proposed design involves constructing a 20-foot round turntable resting on the street with a row of protruding barrier posts perpendicular to traffic to stop vehicles, but allow enough room for pedestrians to pass. When the turntable rotates 90 degrees, it shifts the row of posts to be parallel to traffic, allowing authorized vehicles to pass through. It offers a significant improvement over the Jersey Barrier.

With all the choices for below-ground barriers, it should be noted that these electromechanical units can become submerged in water or mud, creating a major maintenance problem. It is imperative that all underground units be hidden in a hole that has good drainage and that the above ground street is crowned so that there is little risk of standing water draining into the barrier's foundation frame.

Most portable Jersey-type barriers are rented for specific periods of time or for limited-time events and are removed by the company that delivered them. Government agencies rushed to purchase the barriers after 9/11, which doubled and tripled the business to barrier suppliers. The concrete barriers weigh many tons and come in lengths of 10′, 20′, and 40′ which require special 18-wheeler flatbed trucks. A 20′ barrier is so huge that only six to eight barriers can be transported on a truck without going over the weight limit of the highway. The cost to transport these barriers to and from a site is thousands of dollars and finding a final resting place poses another, almost insurmountable problem, since the barriers are rarely ever reused or recycled.

4 CROWD CONTROL BARRIERS

- **SAWHORSES** used mainly as pedestrian control barriers by police are not in the same classification as Jersey Barriers, since they are used mainly to control street crowds. Heavy

FIGURE 29.2 French barriers are much more mobile, storable, and practical than Jersey barriers and are useful as crowd control or deterrents at events, as well as offsets for vehicle traffic – when properly used and staffed. However, with large crowds or highly outnumbered defenses – such as during the January 6, US Capitol insurrection of 2021 or the Israel-Hamas protests in the spring of 2024 – such barriers are ineffective and can even be used against security or law enforcement professionals.

Source: Shutterstock.com.

police sawhorses of rough wood with a stenciled warning "Police Line Do Not Cross" have been a visible staple of most major city landscapes for decades. Since the turn of this century, these symbols of city life have been disappearing and are being phased out by police departments all over the nation. Sawhorses act more as a sign rather than a barrier since they are easy to scoot under, around, or otherwise move.

- **FRENCH BARRIERS** are replacing wooden sawhorses in most cities and are hard to breach once they are locked together; they are also lighter in weight with interlocking aluminum poles that make them easier to move (Figure 29.2). The French barrier was, in fact, invented in France; they are 14 feet long and are very cheap and cost-effective. They can be used at parades, public demonstrations, and special events largely for the purpose of crowd movement and control.
- Over 3,000 French Barriers and wooden horses are stolen in NYC every year, with sometimes as many as 50 pieces taken all at once.

5 POLICE HORSE PATROL

Until the late 1980s, police officers on horseback (mounted police units) had been a familiar sight in many of the nation's urban centers, a traditional sight dating back to 1873. These trained equines typically weigh 1,200–1,500 pounds and are usually taught to work in a group, marching shoulder to shoulder in close formation. They have been effective for crowd control and have become an important element of parades and special events.

Horses are invaluable in controlling crowds or potential rioters since they can get in and out of places much faster than police cars, especially in tight inner-city areas. In spite of their proud history and proven value to any police department, most of these mounted units will probably go the way of the horse and buggy because of budgetary restraints. Meanwhile, in order to slow the trend, several cities are trying to support existing units or start new mounted police units using private donations. Past history has shown that once these animals are gone, it is almost impossible to restore mounted units since most horse stables, once abandoned, will often be sold.

Cities like San Diego, Portland, Ore., Newark, NJ, Boston, Boca Raton, and Tulsa had stopped funding mounted patrols by 2010 and sent their horses to other cities or, in most cases, to horse farms. As recently as the 1980s, there were still over 300 such units, but those departments with mounted units faced budgetary issues that gained momentum during the recession that started in 2008. The high cost of maintaining stables, veterinary bills, and hay forced major cities that supported the mounted units to take them out of service. If the trend continues, the public in only a limited number of cities will enjoy the sight of a policeman on a horse again before they disappear.

Mounted units perform a great public service and have improved relations between residents and the police in cities struggling with high crime rates. In certain crowd control situations, one officer on horseback is worth ten because they form a formidable skirmish line that can easily make a crowd turn around and leave the scene of the disturbance. When police are on foot marching in a skirmish line trying to move the same crowd, it often creates tension and conflict and potential serious problems. As the quoted saying goes: *No one ever asks to pet a squad car.*

As of 2015, it was estimated that there were 75 full-time mounted police units in the nation with the largest consignment belonging to the 64-horse unit in NYC followed by the 32-horse unit in Chicago and 30 in Los Angeles. None of these cities has plans to eliminate these units in the near future. The city of Philadelphia raised private donations to start a unit in 2011 by purchasing four horses from the closed Newark stable that they hope will grow to an eventual platoon of 25 horses. Baltimore's mounted police unit staved off elimination in 2010 after receiving private donations of $100,000 and remains the oldest continuously operated mounted police unit in the country, having started in 1888. Similarly, a fundraising group donated money to keep the mounted police in the saddle in Portland, Ore., which lost public funding. It is now partially supported as an official charity of that city's annual marathon.

At present, as of the time of this book's publication, it seems that such mounted units are making a comeback with over 150 estimated in the United States.

Not everyone welcomes mounted police patrols, including many residents and store owners who complain when the result is streets cluttered with horse manure. Some cities that rely on horse-drawn carriages to attract tourists are better prepared to clean the streets and pay broom and pot cleaners to follow the horses around the city (New Orleans and Charleston SC, for example).

Section XI

Locking Arrangements – Life Safety Rules

We provide an extensive review of the history and changes that have revolutionized the lock and cylinder hardware arrangement since the skeleton key was replaced by the pin/tumbler invention in the 1980s by Linus Yale. The invention made duplicating keys easy and created the locksmithing industry. After 150 years, the introduction of access control systems and the availability of lock hardware at Box Stores like Lowe's have practically eliminated the friendly neighborhood locksmith who has become as useless as a blacksmith. Electronic access control systems, both wireless and hardwired, are not found through a locksmith. New locking arrangements require a knowledge of digital technology and some understanding of being in code-compliance.

The chapters describe the important upgrading of lock hardware, special lock arrangements, and advances in lock technology and include push button shortcomings, electrified locks, electromagnetic locks, and access control devices.

Coverage offers specific details on how the electromagnetic lock became a controversial subject with Life Safety Code Enforcement organizations like the National Fire Prevention Association and local fire departments. The electromagnetic locks for fire exit doors were finally accepted as being compliant with the amended codes specially written to cover special locking arrangements. The controversial process for acceptance is covered for the first time in the section.

A detailed explanation exposes the ease in learning how to pick any lock pin tumbler arrangement and where to purchase lockpicks. Less expensive locks are easier to pick since they are not manufactured within the tolerance of one-millionth of an inch as found in finer-made items like watches. If locks were manufactured to closer tolerances, the cost of each lock would be so prohibitive that there would be no market. To keep prices as inexpensive as possible, the average lock is produced with a relatively loose set of tolerances and that makes it easier to pick.

DOI: 10.4324/9781003091103-41

30 Locking Hardware

1 DISAPPEARING LOCKSMITH

The neighborhood locksmith is following in the tracks of other now obsolete occupations like typewriter repairman, milk delivery man, blacksmith, and others that have been replaced by new technology. The locksmith trade began to fade with the arrival of card access technology. Key grinders and lock pickers, as locksmiths are sometimes referred to, have slowly become as rare as elevator operators, toll collectors, singing telegrams, and the wall-mounted hard-wired house phone. Since major hardware retail outlets started carrying a full array of lock hardware, local locksmiths have become as useful as the radio repairman of the 1930s. The locksmith will remain in business as long as people keep losing their keys, but these mostly small self-employed mechanics are not qualified to install or program any device more complicated than a mechanical lock.

It is not well known, but all locksmiths must not only have a license from the community in which they work but are also supposed to follow a code of conduct, having taken the "Baldomeric Oath," named after St. Baldomerus, a locksmith who died in the 7th century, when entering the profession. An excerpt from the oath:

> That I will maintain this sacred trust, holding myself far aloof from wrong, from corruption, from the tempting of others to vice.

Problem: A newspaper reported that a well-intentioned locksmith was telling authorities at a city-owned hospital that he was getting overtime pay when he was only working the 40 hours required. It seems that after signing in each morning, supervisors were adding extra hours to the length of the workday without his knowledge or approval. He discovered the fraud when he got bigger checks than the amount he was entitled to. Quoting from the Baldomeric Oath, the locksmith reported this action to his immediate supervisors who informed him that since all city employees had had their salaries frozen for three years, the overtime was to make up for this injustice.

The locksmith then went to the inspector general – before he was fired.

Over the last 35 years, while the security industry grew, the locksmith trade has been unable to redefine or reeducate itself as part of the electronic age of door security. The locksmith represents, in many ways, a self-destructive and dormant industry content with making duplicate keys – rather than expanding their knowledge and learning how to install card access or biometric systems, instead of obsolete, conventional locking hardware. Articles often appear in the few locksmith journals still in circulation, urging readers to learn more about the expanding security technology – but it has been a fruitless mission. Compounding the problems created by hardware stores carrying locking devices was the introduction of vending machines located at kiosks inside shopping malls that duplicate keys automatically for a few dollars (Figure 30.1). For over a century, duplicating keys was the main source of income for locksmiths.

Entrance door protection hardware sold through locksmith stores or big box hardware stores is perfectly suited for entrance door protection, but it is not considered a realistic supplier for sophisticated electronic security systems. Until the early 1970s, the only recognized "security specialist" was the neighborhood locksmith (behind a narrow counter in a very small storefront shop). The window displayed several doorknobs, key chains, and heavy-duty locks – but almost never any electronic devices. Until the 1970s and 1980s, there were few electronic security installers, and the only trade shows were Locksmith Conventions. The trade shows continue to be scheduled every year and the exhibits are usually focused on the newest innovations in key-making machines – with little noticeable displays dedicated to the growing electronic security and access control industry. The number of attendees at these meetings has remained consistent over the last three decades,

DOI: 10.4324/9781003091103-42

FIGURE 30.1 A self-service, fully automated kiosk to duplicate keys at a Menards in Gillette, Wyoming.

Source: Wikipedia, public domain.

while during the same period, the alarm detection, video surveillance, biometric, and digital industry experienced a technological revolution and became one of the fastest-growing industries in the American economy. During this same period, the number of locksmiths fell from 35,000 in 1940 to 15,000 by 2015.

The most noticeable advancement in the locksmith industry was the development of locks and associated hardware that were decorative and more stylish but operated with the old pin-tumbler arrangement. New locks are simply variations of the basic designs that go back a century or more. While some locksmiths install complicated electromagnetic locks and maybe a few card readers, they generally avoid anything that involves "electronic" connections. Locksmiths consider themselves skilled mechanics who install hardware rather than electricians who install card access control systems.

The history of mechanical locking hardware started around 1865 following the invention of the pin-tumbler lock cylinder by Linus Yale. The locksmith trade grew slowly but eventually became the main provider for premises protection and could be considered pioneers of the security industry. That same skill and ingenuity that were associated with door locks did not advance with other skilled occupations during the Industrial Revolution. Before the Yale cylinder became the standard in the lock hardware industry, the most prevalent protection device was a wooden beam mounted across the door, which was removed with a skeleton key. Yale also invented the key duplicating machine in the early 1900s, which was followed by key code books published in the 1920s, the time that the modern locksmith shop became an established storefront business in many major cities. Some of those early key-making shops are still in operation with third and fourth generations of the founders still active. Skilled locksmiths work out of storefronts or trucks, having graduated from the $1,500 corresponding "Locksmithing" school that includes a syllabus describing how pin tumblers fall into a "shear" line and how to pick locks for fun and profit. Most schools provide a duplicating key machine as part of the educational course.

In most municipalities, locksmiths are licensed, but if the "shop" moves to another city within the state, it would be illegal to start cutting car keys and repairing lock hardware immediately without a license. Every municipality seems to have their own rules and in Las Vegas, locksmiths must not only be licensed but must also work out of a store or office (no trucks). In general, locksmiths and security alarm installers operate in two different and totally unrelated worlds yet share characteristics such as (a) it takes very little money to enter either business, (b) no particular talent, skill, experience, or education is required, and (c) both trades operate in a mostly unregulated industry.

New York is one of the few states requiring alarm and fire system installers to possess a license that requires experience as well as the passing of a written test. The objective is to protect both the public and institutions against unreliable and potentially criminal firms that install security systems. The idea was to have all government and major organizations place the licensing requirement into the bid package. If the licensee did not perform as promised, the license would be revoked. It did not guarantee that the installer was competent, but it could prevent the vendor from repeating shoddy work, a problem prevalent in the security industry which has practically no licensing restrictions. Since the law was enacted in 1993, no significant impact has been made on enforcing this licensing statute/law.

The small locksmith storefronts may be disappearing but locksmiths have been slowly reinventing themselves by becoming in-house locksmiths, working as employees of hospitals, school systems, or corporate entities with many doors. The freedom of owning a locksmith shop has been traded in for a position that may not pay very much but offers more security than an independently owned store. Almost all major medical centers with over 400 beds have their own locksmith office and one or two skilled locksmiths.

Locksmith Scam. Several years back starting in 2015, a scam came to the public's attention when people sought help from a neighborhood locksmith after they lost a house key or were locked out of their car with the key left inside. In desperation, they would use their smartphone to search for a locksmith, resulting in a list of names and numbers of local companies and the soon-to-be victim would assume that the displayed names were not only in the vicinity but that the search engine's algorithm had presented legitimate addresses.

While a few names may fit that description, most of the names were not actually locksmiths. Such calls were actually being forwarded to call centers – often out of state and sometimes in different countries – that had used a high-tech trick to get Google to present them as nearby physical stores in the neighborhood. This operation is known as lead generators, or lead gens for short. The lead-gen would then contact someone local to the caller within their association, after telling the caller that the cost would be between \$50 and \$90. The subcontractor (working out of his or her car and with no store location) would then show up on site seeking payment of hundreds of dollars to do the work, from the desperate victim who frequently ended up paying the demanded amount. It was a classic bait-and-switch scam.

2 OBSOLETE LOCKING ARRANGEMENT

A security device created in the early 1970s, at about the same time IBM Selectric typewriters became the standard word processor in every office in the nation, was the typewriter lock. The Selectric sold for over \$1,000 when first introduced and instantly became a prime target for office equipment thieves. It may have been bulky and heavy, but it was reported that up to 12 million of these typewriters were stolen within a two-year period between 1979 and 1982. The various manufacturers of electric typewriters took absolutely no action to protect their units by offering or incorporating some sort of locking system for obvious reasons (Every time a unit was stolen, another typewriter was purchased). The lock industry reacted to this problem by introducing a series of rudimentary yet innovative locking devices to prevent the typewriter from being removed, but with questionable success. The simplest system was to attach a cable to the side chassis of the unit and the other end to a drilled hole in the desk. It required a key to release the cable.

The cable was inexpensive and discouraged the amateur opportunist thief (usually an employee) from stealing the typewriter. However, it did little to prevent professional thieves who simply cut the cable and removed the typewriter.

The most innovative and successful typewriter locking system was introduced in the early 1980s when offices had dozens of IBM Selectric models and a lesser number of other makes. The "Anchor Pad" used two-way adhesive tape on a bottom plate that was "pasted" to the desk with an upper plate screwed to the bottom of the typewriter base. The two plates were then locked in place and could only be separated using a key.

The major drawback of this unique unit was that once the bottom adhesive plate was sealed in place, it could not be removed to relocate the typewriter unless a heat lamp was applied to melt the adhesive, though the removal process sometimes left a scar on the furniture surface. The only noticeable change was that the typewriter sat on the two-inch plates instead of the desktop. While the adhesive pad and lock may sound a little impractical, it proved to be the most effective locking device for typewriters for almost ten years and while the cost ($125 at the time) was triple of the cable systems, it provided a much more secure arrangement. The 1980s are now referred to as the end of the typewriter age as desktop computers began to make an appearance. The same technique was initially used to protect desktop computers as well as laptops, but it did not sell for very long for two reasons:

- When the adhesive method was used on the desktop model, thieves simply left the terminal and keyboard in place… and stole the unprotected hard drive.
- Laptops became so inexpensive and were typically carried off the premises by the owner, so the complicated locking arrangement had no practical, cost-effective value.

Laptops are now lightweight and small; thousands are stolen every day from private industry as well as from governmental security agencies. The real difference between stealing a typewriter versus a laptop is that an enormous amount of sensitive information can be stored on a 3–5 lb. laptop. The Veterans Administration lost millions of bits of information when a single laptop disappeared, a problem shared by industry and colleges alike.

2.1 Unconventional Concepts

THE CLUB. Sometimes, a simple locking concept can become a universal locking standard when promoted properly. A case in point is the phenomenal success of the automotive security mainstay simply named "The Club." This simple device was a familiar sight on the steering wheels of millions of cars but was not unique or innovative. The Club was introduced to the public in 1986 through Sears and then heavily advertised on late-night television. The TV ads featured a policeman touting The Club and reminding viewers that "I'm not an actor, I'm a policeman." The Club was hardly a novel device – a similar gadget had been patented in 1919, and other models were already on the market, but by using advertising and a considerable promotional effort, The Club manufacturer became a millionaire.

Law enforcement agencies and product testing groups proved that The Club was easily defeated, either by spraying it with Freon to make it brittle then cracking it with a hammer or by taking a hacksaw to the steering wheel. The Club manufacturer refuted the criticism by claiming that thieves didn't carry hacksaws or Freon.

In spite of its weaknesses, over 10 million Club devices were sold in the seven-year life cycle of the unit and now sells is upgraded, revamped version, The Club 3,000.

3 NO-LOCK NEIGHBORHOODS

The picture of small-town living typically includes the description of a community that is so safe that people don't lock their doors. These homeowners are known as "no lock people" and they exist in large numbers in most of the nation's largest urban centers. Living in a high-rise, inner-city building with a doorman creates a sense of security for tenants who do not always bother to lock their apartment doors. (A study by State Farm Insurance Company reported that fewer than half of those surveyed always locked their front door.) While people who habitually lock their doors are incredulous that others do not, those who don't lock their doors are surprised that anyone would be shocked by the practice.

Historical FBI data shows that home burglaries have dropped over the last decade plus. The last statistics from 2019 recorded 6.9 million property crimes. Of those, there were an estimated 1,117,696 burglaries, a decrease of 9.5% when compared with 2018 data. The number of burglaries decreased 29.6% when compared with 2015 data and was down 48.5% when compared with the 2010 estimate. Additional statistics from 2019 include

- Burglaries accounted for 16.1% of the estimated number of property crimes.
- By subcategory, 55.7% of burglaries involved forcible entry, 37.8% were unlawful entries, and 6.5% were attempted forcible entry.
- Victims of burglary offenses suffered an estimated $3.0 billion in property losses in 2019. The average dollar loss per burglary offense was $2,661.
- Burglaries of residential properties accounted for 62.8% of all burglary offenses (Source: FBI).

The decision to lock or not lock a door is not always logical – often, it is based entirely on emotion. A person's perception of safety is not necessarily tied to statistics but is usually associated with a false sense of security based on such factors as having a doorman inspecting every visitor. But the same doorman has access to a duplicate key that is kept at the front desk. Delivery people are only admitted after being confirmed by a call to the tenant, though there is nothing to prevent these individuals from trying all the doors on the floor after making a delivery.

4 100-YEAR-OLD PIN-TUMBLER LOCK

The common pin-and-tumbler locking mechanism has a series of spring-loaded pins, which are loaded into a series of small cylinders. Each cylinder has a bottom part, called a pin, and a top part called a driver. Upon insertion of the key, the springs will be compressed as the key lifts the pin, pushing the driver into the upper chambers of the cylinder. When the correct key is in the lock, the bottom and top pins align the space between them around a track similar to a ward, which is called the shear line.

When all of the pins and drivers are aligned to clear the way for the shear line, the key will turn. When the key is not in the lock, the pins should be at rest in the chamber, with the bottom pin in the key chamber and the top part of the pin resting on top of the bottom half due to the gentle pressure of the relaxed spring. An incorrect key will misalign one or more of the locking spring-loaded cylinders, blocking the key from turning by placing at least one of the pins or drivers in the way of the shear line (Figure 30.2).

The side-locking bar technology is vulnerable to lock-bumping attacks. Lock bumping is an increasingly popular lock-picking technique in which an intruder would use a specially cut key to "bump" a lock open using skilled strikes to the key in the lock. Kwikset SmartKey deadbolts and locks feature the patented side-locking bar technology, which is also central to the SmartKey's ability to be rapidly and easily rekeyed.

FIGURE 30.2 Cutaway pin-tumbler lock with the correct key inserted. The lock mechanism uses pins of varying lengths to prevent the lock from opening without the correct key.

Source: Shutterstock.com.

A rim lock is a locking device that attaches to the surface of a door, usually the inside surface. A small latch is used to unlock the door from within. On the outside of the door, there is usually a smaller sized rim with a keyhole.

A mortise lock is a type of lock that requires a pocket – the mortise – to be cut into the door into which the lock case is to be fitted. Mortise locks have two different locking devices built into the lock case: the latch bolt and the deadbolt.

31 Lockpicking for Fun and Profit

1 AN EASY TO LEARN SKILL

Purists in the lockpicking field have honed their skills like brave matadors or golf professionals. The main difference is that bullfighters and golfers take years to perfect their skills, whereas lockpicking techniques are not that difficult to learn. There are many books on the subject directed specifically at the amateur, and the special tools needed for lockpicking are readily attainable from catalogs and the Internet by just typing in "lock picking." There are even corresponding schools that offer inexpensive training programs that award certificates that can often be used to obtain a locksmith license from the local municipality. All the courses include an in-depth lesson on the art of "picking locks" with the same level of detail reserved for developing wine-tasting expertise.

Picking a lock requires two tools: a picking instrument and a turning wrench, also called a tension wrench. A typical package consisting of 15 rakes (picks) and two wrenches costs about \$15–\$20 and can be purchased from any of the lock hardware distributors found on the Internet.

A typical lockpicking course (previously correspondence courses, now more likely online courses) usually consists of several lessons with some lock hardware supplied, such as a set of keys, a cylinder, and eventually a key-making machine along with the certificate of completion – but very few complete the course. Each lesson includes an instruction booklet and a simple ten-question test that requires answers that are returned to the school with a check as partial payment. The test is graded and returned along with the next lesson package and so on, until the course is completed. As long as the check clears – no one has been known to fail any of the tests.

The most popular lesson includes the lockpicking instructions themselves accompanied by the "picks." Most students seem to quit the program once they receive the picks, memorize the illustrations or video test lesson plan, and practice for one or two hours. With that brief training, the average student can pick 90% of all the locks in the community, including nearby retail stores.

When following instructions for the lockpicking lesson, the student is given a lock or cylinder that comes with five pins. The student removes four of the pins and practices the fine art of lockpicking by using the picks or rakes and a turning wrench on the cylinder with only one pin. The idea is to move that single pin up and down continually until the pin falls into a "shear line," allowing the tension tool to turn the cylinder into the unlocked position. After mastering the one-pin exercise, the student will dismantle the cylinder and insert two chambers with pins and springs and then practice raising the pins, once again, to the shear line. By adding one pin at a time, the student can slowly build up their confidence to pick the full five-cylinder arrangement and become a full-blown lockpicker.

A cutaway view of the cylinder plug section side shows that none of the pins in a typical cylinder are in exact alignment, because no matter how carefully a manufacturer drills the holes in a plug, they never line up perfectly (Figure 31.1). The cheaper the lock – the greater the misalignment, as seen in Figure 31.1 where holes for the pins are not aligned because of loose tolerances. The loose tolerance measurement is why it is easy to raise one pin at a time to the very critical shear line. When a pin reaches the shear line, the plug holding all the pins can be turned ever so slightly by applying pressure on the wrench. Continuing to raise one pin at a time to the shear line allows the plug to turn, which is a feeling that the lockpicker can actually recognize with repeated practice. The movement can be as slight as two-thousandths of an inch to get the pin in position.

DOI: 10.4324/9781003091103-43

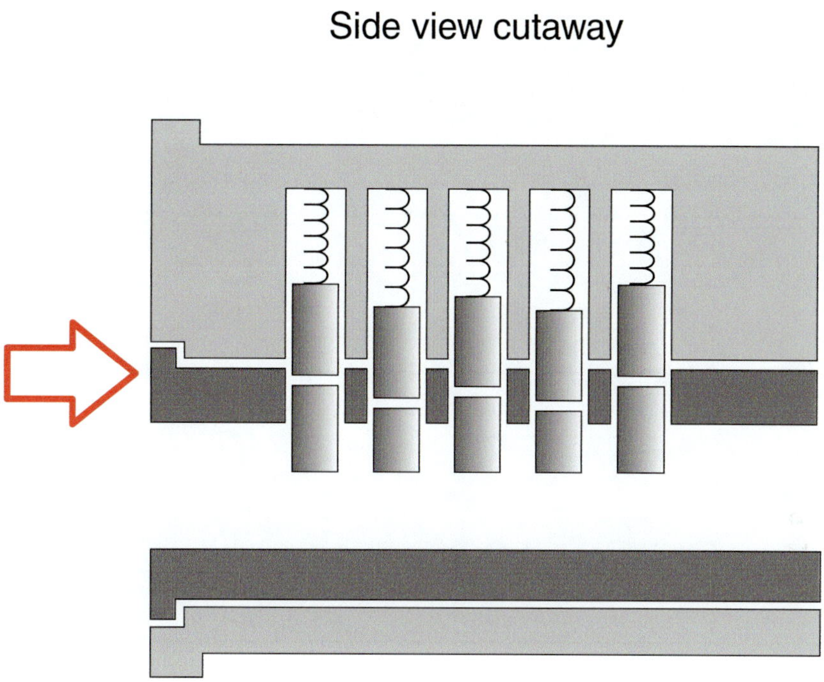

FIGURE 31.1 A cutaway of the lock where you can see the inexact alignment that can allow for lockpicking.

Less expensive locks are easier to pick since they are not manufactured within the tolerance of one-millionth of an inch as found in finer-made items like watches. If locks were manufactured to closer tolerances, the cost of each lock would be so prohibitive that there would be no market. To keep prices as inexpensive as possible, the average lock is produced with a relatively loose set of tolerances and that makes it easier to pick.

Locksmiths must know how to pick locks because customers lose their keys and professional thieves use their lockpicking skills to rob premises and to steal cars without breaking down the door. After a premise has been robbed, and it's suspected that the robber used a pick to enter the premises, police detectives take apart the lock and examine the cylinders, observing any scratches on the pins indicating whether a pick was used rather than the proper key. If there are no scratches – the robbery is considered an "inside" job by someone with a key.

- **Picks:** Come in many shapes to conform to the type of cylinder in the locking hardware. Designed to manipulate the pins inside the cylinder to be aligned to the shear line, which allows the turning of the cylinder – unlocking the door.
- **Turning Tool**: Used to rotate the lock mechanism's cylinder. Many crime scenes in movies and TV shows misinform the viewing public by showing a police officer or bad guy picking a lock with a single tool – not in real life.

The pick and turning tool simulates the shape of a standard key. Picks are used to simulate the pins in the lock cylinder being moved by the cuts in the front section of the metal key – the turning tool simply acts as the rear section of the key that is held by the thumb and first finger, used to rotate the cylinder by applying torsion. After the picks get the pins aligned along the shear line, the cylinder will rotate. The illustration shows that the cuts in the key move the pins into the shear line which must be duplicated by the pick.

FIGURE 31.2 Deadbolt lock device with lockpicking tools.

Source: Shutterstock.com.

There are a number of different types and styles of picking instruments and turning tools; some are more flexible, while others are easier to pick for certain cylinders. Purchasing the proper set of picks is much like selecting any household tool (Figure 31.2).

- **Pick Gun:** An inexpensive picking device that uses a triggering action to rapidly move the pins into a series of positions until the shear line is in alignment. No particular skill is needed except the ability to pull the trigger.
- **Tubular Vending Machine Picks**: Specifically designed to fit into the round opening in the typical vending machine that has the pins set in the rear of the cylinder. The tubular lockpick performs the same operation as the conventional lockpick, but instead of moving the pins up and down – it moves them back and forth until the shear line is in place.

Sometimes, the close positioning of the turning wrench and the pick can cause "crowding" within the cylinder keyway, preventing the pick from working freely. Placing the wrench on the top edge of the keyway may prove to be more successful than using it at the conventional lower position. In working both positions, the lockpicker will be able to tell (or feel) which is the most practical and most comfortable in reaching the desired goal. To become more proficient, some lockpicking instruction manuals suggest that after getting the five-cylinder plug to align with the shear line.

There are lockpicking clubs in 30+ countries dedicated to doing so for fun. The original club started in Belgium in 1999 and spread to over 25 countries, including the United States. The clubs that belong to The Open Organization of Lockpickers (TOOL) meet regularly to share secrets and hold an annual competition – a sort of Olympics of lockpicking – in different countries. This annual event has entrants compete in various categories – padlocks, mechanical locks, and free-style, in which contestants confront a variety of locks with tools they choose, as long as they do not damage the lock. Of course, some of the scheduled meetings and events have occasionally attracted unwanted visitors (those wanted by the police). Unlike burglars who drill into or otherwise damage locks, the club's members follow a cardinal rule that locks may not be damaged in any way.

- **Bumping:** This technique is a rather unskilled method of lockpicking since it involves brute force instead of a careful manipulation of pins and shear lines. It is done by simply inserting a filed-down key into a lock cylinder and then knocking the key with a hammer to momentarily pop the lock's pins into an open position. It takes about ten attempts before the cylinder will turn into the open position. The same lock expert who was responsible for the "bump" technique also exposed the Kryptonite bike lock's vulnerability to a Bic pen and how to "shimmy" opera combination locks. He also published a 300-page book, *Locks, Safes, and Security: An International Police Reference*; it is considered the bible in the locksmith trade. The same author published a book that details a method to pick the allegedly uncrackable Medeco high-security locks.

2 PICKING HIGH-SECURITY CYLINDERS

A typical lock cylinder is "picked" by moving the pins up and down to place them along the shear line – thus opening the lock. The Medeco lock manufacturer introduced a unique cylinder around 1970 that revolutionized the industry. Two inventors who also operated a small locksmith shop in Virginia designed a keyway such that the pins must be "twisted" as they moved up the tumbler; conventional lockpicks could not duplicate this pick-proof feature. The Medeco Company caused a sensation in the industry and offered a $100,000 reward to whomever could pick their twist-pin cylinder within an hour. The NYPD's Chief of Detectives, also head of the robbery squad, took up the challenge. He had already made a name for himself in the industry by developing a technique – as we mentioned previously – wherein law enforcement could determine whether or not a cylinder had been picked. Following a robbery, he would remove the cylinder from the locking hardware to determine if there were scratch marks on the pins – indicating the use of a pick. No marks – no pick – no break-in. It proved to the store or homeowner that someone must instead have used a key and thus it was an inside job.

The detective picked the Medeco cylinder within 20 minutes and collected the prize money. Two years later he retired to become the VP of Medeco. No one else duplicated that feat until 2006, when the lock expert who had exposed the "bumping" method and his new partner discovered a method for opening the Medeco in about a minute. They then distributed videotapes demonstrating exactly how to "Pick and Bump" the Medeco cylinder and published a book titled *Open in Thirty Seconds* that explains how it was done.

3 LOCKSMITHS/BLACKSMITHS/TYPEWRITER REPAIRMEN (ALL OBSOLETE)

1. Most card access systems use some type of electrically powered lock (or door strike) and that type of lock falls outside the talents of traditional locksmiths most content to duplicate keys and install mechanical locking hardware. A device or system that requires electrical wiring or knowledge of computer programming is outside the interest of the "small store" locksmith, which is why they've become obsolete in the digital technology age. Many companies and unskilled individuals entered the security business despite a lack of background in the business. Prior to that, locksmith businesses might have possessed over a century of combined experience passed on from father to son.

Friendly local locksmiths still make keys and install door hardware but rarely graduate to card access systems or surveillance systems. Locksmiths seem to simply enjoy making keys; only a very few install card readers or biometric devices on entrance doors. Locksmiths are also challenged by advanced electrical locking devices that have been gradually eroding the pin tumbler hardware's market, despite potential building and fire code violations.

3.1 Push Button – Digipads

The once-popular push-button pads were introduced into the marketplace to enhance the security of mechanical key locks by adding another level of security or even eliminating the need for a key. They were gradually phased out as more sophisticated controls became available but millions of these numeric keyboards remain attached to door frames – with the highly secret four-digit code clearly written on the adjacent door buck. Some of the better push-button pads are built into the locking hardware but most are the standalone variety that is connected to an electric door strike. Push-button models first appeared in the early 1930s and enjoyed a small level of success early on until the card reader access control systems became the most visible door control device by the late 1980s. Push-button numerical digipads are available in various shapes and sizes, and the only characteristic they have in common is that they can be easily compromised unless combined with a biometric device, like a fingerprint reader. The push button is used in low-security situations like bathrooms, closets, and office doors. Many units include the lock and the "pad" as a single package, which eliminates wiring the coded "pads" to an electronic or mechanical lock.

- These push-button devices have several shortcomings since the four- or three-digit codes are often shared (voluntarily or inadvertently) by someone memorizing the digits through simple observation. The code can often be found in the top drawer of a desk or taped to a computer and is usually the owner's birth date. James Bond would have no trouble breaking the code … and neither would a coworker! Secretly identifying a code can be accomplished by smearing a thin coat of Vaseline over the keypad… the authorized person then uses the pad and fingerprints can be observed in the Vaseline covering the digits. Assuming the numbers to be 4-5-6, the only possible combinations would be 4-5-6, 4-6-5, 5-4-6, 5-6-4, 6-4-5, or 6-5-4. An easier, but more time-consuming, method of identifying the digits would be to observe, after a few months of use, which buttons have more wear, eliminating the need for the Vaseline.

Push-button units are compromised whenever the code is shared. Many times, the numbers are sequential – 1-2-3-4 and the password is actually password, or the numbers are scribbled on the door frame adjacent to the unit. Many people simply forget the password since the average American must now memorize several different passwords for all the online services that require one.

The most popular method to overcome all the shortcomings of a push-button system is to simply keep the door propped open.

There have been a few significant advances in push-button technology, with three notable examples:

- Units can be programmed to allow several unrelated codes to trigger the mechanism.
- Some units retain the numbers used and the time of operation. This stored data can be retrieved with a portable handheld reader unit.
- The most unusual, and by far the most successful, modification to the traditionally fixed layout of numbers is the "scrambler" that automatically and randomly moves the numbers into new positions on the keypad after each use-

SCRAMBLER push-button units prompted a new level of mechanical technology that not only confused unauthorized individuals, who were trained to read codes by standing behind the user, but also confused the very people the system was designed to serve. This innovative device has a standard ten-digit, backlit keypad; however, the numbers are never sequenced in the same position. The locations scramble in random sequences so that anyone looking over the user's shoulder could not identify the number by its position. The top three numbers on other keypads are 1-2-3, but on the scrambler keypad, the top three numbers may be 5-9-1 and, after an activation the top numbers may be 5-2-6, etc. As with many inventive security devices, the public needs time to fully accept any new concept. We are programmed to associate the position on the keypad with the digit; for example, 3 is always on the upper right-hand corner, and that automatic association causes confusion since the user must now read the keypad before every use. Eventually, users became accustomed to the system and appreciated the high level of security it offers.

- **OMNI LOCK.** This standalone digipad allows multiple users to enter a unique PIN for entrance to a secured space, retaining a memory of each authorized user. By connecting it to a portable programmer, it can identify the last series of users who entered their code to gain access to the door and the time of entry. The smallest model can support 2,000 codes per door, track up to 5,000 events, and be used in conjunction with a swipe card or proximity reader. Larger models can track 89,000 events and all the models are programmed using a PC-based management system.
- **TOUCH SCREEN** push buttons were introduced in 2011 by several manufacturers that hid the numbers under a blank screen until touched.

MAGNETIC CAM LOCK was recently introduced to the marketplace that eliminates the one flaw of high-security grade locks that may be pick-proof as well as bump-proof but they still have the vulnerable keyway. A keyway offers an opening for potential thieves to tamper with the lock, possibly gain access through the lock, or cause internal damage. The cam-lock has no keyway and features a flat, seamless exterior and is generally used in display cases.

The locking mechanism within the cam-lock is very similar to that of a combination lock. When the proper magnetic key is used, the brass disks within the lock are magnetically rotated until the notches in the disks are properly aligned into the unlocked position and then the key can then be turned to operate the lock. The lock is made secure by specifically positioning each magnet in a unique sequence that only matches up to the magnetic "key" that has the magnets within it in the same special pattern.

3.2 Vault Combination Locks

Older vault doors were a foot thick and had combination locks that no human could defeat. Bank robbers used dynamite to destroy the locking mechanism and spring the door open. It was a time-consuming and dangerous way to rob the vault area.

Another method to circumvent the bank's round vault door was to drill up from the floor below and avoid the thick vault door entirely.

4 ELECTRIFIED LOCKS

Conventional mechanical locking hardware has rapidly been replaced by various types of access control systems, including card readers, fobs, biometrics, and cyber-lock cylinders. These new locking arrangements operate with electrical power and open when used with the proper programmable device and some operate using batteries, eliminating the need for outside power. Basic electrified locks may be locked to prevent egress under normal conditions but must allow egress in emergency situations. In a growing number of locations, delayed egress locks are being selected since they add additional security to public buildings such as hospitals and schools and eliminate the dangerous practice of chaining panic bars that prevent egress and ingress in complete violation of every safety code.

There are different code requirements that apply to the various types of electrified locks and those requirements can differ from one jurisdiction to the next. The International Building Code has been adopted by most states and is used primarily during construction. The second code, the National Fire Prevention Association (NFPA) Life Safety Code 101, is better known and more important to security practitioners and is used by most municipalities to evaluate the safety of public buildings. A brief description of major electrified locking hardware's operating features:

- Free Egress Control. Allows egress at all times by pushing on the touchpad or panic device or turning the lever handle on a lockset. An access control device is typically on the ingress side of the door and is operated by a card, fob, push button, fingerprint, or another biometric credential to allow access through the door at all times. The locking device does not inhibit egress; it only controls activation of the exit device and possibly identifies the user.
- Electromagnetic Lock with Sensor Release.
- Electromagnetic with Hardware Release.
- Controlled Egress.
- Delayed Egress.

Electric Strikes are embedded in doorframes and operated by a solenoid. This method of unlocking a door makes use of the door lock's strike to hold the door closed until the lock is manually released or the frame-mounted strike is released when power is removed by using the card reader access control system. Most applications use a doorframe-embedded unit since it is more aesthetically appealing and easier to install.

Electric Locks can be either deadbolt or electric doorknob locks where power is provided to the door lock through the electric hinge that carries very fine wire into the door from the frame. From the outside, these locks appear as a standard mechanical lock. They operate by applying and removing power to the locks. This lock has a pin or bolt that is moved into the door when a solenoid is activated in the failsafe version. They are generally used in conjunction with a card access system.

Electromagnetic Special Locking Arrangements (or Delayed Egress) are described in NFPA Life Safety Code 101 under Special Locking Arrangements and are available in two configurations: the direct hold and the less frequently used shear lock. With the direct hold method, a metal plate is mounted on the door frame with the electromagnetic unit bolted to the upper part of the door so when the door is in a closed position – the plate and lock become tightly attached. This arrangement is visible to anyone using the door and some architects consider it an eyesore and dangerous. A subtler application is the shear lock, where both the lock body and the plate are hidden in the upper door jam and in the top edge of the door. When the door is in a closed position – both the plate and lock disappear into the upper frame. Architects like it since it is basically unseen, but it has been difficult to maintain since the door alignment shifts slightly over time and the lock has to be constantly adjusted to be in alignment with the upper metal plate.

The history of electromagnetic locking systems is one of the most controversial in the industry since it was met with a great deal of resistance from the fire code community as well as the local agencies having jurisdiction. Many expressed great concern about allowing the magnetic lock to be used on fire exits that prevented immediate egress and ingress during emergencies. It wasn't until 1980 that the NFPA amended its Life Safety Code 101 to allow this controversial lock to be used on fire exits, and it wasn't until four years later (1984) that this "special locking arrangement" was actually approved by a local fire department for the locking of all the fire exits in a high school in Newark, NJ.

32 History of Electromagnetic Locks

1 FAIL-SAFE DELAYED FIRE EXITS LOCKING SYSTEMS

One man is responsible for introducing a unique lock that eventually became a multimillion-dollar sector within the traditional lock industry. This strange-looking lock, with no moving parts, was introduced in 1972 when a salesman for a lock company, Sumner Saphirstein, hand-molded a few 20-pound ***electromagnetic locks*** and sold them one at a time through a firm he named "Locknetics." The lock was about a foot long (the shape of a cigarette carton) with a coil of wire, embedded inside the casing, to carry the 24 volts that converted the metal shield into a magnet with a powerful holding force. It was bulky and very conspicuous when mounted on the upper door buck, as well as relatively expensive ($300 per lock at the time).

No one in the security industry appreciated its potential value, or the advantages it offered, over the key-cylinder standard lock hardware. The principal feature of an electromagnetic lock was based on the same technology used by powerful and heavy magnetic devices that hang from cranes in junkyards that lift crushed vehicles by powering a large electromagnetic device. A total of 20 of the magnetic locks were made and sold in the first few years by this entrepreneur. It didn't sell very well at first.

In 1975, Saphirstein, under a contract to design a perimeter locking system for NYC's famous Bellevue Hospital, was contacted by the Locknetics entrepreneur. They agreed to participate in the nation's first program to lock fire exit perimeter doors with this heavy gadget. It was the only locking device that would stop unauthorized ingress and egress at selected hospital fire exit doors by overriding the existing panic hardware. The magnetic lock prevented the opening of the door (from both sides) as long as electric power was running through the electromagnetic coil. In order to be in compliance with fire codes concerning egress and ingress, the power cable was connected to the building's fire alarm system (FAS) circuit – which would automatically remove power to the electromagnet whenever a fire alarm was activated; hence, the term "fail-safe." It appeared to offer the ideal solution to the problem of the unauthorized entrance and exit into or out of the hospital premises without compromising life safety regulations. The concept was enthusiastically endorsed by the Bellevue administration, and six doors were locked in this way.

Bellevue Hospital Executive Directors requested that FDNY Commissioners visit the premises to witness a demonstration of this innovative fire exit door lock to gain the fire department's approval, not only to keep the six doors locked but to eventually expand the locks to 45 secondary doors throughout the hospital complex. While the installation was a milestone in the security field, it also conflicted with all the existing fire codes that had been in existence since the Revolutionary War and, of course, the FDNY officials who were invited to inspect the locking system had never heard of a fail-safe lock before their visit. Three fire marshals witnessed a test where the fire alarm was activated and the locks opened immediately. The Bellevue Hospital administrators were overjoyed and the consultant and inventor were counting their potential millions… but these feelings were shattered when the unimpressed fire officials rejected the system as being unsafe and in violation of every fire code in existence. The six locks were removed from the doors… but not forever.

The shocking rationale for the rejection was that the fire marshals did not want another Triangle fire disaster under their watch. In, the Triangle fire disaster which took place in 1912, over 140 garment factory workers died. The workers could not escape the inferno since the exit doors were padlocked to prevent employee theft… that 60-year-old event was the basis for the Bellevue rejection.

DOI: 10.4324/9781003091103-44

After the unexpected rejection by the fire officials, the concept of locking fire exits anywhere in the nation lay dormant until 1980, when the National Fire Protection Association (NFPA) amended the Life Safety Code 101 and recognized that security had become a serious problem that posed as much of a threat to life safety as a fire (especially after the introduction of the "ion smoke detector" in the 1970s by the Swiss firm Cerberus). Until the NFPA Amendment was adopted, the locking of any exit in any building without allowing immediate ingress and egress was considered illegal by every Authority Having Jurisdiction (AHJ) in the nation. The Amendment made the Bellevue rejection appear almost prophetic, if not groundbreaking.

After the NFPA Code Amendment was approved, not one building in the nation attempted to challenge local fire officials by following the Life Safety Code Amendment. It took until 1984 before any public building had the courage to act under this code change, and it did not involve a famous high-rise building or a large city institution. The installation took place at a high school in Newark, NJ. The Newark School District had hired the same security consultant who had been responsible for the Bellevue confrontation a decade earlier, and the consultant's task was to upgrade security in all of the city's 100 schools. That assignment included finding a replacement for the chains that were mounted on all the exit doors, a practice followed by many inner schools throughout the nation at that time. The consultant worked closely with local fire officials to design the same exact fail-safe electromagnetic lock system as he did in 1975, only this time, the locking system received approval from the local Fire Department.

Security in the Newark High School improved dramatically, and the State of New Jersey amended the State Fire Safety Code based on the experience at the Newark School. Soon after the state approval, the famous Paterson, New Jersey, Eastside High School became the second building to replace door chains with the magnetic lock. In 1984, officials from the FDNY visited the Newark School to witness a demonstration (not the same fire officials from the Bellevue Hospital a decade earlier) and they approved the use of the locking system for school facilities in NYC. They recognized the fact that the fail-safe lock was a far better alternative to chaining secondary doors shut.

Brooklyn Technical High School became the third building in the nation to obtain approval from AHJ to lock fire exits. It was the exact same system rejected by the same Fire Department a decade before.

After the locks proved their value, and it was made legal to lock fire exit doors (the weakest part of any school building security system), the fail-safe locking arrangement did not receive overwhelming acceptance by fire department officials, building managers, architects, consultant engineers, or very many public officials in city agencies. Thirty years after the locking of a fire exit became part of the NFPA Life Safety Code, the concept of delaying egress by locking fire exits is not a practice many people feel comfortable with. Most building owners were reluctant to attach devices to fire exit doors as they didn't want to deal with potentially violating local fire regulations. No NYC facility appropriated funds to allow the same locking arrangement on any exit doors for several years because most fire inspectors failed to recognize that security considerations had become as important as the threat from a fire, particularly in a school environment. It seems that old habits never die, and fire officials were reluctant to encourage the use of these controversial locking systems. No one in the architect community would design a new building with electromagnetic locks because the perceived liability risk was deemed more important than the safety of the building occupants. The NFPA Amendment turned out to be a pyrrhic victory, and the concept of locking fire exits again became dormant until 1986.

Sometimes it takes an outside event to force authorities to accept new concepts, especially when dealing with life safety and security issues. There were very few fire codes in existence until the Triangle fire in 1912 and fire doors remained chained and/or unlocked until 1986 when an event took place that changed the history of electromagnetic locks. A student in a Brooklyn High School was shot dead in the corridor by another student who had smuggled a gun into the school through an unlocked secondary exit. The Board of Education suddenly discovered extra funds and immediately authorized the installation of new FASs along with the electromagnetic locks. The timeline of this is outlined in the sections to follow.

2 REPLACING CHAINS ON FIRE EXITS WITH LOCKS

PRE-1980. The nation's leading fire code-writing organization, the NFPA, publishes its well-known Life Safety Code 101 which is used as a guideline by almost every municipality in the nation. This code initially only recognized fire hazards and did not consider any security problems as of equal importance to fire safety, nor did it recognize the common practice of chaining exit doors shut. The Life Safety Code specified that every fire exit door must remain unobstructed to allow immediate egress and ingress to any occupied public building. In some situations, it allowed the locking of doors after the building was closed and empty of occupants. The Electromagnetic Fail-Safe Locking System for secondary doors offered the perfect alternative to the chaining of doors but was illegal under national and municipal fire codes that dictated that no fire exit door could be locked. Free egress from a building is a fire code motto as old as the Bible, but it began to be challenged with the advent of smoke detectors, sprinkler systems, and the installation of a magnetic locking system on fire exit doors in Bellevue Hospital in 1973.

1980. Under a great deal of pressure from the security community, the powers that be at NFPA began to realize that there were serious problems relating to security. Until this time, every fire exit had to remain unlocked to allow free egress from the building in case of a fire or any other emergency. Meanwhile, school officials ignored the Life Safety Codes and chained secondary fire exits shut. Buildings had become less vulnerable to fire damage following the introduction of smoke detectors and the installation of sprinklers; however, maintaining a secure environment had become a growing priority due to the violent behavior of intruders and the threat of terrorism. Following the consensus vote of its members, who finally recognized that security had become as great a threat to individuals' safety as fire, the NFPA issued an amendment to the Life Safety Code 101. The amendment allowed the locking of secondary exits with a delayed fail-safe magnetic locking system, once it was connected to the building's FAS. Occupants could no longer simply press on the panic hardware and exit the building once the magnetic lock was installed and approved by the agency having jurisdiction. The fail-safe feature referred to the requirement that, after the FAS was activated, the lock would automatically release instantly. This concept was fought by fire officials and was not accepted by the fire alarm community. There are still professionals in the field of fire safety, including architects, consultants, and building managers, who insist that exit doors should open freely from both sides. Though they represent a small minority, these opponents have slowed the full acceptance of the locking arrangements.

1984. Four years after the controversial code amendment was enacted, not one electromagnetic locking system was installed on fire exits anywhere in the nation. It was a "lock" associated with security but considered an ancillary part of the FAS. Also, its use was met with a great deal of skepticism and outright rejection:

- Locksmiths refused to get involved with any devices that required electric current to operate
- Alarm system integrators did not understand fire codes or work with FASs
- Electricians are traditionally uncomfortable with low-voltage security alarm systems
- Architects rarely recommend any locks on fire exits for fear of lawsuits and have no understanding that security issues should be balanced against fire safety.

Barringer High School (Newark, NJ) had chains on the doors as did the other 20 high schools in the district (as well as probably additional 30-grade schools in the 100-school district). The fire department marshals regularly inspected the schools and, using a heavy-duty cutting tool, had developed a routine of cutting the chains from the doors. Little did these fire officials realize that they were about to make security history because of a unique combination of personalities. (1) The principal was a strong-willed woman, (2) the school district security director was probably the nation's most knowledgeable expert on school security, and (3) the Newark fire commissioner had an open mind and did not follow the philosophy of fire officials since time immemorial. After cutting the chains, the fire officials left the school building and were on the sidewalk when the principal immediately began chaining the doors shut again. This went on for about two hours until the frustrated inspectors called the police who arrested the principal.

The security director was summoned to the scene and promised everyone that the district would comply with the fire codes and under the amendment to the Life Safety Code would use electromagnetic locks as replacements for the chains. No one was aware of this change to the NFPA Code except for the director and the security consultant who had been hired to solve the exit door problem. Magnetic locks were mounted on the 22 fire exits and hard-wired to the FAS through a relay that released power once the fire alarm was activated. The Newark fire commissioner surveyed the site and gave approval to the school authorities and, soon after that event, the state of New Jersey sent code compliance officials to witness the operation and it became part of the NJ State Fire Safety Code.

This event was the first time any building in the United States was permitted to legally lock their fire exits.

Newark eventually locked the doors in almost all their high schools, but NYC allowed only a pilot program in a single troublesome school with intruder issues. The locks, once installed, stopped all intrusions in that particular school in 1985. No other school was selected for this locking system as the NYC Board of Education was reluctant to spend money on security devices with the exception of metal detectors at the entrances, contact switches on exit doors, and audio detection systems for nighttime security.

1987. The FDNY became aware of this locking arrangement and realized the advantages of the fail-safe lock as an alternative to chaining exit doors. A team of FDNY deputy fire commissioners visited Barringer High School and accepted the electromagnet for NYC school buildings. This NFPA Amendment was incorporated into the NYC fire code for buildings in 1987.

John Jay High School in Brooklyn became the first NYC school to lock fire exits, but the Board of Education was reluctant to expand the locking system to other troubled schools until an event that occurred in another large high school immediately changed the policy. It was a classic example that relates to all public buildings whether they are schools, museums, or hospitals. Protection and security systems are only purchased following a disaster, never as a precautionary measure.

1988. There were districts all over the country that were aware of the Newark and NYC locking systems and these locks gradually became visible in school districts throughout the nation. But not all boards of education automatically approved the expenditure for fear of being accused of violating fire codes. Joe Clark, principal of a school in Paterson, NJ, was continually being arrested for refusing to remove the chains, which he insisted were the only way the school could provide a safe environment for students and teachers. The principal gained a national reputation as an expert on how to control school violence, due to his public displays of defiance against a very uncooperative school board. After visiting the Newark School, his repeated requests for funds to purchase a similar locking system were always rejected by the very resentful Board of Education. Finally, the Board succeeded in firing the principal who then took a position with a national organization dealing with school violence. One week following the departure of this dedicated and controversial principal, the Board of Education approved funding for the locks in the school that the principal had fought to obtain for over a year. Magnetic locks were installed and exit doors were locked against unauthorized intrusion.

1991. A student at Jefferson High School wanted to settle a dispute with another student and could not get past the secured front entrance of the school, so the student entered through an unlocked secondary door. He carried a loaded gun that would have set off the metal detector at the front entrance.

He shot and killed a student in the main corridor. Twenty-four hours later that school district found $30 million to lock 35 high schools with electromagnetic locks and install new FASs to be in compliance with all municipal and national fire codes.

This is another example of the difficulty in getting financial support to upgrade security in private or public institutions unless some disastrous event forces a positive reaction. After all, security is seen as a necessary cost of business but is not a revenue-generating investment.

Finding an installation firm posed a challenge for the Board of Education since the contract involved not only an innovative electromagnetic locking system, but also the system had to be integrated into the new FAS and be in full compliance with codes and regulation concerning life safety. Finding a competent vendor that understood all the restrictions and the method to interconnect with the FAS was not an easy task.

- Locksmiths sometimes mounted these locks on interior office doors which avoided code compliance obligation – or the need to connect the lock into the FAS in some cases. Locksmiths are mechanics – not electricians.
- Alarm Installers were not familiar with the lock hardware since they left that responsibility to the locksmith – unless it involved access control or biometric systems.
- Licensed Electricians avoided getting involved with any security-type devices since they were unfamiliar with building protection procedures and usually focused on high-voltage infrastructure (above 50 volts).
- Fire Alarm Installers rarely worked on security systems and never on a questionable locking arrangement.

The consultant involved in the Bellevue fiasco and the Newark design formed an installation company that was selected to install the electromagnetic locking system in all the schools and the consultant purchased the lock hardware from Locknetics. It was the largest contract the firm had had up to that time… but the inventor had sold his interest in the firm the previous year. An order was also placed for six locks in that same year to be reinstalled on the six doors at Bellevue, which were approved by the FDNY in 1987.

Almost three decades since the initial encounter with fire officials at Bellevue Hospital, and the reluctance by building owners and architects to lock fire exits, the electromagnetic lock industry is now well established (See Figure 32.1). These initially controversial locking arrangements can be

FIGURE 32.1 Kit parts for an electromagnetic door lock.

Source: Shutterstock.com.

found in every type of institution in every section of the nation and in many foreign countries. About 10% of the installations are on fire exit doors. These locks are now found in every public building where security is essential, and the AHJ has understood their value:

- Out of the way door through which inventory might disappear
- Preventing paintings from leaving a museum undetected
- Airports have the locks on doors leading to the flight line tarmac areas
- Schools use delayed egress systems to maintain a safe environment and prevent drugs from entering
- Hospitals lock maternity ward doors to prevent infant abduction

3 FAIL-SAFE LOCKING ARRANGEMENT – CONFLICTING CODES

On the surface, it would appear that there are irreconcilable conflicts between fire safety and physical security requirements. Fire codes have as a principal objective the assurance that occupants of buildings can always promptly evacuate the premises and they specify that there be nothing which prevents any person from gaining instant egress from a building. In many cases, that is about the last thing that the security officer wants since it should be made difficult for persons to leave at will, without being seen – especially in situations where the person is carrying a recently stolen painting from a museum or is kidnapping an infant from the hospital nursery.

The real problem is not so much to prevent persons from leaving the building unnoticed as it is to keep unauthorized persons from entering. For generations, this issue was in conflict with any fire code or security requirements, but in the last 25 years, a conflict regarding exit doors arose, and that issue was addressed by the NFPA in 1980 when they amended the Life Safety Code 101. Panic hardware was almost always installed on exit doors, particularly in schools and hospitals. The problem was that panic-release locking hardware did not provide any level of security. During the 1970s and 1980s, exit door hardware was often deliberately rendered useless since the authorities operating the building had used bars, chains, padlocks, and other means to keep people from breaking in. Removing panic hardware where it was deemed unnecessary and substituting deadbolts or deadlatch hardware, plus providing other security for the door, seemed to remove the problem and resolve the conflict.

Another method to prevent unauthorized egress was to eliminate unnecessary doors, which, for some reason, had been installed because the architect believed that special exit doors should be provided for ingress/egress routes, even if the fire code did not require them. Security would be weakened considerably by having more doors than necessary. Keeping those inside the building from walking out using the fire exit was a challenging problem until the NFPA allowed the use of fail-safe electromagnetic locks in 1980. It was a necessary compromise by fire officials when they realized that the day-to-day security risks (assault, robbery, etc.) due to inadequate control of building exits exceeded the potential risk of injury from delayed egress during a fire.

Fire safety advocates and security specialists cannot strive for their opposing measures to be accepted without regard to the possible effect on the other. Security regulations cannot blandly state that it is acceptable to put bars across doors or use chains and padlocks when fire safety regulations clearly state that their use violates basic life safety principles. For many years, organizations like the NFPA failed to understand that the problem of providing an adequate degree of security was impossible when all exit doors were available for free egress. The 1980 Life Safety Code Amendment was a revolutionary change in NFPA philosophy, and it was the first attempt to resolve the age-old feud between fire safety and security and attempt to balance the requirements of both.

There are situations where these locking systems are an essential component of the security system, an example of which would be in a hospital nursery where it is part of the infant abduction security system. Every door within the nursery area (corridor, stairwells, elevator, etc.) is programmed to lock whenever an infant protection system detects the unauthorized movement of an infant near an exit door. This system is more effective than many facilities that still rely on an audible alarm signal to indicate that the infant protection system was activated.

The Tenth Edition (2012) *Guidelines on Prevention of and Response to Infant Abduction* published by the National Center for Missing & Exploited Children recognizes the importance of locking the perimeter doors leading into the nursery area:

> Install alarms, preferably with time-delayed egress, on all stairwell and exit doors leading to/from or in close proximity to the maternity, nursery, neonatal intensive care and pediatrics units. Establish a policy of responding to all alarms and instruct responsible staff members to silence and reset an activated alarm only after direct observation of the stairwell or exit and person using it. The alarm system should never be disabled without a defined countermeasure in place. A record of the alarms should be maintained and periodically analyzed for cause and potential opportunities for improvement to minimize false or nuisance alarm activations.

Only a small fraction of the fail-safe electromagnetic delayed egress locks in use in healthcare and school facilities have obtained the necessary variance from the municipal Department of Buildings to lock fire exits. For example, in NYC, hospitals had started to use the fail-safe locking system in nurseries as part of infant protection systems, but only a fraction of the installations had been filed with AHJ for a variance, making the locking of their exits illegal. This is probably the same ratio found in facilities throughout the nation. The problem centers on the fact that manufacturers providing magnetic locks may not be familiar with the different municipal fire codes issued by the AHJ. Compounding the problem is that installation firms are unaware of the NFPA Code and have no experience in filing for the necessary approvals since the process is expensive, requiring preparation of plans and schematics signed by a professional engineer or licensed architect, and the installation supervised by a master electrician. As a consequence, most locking systems throughout the nation have not received approval from AHJ.

The overriding codes for installing FASs or any device attached to the system must be in compliance with the Fire Code, Building Code, and Electrical Code. When a magnetic lock is used on a fire exit door, it must be connected to the FAS and operate exactly as described in the NFPA Code under "special locking arrangement." The installation, and all ancillary components, must be approved by both the NYC Fire and Building Departments and meet the city's municipal code even though the locking arrangement is meant to improve security – not life safety. The NFPA Life Safety Code is used as a guideline, but the Municipal Code is considered the law.

4 FIRE CODES/MANDATORY – SECURITY CODES/OPTIONAL

The NFPA Life Safety Code 101 Amendment in 1980 to allow the locking of fire exits was a revolutionary concession by the most important fire alarm code-writing organization in the nation.

Almost all municipalities follow these NFPA Codes when developing their own building codes particularly as they relate to the installation and operation of FASs. At one time, a few of the nation's largest cities, like New York, Chicago, and LA had developed their own building and fire codes that usually followed the guidelines issued in the NFPA Codes, but were modified before being adopted for inclusion into the municipal codes to reflect the particular requirements of their metropolitan area. Now every city in the nation has adopted the NFPA Code completely.

NFPA has published almost 300 Codes and Standards, with their mission to prevent the loss of life and property. Topics include fire prevention equipment, extinguishing systems, inspection, residential and commercial dwellings, the safety of flammable gases, etc. They also publish the National Electrical Code, which is the most widely adopted code in the world used in every state in the nation. Standards include many directives that relate directly to specific facilities or systems:

NFPA 10 Portable Fire Extinguishers
NFPA 70 National Electrical Code
NFPA 90A Installation of Air Conditioners

NFPA 99 Healthcare Facilities
NFPA 101 National Life Safety Code 101
NFPA 704 National Fire Alarm Code
NFPA 1561 Incident Management Incidents
NFPA 1600 Disaster/Emergency Management

Fire safety rules and codes are a century old and follow the well-established rules and regulations of the NFPA (and a few other code-writing organizations). Security follows a far different set of protocols since it isn't guided by codes or statutory requirements. It follows available technology unencumbered by codes or regulations. This lack of any restrictive codes has resulted in the phenomenal growth of the modern security industry since the late 1970s. The separation of fire safety from security had not been breached until 2005 when the NFPA unexpectedly entered this no-man's land separating these industries by issuing a set of controversial security standards that, from all indications, have not necessarily been recognized or adopted by the security community.

5 JURISDICTIONAL DISPUTES

Agencies under the control of large municipalities can become redundant or cause problems with different and conflicting levels of authority – especially in defining the codes, regulations, and standards. Within the boundaries of one NYC Borough, there are several facilities that fall under the jurisdiction of different governmental agencies, each mandating their own special sets of codes and standards, some relating specifically to security. This conflict in the jurisdiction is probably common in the nation's 100 largest cities, and it illustrates how baffling and difficult it can be to obtain approval to install a new concept using technology that is more advanced than any code or standard in existence.

Sometimes, one agency will approve an installation that is rejected by another enforcement agency within the same jurisdiction. An example of a common dispute between agencies is the installation of a fire exit door delay-locking system that meets the NFPA Life Safety Code 101. Problems arise when the locking system is used on a facility that is owned by the state with its own fire code – located in a city with another fire code – and sometimes the facility is under the jurisdiction of the Federal Aviation Authority (FAA) where an airport facility is involved. Other examples include:

- A state-owned mental health institution is authorized to operate under the New York State Code but must meet the New York City Code since the FDNY responds to fire emergencies.
- Federally owned Veterans Hospitals operate according to federal agency fire codes and standards but must also be in compliance with the local municipal fire code.
- Airport terminals (LaGuardia/Kennedy) are under the control of the FAA (federal) but the airports are under an ownership agreement of the Port Authority of NY & NJ (state) and are located in a municipality (New York City). The federal, state, and municipal fire codes vary and expecting that the facility is in compliance with all the fire codes is an impractical and almost impossible task.

If a fire emergency arises, which fire department has jurisdiction, the airport (FAA) fire department or the FDNY? If both respond, there is the possibility of a jurisdictional dispute.

6 OPERATION CHARACTERISTICS

Since electromagnetic locks have no moving parts, they are completely silent even when powered and locked. They are vandal-proof, weather-resistant, and virtually maintenance-free. There are several configurations of magnetic locks with the most common type being the pull magnet. The lock comes in two separate pieces; an armature mounted on the door frame (strike plate) and the magnet bolted to the upper edge of the door. The magnet and the strike plate must align properly to secure the door in a locked position. The locks are powered through the building's FAS so that, in the event of fire alarm activation, the power is removed instantly and the locked door can be opened. According to all the codes, the building must have an approved FAS and the authority that approves the FAS must also approve the locking system.

There may be situations where a fire condition exists, but the FAS is not activated and there are mandatory provisions in every building code that the door shall have power removed after someone manually pushes on the door for 15 seconds (or for 30 seconds under special situations). A nuisance delay period is programmed into the initiating circuit that ignores short pressure on the door caused by someone accidentally bumping into it or a gust of wind, etc. The obvious intent of this nuisance alarm is to minimize false alarms without compromising the integrity of the system. If an exit is required due to emergency conditions, and the FAS was not activated, escape can be accomplished by pressing on the door for a programmed period of seconds. In order to activate this delay egress override, pressure must be maintained on the door until the nuisance delay period has expired. The irrevocable release countdown will continue until the lock automatically releases. The lock will remain unlocked (without power until manually relocked by a reset device).

Other core requirements that every delay egress system must meet are:

- A circuit that ignores the 3-second nuisance "push" but once the 3-second threshold is passed, the 15-second release cycle is continued to the end of the 15-second delay period before power is removed from the lock which solved the annoyance problem.
- Applying no more than 15 lbs of pressure for more than 3 seconds (ignoring under 3 seconds eliminates nuisance delay) will start an irreversible process to unlock the door in 15 seconds. The NFPA Code allows an extension to 30 seconds when the door is located far from the security desk and the long delay is approved by AHJ. The lock must be manually reset at the door to ensure that someone visits the site and that no fire condition exists.

A sign must be applied to the door stating: **PUSH UNTIL ALARM SOUNDS**.

Letters must be at least 1-inch high with a stroke width of at least 1/8-inch stating: **DOOR CAN BE OPENED IN 15 SECONDS.**

6.1 Basic Locking System Definitions

Ingress: The act of entering a building.

Egress: Leaving or exiting a facility.

Fail-safe: Usually relates to a loss in power supplied to an electromagnetic or electric strike or lock, causing the lock to open and will remain in an unlocked condition until power is restored. It defines that an immediate egress after an emergency situation removes power to the lock by activation of the building's FAS.

Fail-Secure: A failure in electric power causes a secure or failed closed condition when power is not sent to the lock, as in a power outage situation. There would be no way to release the lock without a manual override feature. There are very few situations where this type of power configuration can be used and still be in conformance with national fire codes.

Latched: Door can be opened (turning a doorknob) but cannot be pushed open without operating the door handle or releasing the latch.

Access Control: A popular term that usually refers to a card reader system, but not necessarily – since it describes a process that grants or denies an individual access to a specific area based on their possession of a card or a code, a programmable key, or even a physical characteristic of the individual. Since access control systems have the potential to limit a person's ability to escape a fire or other emergency, fire and safety codes must be strictly observed.

Fire Rating: A door assembly consists of the door and the frame and is rated in terms of its ability to withstand a fire. Building codes determine the fire rating in a particular community. When door lock hardware is installed, and the door is then penetrated to allow wiring, this rating may be significantly affected. Access control-locking systems must conform to the door's fire rating so as to not reduce the fire rating of the door.

6.2 Organizations Involved in Fire Exit Locking Systems

NFPA 101 SPECIAL LOCKING ARRANGEMENTS Nuisance Delay: up to three seconds permitted

Release Delay: 15 seconds total or with 30 seconds by approval from AHJ

Relocking: Manual switch, cannot be a switch located on or triggered by door movement.

STANDARD BUILDING CODE (SBC) Nuisance Delay: Not allowed.

Release Delay: 15 seconds or with 30 seconds by approval from AHJ.

Relocking: Reset must occur only on opening, must be a switch located on or triggered by the movement of the door.

6.3 Uniform Building Code

Nuisance Delay: Required and must be set to two seconds.

Release Delay: 15 seconds total without options.

Relocking: Manual switch, cannot be a switch triggered by the movement of the door.

BUILDING OFFICIALS CODE ADMINISTRATOR (BOCA) Nuisance Delay: Required and must be one second.

Release Delay: 15 seconds total or with 30 seconds by approval from AHJ.

Relocking: Must be a switch located on or triggered by the movement of the door.

6.4 General Definitions

REGULATIONS are usually requirements published by local or state agencies and can evolve from a directive set forth in a statutory law. Regulations can have the same authority as if they were issued by a legislature and are often considered more binding than codes or standards.

> **STANDARDS** are a set of voluntary criteria, guidelines, or best practice usually used to enhance the quality, performance, and consistency of a product, service or process. Standards are often interchangeable with "codes" since both become effective only after being adopted as a regulation by some governmental authority (federal, state or municipal). A standard does not contain mandatory provisions, especially when it uses the word "shall" to indicate requirements and is in a form generally suitable as a reference by other standards or codes or for adoption into law. A standard is often adopted into law, but its typically made applicable only when it is referenced in a building code.

CODES are usually modified to accommodate or reflect specific conditions of the jurisdiction before conversion into laws and regulations (An example: Florida stresses hurricane code compliance while Midwestern cities address tornadoes, etc.). The many code-writing organizations that deal with building construction and fire protection traditionally have tried to publish a code for every conceivable situation and sometimes the codes can be either redundant or conflicting. A code becomes a legal mandate after it is adopted by a municipal-, state-, or federal-level agency. The NFPA is, by far, the most recognizable code-writing association in the nation, but there are other organizations that develop codes which usually closely follow the procedures adopted by the NFPA. Codes are not meant to be unchallenged, nor are they based on some scientific discovery – the codes are updated by changes in technology or by events that expose an established code to be inadequate or even obsolete. Codes generally undergo a revision cycle every three years and are often separated into committees that deal with specific categories such as building, fire, electrical, and plumbing codes. When a code is referenced in a state or municipal statute, it will usually specify a particular edition of that code since it takes a significant amount of time for a governmental agency to adopt, and the latest code will override the previous editions. Codes have a long and proven record of success in the fire alarm and building construction industry but have not been universally accepted in the security industry.

The NYC Department of Buildings with its 1,500 employees has hundreds of plan examiners and building inspectors to ensure that the city's nearly 1 million buildings are in compliance with all codes and statutes, a cost that none of the nation's other municipalities can afford.

An example of the complexity and restrictiveness of certain codes is in the construction of a hospital complex. Architects on staff at some of the nation's largest firms consider the process of obtaining proper permits and approvals more exasperating than if the buildings were to be constructed in a small rural community that follows national codes rather than customized local codes and statutes. Negotiating through a maze of federal, city, and state codes is complicated enough, but working with government officials can be a frustrating experience, especially when state and municipal codes conflict or are interpreted differently by various plan examiners within the same agency. In many situations, problems aren't discovered until after the structure is built and occupied. One solution that has been tried to mitigate the disputes between agencies has been the creation of a universal hospital building code which has been adopted in some edition by 43 state agencies.

A classic case in code enforcement was the construction of a neonatal intensive care unit when, just prior to full occupancy, the state building inspector refused to approve a five-foot-wide corridor leading to a storage room. The code required eight-foot-wide corridors, the width of an adult stretcher, while bassinets are only two feet wide making a five-foot corridor more than adequate. Widening the corridor would have cost $100,000 and months of delay before occupants could use the unit. The architect in charge of the project refused to comply with the inspector's demand and the matter was brought to higher government official along with the hospital's threat to initiate a lawsuit to remove the violation. The corridor remained at five feet and the unit was granted a variance allowing it to open on schedule. Codes are obviously important for any society to function, but they can be challenged and a variance issued if a reasonable request is made that would satisfy all concerned parties.

> Many major cities publish their own codes and regulations due to their sheer size and unique requirements. Generally, smaller cities haven't the need nor the resources to write and enforce their own codes and simply adopt the building, fire and electrical codes published by code-writing organizations created for that specific purpose.

FIRE CODES follow the NFPA codes except for a few exceptions where several of the nation's largest cities combine the NFPA Codes with modifications that reflect certain unusual requirements of a particular municipality. For many years, the Fire Code in NYC was customized according to the requirements issued by the FDNY, but since 2009, the NYC Fire Code has more or less followed the NFPA publications including that organization's codes. There has always been a strong interrelationship between fire and building codes which sometimes appear conflicting in goals. Balancing both building construction codes with fire codes has placed a burden on the architect community that has to comply with both. A fire code is considered a "model" code since it is adopted into law by a jurisdiction and enforced by municipal fire prevention officials. It prescribes minimum requirements relating to fire protection, explosive hazards rising from storage procedures, handling the use of dangerous materials or other specific hazardous conditions, focuses on fire suppression equipment (sprinklers and smoke detectors), and specifies the schedules of inspection of the premises and the fire prevention equipment (extinguishers).

LOCAL ORDINANCES are generally produced through a process that involves consensus vote by elected officials (City Councils) which then issue a "local law." Sometimes, the state will mandate minimum requirements that supersede local ordinance (e.g., smoking ban in restaurants, gun control, and medical marijuana dispensaries), but the public usually adheres to local law. A local ordinance is a municipal code; in the larger cities, these are enforced by the Department of Buildings and only slowly revised, if at all. Local laws are periodically issued to amend the basic code. NYC has been the focus of attention since 9/11 for obvious reasons, and while the enormity of that eventful day has caused certain changes in the building code, the code has remained essentially the same since it was last revised in 1972. A newly revised code was issued in 2008 that reflected on the collapse of the twin towers and provisions were added in many areas concerned with factors that addressed emergency situations that might be caused by future disastrous events.

STATUTORY LAW is used at the federal and state levels to cover the building code for fire and safety requirements – never for security purposes. On nonfederally owned property, authority rests with the state and local governing body which issues laws, ordinances, or statutes for specific situations.

RULES should not be confused with "rulings" made by a judge in a judicial courtroom setting. A rule may not be enforceable since that would make it a regulation or a code. A rule is usually an authoritative statement of what to do or not to do in a specific situation and may be issued by an appropriate person or organization, becoming a regular practice even informally, like a "rule of thumb." It could also be used to clarify or interpret a law or policy but, in general, it is just a little more significant than a guideline or tradition. Rules can become important over time simply due to repetition and established practice.

Sometimes certain rules can create problems, specifically when they are old rules and have not been updated to be in line with modern-day society. In the security community, rules are too vague and serve no purpose unless they are in written form, published by a respected organization and, most importantly, there is a penalty involved if the rule(s) is broken. Rules can be divided into four categories:

Folklore – Conveyed by behavior and is implicitly understood.
Guidelines – Commonly published as recommended practices and carry different interpretations.
Mandates – Commands that cannot be ignored and violators are punished.
Policies – Cover predictable behavior and violations can be dismissed under certain circumstances.

Section XII

Fire Safety and Code Compliance

There has always been a close and often controversial relationship between traditional fire safety codes, many of them in direct conflict with security requirements. Some of the conditions and restrictions mandated by fire codes often compromise the security of building occupants and are reviewed and criticized in depth. Fire codes have a long history and, for the most part, have been created and defended by the powerful and often controversial National Fire Prevention Association (NFPA) which uses a consensus vote to change or update codes.

This section focuses on various types of fire and smoke detectors and how they operate using different technologies since being introduced in the mid-1970s. These small, inexpensive, and highly effective detectors have been the basis for the decline in the nation's fire deaths while security threats have grown exponentially. Yet, the security industry has never adopted "Security Codes" to counteract the influence of the NFPA-created codes. This section will also cover the operation of smoke detectors that include Ionization Detectors, Photoelectrical Detectors, Chemical Detectors, and Carbon Monoxide Detectors.

There have been many well-intentioned attempts by a number of national standards writing organizations to establish codes relating to the security industry, which go as far back as 1975. These attempts have had no impact on the industry, mainly due to rapid changes in technology and the lack of any centralized code-writing organization that has a similar influence as the NFPA on the fire prevention community. The technological changes in the security field have been so rapid that any attempt to set a standard acceptable to all concerned parties would fail, since any basic standard that would satisfy everyone's interests would take so long to develop that a security standard would become obsolete by the time it was adopted by the relatively slow consensus vote process.

Fire codes are continuously updated following disastrous fires causing massive death tolls. Four major fires are detailed to illustrate how fires can start after code compliance failures:

Iroquois Theater
Triangle Shirtwaist Factory
Cocoanut Grove Restaurant
Beverly Hills Supper Club

DOI: 10.4324/9781003091103-45

33 Reconciling Fire Safety and Security Protection

1 FIRE PROTECTION CODES AND STANDARDS

The largest and most influential code-writing organization in the nation has been the National Fire Protection Association (NFPA) for almost a century. This organization has more than 60,000 members, encompassing about 200 technical committees and 7,000 voluntary participants in the standards development process. The codes and regulations are adopted by thousands of municipalities all over the world, with some larger cities having their own codes that incorporate the NFPA codes and standards into their municipal electrical and fire codes with minor modifications. Membership is made up of representatives from various technical disciplines and industries, including the American Institute of Architects, National Electrical Contractors Association, Builders Hardware Manufacturers Association, and the International Council of Shopping Centers, as well as thousands of non-affiliated individuals. Anyone who can afford the $175/year dues can join and vote on any newly proposed codes at the organization's annual meeting.

A few security practitioners have become active in recent years, mainly as representatives of the Professional Alarm Services Organization of North America, the Security Industry Association, and ASIS International. Since its founding in 1922, the NFPA's main focus has been working toward having a more balanced membership – but the voting membership has traditionally come from the fire alarm industry and not from the security or other adjacent industries. By comparison, ASIS International, which is in many ways the counterpart to the NFPA, has roughly 34,000 members, almost all of whom are from the security field, with very few from the fire alarm industry. There has never been much commingling between fire safety personnel and security advocates or between the NFPA and ASIS. ASIS was formed almost a half-century after the NFPA; these two highly visible organizations share an interest in life safety, but from different directions and with different goals. While this observation may be too simplistic, the differences between the two organizations became more evident in 2005 when the NFPA issued two security standards, 730 and 732. The publication was met with some hostility from the security community, and ASIS reacted almost immediately by establishing their own standards and guidelines in 2007. In doing so, ASIS started a rivalry with the NFPA (possibly unintentionally) and may expand their standards committee to issue fire alarm system codes.

Fire Prevention Standards developed within the NFPA are based on a consensus vote procedure that allows members in good standing to participate in the development of codes and regulations that relate to the design of fire detection systems. Members who are not present at the meeting cannot vote. While the process may not be scientific, it is a method that has produced fire codes that have been adopted by municipalities throughout all 50 states and have been copied by numerous foreign nations around the world. There has never been an equivalent code-writing group in the security field with influence comparable to the NFPA within the fire safety community. The few attempts that have been made by other code and testing organizations have never produced any significant results, and certainly nothing that could compare to the influence on life safety in building design as NFPA codes. NFPA does no testing, has no laboratory facilities, and has no legal authority to enforce any of the codes it establishes; some of the codes have been criticized over the years for being impractical, unscientific, and too closely aligned with particular segments within the alarm system manufacturing industry.

DOI: 10.4324/9781003091103-46

2 CONSENSUS VOTE IN ESTABLISHING UNSCIENTIFIC CODES

Smoke detectors in hospitals were the subject debated by NFPA membership in the early 1980s, and it illustrated the popular misconception surrounding codes and regulations, how they are developed, and how competent the members of code-writing organizations are. The public, by and large, accepts the codes as if they were scientific discoveries based on lengthy research development procedures. These codes, whether they be fire, building, or electrical, are not based on any testing like those the Food and Drug Administration makes the pharmaceutical industry go through before drugs are approved. Fire codes are issued almost 100% of the time by the NFPA and, it has been argued that sometimes these codes are harmful or based on financial gain rather than directed toward improving life safety. Again, the fire code is adopted by a consensus vote – not after a scientific testing procedure.

3 UNACCEPTABLE SECURITY STANDARDS

In contrast to life safety and fire codes, those who design, construct, and install security systems have shown no desire to establish codes. There have been many well-intentioned attempts by a number of national standards-writing organizations to establish codes relating to the security industry which go as far back as 1975. These attempts have had no impact on the security industry. The technology changes in the security field have been so rapid that any attempt to set a standard acceptable to all concerned parties would fail, since any basic standard that would satisfy everyone's interests would take so long to develop that a security standard would become obsolete by the time it was adopted. On top of that, there is no consensus body or process that could hold a debate or vote on such.

Fire alarms impact life safety with their main focus of protecting against fire damage to a building and injury to occupants, while security systems are primarily concerned with protecting property and occupants at all times – not from fire, but from theft or other crimes against occupants. That separation of goals between the two technologies is sometimes overridden by the federal government's emphasis on terrorism. Commingling of codes between the two factions may never be achieved due to philosophical aspirations and tangible objectives that frequently conflict.

Since the 1980 amendment to the NFPA Life Safety Code 101, which dealt with the locking of fire exit doors, very few other concessions were addressed specifically to the security community by that organization. That indifference changed in 2005 when NFPA 731 and NFPA 730 were issued by the association, developed by one of the many NFPA committees. The NFPA went against tradition by entering the security industry domain, based partly on the previously issued NFPA 5000, which covers the "Building Construction and Safety Code." The wide acceptance of that code appears to have encouraged the organization to move into the whole building environment, which includes topics relating to electronic security systems. Even though the NFPA attempted to recruit the expertise required to develop a security code and solicited comments from security industry practitioners, the resulting standards became very controversial, and it is doubtful that NFPA 731 and 730 will ever be accepted by the security community. It wouldn't be the first failed attempt to establish standards in the security field, and it probably won't be the last:

NFPA 731: Installation of Electronic Premises Security System
NFPA 730: Guide for Premises Security
NFPA 99: Standard for Healthcare Facilities
NFPA 500: Building Construction and Safety Code

These four standards were intended to provide a minimum standard for the installation of several types of security systems. NFPA 731 focuses on eliminating false alarms as much as possible. The code also discusses the training and qualifications required of the installer and other superfluous and extraneous conditions. NFPA 730 focused more on security management in healthcare facilities. These two security-related codes were prepared by a fire alarm organization and have been ignored by the security community. NFPA 99 is a little-known standard that requires the installation of electronic premises security systems that meet the requirements of NFPA 731.

The non-standardized security industry seems to periodically attract a few well-meaning code-writing organizations that, for over several decades, have tried to establish guidelines that cover some of the same objectives that have been used in the century since the NFPA was established. Such code-writing attempts reappear every five years, repeating the same mistakes as their predecessors. There are no organizations that perform for the security industry some of the same functions that the NFPA has done so successfully for the fire safety industry since 1921. Many of the individuals involved in past attempts to establish security standards have retired without leaving a lasting impact on the subject. There are other highly technical industries that have also resisted attempts to establish codes and standards, like the computer industry that changes technology almost annually, or even the aircraft industry, which generally has only to meet safety or hazard standards – not aerodynamic standards. Some of the early adventures in developing security standards occurred in 1973, when two of the nation's largest standard-writing organizations combined forces to develop some standards for a new, evolving security industry. The first private code-writing organization to focus on security standards and regulations was the American Society for Testing and Materials (ASTM). The ASTM formed the F-12 Committee on Security in 1972 and held a series of meetings at their headquarters in Philadelphia that only a few attended; no one paid much attention to their recommendations.

In 1979, the ASTM was still trying to write standards and co-sponsored a symposium on the topic of building security with a federal agency known then as the National Bureau of Standards (NBS) (which became the National Institute for Design and Technology (NIST) in 1982). The NBS meeting was attended by 200 practitioners in the various disciplines that comprise the private security industry. That meeting is considered the very first attempt to establish an industry-wide set of standards in the security industry. It resulted in a series of publications on various security-related subjects, and a book was published by the NBS containing the 14 presentations made by the experts at that historic meeting. Some of the topics discussed at that meeting could still be informative in today's environment, even more than 40 years later:

1. Emerging Concepts in Building Design.
2. Evaluation of the Hartford Crime Control Experiment.
3. No-Force Residential Burglaries.
4. Use of Victimization Surveys in Security Planning.
5. Reducing Vandalism in Naval Bachelor Enlisted Quarters.
6. Crime and Economic Development.
7. Security Systems in Museums.
8. Crime Prevention Through Building Codes.
9. Reconciling Fire Safety and Security Requirements.
10. Conflicting Requirements of Exit Doors.
11. School Building Security – A need for Standards and Guidelines by **Charles Schnabolk.**
12. Design Considerations for High-Security Interior Intrusion Detection Systems.
13. Jamb/Stud Wall Security.
14. Perimeter Lighting for Security.

Nothing ever came out of that well-intentioned meeting, and, in the ensuing 30 years, there have been several code-writing organizations that have attempted to become pioneers in developing security-related codes and standards, including the ASTM, American National Standards Institute (ANSI), and BOCA. These national organizations issued a few basic codes and regulations supposedly to force manufacturers to follow certain guidelines in manufacturing protection and detection devices and systems. All those early efforts experienced one common trait… all the codes related to security systems have been ignored and long forgotten. Even the US government's Sandia National Laboratory, a world-renowned testing organization, had, as long ago as 1980, tested and issued some standards on security devices, but no one still in the industry from those early years can remember exactly what the standards involved and all the devices tested have, of course, become obsolete.

4 LIFE SAFETY AND SECURITY ARE CONFLICTING

Creating a national code or standard often relies on very unscientific methods, and sometimes the rules and procedures can be circumvented by the very people who created them. Certain voting members of the code-writing organizations that rely on consensus vote to develop codes can be influenced to favor vested interests over practical considerations. An example of "stacking" the vote occurred in the early 1980s when the NFPA was considering adding smoke detectors in every patient room in hospitals. The previously established code specified smoke detectors in the corridor at designated separations. The vote was to add a smoke detector to hundreds of interior patient rooms which would be a costly expense to the medical facility. The code was submitted to members prior to the annual meeting so as to promote discussion, and possible revision, of some of the conditions spelled out in the code. There were several hundred meeting attendees, which was unusually high for a meeting on code change. Instead of a calm discussion, there was a heated debate between proponents and opponents that appeared much like a current political debate.

Hospitals within a thousand miles of the meeting location had hundreds of their medical and building staff join NFPA, prior to the meeting, and bused these new members to the meeting. At the same time, manufacturers of smoke detectors had arranged for their employees to join the NFPA and provided dozens of buses for *their* members to attend. Not too many in either group knew exactly what the difference was between a smoke detector and a thermostat. The vote was close and the proposed amendment for installation of a smoke detector in every patient room was defeated. The following year the amendment was approved because manufacturers had gathered more voting members (employees) at this follow-up meeting. This scenario is not unique for a consensus vote. Changes in technology and the threat of terrorism mandate amendments to some life-safety codes. There are now some safeguards in place to prevent one industry from controlling a vote by limiting the right to vote to only those individuals who have been an NFPA member for a year or longer instead of just 24 hours. The passage of the 1980 amendment to the life-safety code to allow locking of fire exits was a major compromise by an organization that held the locking of a fire exit sacrosanct, and it indicated that security considerations would have some impact on how life-safety controls were incorporated into the basic building design. Since that amendment, the NFPA slowly became more involved in attempting to issue security standards. Again, the NFPA and ASIS International were looking to demarcate and defend their domains and in doing so, created a sort of technical Hatfield – McCoy feud.

ASIS International experienced rapid growth along with the expanded security industry and now has over 45,000 members. Their attempts to create security standards and regulations have not been successful in overcoming the NFPA as a source of building codes and standards. The NFPA remains the leader in safety codes with the same level of influence the National Rifle Association (NRA) has on the Second Amendment. In fact, the NFPA issued the NFPA 3,000 Standard for an Active Shooter/Hostile Event Response Program standard in 2018, which, per their website, states that the program:

> … helps entire communities holistically organize, manage, communicate, and sustain an active shooter/hostile event preparedness, response, and recovery program.
>
> The world's first active shooter standard, NFPA 3000 was developed by representatives from law enforcement, the fire service, emergency medical services, hospitals, emergency management, private security, private business, the Department of Homeland Security (DHS), the Federal Bureau of Investigation (FBI), and the Department of Justice.

Again, the NFPA is seeking to fill the void surrounding the lack of security standards for buildings that may be prime targets for active shooters.

The curriculum that undergraduate architects and engineers are exposed to does not include courses specifically focused on security design or related technology, but they are very conscious of the liability to building owners if fire codes are ignored. Furthermore, architects and design engineers traditionally treat security as a task for the building owner or manager to handle after the building is occupied, which adds significantly to the building costs. It would be less expensive if security systems were built into the basic design prior to construction as is the case with fire alarm controls. Critics of this attitude, held largely by the architectural community, describe the typical architect's focus on aesthetics rather than function. And architects rarely add electromagnetic locking arrangements to the fire exit doors, even though the NFPA approved the system in the 1980 amendment.

5 UL SOLUTIONS (PREVIOUSLY, UNDERWRITERS LABORATORIES)

The largest independent testing laboratory in the United States, UL (Underwriters Laboratories) Solutions, with more than 15,000 employees, has conducted tests on systems, devices, and materials to ensure compliance with safety and health standards for nearly 130 years. This laboratory has established 1,000 standards and more than 19,000 unique products are tested each year, including firefighting equipment, lighting fixtures, and access control systems. Examples include the thickness and type of material used as wire insulation, proper grounding, components in a fire alarm system, and electrical appliances – to ensure that there is no danger of overheating or electrical shock, etc. The tests also include many fire and security products found in hospitals such as electromagnetic locks; recently, UL has included infant protection systems that are tested to meet existing standards. UL-listed devices are identified with the UL label which can be found on over 22 billion unique items per year (see Figure 33.1).

FIGURE 33.1 A typical UL tag near the electrical plug for a set of holiday lights.

UL is the most well-known testing laboratory in the nation, and it began its association with the burglar alarm industry in the 1920s. This association was formed when a group of insurance companies requested the assistance of UL in the establishment of a rating system for alarm installations for the few security products available at that time. The result was the development of definitions for different levels of alarm protection and the establishment of priorities of vulnerability and grades of protection, a reference that was lacking until this time. Before a standard is adopted, comments are solicited from interested parties that include manufacturers, consumers, individuals associated with consumer-oriented organizations, academicians, government officials, and industrial and commercial users. Care is taken to avoid having a small group of manufacturers design the testing procedure for their own product.

A manufacturer seeking a UL listing pays a fee to have the unit tested against related standards. In the healthcare security marketplace, the UL label is essential for any device that is connected to any part of the fire alarm system, which means that fail-safe electromagnetic locks must be UL listed under all circumstances. The UL label for other security products is not essential as the devices are upgraded so often to remain competitive that any practical test would be useless because by the time the test was completed, the product would likely be obsolete. One infant protection system had been UL tested, but the test was probably more related to environment tests involving humidity and temperature limits, which translates into proving that the system will operate at temperatures over a hundred degrees and in humidity, but there is no test to ensure that the system will activate the alarm system when the tag is moved near an exit. Once the device is listed, the manufacturer must affix a UL label to each unit, and there is a charge for each label. The public relies on these labels when purchasing electrical supplies like lamp cords and light bulbs, but the UL label is not very important for a complex security system.

In New York City, when any security device is connected to a building's fire alarm system, the device is considered an ancillary component of the system and must be UL "listed," or meet the testing standards of another approved laboratory, before the device will be accepted by the Fire Department of NY.

34 Smoke Detectors and Sprinkler Heads

1 FIRE DEATHS ARE DECREASING – EXPONENTIALLY

The nation has experienced a significantly low rate of deaths and injuries due to fires since the turn of this century. This could be credited to better code enforcement and the fact that buildings with sprinklers have never experienced a death – and that smoke detectors, introduced in the 1970s, have revolutionized residential safety. Some critics of the National Fire Protection Association (NFPA) consensus method of code creation and fabrication claim that since sprinklers are so effective, codes requiring certain facilities to have *both* sprinklers and smoke detectors are overkill. One possible reason is that the NFPA is strongly influenced by insurers and manufacturers who profit when additional detectors are mandated, even though statistics indicate that adding sensors doesn't necessarily provide a safer environment in every situation. If smoke detectors save lives, why add sprinklers and vice versa? Critics feel that property insurance firms are more interested in saving the property by encouraging codes that mandate that buildings have sprinklers rather than in saving lives by adding smoke detectors, but this explanation may be too critical of the insurance industry.

According to the NFPA, US fire departments responded to more than 47,000 home structure fires in 2023. Over the last five years, an average of between 3,700 and 4,400 Americans died in fires each year, with thousands more injured. Reliable studies have shown that sleeping children often do not react quickly enough when a smoke alarm goes off and must be roused by an adult. Very young children can't get out of the building by themselves and need to be rescued. Roughly half of all children who die and are injured in fires are four and younger, according to a federal report issued by FEMA. Children younger than five made up over half of the deaths of children younger than age 16 in fires. As a consequence, death from fire for children younger than 16 was the third leading cause of accidental death after transportation accidents and drowning.

On average, 12 people die each day in US home fires, with older adults being the most likely victims. Cooking equipment remains the leading cause of home structure fires and fire injury; however, smoking materials also persist as a leading cause of home fire death, according to the report. Home fire deaths – from homes in which no smoke alarms were present, or in which smoke alarms were present but did not operate – accounted for 60% of all home fire deaths. Smoke alarms were lacking in 37% of home fire deaths, and at least one alarm was present but non-operational in 23%. Fire sprinklers were present in only 6% of reported home fires.

500 civilians died in highway vehicle fires and over 3,000 died in home fires, while deaths in commercial buildings in recent years numbered 50.

The advocates of stricter fire codes still quote the deaths that occurred during the Great Chicago Fire or the Triangle Shirtwaist Factory fire in New York City (NYC). These two disasters occurred almost a century ago, and many fire officials continue to use those two incidents to justify the need for more sensors and sprinklers.

Local Agencies Having Jurisdiction sometimes find it to be politically safer to reject a new concept (e.g., locking fire exits) than to issue a variance to allow fail-safe locks. Around 1990, ten years after the NFPA amendment was adopted that permitted the installation of a locking system on fire

DOI: 10.4324/9781003091103-47

exit doors, the security industry actually began to install the locks on fire exits in large numbers. Recent statistics indicate that when these delayed egress fail-safe locks are installed in full compliance with municipal codes, they enhance security without compromising life safety.

Traditionally, fire protection codes had been given higher priority than security considerations, an attitude that has slowly changed over the past decade (since 9/11). Thanks in large part to the introduction of smoke detectors and the mandated use of sprinklers, building occupants today are in more danger from potential criminal assaults and terrorist attacks than from fire.

Suppositions and Preconceptions

- Fire equipment manufacturers do not produce security or surveillance controls.
- Fire alarm installers are rarely trained or interested in the installation of security systems.
- Security Directors may or may not share their responsibilities with the building Safety/Fire Directors. Some wear many hats, while others specialize.
- Municipal Fire Inspectors have existed for many years – there are no equivalent Security Inspectors.

A statistical example of the impact that fire codes and smoke detectors have made in the nation is the fact that NYC had the lowest number of fire-related deaths in 2016 since the Fire Department started keeping records 100 years ago. That decline is shared by every municipality in the nation.

2 HISTORY OF SMOKE DETECTORS

In the 1930s, American firms concentrated on upgrading burglary protection and expanding central station coverage, while Europeans took the lead in developing new technology that improved the art of fire detection. It was during this period that a device was developed that would eventually change the way fires were detected throughout the world. The device was an "ionization" smoke detector, which was far superior to existing heat and flow detectors since it could detect a fire before it burst into flames or gave off heat (heat sensors failed to provide early warnings for fires that smolder and produce deadly smoke before bursting into hot flames).

The device was invented in Germany to protect munitions plants in the late 1930s and was refined by a scientist, Dr. Miele, working with a Swiss firm, Cerberus, that was investigating techniques to detect toxic gases in underground mine shafts. Some experts in the industry insist that Dr. Miele was a heavy smoker and would accidentally activate the gas sensor as the invisible products of combustion from cigarettes caused gas sensors to alarm. When it was realized that there was more potential in developing a smoke detector, rather than a gas detector, a patent was obtained. The ionization detector was born and sold throughout the European market following the end of WWII.

Smoke detectors immediately changed the way fires were detected since they could sense a fire earlier than all other sensors used in the industry. Ionization detectors sense "charged particles" which are common to all fires in their early stages. The technology was so innovative that no American firm could be found to distribute the smoke detector. Several fire alarm system manufacturers and installers, such as ADT, Grinnell, and Edwards, were contacted by the Swiss manufacturer (since they controlled most of the fire protection and monitoring market in the United States), but none were interested. These well-established firms did not have the slightest understanding of the enormous market potential for a ceiling-mounted device that could detect a fire before the flame or heat was noticeable. In later years, an entirely new industry was created; in the late 1950s, a very small company in New Jersey sold CO2 fire extinguishers (Pyrene), becoming the nation's sole distributor for the product. They sold millions of detectors, which eventually changed fire codes throughout the nation and significantly lowered the death and injury rate from fires.

The first description of a smoke detector in a municipal fire code was in 1975 when NYC issued the Local Law 5 code for high-rise buildings. The code included many controversial changes and took many years before it was approved. Most of the experiments used to develop this code were conducted during the controlled burning of buildings in downtown New York City (NYC) that were being ripped or burned down to be replaced by the World Trade Center.

Facility fire and safety protection have slowly evolved over the last century, and changes were made, not usually because of technology but more often as a result of fire-related tragedies. There has been little controversy surrounding fire safety since the introduction of smoke detectors in the United States by the Swiss firm Cerberus in 1973 and the installation of sprinklers in all types of structures. Fire safety requirements are strictly enforced by local, state, and federally mandated codes and regulations, but there is no such enforcement governing the security of personnel and property protection.

There is really only one national fire code that is followed by practically every municipality in the nation, and it is not monitored by a federal agency; on the other hand, the Department of Homeland Security has attempted to integrate security requirements into building codes since 2001. This universally accepted set of fire codes is issued by the civilian membership of the NFPA. Even the few large cities that have their own municipal fire codes generally closely follow those codes based on the NFPA Life Safety Code 101, with only minor modifications. There is no comprehensive organization that compares to the NFPA and their manual is considered the bible of fire codes in the nation.

Fire/Heat Detectors

- Heat-sensitive devices react to: (a) preset fixed temperatures or (b) rapidly changing temperatures of 15 degrees/min (rate-of-rise type). These devices have been available for over 100 years as the first line of defense against fire damage. While these detectors are the least expensive, they are also the slowest to react to a fire and are not considered life safety devices since most fire deaths are caused by smoke inhalation, not heat. These detectors are typically used in unoccupied locations as property protection devices. Most national codes indicate that if the space has sprinklers, there is no need to install a heat detector since a sprinkler head is nothing more than a heat detector.
- Flame or gas sensors are designed to "see" the light emitted by a flame or detect flammable gases. They are usually used to detect explosive situations from a long distance.
- Sprinkler flow sensors are activated when the heat sensor on an individual sprinkler head melts and water flows through the system.

Products of Combustion Detectors

- Ionization smoke or products of combustion detectors are the most important of all the fire detection devices and the following history indicates many of the same growing problems that were experienced by the security industry.
- Photoelectric smoke detectors rely on a different principle to detect invisible smoke.
- Air sampling detectors are used mainly in data processing centers and similar environments. Air from a duct is drawn into the detection chamber, where a laser is used to look for a certain number of smoke particles in a given amount of time.

The entire scenario that defines the fire alarm industry is more convoluted than the evolution that defined the security industry. The following list of corporate buyouts, mergers, name changes, and takeovers describes the complicated life cycle of the smoke detection industry:

- **Cerberus** manufactures the high-voltage ionization smoke detector in Switzerland and sells it throughout Europe – 1950.
- Pyrene CO2 becomes the American – Canadian – South American distributor for the high-voltage ionization smoke detector – 1956.
- Pyrene changes the name and forms Pyrotronics Inc. (owned by Baker Industries) – 1958.

- Baker Industries starts an armored car division and buys the Wells Fargo name from bank – 1970.
- Photoelectric and low-voltage ionization detectors introduced – 1970.
- Fire code changes allow its use as a better sensor than fixed-temperature heat detectors and list ionization detectors for high-rise buildings – 1975.
- Borg Warner buys Baker Industries/Pyrotronics/Wells Fargo – 1980.
- Merrill Lynch buys Borg Warner – takes company private – sells off fire divisions – 1988 Cerberus buys Pyrotronics – 1989.
- Siemens buys Cerberus (Cycle Complete) – 1998.

Other mergers and acquisitions that have occurred over the years in the fire detection and protection industry have involved big names in the industry, with G. E., Honeywell, Tyco, and Siemens acquiring such companies as:

Edwards
EST
Gamewell
Notifier
FCI
Thorn Automatic
Simplex Grinnel
Acme
Kiddie
Cerebus
AFA
Faraday

More detailed technical characteristics of fire alarm system devices are available from the NFPA. While the fire alarm industry is usually treated as a completely separate industry from security, there are certain situations where the two technologies overlap. One of these areas concerns the fail-safe locking system for fire exit doors. Another evolving but controversial subject involves security regulations that were adopted in 2005 by the NFPA. That particular set of regulations describing minimum operating standards for security systems caused a negative reaction and a great deal of opposition from many security professionals and security organizations.

3 SMOKE DETECTORS

The two most commonly recognized smoke detection technologies are ionization smoke detection and photoelectric smoke detection.

3.1 Ionization Smoke Detectors

The amount of the radioactive element radium in each detector was so infinitesimal that it would be harmless to humans (unless, of course, 50 detectors were swallowed all at once). The concern over low-level radiation isotopes in the detectors caused some municipalities and agencies to ban their use due to fears as to how to destroy used units without contaminating the atmosphere. The public is exposed to more radiation each time they go out into the sun or board a commercial jetliner than if they carried a thousand detectors in their pocket. However, once a rumor starts, it is hard to stop. Most manufacturers substituted americium-241 for radium in response to the radiation fear.

The ionization detector was an advance over every other detector, but it wasn't perfect; it proved too sensitive for certain areas. For example, it could not be used in the kitchen and was slow to detect fires that smoldered in rugs, couches, and other cloth material. The biggest shortcoming was that the required power source was 220 volts of direct current, which was costly to provide, and it had to be connected to the electrical system. If a fire destroyed the electrical system first, the detector would not be able to send an alarm. The original units sold in the United States cost $110.

In 1964, an American firm refined the device and manufactured an ionization smoke detector that operated off 24 volts and could use alternating current. In the mid-1970s, this company designed a battery-operated unit suitable for the private homeowner marketplace; that unit became the first to be mass merchandized through retail stores at a cost of $10 or less and was described on the front page of Consumer Reports magazine.

Inside an ionization detector is a small amount (perhaps 1/5000th of a gram) of americium-241. The radioactive element americium has a half-life of 432 years. An ionization chamber is very simple. It consists of two plates with a voltage across them, along with a radioactive source of ionizing radiation. The alpha particles generated by the americium have the following property: they ionize the oxygen and nitrogen atoms of the air in the chamber. To "ionize" means to "knock an electron off of." When you knock an electron off of an atom, you end up with a free electron with a negative charge and an atom missing one electron with a positive charge. The negative electron is attracted to the plate with a positive voltage, and the positive atom is attracted to the plate with a negative voltage (opposites attract, just like with magnets). The electronics in the smoke detector sense the small amount of electrical current that these electrons and ions moving toward the plates represent.

When smoke enters the ionization chamber, it disrupts this current, and the smoke particles attach to the ions and neutralize them (Figure 34.1). The smoke detector senses the drop in current between the plates and sets off the horn.

Speaking of alarms, whenever the phrase "nuclear radiation" is used, an alarm goes off in many people's minds. The amount of radiation in a smoke detector is extremely small and is predominantly

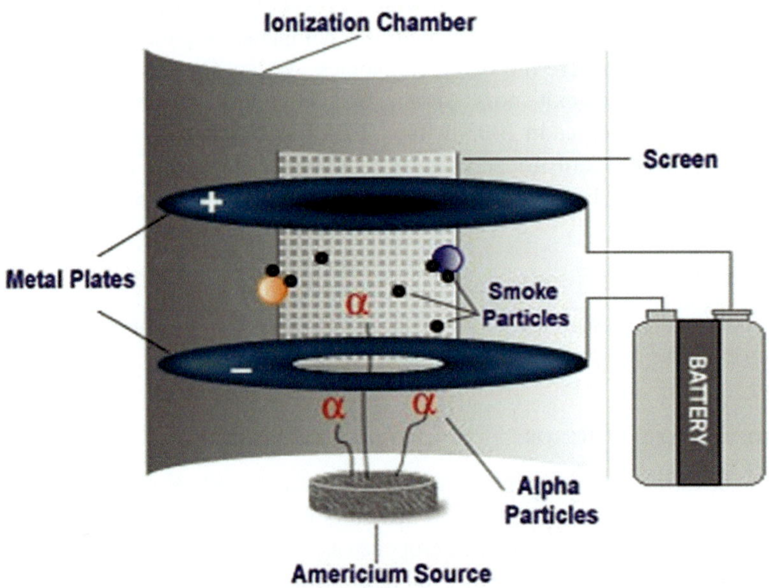

FIGURE 34.1 A diagram showing how an americium source ionizes air particles and makes an ionization smoke detector work.

Source: U.S. Environmental Protection Agency (EPA).

alpha radiation. Alpha radiation cannot penetrate a sheet of paper, and it is blocked by several centimeters of air.

PHOTOELECTRIC detectors operated on different scientific principles than ionization detectors. The photoelectric-type sensor operated more or less like the light beam detectors that were in use as early as 1935. The theory behind this application relies on the principle that smoke can block a light beam just as a human can block the light beam on the intrusion device. The small photoelectric cell allows detection of visible smoke and the operating principle is similar to that of a movie theater projection beam. If the theater were absolutely smoke- and dust-free, it would be difficult to see the beam of light between the projector booth in the back of the theater and the screen.

The beam becomes visible ("lighting up") due to smoke particles being illuminated along its length. The audience in the theater does not see a light beam; it sees the dust particles reflecting the light from the theater projector. In the photocell detector, the projection booth is replaced by a light bulb and when particles of smoke enter the chamber, they are illuminated and detected by the photocell. In order to overcome the need to periodically replace the bulb, a firm developed a chamber with a pulsing LED that lasts for years before the need to replace it (Figure 34.2).

One feature that was added to smoke detectors was that, instead of emitting a screeching sound when smoke was detected in a home, a recording of parents' voices was substituted. The prerecorded voice of a parent proved to be much more effective in alerting children to emergency situations than conventional audible alarm sounds.

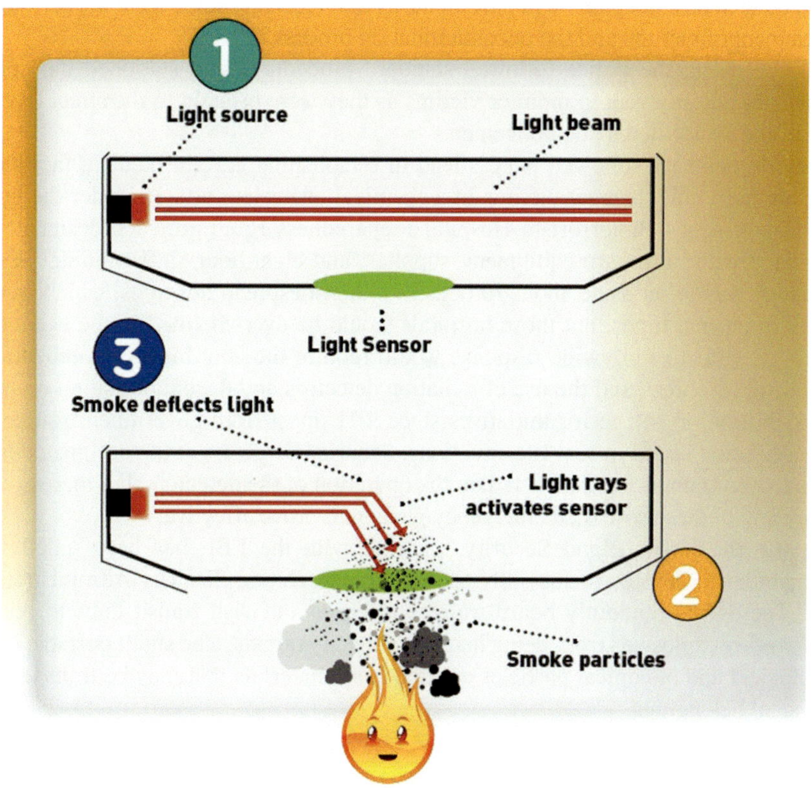

FIGURE 34.2 Diagram of how a photoelectric fire alarm works.

4 CHEMICAL DETECTORS

The possibility of a nuclear, biological, or chemical attack has become the most feared event concerning government agencies due to the potential for mass casualties.

Vacuum cleaner-type chemical detectors have been installed (and tested) in large train stations, most notably (a) Grand Central in NYC, (b) Penn Station in NYC, (c) 30th Street in Philadelphia, and Union Stations in (d) Washington and (e) Chicago. These very expensive systems were first developed in a government-controlled laboratory following the Tokyo subway attack in 1995. The system uses a modified vacuum cleaner-type apparatus to "sniff" the air collected by sensors scattered throughout the station and analyzes it for chemical toxins and lethal gases. The prototype models would falsely trigger the chemical sensor caused by the odor emitted from cleaning supplies even though the sensors were programmed for specific chemicals expected to be used as weapons. The first installation in the nation was in Grand Central Station during the 2004 Republican National Convention. Cameras were strategically placed near the sensors to help security staff isolate real terrorist threats.

The program was developed under Homeland Security sponsorship and was known as PROTEC (the acronym stands for Program for Response Options and Technology Enhancement for Chemical/Biological Terrorism). The proper response to a chemical attack in a train station (or anywhere else) is not clearly defined and the decision to designate police departments or fire departments as first responders is still being debated. In NYC stations, the Metropolitan Transit Authority (MTA) had to develop plans that defined the relationship and chain of command between the MTA police force, the Fire Department of New York (FDNY), the New York Police Department (NYPD), and even the Health Department on how to coordinate their response in the event the sensors detected an attack. It remains to be seen how hospitals will participate as first receivers since there are serious concerns as to the treatment of victims and the decontamination process.

In the Tokyo event, the police followed standard procedures and stationed the control desk directly above the subway vent to monitor victims as they were brought to the street level. The gas coming out of the vent killed some policemen.

Some major cities have followed NYC's lead in establishing certain medical facilities as special emergency burn units to treat victims in a chemical or nuclear attack. Under the guidance of the Health Department's Bioterrorism Hospital Preparedness Program, each designated hospital will be provided with burn care equipment, supplies, and eight-hour staff training sessions. The original strategy was to set aside about 70 beds in four widespread hospitals, which were thought to be enough in normal times, but those hospitals would be overwhelmed in the event of a major catastrophe. Any effective citywide response would require the coordinated mobilization of city, state, and federal resources and the use of radiation detectors and decontamination units. The city spent over $1 billion on anti-terror initiatives since 9/11 (most from government sources) without the system ever being tested in real-life situations. Security Directors at the designated health care facilities have been trained and instructed on the operation of the detectors, but there are questions as to whether any of these so-called emergency plans will prove effective.

The Department of Homeland Security, working with the FBI, has been quietly assigning handheld radiological devices to specially trained Transportation Security Administration agents since 2009. The agents randomly board trains and buses and visit transit hubs to prevent dirty bombs and/or other explosives from being hauled to big city targets. The small portable instruments detect radiological and biological particles emitted from materials that may be transported in baggage or personal belongings.

5 CARBON MONOXIDE POISONING

Carbon monoxide (CO) – often called "the silent killer" – is a colorless, odorless, and tasteless toxic gas produced by incomplete combustion in many devices. According to the Journal of the American

Medical Association, more than 400 die and more than 20,000 visit medical emergency rooms as a result of CO poisoning every year in the United States, making it the leading cause of accidental poisoning deaths in America. Inexpensive CO detectors have been on the market for many years and while all occupied buildings have to comply with fire codes requiring smoke detectors throughout the premises, very few codes mandate the use of CO detectors. The public considers the danger from fire far more predictable and destructive than that of CO poisoning. Most of the deaths caused by undetected CO are in homes (with indoor garages) and in buildings with equipment that produces combustible fumes such as:

- Fuel-burning space heaters.
- Gas stove or stovetop.
- Water heater.
- Fireplace.
- Running car in a garage or enclosed space.
- Furnace.
- Portable generators.

Fires are visible: smoke causes choking, flames are immediately recognized – but the presence of CO is completely invisible, and by the time it becomes noticeable, it's probably too late to control. Fire extinguishers and water are effective in preventing a fire from spreading, but there is no similar device or system to extinguish the insidious flow of CO. It's both deadly and undetectable by humans' sensors. CO detectors are available using two technologies to activate an alarm. One is based on an electromechanical reaction that produces currents when CO is present and the other relies on a semiconductor sensor that changes its electrical resistance when CO enters the chambers.

Severe harm can occur when exposed to high levels of CO in a short period of time (car left running in the enclosed garage under the bedroom) or when exposed to lower levels over a long period of time, as when the basement water heater is defective. People die when their red blood cells pick up CO far faster than they pick up oxygen; CO replaces oxygen, causing tissue damage and often asphyxiation. The symptoms can vary from person to person, but in general, they include headaches, dizziness, weakness, nausea, chest pains, and almost always confusion. Victims must be evacuated to an emergency room where blood samples can determine the level of CO in the blood in parts per million (ppm). When the CO levels rise above 70 ppm, recognizable symptoms of distress can be produced; levels above 150–200 can result in dizziness, unconsciousness, and even death. The common treatment used in the emergency room (ER) is administering pure oxygen to increase oxygen levels in the blood as rapidly as possible.

HOTELS. While CO detectors have been being perfected for many years and have been proven to save lives, there are many committees dealing in building codes that reject installing them in all occupied places. As an example of what is now a controversy, the people who prepare and issue International Building and Fire Codes believe that the danger is not great enough to justify the $50 cost to mount the device on ceilings throughout the premises. In the 1980s, the NFPA found itself in a similar dispute when they tried to install a smoke detector in every hospital room. In 2004, the building code-writing fraternity found itself in a controversy involving hotel safety when some municipalities sought to convince code writers to mandate CO alarms in every room. Hotel officials claimed CO incidents in hotels were rare and existing code requirements already in force were sufficient. Therefore, requiring alarms in each guest room or detectors throughout a hotel was unnecessary and too costly. The hotel lobbying group prevailed and in 2015, the International Building and Fire Code eliminated a 2012 requirement of CO alarms in every room or a detector system in all common areas. The code-writing committees stated that hotel occupants would have sufficient protection by requiring such devices only near fuel-burning appliances. While 14 states currently require CO alarms in hotels, there is no state that requires them in each guest room or in common areas.

6 SPRINKLER HEADS AND WATER SUPPRESSION SYSTEMS

Municipalities throughout the country have adopted the NFPA, *The Standard for the Installation of Sprinkler Systems (NFPA 13),* when establishing fire codes within their jurisdiction. This well-recognized sprinkler standard describes the proper design and installation of water-based sprinkler systems. The original patent for fire suppression systems was issued in 1872 and in 1881, the Grinnell Corporation produced what was to become the first mass-produced sprinkler head, which has changed very little in the intervening century and a half. That company and its acquired subsidiaries held more than 87% of the control station, fire, and burglar alarm business in the United States by the mid-1960s and were forced to split up their main business into four independent firms under a landmark anti-trust action by the federal government. The Grinnell conglomerate was split into ADT, Holmes Security, AFA Protective Systems, and ADT, with Grinnell retaining the fire safety/sprinkler sector. That split actually contributed to the sudden growth of the security industry since hundreds of small "cottage-sized" companies entered the marketplace. Grinnell is still active in the fire and security industry and is now owned by Tyco International, a large conglomerate that owns more than ten interrelated companies and is still growing.

For the first 75 years after their introduction, sprinklers were installed almost exclusively for the protection of commercial buildings but gradually, sprinklers became a mandatory requirement in certain occupancies and, in the United States, they are required in all new high-rise buildings generally 75 feet above fire department access. The five basic types of sprinkler systems operating throughout the world include:

Wet Pipe Systems are installed more often than the others and are considered the most reliable. The automatic sprinkler heads are connected to a water supply so that water is always stored in the pipe and discharges immediately after its fuse is opened by the heat of a fire. Used in areas that do not normally experience freezing temperatures.

Dry Pipe Systems are the second most common sprinkler head and are installed in unheated spaces where the ambient temperature may be cold enough to freeze the water. Water is not present in the piping and instead, it is filled with air at a pressure below the water supply pressure. When the sprinkler head is opened due to the rising temperature from a fire, the air pressure in the piping drops, allowing water to enter the sprinkler system. According to NFPA 13, the time of water delivery cannot be more than 60 seconds.

The perceived benefits are due to the fear that a wet system piping may slowly leak water without attracting notice, while a dry system does not pose this threat. These dry systems are more complex than wet systems and require more maintenance.

- **Pre-Action Systems** are generally installed in locations where accidental activation is undesired or where the objects can be seriously damaged by water, such as in museums where artwork and manuscripts are on display. Data centers are also vulnerable to water damage associated with wet pipe system discharges. There are two subtypes of pre-action systems:
  - Single interlock – similar to a dry system but requires fire detection to take place prior to the introduction of water to the area. The intent is to reduce the time delay of water delivery to sprinklers that is inherent in dry systems.
  - Double interlock – is much like a deluge system except that automatic sprinklers are used and it requires the activation of both the fire detectors and the automatic sprinkler operation to take place. Without the concurrent operation, there will be no water allowed to enter the piping.
- **Deluge Systems** have all the sprinklers connected to the water piping system and the heat-sensing element in the automatic sprinklers is removed. When a fire is detected – or a pull station activated – the deluge valve (controlling the water flow) is opened and all the heads release water simultaneously. These systems are installed in situations where rapid

fire spread is a great concern, as they provide a deluge of water over the entire hazardous area.

- **Foam Water Systems** are used for special applications by discharging a mixture and low-expansion foam concentrate that results in a spray of foam from the sprinkler head.
- Generally, used in special occupancies found in the closed storage area of flammable liquids and in airport hangars.
- **Water Mist Systems** are used to create a vapor that absorbs more heat than water and will cool the room, thus more effectively reducing the temperature of the flame.
- **Water Spray Systems** operate identically to the deluge system but piping and discharge nozzle spray patterns are designed to protect unique areas such as electrical transformers containing oil for cooling and can be used on the surface of tanks containing flammable liquids or gasses, and the spray is cool to prevent tank rupture.

Waterless Systems are designed to protect assets within the protective space that would be seriously damaged if these valuables became wet. Conventional sprinklers are effective for preserving building structure from a fire, but they're not designed for a telecommunications closet, transformer room, flammable liquids, a computer lab, or data center. Waterless systems are considered special hazard fire protection systems, commonly referred to as waterless sprinklers or clean agent systems. These sprinklers use inert gases or liquid chemicals to smother flames. By avoiding a rush of water during a fire, damage to interiors and critical infrastructure is minimized. Special hazard fire protection systems that are considered replacements for halon either use a halocarbon compound (compounds containing carbon, hydrogen, bromine, fluorine, iodine, and chlorine) or an inert gas such as argon or nitrogen.

Unlike water sprinklers, which measure coverage per square foot, clean agent systems are based on cubic volume and are designed to allow gas to be discharged in ten seconds or less.

7 PHOTO-LUMINESCENT EGRESS

Marking the path of egress in buildings with photo-luminescent (PL) egress products has made a tremendous impact on life safety in a relatively short period of time. The motivating factor for its growth can be traced to the NFPA, the state of California, and the city of New York. These peel-and-stick strips of luminescent tape are mainly installed in fire stairwells and other corridors along evacuation paths in high-rise buildings.

A rare earth element allows their molecules to store light like a battery and release it when darkness falls. The basic element is a phosphor, a substance that radiates visible light when energized or exposed to energy like an electrical charge or light source. Television screens and fluorescent light bulbs are the two most common uses of phosphors. In televisions, electrons energize the phosphors and in fluorescent lights, ultraviolet light is used. In both cases, the visible light can be seen.

PL products are almost identical and share certain features:

- Not radioactive.
- No batteries or outside electricity.
- Once installed – maintenance-free.
- Reliable and glows immediately in dark areas.
- Lasts for 16 hours and longer.
- Should last over ten years under normal use.

NYC has had a history of developing the most stringent building and fire-related codes in the nation for many obvious reasons, particularly in this century and following the events on 9/11. A special task force known as the World Trade Center Building Codes Task Force brought together a broad coalition of affected groups, including experts from the public and private sectors and survivors of the World Trade Center (WTC) disaster. The team focused on reviewing building design construction and operating requirements for high-rise buildings to determine if modifications were needed to enhance public safety.

The results of their work were the passage of 13 recommendations into law, including Local Law 26. This law mandated, among many other requirements, the installation of PL (glow-in-the-dark) exit path markings to aid in the safe and speedy evacuation of high-rise buildings in the event that both the primary and secondary power supplies fail. This law did not provide for enough light to illuminate dark egress paths, but it did describe signs and outlines of egress paths, stairs, handrails, and obstacles so that occupants evacuating a building could easily and safely navigate the path in dark conditions. They are required to be mounted at locations near the ground in case of smoke conditions and so as to be easily seen in crowded situations. They are, to a limited extent, found in various locations in hospital complexes, but the law does not specifically cover hospitals and was limited to commercial high-rise buildings. The law will be extended in the near future to include other buildings such as healthcare facilities.

There was a mad rush to meet this code requirement, and many firms entered the market; within two years over 1800 high-rise buildings in NYC were "lighted." By 2007, the NYC market had all but disappeared, awaiting extension of the code to other types of buildings. Other cities and states have followed the national codes passed in 2008 and 2009, giving new life to the industry as it spread throughout the nation. With a PL sign, the only effort required is the simple installation and the need to hold a light meter to the nearest lighted exit sign to ensure that five foot-candles (one lumen per square foot) of light are present. The measurement is done by holding the sensor of the light meter up against the ceiling next to the fixture, then moving the meter away until the meter falls to five foot-candles. The distance between the fixture and five foot-candles is the effective operating distance for the PL signage to illuminate, according to most codes.

8 LIGHT INTENSITY INCANDESCENT BULBS

It is still amazing to realize that in 1880, less than two years after introducing the tin foil phonograph, Thomas Alva Edison, working alone and not in a well-financed laboratory, earned a US patent for the incandescent bulb. That bulb still lights homes worldwide, an especially impressive feat in a world where the iPhone gets revamped every few months, and computers become obsolete in three years. The longevity of incandescent bulbs finally came to an abrupt end when Congress passed the Energy Independence and Security Act in 2007, requiring light bulbs to become at least 25% more efficient, which meant that the popular 100-W bulb could only consume a maximum of 72 W of power; a similar reduction was placed on other wattage bulbs. This was reversed under President Trump's first term, though the durability of that decision is still in question.

This efficiency requirement all but eliminated the production of Edison's bulb and manufacturers began phasing out all sizes of the incandescent bulb in favor of compact fluorescent lamps (CFLs), light-emitting diodes (LEDs) and mercury-free halogen incandescent bulbs, each of which is at least 75% more efficient than Edison's invention. The justification for phasing out the standard incandescent bulb is its inefficiency, wasting 90% of its energy on heat rather than light. That is the reason why incandescent bulbs are so hot to touch – and why the European Union phased out incandescent bulbs in 2009.

The 100-W incandescent was the first to be phased out (2012), with the 75-W (2013) and the 60- and 40-W (2014) following. California banned the manufacture of the 100-W bulb in 2010, two years ahead of the national schedule. The 72-W halogen incandescent gives off the same light (lumens) as a 100-W incandescent bulb and costs about $2 but is estimated to save $3 in electricity over its 1,000-hour lifespan. Another option that has gained favor over other bulbs is LED lighting, which can last up to 22 years and uses less power than either CFL or halogen incandescent bulbs. LED units cost $20, but the price came down to $10 in 2014, and they now outsell all other bulbs.

However, the public has been reluctant to give up the standard Edison light bulb that they have been purchasing for more than 130 years, and many are stockpiling them before they disappear from store shelves.

Incandescent	CFL Bulbs	LED Bulbs
Initial Cost: $50	Initial Cost: $3.00	Initial Cost: $10–$30
Lasts: One Year	Lasts: Six Years	Lasts: 25–30 Years
Energy Cost $72.50	Energy Cost $16.62	Energy Cost $5.00
(over six years)	(over six years)	(over six years)

Three-way bulbs, appliance bulbs, colored bulbs, and a few other specialty products are excluded from the new standards. Halogens and LEDs can be dimmed, but the less expensive CFLs cannot.

Halogens and LEDs don't contain mercury, but CFLs have trace amounts (an average of 4 mg per bulb; older thermometers have 500 mg). No mercury is released unless the bulb breaks but the amount is too small to cause any harm to humans; there's more mercury in an order of sushi than a CFL, which is why the government warns against eating too much tuna.

8.1 Light Intensity – Measurements

Light intensity is measured using different techniques and there are a confusing number of related but different units of these measurements.

CANDELA has its brightness measured by the amount of light emitted from a candle. The candela measurement was known as a candle at one time. It is a term rarely used to calculate light intensity emitted from a bulb or the brightness on a surface. Lux and lumens, and even watts, are more important measurements.

WATTS is erroneously thought of as a measure of brightness, but, in reality, it is a measure of energy consumed by the bulb – not how much light intensity it gives off. The wattage is the sum of the heat generated as well as the energy of the light emitted. A 40-W bulb uses 40 W of electricity, no matter what type of bulb it is, yet the light produced by the different types of bulbs can vary significantly. A 100-W incandescent bulb can light up a room, but at the same wattage, an LED lamp can be twice as bright.

LUMENS – This measure of light intensity refers to the brightness of the source as the human eye perceives it at its source. The light output produced by a bulb is measured in lumens only – heat is not a factor in the measurement. A standard 100-W incandescent bulb produces about 1,680 lumens, whereas a compact 25-W fluorescent lamp produces about 1,750 lumens. While the two bulbs produce virtually the same amount of light, the CFL does so using only a quarter of the energy consumed by the standard incandescent. It is important to determine the amount of light or "lumens" needed for a specific application and use the amount of "lumens per watt" as a measure of light efficiency – not as a measurement of brightness.

LUX – This measurement term is not always understood and is often confused with lumens because they are closely related. The difference between the lux and lumens is that the lux takes into account the area in which the luminous flux is spread. The measurement of lumens should not be confused with the brightness measurement of "lux," which is the brightness on a spot that is the focus of the illumination under the light. If the bulb produces 1,680 lumens and a magazine cover is placed at the bulb's surface – the lux on the page will also be 1,680, but if the cover were held five feet away from the 1,680-lumen bulb, the brightness on the cover would now be 500 lux.

LED – The only way to compare the brightness of the various light sources on the market is by knowing the number of lumens they produce. Everyone has some idea about how bright a 20, 40, 60, or 100-W bulb should be and that frame of reference is automatically applied to LED lamps. But realistically, the way to compare an incandescent with an LED is by comparing the number of lumens per watt (lm/w). Incandescent bulbs generate only about 12 lm/w, whereas LEDs and even halogen sources are more efficient at 40 lm/w and some fluorescent tubes generate 60 lm/w. Another point of consideration with the standard egg-shaped bulb is that only a small proportion (50% at best) of the lumens are delivered as lux to the surface being lighted – the rest is lost. The LED bulb can be focused so that 80% of the lumens can be directed to provide lux to the surface.

LEDs are as bright as halogen lights, nearly as bright as fluorescent lights, and can deliver more lux per watt than either of the other two bulbs onto a surface below the bulb fixture in almost all situations.

9 DISPOSAL OF FLUORESCENT BULBS – A NEW INDUSTRY

The Department of Environmental Protection in most states has found that most retail outlets and commercial businesses were improperly disposing of their spent bulbs and lamps in violation of existing regulations, especially in the handling of mercury found in bulbs. The US Environmental Protection Agency encourages generators of mercury-containing lamps to handle them as universal waste because waste rules are less restrictive than hazardous waste rules. The rules for universal waste were originally established to make it easier to recycle widely used waste such as batteries, pesticides, mercury-containing equipment, thermostats, consumer electronics, and lamps. Private homes are not subject to either the hazardous or universal waste rules, and homeowners take their spent bulbs to local collection facilities.

Generators of mercury-containing lamps (mostly fluorescent lamps) can only treat them as universal waste if they are unbroken. Once lamps are broken, they must be handled under the more restrictive hazardous waste rules. This distinction in disposing of universal and hazardous waste has created the bulb recycling industry. Companies in this business provide pick-up containers for discarded lamps, ballasts, and batteries and report on the tonnage of the collection, which is used by the municipality to collect a refund from the state Environmental Protection Agency (EPA).

Recycling firms charge $1 per bulb for the pick-up from municipalities, but most of the money is made by separating the valuable chemicals found in bulbs and selling those to various factories. The collection boxes are sent through a conveyor system where the bulbs are counted, then to a separation tank where heat chambers melt the glass and collect the liquid mercury and hot phosphates. All three components are then cooled and sold to glass manufacturers, mercury users, and phosphate plants. Everyone makes money, and the municipality is compensated for meeting the hazardous waste rules at almost no cost.

35 Major and Notable Fires in American History

1 INADEQUATE FIRE CODES

Four famous fires that resulted in an enormous loss of life in the early part of the last century were the Iroquois Theater Fire, the Triangle Shirtwaist Factory, the Cocoanut Grove Club, and the Beverly Hills Supper Club. These disastrous fires will remain seminal events that changed the mostly archaic and ineffective fire and building codes to modernize to what are vastly improved, current standards. These fires, along with other major fires, involved the locking of fire exits by the building owners to prevent theft of material and equipment as well as an unauthorized entrance by people who were trying to avoid the price of admission. All these fires occurred well before the 1980 NFPA amendment.

- The **IROQUOIS THEATER FIRE**, Chicago (1903), for nearly a century, ranked as the deadliest fire in American history. The theater had opened just five weeks before the fire destroyed what was considered the most beautiful entertainment center in the nation. The architects who designed the building claimed that the building was "absolutely fire-proof" with the same egotistical flair used by the designers who assured the public that the Titanic was unsinkable when it set sail nine years later. The Iroquois had 25 exits that, assuming an orderly evacuation, could empty the entire orchestra and balcony in less than five minutes. The fire caused a panic, with people piling up in front of the exits, either preventing the doors from opening inward or having those people discover that most of the exits were chained shut. There were no signs designating the location of the exits (which were mostly hidden behind decorative curtains), and the building turned out to be a blazing death trap.

The seats in the theater were impressively designed, but they were made of wood and stuffed with hemp. Much of the precautionary fire equipment that was advertised to have been installed never actually made it into the building because of the rush to open during the holiday season. The theater had no fire alarms, and other safety factors had been forgotten or simply ignored. One of the most tragic aspects of this fire was the fact that the audience was made up primarily of women and children since it was a Wednesday matinee. There were 1,924 people in the theater seats when the fire started – the occupancy limit was 1,724. The fire caused 602 deaths and left 250 people with nonfatal injuries.

At the beginning of the second act, a stagehand noticed a spark descend from an overhead light, along with some scraps of burning paper that fell onto the stage igniting a curtain. Flames began licking at the curtain and while a collective gasp came from the audience, no one rushed for the exits. It had been surmised that the audience merely thought the fire was part of the show. In order to calm the audience down, the lead actor signaled the conductor to continue playing music, which had a temporary soothing effect on the crowd in the auditorium and balcony. A few minutes later, a flaming set crashed down onto the stage and a stagehand tried to lower the asbestos curtain to protect the audience area. The curtain snagged halfway down, leaving a 20-foot gap between the bottom of the curtain and the wooden stage floor. Panic ensued and the audience, with children in tow, rushed for the exits and immediately clogged the aisles as the lights went out. The crowd prevented inward-moving doors from opening and many died due to being crushed or trampled by panicked people trying to get to the exits. Backstage, theater employees, and cast members opened a rear

DOI: 10.4324/9781003091103-48

set of double doors, which sucked the wind inside and caused flames to fan out under the asbestos curtain and into the auditorium. The fire engulfed the supposedly noncombustible asbestos curtain and when it collapsed, it plunged onto the theater seats (Figure 35.1a and b).

When it was all over, 572 people died in the fire and more died later, bringing the eventual death toll to 602, including 212 children. A grand jury indicted a number of individuals, including the theater owners, fire officials, and even the mayor. No one was ever charged with a criminal act and families of the dead filed nearly 275 civil lawsuits against the theater – but no money was ever collected. The Iroquois Theater Company filed for bankruptcy soon after the disaster.

FIGURE 35.1 a. Firemen work to put out the blaze at the Iroquois Theater and b. The stage, seen here from the balcony level, and theater were left in ruins.

Source: Chicago Daily News collection, Chicago History Museum.

Code Change: Some of the changes to building codes following the Iroquois Theater Fire included the use of steel fire curtains, independent lighting systems for stairs and corridors, and requirements to keep passageways free of obstruction.

The Iroquois Theater Fire was not the first large-scale Chicago fire disaster. That city had been the scene of the greatest and most devastating fire in American history, an event that occurred in 1871 known as the "Great Chicago Fire" which wiped out most of the old city, killed hundreds, and left hundreds of thousands homeless and destitute.

- The **TRIANGLE SHIRTWAIST FACTORY**, NYC. (1911) killed 146 people working on the eighth, ninth, and tenth floors of the Triangle Waist Company factory, one of the 500 so-called sweatshops in lower Manhattan and became the most famous factory fire disaster in history. It took just 18 minutes for the fire to consume the entire building, which was ironically and tragically enough named the "Ash Building" and was constructed in the shape of a triangle. The dead included mostly young immigrant women who chose to jump from the high floors to escape the inferno after finding the exit doors to the fire escape locked to prevent theft. The elevator car was halted by the weight of the bodies that had fallen into the shaft and onto its roof. Fire truck ladders, in that period, could only reach up to the sixth floor of a building. Of the 146 victims, 129 women and 17 men burned to death in the loft building on the edge of Greenwich Village; many had no telltale jewelry or clothing to help identify them (Figure 35.2).

FIGURE 35.2 Image of Triangle Shirtwaist Factory fire on March 25, 1911.

At the time, no New York City agencies or newspapers produced a complete list of the dead, and it wasn't until a century later (2011) that the six unknown victims were finally identified. The obscurity of their names is evidence of the times when lives were lived quietly with no television, cameras, or social media and smart devices to record their every step and thought. The Triangle fire is the event that did more to alter fire and building codes than any other fire disaster in the nation's history, even though there were many other factory fires during that same period.

The two factory owners, who had escaped to the roof without unlocking the exit door below or even informing workers that the place was on fire, were charged with manslaughter and brought to trial. The all-male jury acquitted both men – who then collected a huge settlement and restarted their business and a nearby factory location.

Four months prior to the catastrophic Triangle factory fire another factory burst into flames in Newark, N.J. and took the lives of more than two dozen women and girls. History, like life, has its own perverse winner-take-all quality: the biggest moments are remembered; the others simply fade from history. Most of the Newark fire victims jumped to their deaths, and while the fire made national news and more than 100,000 people flocked to the scene the next day – the blaze did not change the fire codes. The 26 dead in Newark couldn't quite galvanize a movement for change the way the 146 dead in New York finally did a few months later. The fire became a footnote in history, and, if mentioned at all. It is always in connection with the Triangle fire.

In 2015, New York State set aside $1.5 million to build a memorial to the victims of the Shirtwaist Factory, still considered one of the worst industrial disasters in US history. The memorial consists of steel panels mounted outside the original factory, which is now owned by New York University (NYU). One panel is etched with the names of the 146 immigrant workers who perished.

As a side note, during the 1970s and 1980s, a time when urban schools became war zones, schools had been forced to chain their exit doors due to safety and security concerns. The chaining was a last resort and condoned by the school principal with full knowledge; it was in complete violation of every fire code in existence at that time. With the security problem in public assembly occupancies becoming so disrupted, the NFPA amended the Life Safety Code 101 to allow magnetic locks after recognizing that the way to replace the chains was with this fail-safe lock. But four years following the Code change, no fire department or city official or architectural firm had shown any foresight by allowing the use of the magnetic lock, which releases the doors when the automatic fire alarm system is activated.

Ironically, in 1984, 73 years after the Newark fire, that same city became the first municipality in the nation to allow the locking of exit doors with an electromagnetic lock at Barringer High School. Soon after the installation, the FDNY visited the high school and approved the locking system for use in New York City. That singular event in Newark started a new industry, and eventually, thousands of magnetic locks were mounted on exit doors in all types of buildings: hospitals, schools, museums, and hotels.

- The **COCOANUT GROVE NIGHTCLUB**, Boston, [1942] killed 491 people in the deadliest nightclub and public occupancy fire in the United States, even though the fire took only 15 minutes from ignition to extinguishment. At the time, the Cocoanut Grove was the largest and most popular nightspot of its kind and was referred to as the poor man's Ritz. The club had a capacity for 600 people within a one-and-a-half-story building and its basement lounge. During the evening of the fire, owners set up additional tables on the main room dance floor and by doing so packed the club with 1,000 patrons putting the club

FIGURE 35.3 The downstairs bar in the Cocoanut Grove where the fire started.

Source: U.S. Army Signal Corps.

25% over its maximum capacity. The fire started in the basement lounge when a busboy was dispatched to replace a missing light bulb and struggled to locate the bulb socket in the dark and lit a match for light. When the match burned his finger, he dropped it on paper decorations, and the resulting fire quickly spread throughout the building. In panic, the crowd sought escape. Many died from crush injuries as people attempting escape piled up at the exit doors. A lucky few in the basement lounge escaped death by sheltering within the walk-in refrigerator. A few unlucky attendees attempted escape through a faux window attached to the brick walls. Many were overcome by toxic gases released by the burning decorations. Some patrons were found standing at the bar or still seated at their tables (Figure 35.3).

Some facts surrounding the fire:

- The rush of bodies to a locked exit door was so enormous that they were jammed into the space with no chance of being able to turn around and escape. The doors were locked because the management had a policy that no one could leave or enter the nightspot without paying.
- More than 100 were burned to death while those in the lounge area were crushed.
- Some main entrance doors were blocked by a staff member in an attempt to stop patrons from leaving before they had paid their bill.
- The rush at one of the revolving doors quickly jammed it shut.
- The two doors leading to the street opened inward and were of little use since those attempting to escape formed a human wall six-feet high in front of the exit, effectively sealing them shut. Investigators later found 200 bodies stacked at this location.
- Arriving firefighters broke the revolving door down and were immediately met with a six-foot pile of human bodies.

The Cocoanut Grove fire would become a major national story that eclipsed news from the waterfront and the club owners would be indicted and sent to prison. After the fire, many changes were made to the Building and Life Safety Codes:

- No place of public assembly should be filled beyond the maximum seating capacity of the structure.
- Combustible materials are prohibited in places of public assembly.
- Every building used for specific purposes of public assembly must have two means of exit access, and the locations must be remote from one another.
- If the occupancy rating is more than 600 people, then there must be three exits.
- If more than 1,000, then provisions for four exits must be made.

The Boston Licensing Board made a ruling that no other establishment could ever call itself the Cocoanut Grove. Within two years of the fire, more than 400 lawsuits had been filed, and, when the final rewards had been divided among the survivors and the families of the dead, the checks issued were for a paltry $150.00.

The **BEVERLY HILLS SUPPER CLUB**, Ky. (1977) killed 165 after the fire began behind a wall and quickly spread throughout the club, feeding on flammable furniture, tapestries, and decorative panels. It rolled rapidly through the building's main corridor. As many as 3,000 people were thought to be in the building, and they bottlenecked trying to flee through what fire officials later determined were an inadequate number of exits. By morning, workers had pulled 134 bodies from that one site and a total of 165 people died. The Beverly Hills blaze would become one of the deadliest nightclub fires in US history. It spawned one of the country's first-class action lawsuits involving disasters, netting nearly $50 million for clients and their lawyers. The local hospital responded so effectively that medical facilities in 33 states and seven foreign countries requested copies of the disaster plans used during the crisis.

Code Change. Fire officials claimed faulty aluminum wiring was the cause of the blaze, which resulted in a shift in the code that mandated copper as an alternative. Other factors included overcrowding, delayed notification while employees tried to put out the fire and a lack of sprinklers in the building. The changes in the national fire code required new and existing nightclubs (facilities that hold more than 300 people) to have alarms that include voice messages giving evacuation instructions and new facilities to be outfitted with sprinklers. This requirement was extended to all new nightclubs and any existing ones that hold more than 100 people, after a 2003 fire at the Station nightclub in Rhode Island killed 100 and injured another 230 people.

The **STATION NIGHTCLUB FIRE** West Warwick, RI. (2003) was one of the deadliest nightclub fires that resulted in 100 people dead and left 230 injured as a result of it. The band Great White was performing and pyrotechnics ignited highly flammable polyurethane soundproofing foam insulation in the venue, which quickly spread and resulted in a massive blaze. The resulting stampede to flee and exit the building caused a blockage that exacerbated the resultant deaths and injuries.

In the wake of the fire, in July of that year, the National Fire Protection Association issued Tentative Interim Amendments to the national "Life Safety Code" (NFPA 101), which required automatic fire sprinklers in all existing nightclubs and similar locations. Rhode Island's fire code is now considered among the most stringent in the country as a direct result of legislative changes following the Station nightclub tragedy.

Section XIII

Explosives in Major Conflicts

The greatest public anxiety in areas of safety and security has been the potential for another terrorist attack (WTC) and natural disasters (Hurricane Katrina). In response to these concerns, the federal government, along with every State and many Municipal authorities, created antiterrorist or counterterrorism agencies, often with redundant responsibilities and almost always staffed with many politically appointed officials with very little background in areas relating to public safety. Most of these similar-sounding agencies seem to keep busy just trying to explain the relationships and their specific functions and sending messages to each other. They also spend an inordinate amount of energy in trying to obtain allocated emergency funds distributed by the Department of Homeland Security to justify their existence and to protect their individual domains.

The National Asset Database originally issued in 2006 by Homeland Security had been used to divvy up the hundreds of millions of dollars in antiterrorism grants each year. Some security directors believe that the Homeland Security designated list of potential targets, barring some obvious choices, was questionable at best.

Another report on the same topic was prepared by security directors, which may be a more realistic set of targets than any other list. Once the director's list was assembled, there were questions regarding the advisability of identifying a list of prime targets. The thought was that it may inspire a would-be terrorist to use the information as a guideline in planning another attack in NYC, which, on the surface, is a laughable premise.

Confidential list Compiled from a Cross-Section of NYC Security Directors:

Probable terrorist targets in NYC	Unlikely terrorist targets
Brooklyn bridge and the Verrazano bridge	Empire state building
Train terminals (Penn Station or Grand Central station)	Path tunnels
Tall wall street building (in or near the stock exchange)	Airports
Bio-terrorist attack at three large hospitals in Manhattan	Major retail stores
Stock exchange on wall street	Municipal building
Subway tunnels under east river (chemical attack like Tokyo Subway attack)	NYC times square
Underground electric utility and distribution network	Apple festival – Illinois
Municipal building in NYC	Flea Festival in Tenn
Entrance to Lincoln Tunnel	Mule Parade in Columbia
Waldorf Astoria hotel	Petting Zoo in Alabama

DOI: 10.4324/9781003091103-49

With the National Assets Database mostly abandoned at this point, the government has taken a more holistic approach to threats. And now, securing information systems is being emphasized as more vital to the nation's telecommunication system, its transportation network, the power grids, and to many other vulnerable assets, as expressed by the Cybersecurity and Infrastructure Security Agency (CISA). Per their website:

> The Infrastructure Data Taxonomy (IDT) Table of Contents serves as a quick reference of the taxonomy, in which infrastructure assets are categorized up to five levels. The first level is determined by broad infrastructure categories, which are then detailed further by increasingly granular differentiators as needed. … Some infrastructure assets fall into multiple categories of the taxonomy. For this reason, some concepts may be found listed within multiple sectors.
>
> *Source: https://www.cisa.gov/resources-tools/resources/infrastructure-data-taxonomy (April 2024)*

36 Nuclear Threats

Efforts to prevent a nuclear attack may keep that day from becoming a reality, but as long as the threat persists, it is important to formulate contingency plans that will save thousands of lives, billions of dollars, prevent public panic, and promote recovery. The plans will also help preserve America's constitutional government, something that terrorists, even if armed with nuclear weapons, should never be allowed to take away.

The atomic bomb "Little Boy" was released from the belly of the B-29 bomber "Enola Gay" at 8:15 a.m. on August 6, 1945, 31,000 feet over Hiroshima; it unleashed a chain of events that altered the face of warfare and man's future capacity to inflict damage upon its enemies. Detonating 43 seconds later 1,900 feet over the center of the city, with the force equivalent to12,500 tons (12.5 kilotons), Little Boy killed 200,000. 80,000 of the victims died instantly.

1 NUCLEAR EXPLOSION

When the first nuclear bomb exploded on US soil, it may not be a single incident. A terrorist who can fit a device within a small suitcase could conceivably wield many more. Even if the bomber had no other weapons, they could easily claim they did as a terror tactic. While the instinct is to evacuate on day one (at least evacuate the children as the English did in World War II [WWII]), the safer choice is to shelter in place, seal the windows with duct tape, and stay inside for at least seven days to avoid the worst of the fall out. The government would immediately attempt to trace the bomb's source and assaying the bomb material may provide information that points to a particular government's nuclear arsenal or power program. Even if the most sophisticated terrorist groups cannot make sufficiently pure plutonium or enriched uranium, they would need to get their weapons from a government stockpile. Generally, a dirty bomb is the most likely terrorist attack method for a terrorist outfit. Per the US Nuclear Regulatory Commission (NRC): A 'dirty bomb' is a type of "radiological dispersal device" that combines a conventional explosive, such as dynamite, with radioactive material.

And while it would be tempting to retaliate against that state, it may not even be aware that its bombs were stolen, sold, or even deliberately given to terrorists. In fact, it might be counterproductive. Cooperation will yield a better chance of figuring out who did it and how much material they have.

A private foundation that studies the threat of terrorism involving weapons of mass destruction argues that the vast amount of damage at ground zero would be dwarfed in scope by the global economic aftershocks. The explosion itself would trigger a predictable set of responses that a president would almost certainly have to make in grappling with the possibility of a follow-up attack. Assuming that such a weapon would have to be smuggled into the country, the various federal agencies dealing with homeland security could be expected to close the nation's borders, manufacturing would cease, freight movement would stop, and a direct search of virtually any moving conveyance with the ability to transport a nuclear weapon would begin. If US intelligence determined that one or more countries had somehow aided and abetted the attack, the public would demand a level of retaliation at the suspected locus of the attacking group that would dwarf the post 9/11 military response and initiate the prospect of a full-scale war greater than those in Iraq and Afghanistan.

DOI: 10.4324/9781003091103-50

2 SHELTER IN PLACE

In 2008, Congress requested that the Department of Homeland Security create accurate models of a nuclear aftermath in any one of six target cities (L.A., New York, Washington, D.C., San Francisco, Houston, and Chicago). The study produced a 40-page guide, Nuclear Detonation Preparedness: Communication in the Immediate Aftermath. It was meant to act as a guide to a frightened populace in the aftermath of the unthinkable. The overall message: it is better to shelter in place rather

FIGURE 36.1 FEMA's "Be Prepared for a Nuclear Explosion" instructions from Ready.gov.

than try to outrun the fallout cloud from the blast effect. It's the best way to save lives and prevent radiation sickness. The "the remain-in-place reaction" runs counter to a person's basic instincts, which are to get away and reunite with family members. Figure 36.1 provides a more updated preparation guide from FEMA.

In any nuclear bomb scenario, the detonation of an atomic bomb anywhere from 25 to 500 miles above the earth will electrify, or ionize, the atmosphere about 25 miles up, triggering a series of electromagnetic pulses (EMPs). A bomb of the proper size detonated over Omaha would eliminate most of the electrical production in the United States. Accomplishing such devastation would only be possible if the bomb were the size of a powerful Soviet H-bomb. One complication for rogue nations, or terrorists contemplating a high-altitude nuclear blast, is that such an attack requires a missile to take the weapon at least 25 miles high to trigger the EMP. Attempting such a launch would invite massive nuclear retaliation from the US's current stockpile of 5,000 warheads, many riding in submarines far from any pulse effects. Any attempts of a preemptive attack would be considered a suicidal act since US government counter-terrorism labs would quickly trace the origin of the bomb, which will be destroyed within minutes.

According to some of the theories pursued in the government study, it seems that the greatest danger to citizens living in city centers near an explosion site would occur in the 6 to 24 hours immediately following as fallout's radioactivity dwindles. Using L.A. in one scenario, there would be 285,000 people caught in the initial blast who would develop radiation sickness or die. Just getting into a wooden framed house could save 160,000 lives and remaining in a shallow basement or in a two-or-three-story building could save 240,000 of the 285,000, the rest would get sick – but survive. The ideal space to avoid exposure to the radiation cloud would be an underground garage or the core of an office building where there would be no significant exposure at all. In the second scenario, the New York City response in the wake of a nuclear detonation indicated that as many as 5–6 million people would scatter randomly through at least eight states into relatively small communities that would be confronted with thousands of injured, sick, and terrified people who had been evacuated in the face of great danger and anxiety.

There is nothing more potent as a symbol of terror than a nuclear explosion. According to the White House's Planning Guidance for Response to a Nuclear Detonation posted on the Internet, bystanders' miles away would witness a huge fireball shooting five miles into the sky. Heat, hyper-explosive pressure, and 900 mph winds would level buildings for half a mile and between 50,000 and 100,000 people would vanish in smoke and flames. Flash-blind drivers 10 miles away would crash, blocking evacuation routes. Fallout would rain down for hundreds of miles. There would be no opportunity to call 911, and there would be no response if the call were made, since all the cell towers would have disappeared. A nuclear attack would be infinitely more catastrophic than 9/11 and would be a challenging moment in world history.

Nuclear proliferation experts are divided on terrorists' odds of success, with some believing that the probability is so low that it's not worth spending a lot of money and effort to deal with it. It's highly unlikely that terrorists could get enough highly enriched uranium, build a bomb, sneak into the United States, and trigger it without getting caught. Other experts believe that al-Qaeda has been trying to acquire or make nuclear weapons since before 9/11. Since that date, the International Atomic Energy Agency has collected over 400 reports of lost or stolen nuclear material from member states. The more frightening statistic is that the United States has lost at least 11 nuclear weapons, according to the Belfer Center for Science and International Affairs at Harvard.

Balancing all the research, studies, and predictions – and despite the fact that nations like Iran, North Korea, and Russia repeatedly speak of and threaten with nuclear capabilities – the detonation of a nuclear device still seems unlikely.

3 NEUTRON BOMB

Conceived by one of the scientists who worked on the Manhattan Project who developed the first atomic bombs that wiped out Hiroshima and Nagasaki in 1945, the neutron bomb never became part of America's weapon arsenal. In contrast to strategic nuclear warheads, which can kill millions and level cities, the neutron bomb minimized blast and heat by maximizing a barrage of infinitesimal neutrons that would zip through tanks, buildings, and other structures, to kill people inside the structure by destroying the central nervous system of all life forms.

The basic design concept of this deadly nuclear weapon was that it would kill people with neutron radiation without an explosion that destroyed buildings for many miles away from its epicenter. After WWII, it took scientists almost a decade to finally test this very controversial weapon (1963), but its potential was never fully embraced by the military or by the various Presidents and members of federal agencies dealing with advanced weapon technology during the 1960s and 1970s. This fusion nuke produces large amounts of lethal neutrons that could penetrate solid objects to kill the inhabitants with the potential to destroy armies but not cause serious damage to the surrounding countryside. Scientists who helped develop the neutron bomb believed it was an ideal battlefield weapon that would kill fewer innocent people leaving the environment relatively unscathed without any lingering radiation, which would take a whole generation to dissipate.

While nuclear strategists doubted the usefulness, logic, and ethics of killing people while sparing property, the developers of the neutron bomb believed that its unique characteristics of radiation containment were a sane and moral alternative to the conventional nuclear bombs that caused mass destruction and radioactive contamination. President Nixon authorized the initial development of the bomb, but President Carter ordered that the program be shut down. Then President Reagan ordered production to start up again, in part because the weapon's enhanced radiation was thought to be capable of disabling incoming nuclear warheads. President W. Bush dismantled the program in 1992 after only a few bombs had been constructed and stored in an arsenal. Its final demise was due, in part to a lack of government backing, the fall of the Soviet Union and the development of precision-guided conventional missiles, all of which contributed to the lack of need for a neutron bomb. The few remaining bombs were dismantled in 2000.

4 NUCLEAR LEGACY – THE CONTAMINATION AFTERMATH

On the eve of America's entry into WWII, scientists isolated plutonium in a California university laboratory. Years later, in 1987, it was discovered that a radioactive residue had been left behind, forcing the university to rip out an entire adjacent room while students and professors continued to use the building's classrooms. The contamination that may still be in the building is part of a legacy of one of the most important scientific and industrial undertakings in US history. During the atomic bomb development and research period, the government turned to the private sector to help produce nuclear weapons and other forms of atomic energy involving hundreds of companies and thousands of workers. The effort helped defend the country, but this enormous endeavor in the early 1940s left an equally enormous cleanup job that was spread throughout three dozen states.

The California classroom where the plutonium was first isolated is now a National Historic Landmark. An investigation of the contamination by a newspaper discovered that record-keeping by the government had been so spotty that the Energy Department didn't have enough documentation on several dozen sites to determine whether a cleanup was necessary. In fact, there is such a considerable lack of data that it has become impossible to track the exact addresses of the sites that likely still contain traceable residue of the radiation material. There is no precise information to identify the level of radiation that can be harmful or even deadly. A government report contends that risk rises as exposure accumulates over time and that 300 millirems a year, the equivalent of the average background dose could – over 70 years of exposure – cause one extra cancer among about 50 people exposed.

5 DIRTY BOMBS

Dirty bombs are Radiological Dispersion Devices (RDD), also known as "dirty nukes." An attack of this type is considered far more likely than the more destructive nuclear devices we associate with atomic weapons. An RDD combines a conventional explosive, a bomb, and a payload of poisonous, radioactive material. It is designed to scatter dangerous sublethal radioactive material over a large area. This device appeals to terrorists because they require limited technical knowledge to build and deploy compared to the full-blown nuclear bomb. Also, the radioactive material is easier to obtain than uranium and plutonium since the material is frequently used in medicine, agriculture, industry, and research.

The favored isotope for radiological attacks is cesium-137, which emits energetic gamma radiation capable of traveling great distances and can penetrate lead shielding. An inhalation attack – sometimes called a smoky bomb – employs radioisotopes that become aerosolized when burned and are inhaled by victims. Terrorists could also use an insecticide sprayer or similar device mounted on a truck to disperse radioactive materials dissolved in water. Even a few drops are enough to cause fatal radiation poisoning. The first instinct when soaked with water is to wipe your face, which transfers the poison from hand to mouth. Even if the victims avoided swallowing, the solution would still cause severe radiation burns.

Dirty bombs only require dynamite or a similar explosive and a radioactive source like cesium or cobalt, common in widely available medical devices. The initial blast would cause injuries or death, but radioactive residue could cover two or three blocks. Though not an immediately lethal threat, the resultant panic and economic consequences would be disastrous. Present detection devices are intended to stop radioactive threats as far away from the city center as possible.

Dirty bombs, unlike conventional weapons, involve a complex nuclear-fission reaction many times greater than a conventional blast. Detection is not easily accomplished and while various systems have been tested, the results are far from satisfactory. The establishment of Homeland Security's Domestic Nuclear Detection Office in 2005 was a response, in part, to criticism that nuclear safeguards were too disorganized.

6 RADIATION: THE ENDURING AFTERGLOW

In 2011, a 9.0-magnitude earthquake wiped out entire coastal cities and damaged Japanese nuclear reactors causing more devastation than millions of Hiroshima bombs. Explosive power usually comes and goes in an instant but with radiation, the terror lies in the abstraction because it kills incrementally – slowly, diffusely, invisibly like a smoldering fire that can't be put out. The terminology used to set a level for the radiation is very confusing. When traces of radiation were found in water near a leaking nuclear power plant, the danger was reported in becquerels, the number of nuclear disintegrations per second: 210 per liter, safe for adults but high enough to warn that infants should not drink it.

William Roentgen produced the first artificial X-rays in 1895 and got the world's attention by showing the see-through images of his wife's hand. Henry Becquerel found similar emissions coming from uranium. The field advanced further with Madam Curie who made radium famous; 1 curie became equal to 37 gigabecquerels. Radiation was not thought of as something dangerous until Hiroshima, and in earlier years, it was supposedly good for humans. Vacationers soaked their sunburned bodies in radium hot springs, magazines carried advertisements for radium suppositories, radium toothpaste, and radium bread – products ranging from useless to harmful. As late as the 1950s, customers could peer inside their own feet at shoe store X-ray machines. In the 1920s, women who had painted glow-in-the-dark radium watch dials by placing the small brushes in their mouth to sharpen the bristles into a point began to sicken and die, something depicted and outlined in detail in the book *The Radium Girls* by Kate More, as well as a play and movie.

These rudimentary radiating devices eventually led to a medical revolution – X-rays for medical imaging and radium for killing rapidly dividing cancer cells. Sieverts are units of measurement of the levels of exposure, and thus risk, after it became obvious that these amazing rays were not harmless and could accumulate from overuse and become injurious and even deadly.

According to the NRC, the effects of radiation exposure are largely determined by:

- The amount of radiation absorbed by the body;
- The type of radiation (gamma, beta, or alpha);
- The distance from the source of radiation to an individual;
- The means of exposure – external or internal (absorbed by the skin, inhaled or ingested); and
- The length of time exposed.

The health effects of radiation tend to be directly proportional to radiation dose. In other words, the higher the radiation dose, the higher the risk and severity of injury.

7 RESPONSE TACTICS

A terrorist attack on any major city would kill and injure so many that disaster plans previously published on emergency preparedness procedures would have to be re-written or ignored. There is no logical way to anticipate the response to a major nuclear attack. Providing basic medical care after the explosion of a 10-kiloton nuclear bomb, which is equal to the same explosive force of the Hiroshima blast or 5,000 Oklahoma City truck bombs, would be inconceivable. In Washington, D.C., a city with 38 ambulances, and neighboring communities, at least 930,000 people would seek medical care and more than 70,000 would require hospitalization, vastly overflowing the city's 3,600 hospital beds. There would be 1,000 severely injured trauma patients for every available operating room, according to reports published in the journal "Disaster Medicine and Public Health Preparedness."

8 ECONOMICAL AND POWER GRID EFFECTS

Our system of power runs on electricity and an attack on our grid would cause businesses to grind to a halt, immediately resulting in monetary collapse. While some of the consequences are obvious, others are not. It might transform the world – and not for the better.

An EMP powered by a nuclear blast would create an intense magnetic pulse, which would induce ground currents strong enough to burn out power lines and electric equipment within the explosion's line of sight. At risk, according to the Harvard Smithsonian Center for Astrophysics, are the more than 200,000 miles of high-voltage transmission lines that crisscross North America, supplying 1,800 utilities that power TVs, lights, refrigerators, and air conditioners in homes, businesses, hospitals, and police stations.

A realistic taste of what power grid failure looks like occurred in the famous 2003 blackout from Michigan to Massachusetts and southeastern Canada. The two-day shutdown exposed the grid's vulnerability as it idled 100 power plants and cut power to about 55 million people. $6 billion covered repairs and lost work time. Television and radio stations went off air, drinking water had to be boiled, and everything from traffic lights to trains stopped running. Manhattan became the world's largest pedestrian mall.

9 TOO MANY NUCLEAR WARHEADS

Some military experts maintain a belief that the United States has far too many nuclear warheads since there is no national security rationale for supporting a stockpile of over 8,000 warheads, with nearly 2,000 armed and ready to use on short notice. The threats from North Korea and Iran aren't justification for such a huge arsenal. The only conceivable use of so many nuclear weapons would be a full-scale nuclear war with Russia, which is the only nation, besides the United States to have more than 500 bombs. A private organization, the Federation of American Scientists (FAS), records the number of nuclear bombs in the nine nations that possess them as of 2024 (Incidentally, Iran is not listed as one of them) (See Figure 36.2).

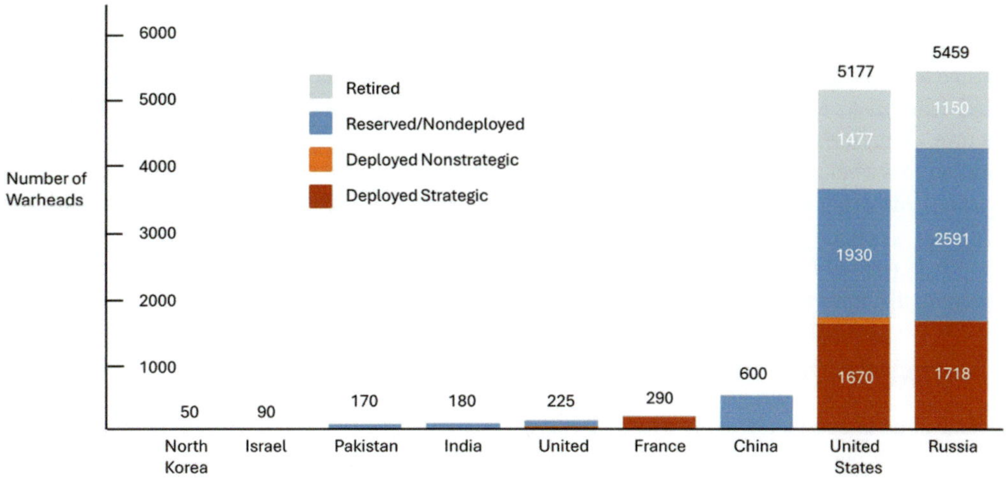

FIGURE 36.2 2024 statistics on the estimated global nuclear warhead inventories as per the FAS.

Source: FAS, https://fas.org/initiative/status-world-nuclear-forces/.

10 UNDERWATER THREATS

It has been 60 years since the Battle of the Atlantic against German submarines that were waging a guerrilla war at sea. German submarines had pursued a form of underwater warfare known as commerce raiding, attacking the enemy's economic assets that traveled between America and Britain. America eventually won that battle, but memories of that time when access to the seas was in jeopardy have faded, and there is very little preparation by government agencies directed toward possible terrorist attacks beneath the waves, an indifference that could prove to be a serious mistake. There is a critical need for the United States to renew their policy regarding defending maritime commerce because of two new developments that are unrelated to German U-Boat-type threats: The first was the emergence of an undersea economy based on the discovery of oil in 1947 at offshore, out-of-sight of land deposits. Nearly, 30% of US oil production and 15% of gas production come from wells on the Outer Continental Shelf. Globally, over 30% of the world's oil output comes from offshore production. Much effort and money have gone into undersea energy infrastructure, including the platforms, pipelines, and pumps to transfer the oil and gas to shore. This vast infrastructure was built with the assumption that while it would have to weather natural disasters, it would not be a target in war. This policy has left offshore platforms vulnerable to attack by explosives that could cause significant and perhaps catastrophic damage. Fortunately, many of these targets were not easy to reach unseen – until now.

The second development in this scenario involved a form of commerce raiding copied from WWII. In recent years, Latin-American narco-cartels have been moving drugs by a submarine that has a range of 2,000 miles and cost a few million dollars to construct. These submarines can submerge to depths of a few dozen feet, which is sufficient to make detection difficult, allowing the object to approach offshore oil platforms with little or no warning. Even more threatening is the proliferation of Unmanned Underwater Vehicles, or UUVs, which were once almost exclusively operated by Western militaries and now widely used for a variety of commercial and scientific purposes. They are probably best known for their role in locating sunken ships like the Titanic. Unlike the small submarines operated by narco-cartels, UUVs can descend as deep as two miles to the ocean floor. When adopted for military purposes, they could carry mines and other explosives, as well as cameras and electronic sensors. To counteract the submerged threat, several countries are developing mines with better sensors, target-recognition systems, and self-propulsion systems to enable them to move about – like an underwater drone. But mines don't need to be sophisticated to be effective, especially against the thousands of soft targets populating the continental shelf.

Some terrorist groups seek to achieve victory not by defeating their enemies on the battlefield but by inflicting unacceptable pain or damage, either against defenseless civilians or economic infrastructure. America's enemies could potentially impose enormous costs on its undersea economy, including loss of energy resources, damaged infrastructure, and environmental degradation. Given the stakes involved, just as the United States and its allies developed the technology, capabilities, and methods needed to defend their economic assets at sea during the Battle of the Atlantic, a similar effort is needed to protect the undersea economic interests of America's allies who depend on offshore drilling for energy resources.

11 SUBMERGED POLICE VEHICLES

The fear of an attack by terrorists using submerged vehicles in cities that have commercial shipping docks has prompted the purchase of remote-controlled devices that can be moved around waterways to look for suspicious containers or bombs attached to ships or docks. The need for underwater security has resulted in an increasing reliance on relatively new tactical weapons for police department anti-terrorist units: unmanned submersible drones, referred to as a remote-operated vehicle, or ROV. As of 2020, nearly 15,000 vessels per year pass through the Port of New Jersey and New York, with cargo ships docking in the various sections of New York City and the surrounding areas where millions of containers reside and are shuttled on and of the piers.

The ROV used by the Harbor Unit acts in a similar manner to the mirrored devices used by police to check for explosives under the chassis of the vehicle. The New York City police purchased six of these units at $75,000–$125,000 each and are identical to those used by the military and oil companies with offshore platforms. The drones weigh 16 pounds and are, of course, watertight, and have lights, and are tethered to a 100-ft cable running directly from the control room on the police boat. The underwater camera is controlled and maneuvered through the tether or cord that carries 12 volts of power to the unit, and transports data and video images (in color) from the depths to the control room.

37 Explosives, Chem-Bio Threats, and Drone Warfare

1 BOMBS, IEDS, AND FANATICS

Explosives planted in public places have caused substantial destruction and fatalities for over a century, including suicide bombers, planted improvised explosive devices (IEDs) remotely detonated, and vehicle-borne IEDs.

Scattered bombing incidents have occurred from time to time but never in any organized pattern and almost all involved homemade bombs. The bomb material and the method of construction have been readily available in books and magazines since the early 1960s and can be found on the Internet; however, the 20th century has seen incidents in war zones, and attacks on civilians in urban areas reach epidemic proportions in becoming a worldwide problem.

Explosives fall into two major categories:

High-order (HE) explosives which include TNT, C-4, Semtex, nitroglycerin, dynamite, and ammonium nitrate fuel oil.

Low-order (LE) explosives create a subsonic explosion and lack the HE over-pressurized wave. Pipe bombs, gunpowder, and pure petroleum-based bombs such as Molotov cocktails or aircraft improvised as guided missiles.

The explosive that nearly doomed two airline flights (landing at Detroit in 2009 and the shoe bomber Richard Reed in 2001) was a white powder that can deliver powerful blasts in quantities as small as tenths or hundreds of a pound. It is known as pentaerythritol tetranitrate (PETN) which was introduced after World War I and is in the same chemical family as nitroglycerin and is among the most powerful of explosives. It is also impossible to detect with X-ray equipment commonly used at airport security checkpoints. Luckily for the passengers on those two flights, it can't be ignited with a flame or by dropping it, or otherwise set off without using a detonator or mixing it with a chemical to cause an explosion. A suicidal airplane bomber needs a chemical reaction that would be hot enough to initiate the PETN and cause it to explode. PETN is a legal explosive used in mining and other industries and is available from its numerous manufacturers.

- TNT (trinitrotoluene) is probably the most important explosive compound in use and has an explosive size that is considered the standard measuring strength of bombs and explosives.
- Dynamite is an explosive based on the power of nitroglycerin and other chemicals. It was invented by Alfred Nobel in 1867 and is the red cylindrical tube that is used in many movies.
- Nitroglycerin is a high explosive with an incredibly unstable nature. It can explode for the most minor reasons, such as a change of one or two degrees in temperature, or a minor shock.

2 EXPLOSIVE DETECTION TECHNOLOGIES

There are two phases in explosive screening: detection and identification. In baggage screening systems, specially equipped X-rays are most often used. Ion mobility spectrometers are also employed for trace analysis of cargo shipments. In some circumstances, when the risk of touching or moving a suspicious substance is too high, Raman technology is used for non-contact identification of bulk solid deposits on a surface.

DOI: 10.4324/9781003091103-51

Threat	Threat Description	Explosive Capacity	Building Evacuation Distance	Outdoor Evacuation Distance
	Small Package/letter	1 lb	40 ft	900 ft
	Pipe Bomb	5 lb	70 ft	1,200 ft
	FedEx Package	10 lb	90 ft	1,080 ft
	Vest/Container Bombs	20 lb	110 ft	1,700 ft
	Parcel Package	50 lb	150 ft	1,850 ft
	Compact Car	500 lb	320 ft	1,900 ft
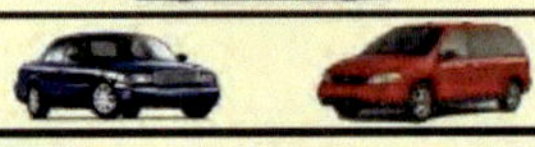	Full Size Car/Minivan	1,000 lb	400 ft	2,400 ft
	Van/SUV/Pickup Truck	4,000 lb	640 ft	3,800 ft
	Delivery Truck	10,000 lb	860 ft	5,100 ft

FIGURE 37.1 Details the range and effects of IEDs and is excerpted directly from the well-known fact sheet from the National Academies and the Department of Homeland Security (Graphic developed by US Technical Support Working Group (TSWG)).

X-ray. When sent through a baggage scanner, complex software algorithms help operators identify contents based on the atomic number which is proportional to material density. When an X-ray penetrates, denser objects allow less energy to pass through. They also look different. A piece of paper is organic, so it will appear light orange. An item containing aluminum appears green, and steel will show up as blue. With multiple X-ray generators, better characterization is achieved. Placing one camera on top, and one on the sides or in front also gives a more complete, accurate picture.

Ion Mobility Spectrometer. This is another detection technology sensitive enough to detect parts per billion of material and works by introducing an electrical charge to tested molecules and then pushing them through a drift tube. The time that it takes those molecules to reach the end of the tube – its drift time – reveals its atomic number. (A large molecule will be slower than a smaller one). This technology can measure particles smaller than nanograms, that is, one one-billionth of a gram. Even tiny amounts of residue left on suitcase handles or other objects at airport and cargo facilities could use this system since it allows swabs of hands that may have been touched by the suspicious material to be analyzed by the Spectrometer.

Raman Spectroscopy. A technique that uses a laser beam directed onto a sample to identify unknown materials based on their molecular structure. The wavelength of the beam is scattered into different frequencies by the atoms and molecules that make up the matter. The resulting Raman scattering is detected and the information is processed by an integrated computer to generate a "Raman spectrum" a graphical representation of the intensity of the scattered light as a function of its frequency.

3 HOME-GROWN EXPLOSIVES

Assembling a bomb can be accomplished by purchasing ingredients from a local farm supply warehouse. All the purchases are legal and can fit into a truck. A bomb of this size can have explosive power greater than the device used on the World Trade Center garage in 1993. Combining ammonium nitrate (terrorist weapon of choice) with fuel oil (known as ammonium nitrate fuel oil (AMFO)) and detonating it with blasting caps produces a powerful, slow-moving explosion. This type of blast has enough power to level a skyscraper larger than the Oklahoma Federal Building. Fertilizer distributors are now required to report suspicious purchases of ammonium nitrate to Homeland Security or face a $50,000 fine.

4 CHEMICAL AND BIOLOGICAL WEAPONS AND AGENTS

The Federation of American Scientists provides a useful rundown of the most dangerous and likely biological agents that could pose a threat if used as terrorist weapons (Table 37.1).

Deadly Chemical Agents are listed in Table 37.2.

Pound-for-pound, the deadliest weapon of them all isn't nuclear but biological; a single gallon of anthrax, if suitably distributed, could potentially end all human life on earth.

5 DRONE WARFARE AND THE AGE OF DRONES

Partly in response to the Iraq war and operations in Afghanistan (in the wake of the 9/11 attacks) – and in large part due to advances in technology – the US government shifted from engaging in further ground warfare operations, and the resultant US solider casualties, to drone operations. With IEDs from old and modified munitions from the regions, for years opposing forces in Iraq and Afghanistan carried out numerous attacks on military forces in both regions.

TABLE 37.1

Table of deadly biological agents

Anthrax	Ebola hemorrhagic fever	Saxitoxin
Botulinum toxin	Melioidosis	Smallpox
Brucellosis	Plague	Staphylococcal enterotoxin B
Cholera	Q Fever	Trichothecene mycotoxin
Clostridium perfringens toxin	Ricin	Tularemia
Crimean-congo hemorrhagic fever	Rift valley fever	Venezuelan equine encephalitis

More details information on each can be found on their Web site: https://biosecurity.fas.org/resource/agents.htm.

TABLE 37.2

Highly deadly chemical warfare agents

Nerve Agents	Pulmonary (Choking) Agents	Blood Agents
Taburn	Phosgene	Hydrogen cyanide
Sarin	Diphosgene	Cyanogen chloride
Soman		Arsine
Fluoride-containing organophosphate		
Standard V-agent		
Vx or V-gas		
Binary nerve agents		

This led to a shift in policy across several administrations – including each of the Bush, Obama, Trump, Biden, and Trump's second administrations – in which drone warfare and operations being utilized were the first and preferred option against military targets and terrorist targets.

6 SECURITY WEAPON OR PUBLIC CONVEYANCE

6.1 Military Drone

The Central Intelligence Agency (CIA) had been flying unarmed drones over Afghanistan since 2000 and, following the 9/11 attacks, it began flying armed drones.

Air warfare has been with us for a hundred years, since the Italian invasion of Libya in 1911, and the development of drones was in the works from the start. The reason is simple: even with all the advantages offered by air power, humans were still needed to fly the devices. There were limits to the risks that could be taken and whatever purpose an airplane was used for, it ultimately had to return to base with its pilot. Not surprisingly, from the start of the development of airplanes for use in war, engineers labored to circumvent this limitation.

During World War I, the Navy hired the inventor of the gyroscope to develop a fleet of "air torpedoes," unmanned biplanes designed to be launched by a catapult and fly over enemy positions. A secret program was run out of a small outfield in central Long Island, New York and when the secret was revealed in 1926, a newspaper reported that the planes were "automatically guided with a high degree of precision." After reaching a predetermined distance, the plane was supposed to suddenly turn and fly vertically downward, carrying enough TNT to "blow a small town inside out." The program ran out of steam when the war ended in 1918. In reality, according to a Navy history, the planes rarely worked: they typically crashed after takeoff or flew away over the ocean, never to be seen again.

7 HISTORIC SIGNIFICANCE

In 1935, the British Royal Navy demonstrated a remote-controlled aircraft for target practice for visiting American military officers and called the Unmanned Aerial Vehicle the Queen Bee. When the American Navy developed a similar vehicle, it was named "drone" in reference to the British model. Drone has for many centuries been used to describe a male honeybee. The word began branching out as a verb, meaning to buzz like a bee or to speak in a monotonous fashion reminiscent of a bee's persistent hum. Drones, unlike worker bees, need not worry about gathering nectar or pollen, they have often been seen as idlers. By the 16th century, a "drone" referenced lazy humans. Since the weaponized drone could only function when controlled by an operator, either on the ground or in another plane, the word was adopted to describe any remote-controlled aerial vehicle.

In World War II, a different approach was taken. The Navy had launched a new program called Operation Anvil to target deep German bunkers using refitted B-24 bombers. The bombers were filled to double capacity with explosives and remotely guided to selected German targets both in the Motherland and Nazi-controlled France. Remote control technology was limited to crude, radio-controlled devices and actual pilots were needed for takeoff. After guiding the plane to a cruising altitude, the pilots would parachute to safety in England. Thereafter, a "mothership" would guide the plane to its target. The program was a disaster. Many planes crashed, or worse. John F. Kennedy's older brother, Joseph, was one of the program's first pilots: he was killed in August 1944 when a drone-to-be that he was piloting exploded prematurely over Suffolk, England.

The target of Kennedy's mission was Nazi scientists working on competing drone technology: the world's first military rocket program. Indeed, German engineers had switched to rocketry given the difficulties of building full-scale pilotless airplanes. They worked extensively on rockets both during and after the war, continuing under the auspices of US and Russian government programs. In the late 1940s and 1950s, hundreds of former German rocket engineers and other Nazi scientists were brought to the United States and granted citizenship in exchange for rocket science – often despite clear ties to Holocaust-related atrocities. Stanley Kubrick's character Dr. Strangelove was a caricature of an expatriate Nazi scientist.

The development of drones stagnated for decades because there was little need for them, thanks to developments in rocketry. By the late 1950s, the US military had developed, in addition to many rockets, a slew of slower but more guidable "cruise missiles" which, in their own way, were like little airplanes. Cruise missiles maintain airplane-like "lift" on stubby little wings, unlike ballistic missiles, which move through a long curve of flight comprising a launch and rise followed by a guided fall.

Cruise missiles were, in a sense, proto-drones, miniature versions of what the military had attempted as far back as 1917. They could be dispatched and guided in flight; some had cameras and, in some incarnations, could even change target mid-flight. But cruise missiles could not linger over a battlefield in a holding pattern nor could they return to base. Also, their weapons delivery was blunt and inflexible; the missile itself was a single warhead. So, in the 1960s and 1970s, Air Force engineers continued to tinker with unmanned aircraft – in particular for use in surveillance flights, which don't require complex flight maneuvers hence less sophisticated piloting.

Drones are disturbing for an unusual combination of characteristics: the distance between killer and killed, the asymmetry, the prospect of automation, and, most of all, the minimization of both pilot and political risk. Taken together, these issues captivate the attention of journalists, military analysts, human rights, and researchers. A hint at what human-on-human violence may become, the unique technology allows the mundane and regular violence of military force to be separated further from human emotion. Drones foreshadow the idea that brutality could become detached from humanity – and yield violence that is, as it were, unconscious.

The weapon's names suggest ruthless and inhumane characteristics. The first drone aircraft deployed by the CIA and Air Force after 2001 was the Predator, a rather coarse name even for a weapons system, suggestive that the enemy was not human but merely prey, those military operations were not combated subject to the laws of war but a hunt. The Predator's manufacturer, General Atomics, later developed the larger Reaper, a moniker implying that the United States was fate itself, cutting down enemies who were destined to die. That the drones' payloads were called Hellfire missiles, invoking the punishment of the afterlife, added to a sense of righteousness.

But the real issue is the context of how drones kill. The curious characteristic of drones – and the names reinforce this – is that they are used primarily to target individual humans, not places or military forces. Yet they simultaneously obscure the human role in perpetrating the violence. Unlike a missile strike, in which a physical or geographic target is chosen beforehand, drones linger, looking precisely for a target. But the perpetrator of the violence is not physically present. Observers are drawn to thinking that it is the Predator that kills Anwar al-Awlaki, or rather, the Predator's Hellfire missiles, not the CIA officers who order the weapons' engagement. On one hand, we have the most intimate form of violence, targeted killing, which in some contexts is called assassination, and on the other hand, the least intimate of weapons.

There are reports of military drone operators suffering from post-traumatic stress disorder, and studies have shown that those who conduct strikes or watch videos of strikes suffer from "operational stress," which officials believe is the result of operators' long hours and extended viewing of video military operation feeds, i.e., dead bodies. Still, these reports pale in comparison to those of PTSD among combat veterans and there is no public information regarding stress among those ordering the strikes.

There is serious concern over the larger-than-expected number of civilian casualties resulting from drone attacks on terrorist sites. The growing number of casualties of innocent civilians (euphemistically known as "collateral damage") has caused a significant portion of the population in many foreign countries living in war zones to turn against the US government. Anti-American backlash following each drone attack and the collateral damage argues strongly for concentrating future attacks only on dangerous and high-ranking leaders who cannot be realistically captured or killed any other way.

One of the problems with drone strikes is that the accuracy of the attack cannot always pinpoint the target and opposition to the strikes is based primarily on reports of civilians killed unintentionally. While proponents argue that missiles fired from unmanned aircraft are the most precise way to eliminate terrorists, intelligence agencies do not have enough detailed information about who is in a strike zone to be certain that all are militants posing a threat to the US or to Americans overseas. A central role in identifying suspects and in determining their exact location is the use of eavesdropping and the tracking of electronic signals from mobile cellular phones. Such sophisticated technology may improve the odds of finding the intended target, but the data may be inaccurate.

A smartphone carried by a target can be easily tracked by an NSA or CIA operator and can provide a positive identification. But phones can be passed from person to person, leading to mistaken identifications.

8 SPY IN THE SKY

Thousands of drones have been flying over war zones in the Far East almost daily since 2010 and there have been drones flying in the United States for several more years, but these unmanned flights had been mostly limited to restricted airspace near the borders shared with Mexico and Canada. Some local law enforcement agencies have shown an interest in using unmanned aerial vehicles to find lost children, hunt for illegal marijuana crops, and ease traffic jams during the evacuation of cities before a hurricane or after a terrorist attack. Drones used by police departments are considerably less expensive than regular aircraft and while there are barriers to training, the benefits seem to outweigh any time, energy, and expense to get a program up and running.

Just as there are different types of manned aircraft, such as fixed-wing and rotorcraft, there are fixed-wing and min-rotor drones – each type serving different purposes. Fixed-wing drones resemble remote-control airplanes and are often used for farming and large surveys, whereas multi-rotor drones work better for aerial photography and inspections.

The three broad categories for drones include:

- Micro drones, which are the best for learning how to fly
- Consumer drones geared toward those who fly for recreation but are considering moving into the commercial area
- High-end commercial/professional/military drone

9 DRONES DARKEN THE SKIES

The surge in worldwide demand for drone surveillance comes as electronics and communications gear prices continue to drop. Suddenly, sophisticated surveillance platforms can be afforded on a modest budget. Increasingly, drone surveillance poses a significant threat to privacy and civil liberties. Millions of Americans and people around the world face daily technological intrusions. The law has lagged behind technology – it always does – and fast-changing policy must focus on basic principles if we are to protect even a modicum of our privacy rights.

Congress, in recognition of the drone revolution, ordered the Federal Aviation Administration to integrate these aircraft into the national airspace. With only a GPS sensor and autopilot feature, a drone can be programmed to fly through a pre-programmed sequence between coordinates at various altitudes in much the same way as an airliner. A system developed by the FAA called "sense and avoid" allows each plane, whether manned or unmanned, to communicate with a ground computer and map all aerial moving targets. That information is relayed to both aircraft and ground controllers.

The ability to use satellites to capture data within the US' borders has raised privacy concerns that the information is being abused by government agencies. The imagery gathered by drones could be analyzed by the private sector and the intelligence community for purposes that have never before been possible. Many privacy advocates are concerned that the data can be compromised regardless of all the safeguards in place. The civilian use of drone surveillance for law enforcement purposes is still in the premature stage. The intelligence community is already swamped with data from foreign surveillance and the various government agencies involved in this type of drone surveillance have little time for lower priority domestic law enforcement work. Since the drones have long wings, it's possible to strap all kinds of surveillance devices on them, including cameras, eavesdropping equipment, and, of course, even munitions.

Section XIV

Deadly Weapons of Apparent Mass Appeal

This section briefly describes several of the more popular handguns in circulation to those licensed to carry a concealed weapon and those who intend to use the weapon for a criminal purpose. The guns are the derringer, the Glock 17 and the AK-47.

About 20 children and teenagers are shot daily in the United States, according to a study by the *Pediatrics* journal. According to the FBI and the Centers for Disease Control and Prevention, guns

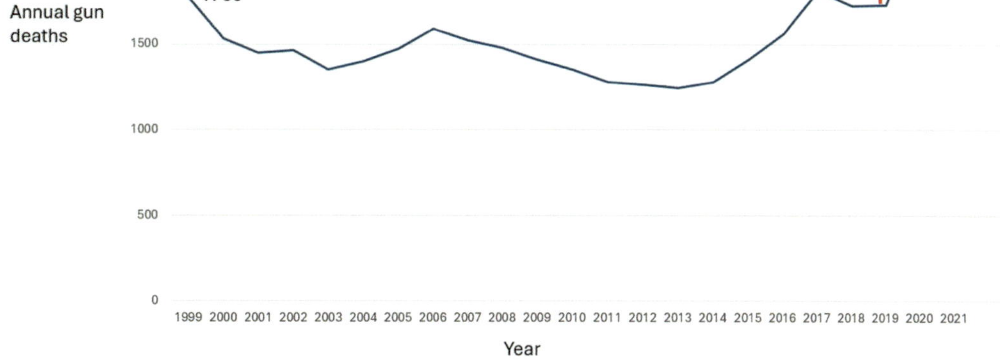

FIGURE S14.1 Annual gun deaths in the United States of children and teens under the age of 18.

Source: Pew Research.

DOI: 10.4324/9781003091103-52

kill more preschool-age children (about 80 a year) than police officers (about 50). In about 65% of annual cases in which children die from accidental shootings, the fatal shot was fired by another child. See the statistics in Figure S14.1.

Toddlers have the highest risk of unintentional shootings of themselves or another toddler, which doesn't happen in many other developed nations. The technology to eliminate these senseless deaths already exists and the proposed Handgun Trigger Safety Acts of 2017 and 2021 would have achieved this goal by mandating that all new handguns manufactured in the United States include "smart gun" technology if the law is passed.

There have been a few attempts that were funded by states or by private sources to develop a child-safe handgun to keep accidental deaths to children under the age of three from being killed by children under the age of three. The two major techniques that were tested include the Dynamic Grip Recognition System and the RFID gun release system. Neither concept has made an impact on the gun industry since the cost to convert a conventional pistol or handgun would be so expensive and to incorporate the safety device into new guns economically impractical. The techniques are covered in the next Chapter paragraph I. CHILD-SAFE HANDGUN

- Gun violence in America is no longer a phenomenon or an infrequent aberration; it has become almost a way of life with gun violence occurring almost daily
- People are shot in homicide assaults 300 times every day
- In 2018, people died as a result of gun violence at the rate of 90 times a day on average
- People who are shot but survive number over 200 every day

Other topics in the Section involve:

Construction of a gun using 3D printing technology Body Armor
Kevlar diapers
Armored Car Commercial and Private Service

38 Guns and Active Shooters

Gun homicides have become a common cause of death in the United States, consistently killing more people than car crashes in any given year. According to the Centers for Disease Control (CDC), in 2023, nearly 47,000 people were killed as a result of guns and gun violence.

To put that in perspective as compared to other countries, about ten people in Germany, the Netherlands, and Austria out of every million are shot to death by another person. That's about the rate at which people in the United States are killed by falling objects or plane crashes. In Poland and England, one person per million dies in gun homicides each year – about as frequently as an American dies in an agricultural accident or falling from a ladder.

1 CONCEALED AND OPEN CARRIED GUNS

In 2001, a credit card company in Florida (Merchant/Service.com) offered a free AK-47 rifle as a bonus to anyone signing up for their card. It was an unusual bonus in the very competitive credit card business and was worth $750, redeemable at any authorized gun shop in the state. Over 5,000 people signed up, but there is no data available as to how many recipients used the gun, or the card, to acquire their goods and services.

- **Handgun** originally referred to any firearm that could be carried and used by an individual. Today, it refers mainly to a pistol but can also be a revolver, a rifle, or even an air or paintball gun.
- **Revolver** describes three types of handguns: revolver, self-loader, and derringer.
- **Single-Action Revolver** has to be manually cocked before each shot, while a semiautomatic revolver requires cocking only before the first shot is fired.
- **Double-Action Pistols have** a long trigger pull that fires the weapon and brings the cartridge into the chamber in readiness for firing. Generally, a semiautomatic handgun.
- **Six Shooter** usually refers to a Colt Peacemaker six-shot revolver made famous in Western movies. While a six shooter is a revolver, not all revolvers hold six shots; the number of shots in various revolvers ranges from two to twenty.

Some of the definitions for the family of handguns can be a little confusing.

- A **pistol** is usually loaded with a magazine inside the handle, which holds bullets, while a **revolver** has bullets loaded in a rotating section in front of the trigger, which "revolves" as each bullet is fired – placing another bullet in the firing chamber.
- A **derringer** is a small pistol that holds only one or two bullets.

Mechanical Semiautomatic Pistols are popular handguns that align bullets in the chamber in a very unique way. Bullets are housed in a separate, spring-loaded container called a "magazine" (also known as a *clip*) that is inserted into the hollowed-out handle or grip of the pistol with the top bullet nearly aligned with the chamber. In order to fire, the slide is manually pulled back under spring pressure and released. The backward and forward motion of the slide does two things: first, pulling back the slide lifts or cocks the hammer, which remains cocked by engaging a catch. Next, the slide is released and moves forward to its rest position. This action strips the top bullet from the magazine and pushes it into the chamber.

DOI: 10.4324/9781003091103-53

When the trigger is pulled, it trips the catch, releasing the hammer. The hammer then hits the firing pin, firing the bullet. The resultant force of the explosion of the bullet out of the barrel in one direction causes a force in the opposite direction. The force drives the slide assembly backward – pulling out the empty shell, ejecting it from the chamber area, and cocking the hammer. The spring in the magazine advances the next bullet for feeding, and again the slide returns to its rest position, strips a fresh bullet, and feeds it into the chamber.

2 WEAPONS OF MASS SIGNIFICANCE

- **DERRINGER** was first introduced in 1825 and named after its inventor (who had one "r" in his name), who neglected to file for a patent. The short-barreled gun had instant popularity since it was small enough to be concealed in a pocket or even in the palm of a hand. It received notoriety when, in 1865, John Wilkes Booth used a derringer to assassinate President Lincoln. This small pistol was copied by many other manufacturers and the "Derringer" (with two r's) became the generic name for any short-barreled, single-shot pistol.
- **GLOCK 17** has become a household name, but it wasn't invented until 1982 when a machinist in Austria constructed the first model. The Glock was an immediate success with both law enforcement and criminals. Its adoption in the US marketplace can be traced to Bruce Willis's role as Lt. John McClane in the 1990s movie, *Die Hard 2.* Within the movie, he yelled to a police officer in the airport, "That punk pulled a Glock 7 on me! You know what that is? It's a porcelain gun made in Germany. It doesn't show up on your airport X-ray machines, and it costs more than you make in a month!"

There was never a model called the Glock 7; the gun was made in Austria – not Germany; it did show up on X-ray machines, and the Glock didn't cost more than the monthly salary of a police officer. While his description of the gun may have been exaggerated, it did help make the manufacturer a millionaire in the same way Clint Eastwood's performance as "Dirty Harry" Callahan did for the Smith & Wesson company and its .44 Magnum. The sudden mass acceptance of the Glock was also helped by the infamous shootout in Miami in 1986 between two bank robbers and eight FBI agents. Despite being outnumbered four to one, the robbers, armed with semiautomatic weapons, killed two of the agents and wounded five others before being killed. The agents, like most other federal and local law enforcement officers at the time, were armed with standard six-shot revolvers. The Glock immediately became the choice of most law enforcement personnel since the semiautomatic Glock 17 9-mm was lighter weight, had fewer and interchangeable parts, could fire as many as 17 bullets from its large-capacity spring-action magazine without reloading, and could be reloaded more quickly than most other semiautomatic guns. Two-thirds of all US police departments use the Glock 17 as their standard firearm.

3 ARMALITE ASSAULT RIFLES AK-47 AR-15

AK-47, as the world's most popular semiautomatic rifle, is more portable than a machine gun and has a longer range than a submachine gun. The Russian-made firearm was introduced following World War II and is by far the most widely distributed firearm in history. The name, appearance, and even the muzzle sound are instantly recognized worldwide. The rapid-fire AK-47 represents the final stage in the development of automatic weaponry dating back to the Civil War era's Gatling gun, which the US Army was slow to adopt once the war ended. European armies used Gatling guns in numbers far greater than those used in the United States to devastating effect in many of the colonial wars. It was almost 80 years after the Gatling gun was introduced that the AK-47 was developed, but this particular rapid-fire gun is far superior to any of its rapid-fire predecessors and revolutionized the world's gun culture. An estimated 75–100 million AK-47s are in circulation – one for every 80 people in the world.

The former Soviet Union's greatest engineering triumph may not have been the construction of an atomic bomb in 1949 (based on stolen Western secrets), the launching of the first satellite into orbit (1957), or even the first man in space (1961). Far more enduring, if low-tech, it was the world's most ubiquitous firearm – the AK-47. The automatic rifle was named after the designer, an Army Sergeant (**M**ikhail **K**alashnikov – MK) who, according to Russian government information, had no formal engineering or weapons design training. The year the gun was designed became part of its name (1947) and the sergeant was subsequently treated as a national hero in Russia and died in 2013 after retiring with the rank of Lieutenant General.

Western countries initially dismissed this automatic rifle as crude and simplistic compared to guns made by industrial nations with precision tools that allowed assembly lines to work within tight tolerances and mill parts to an exacting fit. But tight tolerances were not the primary goal in the production of the AK-47. It turned out its defects proved to be its virtue and made the AK-47 the winner in a global arms race that had been going on for more than 60 years because of how it performed in the field. The gun has a minimal recoil effect, making automatic fire possible in a one-man, handheld firearm (unlike the Gatling gun). It also used a minimum of parts, is built far more strongly than necessary, and its low tolerance during construction resulted in a relatively loose fit between its few moving parts, allowing the AK-47 to continue firing even when clogged with powder, residue, and dirt. It is a practically indestructible rapid-firing rifle that is simple to manufacture and considered a foolproof gun that works in the most extreme conditions despite neglect and abuse.

Over the next 50 years, it became the universal weapon, and by the late 20th century, it was used by armies, militias, and terrorists on all sides in practically every armed conflict, as well as being the weapon of choice of many criminals. The United States had purchased mass quantities of AK-47s for friendly forces in Iraq and Afghanistan, as well as for its own armed forces. Practically anyone, even a child soldier, could use this compact marvel, which was less than three feet long, weighed about ten pounds, and pushed out a blistering fire for the length of two or three football fields. Mass production of the AK-47 in Russia started in 1950, 15 years before the United States introduced its own automatic rifle, the M-16, which was far less popular. The AK-47 appears on the flags of Mozambique and a number of terrorist groups, including Hezbollah (Figure 38.1).

In recent years, the AK-47 and AK-15 semiautomatic assault rifles have become the weapons of choice for warring gangs, major illegal drug distributors, and foreign warlords throughout the world due to their easy accessibility, low cost, and the fact that they can hold up to 100 rounds in a magazine clip.

AR-15 is similar in configuration to the AK-47 and was less known until the mass murder of 20 first graders and 6 teachers in the Sandy Hook Elementary School massacre in Newtown, CT, in 2012 (Figure 38.2). It was the incident that prompted federal and most state governments to review their gun laws and outlaw assault weapons, without providing an adequate definition of exactly what differentiated an assault weapon from other guns that shoot bullets just as rapidly.

In 2017, the worst massacre in the nation's history involving automatic rifles occurred at a music festival in Las Vegas, where a gunman killed and injured hundreds one year after the second worst gun massacre in the nation's history took place in Orlando, FL. There a 29-year-old who had been employed as a security guard for the world's largest security company for nine years entered a crowded nightclub and killed 49 patrons and injured another 50. He used an AR-15 automatic gun.

FIGURE 38.1 The Hezbollah flag featuring, among other things, an AK-47.

FIGURE 38.2 Sandy Hook in the aftermath of the December 14, 2012, shooting.

Mass shootings in the United States that involved the use of the AR-15-style rifle:

1984 – Killed two and wounded 12 at an elementary school in Los Angeles. **2007** Six were killed and one injured in an apartment in Crandon, Wis. **2012** 12 killed and 58 injured at a movie theater in Aurora, Colo. **2012** Newtown Conn. Sandy Hook School, where 27 people were killed.

2013 – Killed five and injured three in a Santa Monica, Calif., home.

2015 – Killed two and injured one in Little

2015* – Killed nine people in a South Carolina church.

2015 – Two killed and two injured in store in Conyers, Ga.

2015 – Killed 14 and injured 21 in San Bernardino, Calif.

2016 – 49 killed and 53 injured in Orlando nightclub.

2017 – Killed 58 people and injured hundreds at a music festival in Las Vegas.

2017* – Killed 27 people at a church in Sutherland Springs, Texas.

2018* – Killed four people in a Waffle House in Tennessee.

2018 – Parkland, Fl., high school killing 17 and injuring over 50.

2018 – Santa Fe, TX, high school killing ten and injuring ten.

2019 – Virginia Beach, VA Kills 13 and injures four others at a municipal building.

2019 – El Paso, FL A gunman kills 23 and injures 22 others at a Walmart Supercenter.

2022 – Uvalde, TX A local man wounds his grandmother. Drives to his former elementary school, killing 19 children and 2 adults and injuring 18 others. Note: Police response was widely and resoundingly criticized.

2023 – Monterey Park, CA, 11 killed and nine injured when a gunman opened fire at a dance studio.

2023 – Lewiston, ME A shooter opens fire at a local bowling alley and bar, killing 18 and injuring 13 others.

*The shooters in the above-noted massacres had gun licenses and owned guns even though they had been convicted of criminal violations that, in eight states, would have allowed the police to confiscate the guns under the laws in those states. Those required to turn in mandated weapons to authorities include people with felony, domestic abuse, and drug convictions; fugitives; people adjudicated mentally ill; dishonorably discharged military veterans; and illegal immigrants.

Those eight states with gun seizure laws rarely enforce the statute and are, for all practical purposes, ignored as if the law never existed. The police cannot simply walk into the house of someone in violation of the statute without a court order.

4 GUN VIOLENCE IN AMERICA

In the two years following the Las Vegas shootings, there were at least 700 mass shootings, defined as involving four or more victims, across the United States. Gun violence has become so routine that the public is no longer shocked by the frequent deaths – the reaction by the government is short-lived, and very little legislation has been passed to address gun violence.

According to the CDC, on an average day, 132 Americans die by firearms, according to the Centers for Disease Control and Prevention with over half of those deaths by suicide.

The two major shootings in Florida within a year of each other (the Pulse Night Club and the Parkland High School) resulted in the state of Florida passing laws to raise the minimum age to purchase a firearm to 21 and extend the waiting period to three days. The two restrictions had little effect but were, at least, a start in recognizing the active shooter threat and the ease in purchasing weapons of all types. As expected, the National Rifle Association filed a lawsuit shortly after the bills were enacted, claiming that they were in violation of the Second Amendment.

5 HIGH-CAPACITY MAGAZINES

There was a public outcry in 2011 after six murders and 14 injuries (including a US congresswoman) in Tucson, AZ, not over banning guns, a reaction that usually follows a mass shooting, but against the selling of high-capacity magazines. While advocates (National Rifle Association (NRA), etc.) of the Second Amendment promote the right to carry arms, the nation was outraged that it was so easy to purchase high-capacity magazines and thought there was no justification for such magazines to be sold. Having a gun for hunting or to defend oneself cannot be used to justify what is, in effect, a rapid-fire machine gun. The deranged killer had a 33-round clip in his Glock 19 and was ready to reload with another clip before being subdued. That particular clip is banned in six states, including New York and New Jersey, but the extra-large clips continue to be sold freely throughout the rest of the country.

The most startling fact is that high-capacity magazines have been involved in virtually every mass shooting in the United States and, in general, ammunition is completely unregulated nearly everywhere. It's impossible to estimate how many magazines of all capacities are in circulation since there are no reporting requirements or background checks involved in the purchasing process. Most high-capacity magazines were once illegal under the Federal Assault Weapons Ban that was part of a law enacted in 1994 but expired in 2004, after which large clip magazines re-emerged on the marketplace.

The large clip was readily obtainable at any gun shop or at most large container discount outlets, including Walmart. Congress attempted to limit the number of bullets in a clip in 2011 following the Tucson shooting, but it wasn't until the Newtown shooting in 2012 that states like New York and Connecticut passed laws that included background checks for gun purchases and limited magazine clips to ten bullets. Congress soon followed with a similar federal law.

- The Tucson shooter was a man rejected as unfit by the Army, deemed unstable for college (thrown out), and thought to be inclined to violence by neighbors and school officials. He was nevertheless able to walk into a store and buy a gun and as many magazine clips as his credit card would allow. The gun he used was a Glock semiautomatic pistol.
- That event severely wounded Arizona US Representative Gabby Giffords and killed six innocent bystanders, stunning the nation and putting the focus on the Arizona gun control law that allows citizens to purchase and carry concealed weapons after a background check. The deranged shooter was turned down for the gun at one Walmart but purchased the gun at another Walmart the same day. The state law does not affect the purchasing of bullets and that fact caused a public outcry across the nation. The outcry soon abated, only to rise again after the next massacre a year later in the Newtown Elementary School.

The mass shooting in December 2012 at a small Sandy Hook Elementary School in Newtown, Connecticut, shocked the nation like no other mass killing in its history. The event placed considerable pressure on Congress to amend gun laws to prevent such horrendous killings. The deranged 22-year-old murderer got his semiautomatic AK-15 weapon from the inventory of guns kept by his mother in the home they shared. After killing her in her bed, he went to the school and walked around first-floor classrooms, randomly unloading 154 bullets from the 22-caliber rifle, killing 26 people, 20 of whom were under the age of 6. The carnage took all of five minutes and ended when the shooter committed suicide.

6 BANNING RAPID FIRE BUMP-STOCK GUN ATTACHMENTS

The bump-stock magazine used by some active shooters attracted a great deal of criticism, with many government officials demanding that bump stocks be banned for civilian use. The inexpensive attachment allows anyone with an AK-15 to turn the semiautomatic rifle into an automatic weapon. The Las Vegas shooter fired from the 32nd floor of the Mandalay Bay hotel on the Las Vegas Strip, having affixed bump stocks to his rifles for his attack on the Harvest Music Festival.

More than a half-million bump stocks had been purchased between 2010 and 2018 alone, and a law passed after the mass shooting in Las Vegas simply stated that the attachment should be turned in to authorities. Not many owners complied, and there was no effort made toward enforcing the ban by confiscating them.

According to the Bureau of Alcohol, Tobacco, Firearms and Explosives (ATF): On June 14, 2024, the United States Supreme Court in Cargill v. United States, 144 S.Ct. 222 (2024), held that a semiautomatic rifle equipped with a bump stock does not meet the definition of a "machinegun" as defined under 26 U.S.C. § 5845(b) of the National Firearms Act because the Court determined it cannot fire more than one shot by a single function of the trigger and, even if it could, it does not do so automatically.

A bump stock combines two devices, a plastic stock and a firearm. As the rifle fires, the bump stock absorbs and then rebounds the recoil, pushing the gun slightly forward into the shooter's otherwise immobile trigger finger, causing the gun to fire again and again and again without an additional "pull" of the trigger. Because a bump-stock device "automatically re-engages the trigger finger," the device allowed the Las Vegas shooter to fire more than 1,100 rounds of ammunition in ten minutes. The ATF deemed it to be a machine gun, but there were no provisions in the new designation that called for compensating the owner for either destroying or surrendering their bump stocks. "Options for destroying the devices include melting, crushing, or shredding," the rule advises. With plastic bump stocks, "individuals can use a hammer to break apart the devices." A cost estimate, included with the rule, anticipates that owners will have to destroy as many as 500,000 bump stocks, worth approximately $100 million.

The ATF provides a series on diagrams illustrating how to destroy various bump-stock devices at this URL: https://www.atf.gov/rules-and-regulations/bump-stocks/how-to-destroy.

The NRA believes that devices designed to allow semiautomatic rifles to function like fully-automatic rifles should be subject to additional regulations. The organization has long touted the distinction between civilian semiautomatic rifles and military guns with fully-automatic capability as the reason civilians should be able to own weapons like the AR-15. The sale of devices such as bump stocks, which put automatic-fire capability in the hands of the masses, threatens to undermine decades of public marketing and court arguments. The unpleasant truth is that semiautomatic fire is more accurate than automatic fire and the military instructs soldiers to use semiautomatic fire in nearly all combat scenarios. Some experts believe that the Las Vegas shooter could have killed more people had he not used a bump stock.

7 ACTIVELY RESPONDING TO MASS SHOOTERS

Besides the passage of gun-banning laws, there was another significant outcome of the Newtown massacre, which is the way in which first responders (and those caught in the line of fire) should react when confronted with a mass shooter in a confined public place, particularly in a school environment. The speed and deadliness of high-profile shootings prompted police departments to recommend fleeing, hiding, or fighting in the event of a mass attack, instead of past recommendations to remain passive, play dead, and patiently wait for help. The wait for a police response has averaged five minutes – the same time it took the Newtown shooter to unload 155 bullets from a semiautomatic rifle.

The recommendation that potential victims in a mass shooting event take an active role instead of waiting is largely due to the fact that "waiting" can prove deadly. Law enforcement agencies have begun to recognize that in active shooting situations, there may be a need for citizens to act as first responders even though they haven't been formally trained. This change in procedure accelerated after Newtown but began after the shooting at Columbine High School in 1999, when some police departments instructed officers who were first to arrive on the scene to act immediately rather than waiting for backup. Previous reaction to mass shooting incidents was to sit outside and set up a perimeter and wait for the SWAT team to arrive.

Since 2000, research on mass shootings has bolstered the idea that people at the scene of an attack have a better chance of survival if they take an active stance rather than waiting to be rescued by the police, who, in most cases, cannot get there fast enough to prevent the loss of life. The average time it took for the police to respond was four to six minutes, which means that half the attacks are over before the police arrive. In 16 of the deadlier attacks studied by the researchers, civilians were able to stop the perpetrator, subduing them in 13 cases and shooting them in three of those cases. Initially, the suggestion that victims should fight back as a last resort caused some controversy, and most did not think it was a wise idea, but in some cases, it has become apparent that fighting back might be the only option.

As part of the research noted above, actions taken by people in classrooms under attack during the Virginia Tech massacre, in which 32 students and teachers were killed before police arrived at the scene, varied and indicated that action may be far wiser than passively waiting for responders:

- In two classrooms, the students and the instructor tried to hide or play dead after the killer entered the room. Nearly all were shot, and most died.
- A professor in a third classroom told the students to jump out the second-story window while he tried to hold the classroom door shut, delaying entry of the killer. The professor was killed, but many of the students survived, and only three were injured by gunfire.
- A joint effort was made in a fourth classroom, where all the students and teacher helped block the door with a heavy desk and held it in place. It kept the killer out of the room. Everyone lived.

As a result of these now commonplace mass shooting events, some people have started an industry to educate the public in the ways in which they should react when confronted with an active shooter situation.

8 SURVIVING A MASS SHOOTING

The old adage of ducking and covering under a desk when a shooter is rampaging through a school hallway and enters a classroom is no longer the way to respond to a potential massacre. While it may seem shocking to most, there are now training consultants who make a living by providing survival courses to schools around the country. Attendees consist of school administrators and law enforcement officials and the training is based on fighting back instead of hiding, which is what people were told to do in case of a nuclear attack over 60 years ago. The instruction consists of staff members roaming the hallways with paintball guns to simulate an active shooter while the students try to hide in classrooms.

The school attacker is not typically a hulking 250-pound trained killer. The average male shooter is between 15 and 19 years old and in non-school shootings (Orlando Night Club, Aurora Theater), the shooter is most likely to be between the ages of 35 and 45 and of average build. Other facts presented at the group sessions indicated that almost half of all guns used in mass shootings are semiautomatic handguns and wounds from these weapons are survivable 85% of the time, but 95% of headshots are fatal.

Most would agree that trying to get seventh graders to react forcefully against an attacker may be unobtainable, yet it may prove to be the best way to handle an active shooter situation. The current protocol to rely on law enforcement to help contain or end the situation is not feasible, and calling 911 has been unsuccessful since the fastest response in a large city is ten minutes, and in smaller municipalities, the time is four minutes or longer. The average shooting spree lasts six minutes which, in effect, makes the student and teaching staff the first responders.

These survival training classes recommend fortifying the room or evacuating if a safe route is available and, if all else fails, throwing objects at the shooter such as a chair, laptop, or anything available. The idea is distraction. Try to prevent the bad guy from entering the room by looping an extension cord around the doorknob and pulling it to the side since most shooters don't spend a lot of time trying to open the door. The worst-case scenario is when a man stands up and pulls a gun and starts firing, a situation that occurs 52% of the time and the only defense is to lunge at the killer, an instinct that is not found in most people.

9 BACK TO NORMAL: A MISNOMER

Gunfire erupted in schools, malls, concerts, theaters, and army posts, and innocent people have been killed and many injured. But after a few days, newspaper reports are back to normal, with wording that is not only inappropriate but also ignores the fact that being in the presence of gunfire and watching another human fall victim to gun violence is far from normal. Individuals who are unfortunate enough to be in the same area as the active shooter will never *feel normal* again. Some will need therapy for the rest of their lives; some may jump at the slightest noise, and most will likely suffer from post-traumatic stress disorder (PTSD).

Our government's problem is not at the border; it's in our schools, churches, and homes. Gun violence does not discriminate; its victims are colorless and ageless, and it can affect anyone at any time no matter how affluent the community.

While the rest of the world has had its share of active and mass shooter scenarios, this phenomenon is by leaps and bounds an *American* problem first. For better – and often for worse – no other country has the freedom, or access to guns, quite like the United States.

10 SILENCERS (A.K.A. SUPPRESSORS)

The use of gun silencers automatically brings up images of the well-dressed hitman portrayed in Hollywood gangster movies. Silencers were not easy to obtain because of a federal law enacted during the 1930s depression, the National Firearms Act, which required a month-long screening process before a silencer could be attached to a weapon. It was about that time that gun manufacturers started using the less sinister name "suppressor," a successful effort that resulted in getting 42 states to allow the purchase of a treated silencer after passing a criminal background check.

In 2016, gun advocates began a campaign to get all the states to accept silencers as a legal weapon. They focused on getting Congress to repeal the Depression-era regulation and allow silencers to be obtainable as easily as a gun, which has far less scrutiny with almost no waiting period. The proposed law is better known as the SHARE Act (Sportsmen's Hearing And Recreation Protection Act) that would make it easy to purchase a silencer without a background check. The noise of the gunshot is important to police and first responders since it helps to locate the source of the firearm noise.

Even with the difficulties in obtaining silencers, they were being purchased at an all-time record high and the number of silencers registered with the federal government more than doubled between 2012 and 2016 to almost 800,000, according to the Bureau of ATF. Despite its name, the silencer, or the suppressor, can only quiet a gunshot to the level of a jackhammer – not much use for a James Bond assignment as a hitman when told to do so by G2. The surprising fact is that silencers are rarely used in crimes, according to a decade-long study by researchers that estimated that silencers were involved in 30–40 of the 75,000 federal criminal cases filed each year.

Having said this, the number and percentages of silencers or suppressors involved in shootings may jump shortly. In June and July of 2025, a provision introduced by Republicans in Donald Trump's "Big, Beautiful" tax reconciliation bill sought to deregulate silencers and numerous weapons. While it was ruled as violating Senate rules under the National Firearms Act, another provision was quickly introduced, attempting to eliminate taxes and, thus, making it easier for the proliferation of silencers and short-barreled rifles and shotguns.

11 NEWER SMART GUN TECHNOLOGIES

11.1 Dynamic Grip Recognition

A few years before the law was passed, the New Jersey State legislature had appropriated $1 million (1999) to fund a research program at the New Jersey Institute of Technology (NJIT), a state engineering college, that would lead to the development of a child-safe handgun locking device. This device would prevent accidents resulting from children finding guns in their home and accidentally shooting themselves or someone else. There are many mechanical trigger locks commercially available, but placing a locked gun on a shelf in the closet may slow a child's natural curiosity, but it won't prevent a child from finding the key, usually conspicuously hidden in the gun owner's sock drawer. Any type of trigger lock that uses a key to open the mechanism is more or less useless. A far better alternative would be a coded sensor pad built into the handle that could be programmed to be operated by only the gun owner.

The Dynamic Grip Recognition biometric ID system was developed by NJIT to overcome the shortcomings of the conventional trigger locking mechanism. After the first appropriation from the state ran out after ten years, the Department of Homeland Security funded another $1 million to continue the research to improve reliability and ensure that the trigger would operate only when the authorized person was using the gun.

American gun manufacturers were unified in their reluctance, *rejection* may be a better word, to sponsor a gun grip research project due to the limited sales potential and the fear that, if successful, a device like the Dynamic Grip Recognition device would conceivably encourage mandatory legislation by federal and local governments, forcing the gun manufacturing industry to redesign all its guns.

Many techniques were examined in the NJIT research lab, including fingerprint comparison on the grip, which was rejected. The researchers then tested radio frequency identification (RFID) technology, which required a remote transmitter to read the "grip" RF signals, etc. All were found to be susceptible to false readings, and after 12 years of testing, a technique was developed that involved a series of embedded sensors built into the pistol grip that measured several characteristics of the hand, grip size, and the pressure that was applied as well as the timing between aiming and pulling of the trigger. (Closely associated with signature verification and keystroke identification devices) The grip recognition data was collected in a memory circuit database to identify and enroll authorized users.

Each person in the pilot program displayed a distinct and measurable physical pattern in the way the gun was held when preparing to fire. Identification was the first of two challenges assigned to the research team; the second was sending the signal to release the mechanical trigger operation. The grip recognition system could not operate a mechanical mechanism; instead, the system required the trigger action to be electronically controlled, much like the rapid-firing machine guns that shoot thousands of rounds a minute. Converting standard operating mechanical triggers to electronically controlled rapid-fire guns would cost millions in retooling costs. Of course, there was strong opposition from the NRA, which did not support the concept, and law enforcement had serious concerns, since there was the possibility that once the gun was programmed, it could reject the owner in an emergency situation, a problem found in almost all biometric ID devices used in access control systems. The program at the college ended with success in 2012, just before a German company, using a different technology, succeeded in restricting the operation of the gun to its rightful owner.

12 TOY GUN REPLICAS FOOL POLICE

A teenager in a Texas school was carrying a gun tucked into his pants; the gun appeared to be a Glock semiautomatic handgun but was actually a perfect replica. It turned out to be a harmless air gun purchased on the Internet for $60. School officials believed it was a real gun and responded by calling 911 and placing the school on lockdown. The responding police officers arrived on the scene within minutes and confronted the student, repeatedly shouting, "Put it Down!" The student refused to comply and pointed the weapon at the officers. Two officers fired their weapons, hitting the student, who was taken to the hospital and died an hour later. It wasn't an isolated case, but the latest in a series of shootings involving realistic-looking fake guns that have appeared in schools and in attempted robberies since replicas became so realistic and inexpensive. Nationally, on average, about four children are killed each year in episodes involving BB guns or pellet guns, but that statistic doesn't include those deaths stemming from police shootings.

The gun at the Texas school had no markings indicating it was an air pistol. A federal law requires toy firearms and so-called airsoft guns – low-impact weapons used by police in training and for outdoor games and hobbies – to have an orange tip at the end of the barrel. The law doesn't apply to BB guns.

13 BUILD A GUN ON A $250 3-D PRINTER

According to an annual report released by the ATF, more than 5.5 million guns are manufactured in the United States each year, with millions more imported. When this total is added to the fact that it's so easy to purchase a firearm at gun shows without having to undergo a detailed background check, it seems a waste of time to make a replica plastic gun in the privacy of a person's home. Under most circumstances, it has not been illegal in the United States to build your own gun with a 3-D printer, but it had been fairly difficult using the limited software available until a 25-year-old law student posted designs for a 3-D printed handgun online in 2013, causing an immediate worldwide reaction.

An increasing number of crimes and homicides have been committed by individuals using these printed guns, which have been dubbed "ghost guns."

At the time, the State Department demanded that the online instructions be removed after discovering that they had been downloaded over 100,000 times, with Spain leading the downloads, followed by the United States, Brazil, Germany, and Britain. In Germany and most European countries, the possession of an unregulated weapon, even if manufactured at home, is illegal and punishable by law – however, trying to enforce that law may prove an insurmountable challenge. Despite the State Department's attempt to block the printed instructions, they are still available for free on websites like Pirate Bay, a popular file-sharing portal. This so-called "Liberator Gun" was the first successful firearm made entirely on a computer printer. It was fabricated on an $8,000 3-D printer that combined 15 printable plastic components, requiring only a single roofing nail as a firing pin to become a lethal weapon; it shoots a single bullet before shattering from the explosion.

An increasing number of crimes and homicides have been committed by individuals using these printed guns, which have been dubbed "ghost guns." While it has been determined that the weapon was partially 3-D printed, this proliferation of such weapons includes the 2024 killing of UnitedHealthcare CEO Brian Thompson in New York City on December 4, 2024 (Figure 38.3). CRC Press/Taylor & Francis has one of the few books published on the subject in Robert Walker's *A Field Guide to Ghost Guns: For Police and Forensic Investigators* (2021).

HISTORY OF 3-D PRINTER. An engineer in the early 1980s realized that UV light could be used to etch plastic layers into a desired shape; then, by stacking these layers together, a 3-D object could be created. The next step was to duplicate the feat on a printer, and eventually, a working model was produced that cost a buyer hundreds of thousands of dollars. Because the early models were so heavy that they couldn't be lugged to demos at prospective clients' offices, the developer made home movies to show executives at the major automobile manufacturing plants. The US automotive industry had fallen behind Japan, and car companies were looking for a secret weapon to better compete. The 3-D printer made it easy to create the prototypes for door handles or gear shift knobs, rather than sending blueprints to a tool and dye shop, shaving months off the design process. By 2014, 3-D printers had become more affordable at $1,300 and people caught the highly

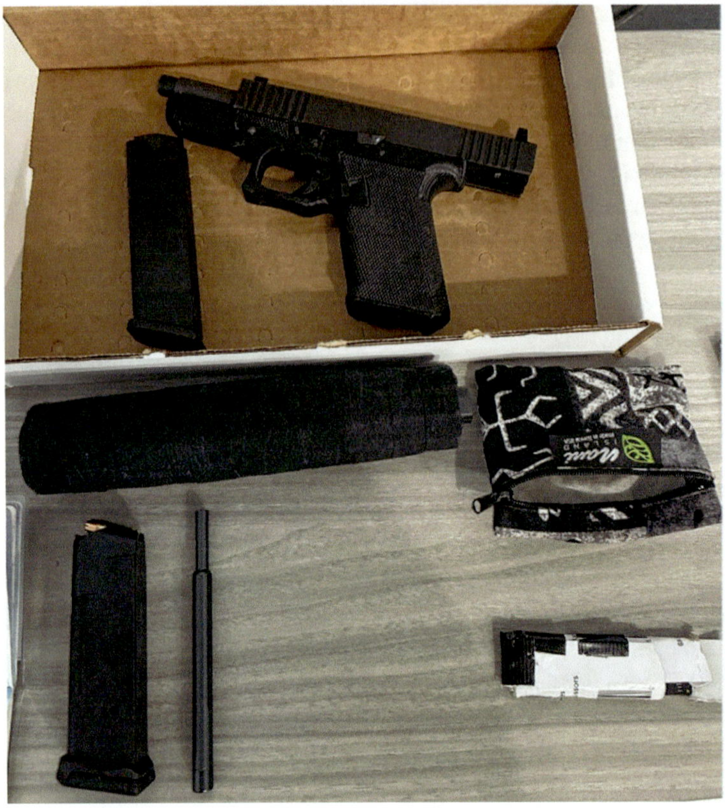

FIGURE 38.3 Portions, but not the entirety, of the weapon carried by the accused killer were 3D printed.

contagious 3-D fever, especially after all the publicity about making a pistol on the printer in a matter of hours.

Police agencies across Europe are very concerned over the proliferation of gun-making software that is easily found on the Internet. Countries in the European Union have much stricter gun control laws than the United States and officials in those nations worry that it has become much easier to covertly obtain and carry potentially lethal weapons. Experts in the Austrian Interior Ministry downloaded the Liberator gun's digital blueprints from Google, a gun that could be built on the type of 3-D printer that anyone could buy. It took the Austrian authorities 30 hours and about $68 worth of plastic polymer, building up layer by layer according to the software instructions, to make the gun. The next generation of 3-D printer-made guns is already available and is continuing to evolve to be easier and cheaper to produce.

14 SPOTTING CONCEALED (REAL) HANDGUNS

Many fatal encounters have occurred when police officers assumed that someone was carrying a gun and responded accordingly, only to find that the bulge in the pocket was not a gun. Police officers have always relied on training, observations, and street smarts to spot a concealed gun, but those methods are not scientifically acceptable. A person with a gun on their hip might grab the front of their coat to keep it from flapping open to reveal the pistol, or when getting out of a cab, they might hold their side to keep the gun secure.

All these gut feelings and streetwise techniques may become an obsolete practice. The New York Police Department (NYPD) was testing a gun scanning system capable of detecting concealed firearms by operating as a sort of reverse infrared mapping tool. It is programmed to read the energy people emit and pinpoint where the flow is blocked by some object, like a gun. Some objects obstruct the flow of radiation, e.g., a gun, which the device will detect without the need for a physical search. One of the early test models detected objects from five to eight feet, and further modifications will extend the range to 80 feet so that the device can eventually be placed in the window of the police patrol car as it moves through the streets. In some ways, it will operate as a long-distance metal detector like those used at airports.

15 GUN INDUSTRY CONGLOMERATE

In the not-too-distant past, the manufacturing of security systems was controlled by a single enterprise, until 1968, when anti-trust laws split the industry into a large collection of small firms that specialized in a few devices. By the turn of the 21st century, conglomerates began changing the industry, reverting it back to what it was in the 1960s, creating a few large corporations controlling much of the market. The same process seems to have happened in the gun manufacturing industry since about 2005.

In recent years, many recognizable, top-selling brands have quietly passed on to the control of a single privately held company, The Freedom Group. The company was formed in 2007 by the investment firm Cerberus, which had sales of over $40 billion from its many resources. The Freedom Group, with the financial backing of Cerberus, slowly became the most powerful force in the US commercial gun industry. They purchased companies like Bushmaster, DPMS Firearms, Marlin Firearms, Dakota Arms, and Remington, one of the most famous names in the industry. Remington has been producing guns since 1816, when, according to lore, Eliphalet Remington made a flintlock rifle in upstate New York. By the 1970s, the brand was so popular that the company diversified into typewriters. It is estimated that roughly 20% of long guns and about a third of gun ammunition come from The Freedom Group companies. Famous gun manufacturers not owned by The Freedom Group include Smith & Wesson, Ruger, Colt Manufacturing, Winchester, and Browning.

Given the prominence of guns in American culture and the continuing challenges made by gun lobbies over the interpretation of the Second Amendment, it is surprising to note that the firearms market in the United States is actually relatively small. As of 2025, sales of guns and ammunition totaled just under an estimated $20 billion annually. 84 million Americans own nearly 400 million firearms.

16 CONCEALED BURIED BOMBS

Improvised Explosive Devices (IEDs) are buried all over the world and have killed more people than bombs dropped from planes. Underground explosives have been around since WWI and some of those landmines used in that war which ended in 1918 remain buried. Insurgents in Afghanistan have used these IEDs to injure and kill military troops as well as thousands of civilians and during the course of that war (2001–13), IEDs remained the number one killer. The US government has spent billions of dollars in an attempt to find these devices before they explode. Some new devices and methods have come from soldiers in the field who adapted to rapid changes in enemy tactics and who understand the need for the United States to react quickly to these ever-changing threats.

MINOTAUR: a small remote-controlled front-end loader with attachments that detect bombs sensitive to weight. It also contains radar to penetrate the ground to detect buried IEDs.

LINE-CHARGE: consists of a small rocket that propels a rope filled with explosives about 80 feet. The ensuing blast clears a footpath about a foot wide, exposing trip wires and buried bombs or disabling them.

AERIAL MUNITION SYSTEM: is a camera-guided missile that can be fired at insurgents hiding on rooftops, inside buildings, or in fleeing cars. About the size of a man's forearm, it weighs about five pounds and has a range of over six miles.

17 BODY ARMOR

A $500 research project initiated by the DuPont firm in the early 1960s sought to find a lightweight fiber strong enough to replace the steel in radial tires. One of the few female chemists working on that research team discovered a poly fiber that was five times stronger than steel, of equal weight, and just as resistant to fire. The material was named Kevlar and in 1975, about a decade after its development, it appeared in the form of a vest resistant to bullets fired by handguns. In the 1990s, the vest was further strengthened by being reinforced with ceramic plates to withstand rifle fire. Since then, Kevlar has been used in military helmets; by applying up to 24 layers of the fiber, helmets have been made less vulnerable to penetration by shrapnel. This amazing fiber is now used all over the world, in boots for firefighters, hockey sticks, cut-resistant gloves, fiber-optic cables, fire-resistant mattresses, armored limousines, and even canoes. It is also used in building materials, making them bomb-resistant.

The development of Kevlar took place less than 40 years ago and has proven to be a financial windfall, generating several billion dollars in revenue for the company since its discovery by the chemist who had assigned all the patents to DuPont. The chemist died at age 93 in 2014, the same week that DuPont announced that the one millionth Kevlar vest was sold to a police officer.

It took many years for the so-called "bulletproof vest" to be fully accepted by law enforcement, and by the turn of this century, police officers no longer treated it as a novelty. The bulletproof vest has become an integral piece of a police officer's uniform, worn to work just as a shirt and pants. Kevlar will stop a bullet much like a net stops a tennis ball – by turning the bullet, slowing it down and dispersing its energy throughout the vest. (High-velocity and hardened-tip bullets like those fired from rifles can cut through the fibers and punch through the vest.) In actuality, the body armor worn by most police officers is not bulletproof – it is bullet-resistant and does not protect against all possible weapon threats. Also, some bullets can go through a vest, even if the armor is rated to stop that particular type of round.

36% of police officers shot to death from 2003 to 2014 were not wearing body armor, according to FBI statistics. The largest police union in the nation, the Fraternal Order of Police, was the unifying organization that made the use of body armor mandatory for all law enforcement agencies in the nation. This accord came amid persistent reports that large numbers of police officers were dying in shootings without the protection afforded by bulletproof vests. Disciplinary measures for those found without the vest while on duty include additional training, suspension, and being docked pay. Many police departments have instituted a policy of inspecting officers during roll call to ensure compliance with their departmental rules relating to armored vests.

There are different categories of vests that are certified by the National Institute of Justice, which can be a little confusing. Some police departments wear Level II-A vests, others wear Level II and others wear Level III-A; it has been a challenge for law enforcement agencies to anticipate which handgun rounds will be popular at the time. While Kevlar is still the most popular bullet-resistant fiber, there are other fibers, some of which are used with Kevlar in the same vest. Some of the more popular materials include Twaron, Goldflex, Dyneema, and Spectrashield.

It is recommended that armored vests be replaced after being worn for five years, according to the standards issued by the Justice Department. If the vest is treated like an old T-shirt and not cleaned in compliance with the care instructions provided by the manufacturer, it will not last more than the five years warranted by the manufacturer. Interestingly, a sizable proportion of police officers' lives are saved each year from wearing vests – not from the vest's protection against bullets from perpetrators, but from the cushioning effect of the padded vest when the officers hit the steering columns of their patrol. The vest is useless if not worn at all times. Far too many police officers have been killed in the line of duty by a bullet that could have been stopped by a vest that was stored away in the trunk of the police car.

Once a vest is hit by a bullet, it's done. It also cannot protect against a knife attack; stab vests are made from a different fiber and with a different weave than bullet-resistant vests. Multi-threat stab vests are available, but they are heavier, more expensive, and very few police departments use them.

"Backface deformation" is the resultant bulge that appears at the rear of the ballistic vest. The bigger the bulge, the more trauma the wearer had experienced. The trauma rating of a vest is determined by placing the panel up against a tray of clay and shooting a bullet into the vest. The size of the impression in the clay is then measured. One threat that can be anticipated is the police officer's own firearm. A large number of police officers are shot, not only by their own weapons but also after gun grabs, friendly fire accidents, and other internal risks.

The price of a Kevlar vest starts at about $560, with one of the more popular models costing $925 each. Under a government-funded program approved by Congress in 1998, the cost to local law enforcement agencies had been subsidized where cities with populations of over 100,000 paid about half the cost of the vest. In 2010, the program helped purchase almost 200,000 vests nationwide, but that number dropped to 60,000 in 2013 due to federal budget cuts. Since then, many states have funded the vests, and all officers in the country will have been issued lifesaving vests as of 2016.

18 BODY ARMOR UNREGULATED

When an investigation uncovered the background of the lone shooter who killed more people in the nation's history by a single person (Orlando nightclub in June 2016), it was discovered that the killer tried unsuccessfully to purchase body armor capable of stopping police handgun rounds. The gun shop he had visited didn't carry armor; however, if he had searched online, he would have found dozens of online retailers selling bulletproof vests capable of stopping not only handgun fire but also bullets from SWAT team snipers or even armor-piercing rifle ammunition, all of which would have been shipped right to his door with few questions and no background checks.

Many law enforcement officials have pressured Congress to pass laws restricting the sale of military-grade protective body armor to civilians. Congress considered such a law but never passed a body armor law except for a little-known federal regulation that forbids felons from buying or possessing the armor. There are no procedures in place to enforce that restriction at the point of sale.

Not too many years ago, body armor was a niche product and difficult to obtain outside of law enforcement circles. In recent years, manufacturers and suppliers have appeared on the market selling the equipment to a growing group of people. There are two major categories of bulletproof equipment sold in the loosely regulated civilian market: soft armor and hard plate.

19 KEVLAR DIAPER

More than 1,500 servicemen and women lost limbs in the wars in Iraq and Afghanistan, according to the Army and according to data from the Joint Improvised Defeat Organization, in 2011 **alone** nearly three in four service members who lost legs to IED blasts also suffered genital injuries. That percentage dropped to less than half for those wearing Protective Undergarments (PUGS), like a Kevlar diaper. It was common practice in the counterinsurgency effort in Afghanistan to have little more than boots and pants protecting their lower body from landmine blasts before the armored PUGS were issued. The shorts prevent dirt and fine debris from bomb blasts from piercing the skin and a protective cup shields troops from larger fragments.

Besides physically protecting American soldiers in the field, PUGS had a great effect on morale by conveying a strong message of support from the government. The Defense Department spent more than $40 billion to buy Mine Resistant Ambush Protected vehicles to protect troops from roadside bombs. The same research effort is also being expended in developing new gloves with better protection after reports showed soldiers losing their fingers when bomb debris pierced the fabric.

Such technological advances and solutions can potentially prevent – or at least limit and mitigate – permanently debilitating injuries to soldiers in the future.

20 AMMUNITION VENDING MACHINES

With the controversy about guns falling into the wrong hands or into a child's grip, it was reported that not only is it still easy to purchase a firearm, but ammunition can be purchased using vending machines. The first such machine appeared in 2015 at several public shooting ranges in one eastern state and has now proliferated to a number of states to the present (Figure 38.4). It was built and installed to cut down on the long lines to purchase the ammo at the range. The machines sell every standard caliber bullet between a. 22 long rifle and a .45 Long Colt. The ranges are considered private gun clubs, and the use of such a dispenser is completely legal according to the lawyers for both the local municipality and the gun club.

FIGURE 38.4 An image of ammunition vending machines (Which might cause one to ask what we've come to! How convenient do we need to make ammunition acquisition?).

Section XV

Nonlethal Weaponry

1 STRANGE AND CONTROVERSIAL WEAPONRY

- Taser/stun guns
- Laser beams
- Reconnaissance robots
- Fog machine
- Powerful loud sirens
- Penetrating police sirens
- Beer bottle weapon
- Blinding strobe light
- Chemical spray
- Hazardous toilets
- Breathalyzer detectors
- DNA mist spray
- Dummy cameras
- Dummy safes
- Dummy containers
- See-through gates

Nonlethal handguns that are called "traumatic weapons" can fire rubber bullets at high velocities. In some foreign countries, they are used extensively for riot control and sometimes referred to as "compliance weapons" or "less-lethal launchers." Most of these rubber bullet guns can be retrofitted to fire metal projectiles (real bullets) except in certain countries which specifically mandate that the guns be designed to prevent the use of deadly bullets. Also, there are laws in some European nations that have made it a crime to use rubber guns at distances closer than three feet, since at close range their use could not be defended as in self-defense.

A low-powered propelling charge gives the rubber bullet gun a muzzle velocity that limits its accuracy to about 100 yards. The intended use is to fire at the ground so that the round bounces up and hits an individual in the legs, causing pain but not injury. The bullets can be fatal if the person is shot in the head, but overall, records indicate that rubber bullet gun fatalities are rare. When used

DOI: 10.4324/9781003091103-54

in target practice with suitable backstop, rubber bullets can be recovered undamaged after firing and reused several times.

Russia, like many European countries, has equipped its riot police with water cannons, tear gas grenades, and electroshock devices. Experience with these nonlethal weapons during actual riots has demonstrated that they can sometimes be lethal. Tear gas can choke people to death, stun guns can cause a heart to stop, and, in the winter, water cannons used in colder climates can cause wet protesters to suffer from hypothermia.

Other less-lethal riot control launchers from rubber bullets to grenades:

- Sage weapons
- Beanbags consisting of a tough fabric bag filled with birdshot flattens out on impact
- Stinger rounds consist of round rubber balls of about ½ inch in diameter around
- Teargas grenade cartridges
- Pepper spray called pepper balls, capsules filled with a solution of capsaicin
- Stun ammo
- Smoke rounds
- Shuttlecock machine guns introduced in Russia, using pressurized cartridges to shot plastic
    - birdies
- Sound projectiles that emit 180 db noise level that causes pain and injury to the eardrum
- Shotgun used in riot control, usually 12 gauge
- Paintball guns loaded with paint to mark rioters for later identification

39 Temporary Debilitating Weapons

1 STUN GUNS

Stun guns have grown in popularity among law enforcement after many years of being a very controversial device. A type of conductive energy device that delivers a shock from short distances, these electroshock guns were better known as "Tasers," a name used by the firm that, for all practical purposes, controlled the stun gun industry from its beginning. The firm was, for many years, known as Taser International, and the Taser was manufactured by Axon since 2017. The so-called stun gun is legal in most states as a self-defense device that administers a high voltage, low amperage shock to individuals. Taser is accepted as the proper name for this type of unit much like Kleenex is used as a generic product name in the facial tissue market (Figure 39.1). The one-pound Taser delivers one to four milliamps, far below the amperage that could do any lasting damage to an individual (100–200 milliamps can kill a person).

The stun gun can shock through clothing, sending hundreds of volts of electricity through metal prongs. Most work only at short range, while others shoot their prongs from longer distances. The three most popular stun guns from Axon were the handgun-shaped Pulse, the Bolt, and the

FIGURE 39.1 A typical Taser frequently utilized by police.

Source: Shutterstock.com.

DOI: 10.4324/9781003091103-55

baton-shaped Strikelight, which can double as a flashlight. Another brand is Vipertek, which offers devices in a variety of colors. After being hit by the stun gun, the current causes skeletal muscles to contract. Although the shock is painful, it is far less lethal than a bullet.

Basic Taser Facts:

- Range of up to 21 feet
- Delivers a five-second blast that briefly incapacitates most humans
- Produces a jolt of 1,500–5,000 volts

There were only four states and the District of Columbia that enforced an outright ban on stun guns until 2017, when New Jersey lifted the ban that had been enforced since 1985. That state's Supreme Court found the ban illegal for being in violation of the Second Amendment. According to statistics compiled by the NJ Office of the Attorney General, in 2016, there were 60 incidents where police deployed stun guns. According to the data provided by Amnesty International, 70 people died after being shocked with a stun gun.

Dart guns (Tasers) and touch guns are often misidentified by the general public since they both operate under the same electroshock principle. The more popular Taser works with two charged electrodes that aren't permanently joined to the housing like a conventional stun gun. Instead, they are positioned at the ends of long conductive wires attached to the gun's electrical circuit. Pulling the trigger breaks open a compressed nitrogen gas cartridge inside the gun, propelling the darts into the subject. The expanding gas builds pressure behind the electrodes, launching them through the air, and the attached wires trail behind using the same basic firing mechanism as in a BB gun. The electrodes are affixed with small barbs so that they grab onto an individual's clothing. When the electrodes are attached, the current travels down the wires into the individual, causing a similar stunning reaction as with the conventional stun gun. The obvious advantage of the "wired" design is that the unit can transmit the shock from a distance of 15–20 feet. The disadvantage is that there is only one shot – then the unit must be wound up, the electrode wires repacked, and a new gas cartridge reloaded.

Electrodes are simply two plates on conductive metal positioned in the circuit with a gap between them. Since the electrodes are positioned along the circuit, they have a high voltage difference between them. If this gap is filled with a conductor (the individual's body, for example), the electrical pulses will try to move from one to the other, dumping electricity into the person's muscles, causing them to do a great deal of work at a rapid pace, instantly depleting the person's blood sugar by converting it into lactic acid. The individual is unable to produce energy from his or her muscles, and the body is then unable to function properly. The best places to strike a person are in the upper shoulder, lower rib cage, or upper hip. When the neuromuscular system is overwhelmed and controlled by the stun gun, it causes an immediate loss of balance. The current does not pass from the individual to the Taser holder if their bodies touch.

Stun guns vary in power from 20,000 volts up to 150,000 volts, and while all stun guns, regardless of their voltage, do essentially the same thing, the higher the voltage, the quicker the effect. Stun guns can be purchased quite inexpensively and come in various disguises, including flashlight shapes, cell phone configurations, and possibly the most popular, a compact model that resembles a disposable razor and is available in several colors. Most operate off a 9-volt or 8 AA batteries that can send a high voltage and low current (amperage) to disable an individual for several minutes. The stun gun will penetrate through clothing, does not rely on pain for results, and it is completely nonlethal.

2 HISTORY

In Mississippi, a pair of Freedom Riders was zapped while in custody in 1961 and two years later, Alabama police officers used cattle prods on African American demonstrators. This type of treatment to mostly nonviolent civilians rightly caused widespread public denouncement and outrage and the cattle prod – or "shock baton" – fell out of favor as a method of crowd control during the 1960s and 1970s. In August of 1965, just before the riots in Watts, California, the US president at the time ordered a study on crime in a "free society" and among its many recommendations was the development of a technology for handling unruly crowd gatherings. The study went on to explain that a police officer should have a way to incapacitate criminals that worked quickly and had little risk of lasting injury. For example, "darts have been used for injecting tranquilizing drugs into animals"; an observation that started a cottage weapons industry.

Inventors rushed to find replacements for prods and darts, and in 1965, a patent was filed for a long-range weapon that worked like an electrified super-soaker. In the following year, a Japanese inventor proposed an air gun that fired two electrode needles that would "stun the criminal to a temporary state of false epilepsy, which rendered him helpless." Hundreds of patents soon followed, but none caught the attention of law enforcement professionals. The Taser was patented in 1974 and wasn't very different from the ones that came before. It had a pair of electrode projectiles tethered to the gun with wires, delivering a high voltage, low amperage pulse of electricity that could subdue individuals without killing them. The first improved model of the Taser was manufactured in 1993 after the owner was inspired by an incident in which a man inadvertently walked into an electrical fence and survived – though he was temporarily immobilized. The invention was named after a Tom Swift science fiction novel, one of which was entitled "Tom Swift and His Electric Rifle"; an "A" was added to the middle, forming "TASER." Initially, the newer model Taser was sold through only one outlet, The Sharper Image, which seemed logical at the time in terms of capturing the gadget-lover audience. However, it was soon discovered that this limited market was not attracting large numbers of consumers interested in buying self-defense devices.

Another factor that hurt sales was that the government classified the invention as a firearm because the product looked like a gun, felt like a gun, and, at a cost of $1,000, only a few hundred units were sold to police departments. The few units that were purchased by police were mostly ignored and treated as a boutique item and stored in the trunk of supervisors' vehicles. From 1976 to 1995, Taser-type devices were considered firearms because the darts were propelled by gunpowder, but in 1998, the unit was modified to use compressed nitrogen, allowing the Taser to be sold to the general public. Because they propelled the barbs with compressed nitrogen instead of gunpowder, Tasers were no longer considered firearms and are not regulated by the Bureau of Alcohol, Tobacco, Firearms, and Explosives or any other federal agency.

Currently, over 7,000 of the 18,000 law enforcement agencies in the United States have distributed this "gun" to their entire force and since the founding of the company, more than 17,000 law enforcement and military agencies in 45 countries around the world use Tasers, as do more than 400,000 individual officers and 200,000 private citizens. A great deal of concern still surrounds the worldwide use of Tasers by police officers; a well-publicized death of a passenger at a Canadian airport in 2007 led to Tasers being outlawed in that country. The proliferation of Tasers has made them controversial, and their frequent use has led to fears of overuse. The human rights group Amnesty International seeks to have the use of Tasers by private citizens prohibited and use by police curtailed pending further study. The manufacturer has had 52 product liability lawsuits that went to court – and has won them all.

CATTLE PRODS are similar to stun guns in design in that they apply an electrical current across two electrodes, but they serve a completely different function. A conventional stun gun uses an electrical charge to incapacitate someone, while a cattle prod applies a charge to get an animal (or person) moving. While a cattle prod causes pain, it does not significantly affect muscles or the nervous system. The voltage in both the Taser and the conventional stun gun is high enough to deposit electricity into the entire body. The lower voltage in a cattle prod can only shock someone at the point of contact.

3 TASER POPULARITY

Regardless of the potential danger, millions of police officers, soldiers, and ordinary citizens carry real-life stun guns to protect against personal attacks. Only since 2001 have public buildings, such as hospitals, purchased these electroshock guns, mainly for sensitive areas like the ER or psychiatric wards. Three NYC hospitals that report possessing stun guns have not yet reported if the guns have ever been used.

A landmark decision that set broad judicial standards for the use of Tasers occurred in a US Court of Appeals in California (2009), ruling that police officers can be held liable for using one of the devices against an unarmed person during a traffic stop. The ruling specified that the electrically disabling device constituted excessive force when used against an unarmed individual who did not pose a threat, and it refused to allow a police officer immunity for its use. The lawsuit involved a man stopped twice on a highway, once for speeding and once for not wearing a seat belt. After the second stop, the driver became agitated, standing outside the car, clad only in his boxer shorts and tennis shoes. It was noted by the judges that the driver did not threaten the officer and was not trying to flee – two of the elements of a three-part test that the US Supreme Court has used to determine when significant force is justified. The third element is the severity of the offense, and a traffic violation did not support the use of a significant level of force. The case was allowed to proceed to court, and the jury found the police officer guilty of using excessive force, awarding the driver more than enough money to replace the four front teeth he lost when he fell to the pavement after being hit by the Taser.

WARNING: The Justice Department has sponsored studies at university research laboratories on stun guns and their implementation and usage over the last several decades. A mid-2000 DOJ study concluded that these electric shock guns may be more dangerous than previously claimed by manufacturers. Controlled lab tests on human volunteers did not find any changes to heart rhythms, but other studies conducted on pigs (which have hearts similar to humans) found that the shock from the guns could cause the hearts of healthy pigs to stop beating due to ventricular fibrillation if the barbs landed near the heart. An electrical disturbance that causes the heart to beat irregularly and leaves it unable to pump blood is known as ventricular disturbance and, unless the heart's normal rhythm is restored, death can occur within minutes. According to data compiled by Amnesty International, more than 500 people have died after being shot by Tasers since 2001. However, this fact hasn't stopped police departments from purchasing them.

The Police Executive Research Forum, a nonprofit group dedicated to improving police tactics and strategies, recommended that officers be allowed to use the devices only on suspects who are actively resisting arrest. Some police departments have allowed officers to use Tasers on people who are simply refusing to follow orders, which was not the purpose of the stun gun. In general, law enforcement officials have often taken a dim view of the average consumer having access to Tasers because they feel that the Taser is better suited for well-trained individuals than in the hands of average citizens. The debate goes on, but private Taser sales continue to rise as average citizens become more concerned with defending themselves.

Several public interest groups have been concerned about how these devices are used and have pressed law enforcement officials to limit the use of Tasers to instances in which the only other option is gunfire and to not use it to subdue targets who are simply resisting orders. Another source of concern is the Taser being used in the commission of a crime. There is no way to prevent criminals from obtaining the device from various sources, and manufacturers have attempted to discourage such purchases by incorporating an elaborate system to activate the gun, including requiring a background check over the phone. The newer units, when fired, discharge a spray of confetti-like pieces of paper printed with the gun's serial number.

Another problem associated with stun guns is the confusion over the term, which can refer to the operation of either Taser guns or conventional handheld contact devices that require touching the individual's skin. Laws regarding both the Taser and conventional stun guns seem to consider both techniques to be the same. Laws that vary by state and local jurisdiction apply to both technologies; those same laws change from year to year, and even suppliers have no single policy concerning the distribution of the product. It is against eBay policy to sell or ship stun guns outside of the United States and to areas where stun guns are restricted, but the Internet lists dozens of suppliers.

The private sector has slowly accepted this device as an alternative to pepper spray for defense against criminal attacks since it is less lethal than a gun. Stun guns are supposedly more reliable than pepper spray, but less reliable, in a manner, than a gun. While stun guns have proven to be effective in many situations and can save lives, they have been criticized by some experts in the criminal justice field as being a cause of death. The manufacturer that controls most of the industry claims that extensive testing found that the Taser (name of manufacturer that produces stun guns) did not change heart rhythms or damage hearts. A 2012 study published by the American Heart Association (AHA) concluded that Tasers can result in "ventricular arrhythmias, sudden cardiac arrest, and even death (Source: *Circulation,* AHA's journal)."

4 HUMAN ELECTRICAL SYSTEMS

In general, electricity is considered a harmful force to bodies, for example, being struck by lightning or sticking one's finger in a light socket, two situations where the current can maim or even kill a person. In smaller doses, electricity is relatively harmless and even essential to everyday human functions, even the simple task of making a sandwich. This activity requires the brain to send electricity to a nerve cell in the muscle in the arm. Then the signal activates the nerve cell to release an impulse to put the sandwich together. In this way, the different parts of the body use electricity to communicate with one another. In a sense, the body operates like other communication systems, and the basic idea behind stun guns is disruption of this communication system. The charge from a stun gun has a lot of pressure but not much intensity and will confuse and unbalance the individual, partially causing a temporary paralysis. Since there are muscles and nerves all over the body, it doesn't particularly matter where the voltage charge hits an individual.

5 PARALYSIS

Stun gun effectiveness varies depending on the particular model, body size, and the level of determination of the individual being stunned. It also depends on how long the gun is applied to the person. When the gun is applied for half a second, a painful jolt will startle the individual. If "zapped" for one or two seconds, the muscle spasms cause a sort of daze; if zapped for more than three seconds, the individual will become unbalanced, disorientated, and lose muscle control. Even with all the voltage applied, it is possible that determined individuals with certain physiological characteristics may stay in control despite any shock.

6 TASER CONTROVERSY

The Taser or stun gun has been labeled a health hazard by some, but in many cases, it can be an effective enforcement tool that may ultimately reduce the number of violent assaults and even save lives.

Some police departments in major cities initially soured on Taser use after the manufacturer, Taser International, issued a statement in 2009 stating that the handheld zapper posed an "extremely low risk" of causing a heart attack if fired into the chest. This memorandum prompted several law enforcement agencies to issue directives telling officers to aim for the back of the legs and adding warnings of possible claims if targets were hit in the chest. Some police departments use the Taser gun as many as 300 times every year and claim that: "if the Taser can't be aimed where it is most likely to work, the device is rendered useless." Firing at the legs was impractical since both Taser prongs must stick for the system to work and the darts veer away from each other as they propel.

It appears that while stun guns are intended to be used only for self-defense or incapacitating an unruly person, many are, unfortunately, commonly used as torture devices in many parts of the world. A number of governments routinely use stun guns to extract confessions from political rivals, since electrical torture leaves less evidence than other methods of torture. Shocks from stun guns are extremely painful but don't leave an obvious wound, so while they are relatively safe when used correctly, they can be quite dangerous in the wrong hands. It should be noted that stun guns cannot be shipped outside the United States without an export license from the US government.

In 2006, a specially made video camera that attaches to a Taser stun gun became available; more than 2,400 law enforcement agencies bought 45,000 of these cameras. The so-called "Taser Cam" is activated as soon as a Taser is unholstered and the officer turns off the safety. There is no way to deactivate the camera without disabling the gun. The camera is meant to provide accountability for police departments using the guns, but it has also aided police departments in proving that the guns are being used properly. Another reason that the camera has been accepted by a growing number of law enforcement agencies is that, following a police encounter, a video of the encounter is often posted on social media by witnesses who recorded the action on their cell phones. The Taser Cams have been used in contentious situations to show that the police were justified in using the stun gun, avoiding lawsuits from individuals who were subdued with a stun gun. There has been a marked increase in the use of dash cams, or cameras on uniforms, for similar reasons.

New Jersey became the 46th state (2017) to allow residents to carry stun guns as a means of self-defense. The legal requirements to obtain a license to carry the stun gun in public areas are identical to those governing the carrying of a concealed pistol, training, and background check. The three states in which it is still illegal for residents to possess stun guns are New York, Hawaii, and Rhode Island, according to Taser International, which has more than 95% of the US stun gun market. Even the states that allow the carrying of these guns generally ban them from schools and courts and some states limit their use to homes and businesses or require a permit to carry one in public.

All 50 states allow law enforcement to carry them.

Connecticut was the first state to require law enforcement officers to document Taser gun use and a state agency issued a report in 2016, the first nationwide study, on how their police officers use them. Police officers in Connecticut used Tasers 650 times in 2015 and listed the following findings:

- 33% of the people involved in Taser encounters were emotionally disturbed.
- 50% were identified as either intoxicated or under the influence of alcohol or drugs.
- Taser use appeared to be ineffective against those under the influence of certain drugs.
- 13% of incidents involving Tasers were described as a person attempting suicide.
- 94% of the victims were male.
- About 30% of those who were the target of a Taser received multiple shocks.

The city of Miami Beach changed how their police officers use Tasers following the death of a teenage graffiti artist in 2013. Officers can no longer aim the Taser at a suspect's chest, and can't use the stun gun devices if the suspect is handcuffed or is not an immediate threat.

WIRELESS STREAM CHARGE falls into the category of stun guns and works on the same principle as the Taser guns except it uses a liquid stream to conduct electricity rather than extended wires. The gun is hooked up to a tank of highly conductive liquid, typically a mixture of water, salt, and various other conductive elements. When the trigger is pulled, an electrical current travels from the gun through the liquid stream to the individual. These guns have a longer firing range than Taser guns and can be "shot" many times in succession. This repeat firing requires water to be carried in backpacks or in vehicle-mounted water cannons. It hasn't replaced the conventional Taser gun, and less than 500 of these "guns" have been in the police department inventory.

7 CHEMICAL SPRAYS

Pepper Spray and Tear Gas are sold in pressurized cartridges containing one of the two gases and are the first line of defense for individuals who treat the handheld device as a weapon of last resort. Tear gas and pepper spray are probably the two most popular nonlethal chemical agents. Tear gas falls into the category known as a lacrimator, a chemical substance that incapacitates by inducing an upper respiratory irritation causing some degree of breathing difficulty, plus some eye irritation. The oil taken from the placenta near the stem of a pepper is called Oleoresin Capsicum (OC) and is the active ingredient used in most self- defense sprays. The expected effect of tear gas is not death since a full and rapid recovery is achieved by changing clothes and washing with soap and water right after the spray is administered.

Aerosol sprays will eventually lose pressure, but canisters generally have a shelf life of at least two to four years, capable of providing adequate pressure for the OC to spray for 15–20 seconds. The pepper or tear gas spray will temporarily blind an individual and restrict normal breathing as long as the human or dog is within six to ten feet of the spray can. Even if the attacker is under the influence of alcohol and/or drugs, the OC spray causes incapacitation.

One useful published reference on the subject of various less lethal options is *Risk Management of Less Lethal Options: Evaluation, Deployment, Aftermath, and Forensics* (2014) by Wyant and Burns, also published by CRC Press/Taylor & Francis.

40 Armored Courier Services

1 LONG-ESTABLISHED TRADITIONS

The familiar scene of a large vehicle with small windows pulling up to a bank entrance or retail establishment, with two uniformed guards jumping out with guns at the ready, epitomizes the specialized armored courier service. More generally, these bulletproof armored vans that move valuable commodities between locations have been in existence since Butch Cassidy and the Sundance Kid began robbing banks. Armored vehicle services do business under a variety of operating authorities, including the Interstate Commerce Commission, the Department of Transportation, and individual state regulations.

Insurance companies can be considered the biggest regulators of the industry; they insure armored car companies and their customers against loss from theft and other casualties. In return for providing insurance, insurers require armored car companies to meet certain prescribed standards. They include the size and construction of the vehicle to be insured, the number of guards per vehicle, the procedures to be followed at the client's facility, and the steps to be taken in the event of a loss.

Armored car companies are responsible for assets entrusted to its guard staff the moment the guard signs for and receives the assets; the company is relieved of liability the moment the party at the other end of the transaction receives the assets and signs the appropriate documents attesting to that fact. This distinction has importance in the law. Several armored car carriers have received and signed for bags of money within the secured areas of bank and other customer facilities and have been robbed only minutes later while on the client's property. The armored carrier and the insurers were responsible for the total amount of the loss.

In the mid-1980s, four companies dominated this segment of the carrier protection field. That has all changed with acquisitions and mergers over the last 30 years. The firms have evolved to include five major carriers, which, when combined, control about 75% of the industry. The remaining 25% of the industry is shared by 125 other companies, some with substantial growing routes, while others have only two or three armored vehicles. The insurance rates have become so high that only those firms with a considerable volume of activity survive.

In 2023, the industry was worth \$4.4B in the United States alone. The top three services, in order, are Brink's, Loomis AB, and GardaWorld Security Corp. GardaWorld Security Corp. made the list despite a high-profile heist in April 2024 at their money storage facility in Sylmar, California, in which approximately \$30M US was stolen (Figure 40.1).

1.1 Largest Armored Car Carrier Services

Armored cars are highly visible; therefore, there is a need to give the impression of being impenetrable. As a result, most robbery attempts occur when the guards stop for a coffee break – or while transporting money between the customer's premises and the vehicle – or, in the case of the GardaWorld incident, the warehouse facility itself. Due to this constant threat and the vulnerability of the guards, they are armed. Insurance companies vary in their policies, but generally, they require three persons with each vehicle: a driver who remains in the vehicle, and two guards who transfer the money to and from the customer's cash location. Another policy followed throughout the industry is that scheduling must be randomized so that potential robbers, either casing or planning a theft, will not be able to anticipate the carrier's timing or route.

DOI: 10.4324/9781003091103-56

FIGURE 40.1 A GardaWorld armored security truck.

Source: Shutterstock.com.

The industry is highly competitive, so most of the firms operating in the money transfer business are unable to pay personnel considerably above, if at all above, the prevailing average industrial wage. Turnover is a problem, but not as pronounced as in the contract guard service industry, which has a 300% turnover rate.

2 PRIVATE BULLETPROOF ARMORED PROTECTION CARS

The rapid growth of the armored car industry was not due to an increase in money from a business entity to a safe bank vault but was the result of the demand to protect individuals from being kidnapped. In many sections of Mexico, particularly in Mexico City and the small towns bordering the United States, carjacking, ransom abduction, and extortion have made the simple task of driving a hazardous journey, and sometimes a fatal venture. Extortion is now almost commonplace in some regions of Mexico, and kidnapping has soared by more than 300% and between 2006 and 2010, more than 28,000 people have died as a result of drug cartel violence. Automakers such as Mercedes-Benz, Volvo, and Volkswagen sell bulletproof models in their Mexican dealerships. Such vehicles can also be obtained anywhere in the world, of course, as such transport is also deemed necessary for VIPs as well as in volatile regions of the world.

Twenty years ago, armored cars barely existed, and only the wealthiest business professionals and most powerful politicians had them; but in the current decade, even Kia models are being bulletproofed, since the less expensive cars are considered "low-target cars." It's this niche armored car industry that experienced a surge in 1994 after a banker was kidnapped in Mexico City. His release was negotiated after a payment of $30 million was conducted by ransom negotiation. That same city has become No.1 in the world in kidnappings for cash ransoms where a victim is forced to withdraw from ATMs, eliminating the need for lengthy negotiations that used to take weeks before money was exchanged. As wealthy families invested more heavily in security, making them more difficult targets, kidnappers began expanding out to target victims paying smaller ransoms.

The armored car industry may have been focused on honest citizens, but it also serves another type of client: drug cartels. The Mexican government has tried to pass laws to set up a customer registry and function like the gun permit regulation enforced in the United States. The law would restrict the improper purchase of bulletproof items such as jackets and body armor. The armored car industry tries to self-regulate by not selling to criminals, but drug cartels often "trick out" their cars with plating on their own, likely due to some of them having direct access or connections to many armor-plating and bulletproofing companies in Mexico.

Section XVI

Lie Detectors and Truth Deception

In this section, we focus on the various techniques used by law enforcement and the criminal justice system to determine the honesty of suspects prior to arrest. This includes descriptions of the lie detectors from the Polygraph to Thermal Lie Detectors to the often-discredited Stress Analyzers.

At the risk of oversimplification, lie detectors like a polygraph convert a spoken response into a chart that records deviations from the norm. Regardless of how nervous, emotionally upset, or distraught a person may be at the time of such an examination, the emotional state of the person being examined can be considered a norm for that person at that time. The basic concept behind lie detection, by either a polygraph or a voice analyzer, is that people who are lying will normally show more stress or emotion disturbance than when they are telling the truth. It is not an exact science, but many criminal suspects often confess when confronted with a lie detection device once they are convinced that the instrument is so accurate that it will detect a lie.

In general, all lie detector techniques have demonstrated an inaccuracy rate of at least 10% and possibly more than 20% for pre-employment screenings, meaning that the federal government alone – by testing 50,000 applicants a year – will unfairly reject perhaps 5,000 people each year for entry-level positions.

Regardless of the controversy surrounding the different lie detector systems, they will remain a factor in the criminal justice system for many years to come, or until new technology is developed that can perhaps detect people lying by recording the changes in their facial expression. In the meantime, the existing lie detection techniques will remain highly controversial.

The practical applications of using a computer programmed to read facial expressions as an indication of someone lying are being tested at sites such as border crossings and at airport security checkpoints. Researchers have written software that not only detects whether a person's face reveals genuine or fake pain but also does so far more accurately than human observers. People generally excel at using nonverbal cues, including facial expressions, to deceive others (the so-called poker face). Some individuals instinctively know how to contort their features to convey physical discomfort or to overcome a sign of nervousness that indicates a lie. In most facial recognition tests, the detection of a lie or faking it is effective 50% of the time, meaning that detecting a liar is like guessing or a coin flip.

It is simpler to use the voice analyzer than the polygraph since the analyzer requires no wires to be attached to the subject being tested and uses only a microphone plugged into a multifunctional notebook computer to analyze the subject's response.

DOI: 10.4324/9781003091103-57

41 Truth-Seeking Techniques

1 POLYGRAPH CREDIBILITY AND CONTROVERSY

The history of civilization has long been filled with attempts to detect lies and verify the truth by strange and often very harsh interrogation methods. Ancient methods of detecting lies included the ordeal of the red-hot iron, boiling water, throwing a knife, or the pattern assumed by a handful of tossed pebbles. Some of these ancient rituals reflected a shrewd understanding of psychology or physiology.

There are several reliable written tests, mostly used for pre-employment evaluations, which claim to spot exaggerations and outright lies with trick questions. Early scientific attempts to detect a person who is lying occurred around 1895 through the use of crude devices that measured human responses but were completely without merit and so unreliable that they were never produced for commercial applications nor recognized as a lie detector. It wasn't until the beginning of World War I that more scientifically based experiments were conducted using an unscientific device to measure and record the rate and depth of the subject's respiration. The first American to become involved in the field of lie detection was a psychologist who used a sphygmomanometer (the same device physicians use to take the patient's blood pressure). Soon after, a more sophisticated model was developed that measured three phenomena: blood pressure, pulse, and aspiration, and was considered the first polygraph measuring instrument. Eventually, another feature was added that measured changes in skin resistance to electricity, commonly known as galvanic skin response.

In 1915, the first person credited with the creation of a lie detector device that measured blood pressure, as relates to truthfulness, was William Marston, who received little acclaim for the technique. He later became wealthy in 1941 by publishing the comic book "Wonder Woman."

Detectors like the polygraph convert a spoken response into a chart that records deviations from the emotional norm. Regardless of a person's emotional state at the time of such an examination, that state will be considered the norm for that individual. While detecting lies may not be considered an exact science, many criminal suspects often confess once they are convinced that the lie detection device is accurate, especially those individuals with a guilty conscience.

Theoretical Concepts: While the acceptance of lie detection devices cannot be considered overwhelming, since even the most loyal and determined advocates admit that the best accuracy rate is less than 90%, the polygraph continues to be purchased by a significant number of law enforcement agencies, particularly those associated with federal employees. A number of theories exist that explain why the polygraph can be effective, with the most logical theory being that the fear of punishment is a mental attitude that makes it easier to detect a lie. A popular hypothesis is that the greater the consequences of being detected, the more noticeable physiological changes will be triggered, which creates a greater likelihood of detection. It is, therefore, not so much the lying or guilty feelings that alter the subject's physiological responses, but rather the fear of punishment if caught.

To substantiate that theory, several laboratory experiments used volunteers – none being a criminal suspect – who had very little punishment to fear. Once fear and stress are removed, the physiological changes associated with lying are reduced, resulting in lowered accuracy levels for people under no stress.

Advocates claim that the accuracy of a polygraph is 90%, while opponents contend that there is no scientific method to verify any claim of accuracy. This debate has been going on since the polygraph was first introduced into the criminal justice system in the 1920s. Opponents also believe that any use of the polygraph violates a subject's constitutional protection against self-incrimination – that is, a subject should be able to take the Fifth Amendment when a question of

 DOI: 10.4324/9781003091103-58

honesty arises. Furthermore, opponents also believe that a polygraph test violates the subject's right to privacy, including his right to conceal past criminal activities. Under common law, the right to privacy protects the individual only against unwarranted intrusions by the state, but since 9/11, the term "unwarranted" is being defined differently than it was prior.

Privacy Issues: The right to privacy is not only guaranteed, but it is indeed sacred, and it must be guarded and respected. But in a free society, no person's right is absolute, and it must give way, to a degree, to the rights of others. The obligation to protect airplane travel means that passengers must submit to searches of their person and luggage; many office buildings and hospitals reserve the right to search visitors' packages in order to maintain public safety. Employers in that same building have the right to investigate the background of all applicants. With the threat of terrorism being so real, lying on a job application by concealing past criminal behavior, including acts of dishonesty and irresponsibility, should not be kept secret from the employer. The employer has not only the right but also the obligation to utilize the most effective methods available to verify applicants' background information.

The most ardent defenders admit that the polygraph, an often-effective tool for interrogating criminal suspects, is no better than guesswork when it comes to weeding out liars or potential spies in large pools of people. Yet, the government is increasingly using polygraph tests for screening applicants for a variety of positions. Generally, the questions on a typical test cover drug and alcohol use, personal finances, foreign travel, possible criminal violations, and relevant background information. The most famous case involving the inaccuracy of the polygraph was when a high-level Central Intelligence Agency (CIA) official, Aldrich Ames, passed the lie detector test while spying for the Russians. That agency has a reputation for having the most qualified polygraph examiners in the nation.

Accuracy is an elusive term that polygraph experts try to achieve. They acknowledge an inaccuracy rate of at least 10% and possibly more than 20% for pre-employment screenings, meaning that the federal government alone, by testing 50,000 applicants a year, will unfairly reject perhaps 5,000 people a year. This type of failure caused President Ronald Reagan to sign legislation in 1988 prohibiting private industry from using polygraphs to screen job applicants, but an exception was made for government employment (See Figure 41.1).

New Mexico is the only state in the nation that allows polygraph results as evidence in court. Even though there is strong opposition to any use of the polygraph, since it supposedly compromises a person's privacy, it is difficult to explain why the public doesn't question other infringements such as mandatory blood tests, alcohol and drug tests, and body searches at airports. These tests have always stood the test against self-incrimination and have been accepted by the public, unlike the public's response to the use of the polygraph.

2 DEFEATING THE POLYGRAPH

Licensed polygraph examiners certified by the National Association of Polygraph Examiners must pass an exam after attending approved classes. There are about 3,300 licensed polygraph examiners in the nation, 70% of whom are possibly employed by government agencies. The independent polygraph consultants provide polygraph examinations for government and private organizations for an hourly fee or a set fee for each test conducted (usually during the pre-hiring process or after a theft of merchandise is discovered).

A few former examiners offer special services to suspected criminals and teach them techniques that could be used to defeat the polygraph by disguising physiological reactions when lying. The first rule in compromising the results of a polygraph test is to control physiological responses because any physical sign of either anxiety or panic can result in a jail term. Sweating must be kept under control since it is a sign of anxiety that is easily detected by the polygraph and a sign of nervousness that may be interpreted as a lie. Maintaining a steady heartbeat, breathing deeply with long and

FIGURE 41.1 Several federal government agencies have relied upon the use of polygraph detection for decades, as evidenced by this photo from the 1970s.

Source: FBI.

steady breaths, applying a liberal swab of antiperspirant, or covering the body with absorbing body powder can help.

Examiners are trained to make the individual being tested relax. The person being tested should beware of projecting certain telltale signs that an experienced examiner can translate into an attempt to cover up a lie and avoid the following movements since they are gestures that can sometimes indicate that the person is lying:

- repeatedly crossing the arms
- constantly shifting in the chair
- looking up and to the right while being questioned
- rubbing the eyes too often
- playing with the hair
- touching the face using unnatural movements
- jiggling the foot
- touching the ear lobe too often

Baselines: There will be many different questions asked by the examiner, only a few of which are relevant to the situation, and there will be some important questions specifically designed to catch a lie. Keeping all responses as brief as possible – "yes" – "no" is the best as the more a person talks, the more likely a cover-up will be suspected, so it is best to avoid elaborate responses. There is an essential need for the examiner to establish a "baseline" that defines a normal reaction to a seemingly innocuous question. A lie on relevant questions will move the polygraph pen out of the baseline range – a lie! Therefore, it is important to trick the device into causing the baseline to display erratic movement. This can be done by biting down hard on the inside of the cheek, digging nails into the hand, etc. These self-inflicted actions must be done when answering the irrelevant questions, which will confuse the machine when uncontrollable anxiety is displayed in response to the important relevant questions.

The practice of teaching people to lie on the polygraph test without being detected has been going on since the early 1980s and has been treated as just another educational program like gamblers being taught to count cards or students being tutored for the SAT examination. In 2013, a Virginia court sentenced a man to eight months in prison for teaching government job applicants how to beat lie detectors. The man was found guilty of providing information to child molesters, intelligence employees, and law enforcement applicants. His lawyers claimed that his classes were an expression of free speech that was protected under the First Amendment. The instructor was only giving advice and had no knowledge that the students would use the techniques discussed in class. Furthermore, they argued that the instruction shouldn't be considered illegal when the results of a polygraph exam were inadmissible at a trial because of the unreliability of the test.

3 COMPUTER VOICE STRESS ANALYZER

Seeking greater versatility in the lie detection field than the controversial polygraph, a group of retired army officers developed voice stress analysis technology in the early 1970s. The first system created a great deal of interest, especially from the advocates of the more established polygraph. It first gained national attention when the only manufacturer of this device claimed to have analyzed the voice patterns of Richard Nixon when he made the claim "I am not a crook" on TV. Based on the analysis of the voice fluctuations, he determined that the President was lying (he was!).

It is easier to use the voice analyzer than the polygraph since the analyzer requires no wires to be attached to the subject being tested and uses only a microphone plugged into a multifunctional notebook computer to analyze the subject's response. As the subject speaks, tiny frequency modulations in the human voice, called micro-tremors, are graphically displayed on a laptop programmed with special software. When a test subject is not being truthful, the automatic, or involuntary, nervous system causes an inaudible increase in the micro-tremor frequency, which is recorded. By analyzing the patterns, the examiner can learn when the subject was likely to be deceptive in answering questions. This process removes any possibility of examiner error and provides a completely objective examination. A voice stress analyzer costs about $10,000, significantly less than the polygraph, and about 1,000 domestic law enforcement agencies have purchased the system despite the efforts of several federal government agencies to discredit the technology in favor of the polygraph.

Of the 50 states, there are only a few that do not recognize or approve of the use of computer voice analyzers: Illinois, Oklahoma, Michigan, Texas, Vermont, Virginia, South Carolina, and Kentucky.

4 EYE DETECTOR

Researchers at a Midwestern University have tried to find an alternative to standard lie detectors and focused their efforts on eye-tracking technology. Their studies were based on hard scientific evidence that liars really are shifty-eyed, as the old adage claims. The researchers took measurements of certain physical responses made by subjects as they were required to answer true or false to a series of questions displayed on a computer screen. Among the measurements taken were pupil dilation, response time, reading and rereading time, and errors. The researchers determined that lying requires more effort on the part of the person being examined than telling the truth. A person who is being dishonest in their responses may have dilated pupils and will take longer to read and answer questions. Eye-tracking tests appeared to have yielded results that are as good or better than polygraph tests, but more data will be required before criminal justice professionals accept this technique. Eye detectors require more sophisticated measurement devices as compared to the sensors used during a polygraph test and may be limited as an alternative only to the voice analyzer.

There is a theory about body language, and its role in revealing a lie. According to this theory, when right-handed people move their eyes up and to the left in response to a question, they are picturing a real memory. When the eyes go up and to the right, they are accessing the creative centers of the brain and visualizing an imagined event and therefore concocting a lie. The theory dates back more than 50 years and is frequently taught in linguistic training courses, but the results have never been conclusively substantiated. In a controlled study by British researchers in 2012, the eye movements of a selected group of right-handed people were monitored as they told lies and truths about recent events to an interviewer. They found that it was impossible to accurately predict who was lying from roving eye movement. The researchers also viewed news videos of people making pleas for the safe return of a missing relative with half of them being told to lie. The researchers could not identify any pattern that correlated eye movement with predicting lies or truths.

Likewise, dilating pupils or moving eyes to the left should not be treated as an uncontrollable response to a question and can be regarded as pseudoscience.

5 THERMAL LIE DETECTOR

Research being conducted in Europe successfully detected liars using thermal imaging cameras. People have been known to beat the polygraph by controlling their breathing and heart rate; however, thermal imaging cameras and computer software overcome the shortcomings in polygraph tests by measures flushing and blood-flow patterns around the eyes while picking up clues such as nail biting, slips of the tongue, nose-wrinkling, and excessive blinking. Features that can be almost impossible to fake.

When a person is being deceitful, there is an increase in brain activity, which is reflected in the face through involuntary facial expressions and blood flow. Thermal cameras focus on the region around the eyes where the skin is thinner, making temperature changes easier to spot. There are cameras available that can detect a warming of the face by as little as 0.1°C. Humans have a limited number of facial expressions known as Facial Action Units, and software is available to detect all the 46 expressions.

- Researchers working on the technology believe that thermal imaging will never be 100% foolproof, but it should eventually achieve the same 90% accuracy rate of the polygraph.
- It will be an effective device for the initial assessment at check-in at immigration centers.
- It will never be a fail-safe method of spotting potential terrorists but could provide a useful extra test.
- In many sensitive situations, it can add to other practices since it can be used covertly.
- Some subjects will be difficult to monitor like Parkinson's patients as they have no ability to smile.
- In the future, the thermal imaging technique could detect dementia in its early stages.

6 TEACHING COMPUTERS TO SPOT DECEPTION

A small group of linguists, engineers, and computer scientists, among others, have been programming computers to recognize emotional speech patterns that reflect deception, anger, friendliness, and even flirtation. Much like the software developed to spot facial flushing, there are programs that have succeeded at spotting emotions that may eventually have many practical uses.

The scientific goal is to understand how emotions are reflected in our speech. The programs that researchers are developing are not likely to be used as evidence in a court of law, just as polygraph test results are inadmissible in court. Algorithms have been developed that spot a liar 70% of the time in test situations by analyzing speech patterns without the need to strap someone to a machine. However, a person's speech provides many clues – loudness, changes in pitch, pauses between words, *ums*, nervous laughs, and dozens of other tiny signs that can be suggestive of a lie.

The use of the tiny pronoun "I" says more about a person than they realize. Frequent "I" users subconsciously believe they are subordinate to the person to whom they are talking, according to a research study published by the *Journal of Languages and Social Psychology*. People who often say "I" feel less powerful and less sure of themselves than those who limit their use of the word, another finding in the study. This theory regarding the use of "I" is counter to the long-held misconception that people who are confident tend to use "I" more than those who are not. On the other hand, the study also found that people who avoid using "I" are not limited to confident individuals but also those who may be hiding the truth.

Avoiding using "I" altogether is not always beneficial. Marriage therapists have long held that people should use "I" instead of "you" during discussions with a partner or when speaking about something emotional. The theory is that it is better to say "I feel unheard," rather than "You never listen," as the use of "I" in this case is considered less accusatory.

With the outgrowth of AI applications and capabilities, this is an interesting area that will no doubt be researched and analyzed further in the coming years.

Section XVII

Odd Protection Devices

The newest and most annoying recently developed systems and devices described in this chapter are not always the most effective, but each was meant to solve some security breach.

- Blinding strobe light beams
- Laser beams
- Reconnaissance robots
- Fog machine
- Powerful loud sirens
- Sirens penetrate car doors
- Beer bottle mass destruction
- Various innovative measures
- Hazardous toilets
- Alcohol detectors
- Drug scanners
- DNA mist spray
- Dummy TV and hidden cameras
- Dummy safes
- Dummy bombs
- Lobster eye detector
- See around corner detector
- See through rolling gates
- Fake proof driver license
- Protective glass barriers

One of the most neglected areas that is rarely addressed from a security standpoint is public toilet facilities. These seemingly safe havens are far from safe and a cause of great concern for building owners. People are often hesitant to use public toilets unless an emergency situation necessitates it.

DOI: 10.4324/9781003091103-59

While drunken driving fatalities have dropped steadily over the last 30 years due to educational campaigns, almost roughly 11,000 deaths each year result from accidents caused by legally drunk drivers, and roughly 1 million driving under the influence (DUI) offenders are arrested annually. Some convicted drivers are sentenced to jail or are required to use an ignition interlock Breathalyzer that the driver blow into to detect if alcohol present and, if so, if it is above a preset level prior to starting the car. One state requires a driver to rent one of these devices after the first arrest for DUI.

42 Devices of Mass Annoyance

1 BLINDING STROBE LIGHT BEAMS

A bright light can be considered a nonlethal weapon when packaged in the form of a flashlight that emits a dazzling strobe capable of temporarily blinding criminals, terrorists, and even unruly airline passengers. Initially developed under a million-dollar research program by the Department of Homeland Security, the device was designed to make an attacker turn away or shut their eyes, providing authorities with enough time to subdue the suspect. The device, "Light-Emitting Diode (LED) Incapacitator," which emits a series of light pulses that blinds and disorients when aimed at the eyes, underwent testing at a government-sponsored college research laboratory. It is similar to the magnifying effect of camera flashbulbs going off every few seconds. The developers selected a diode instead of a laser in order to prevent any permanent blinding; it is debilitating within a range of up to 20 feet. Among its several advantages is that there are no residual effects, and there is no need for the decontamination that occurs when pepper spray is used. Laser-based blinding weapons, called "dazzlers," actually blind people with intense light, and the laser cartridge has to be replaced, while the LED uses rechargeable batteries, eliminating recurring replacement costs.

This bright beam could become the weapon of choice by law enforcement since it is one of the better defensive devices supported by the government and private companies that provide equipment for the military, law enforcement, security, and others for use in potentially dangerous situations. The wavelengths and frequencies chosen for the LED device have certain psycho-physical effects that can range from vertigo to nausea. Sales are strictly controlled by the sponsoring agency, but the devices are available on the black market.

Strobe light exposure can induce fear, but there are other effects that are just as desirable insofar as disabling a potentially threatening individual. Disorientation and confusion caused by the strobe

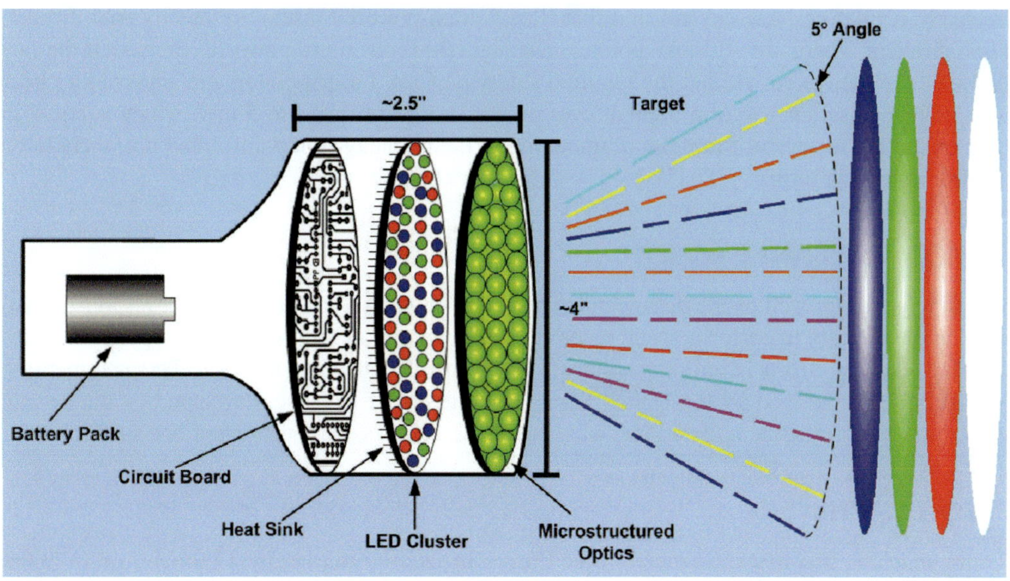

FIGURE 42.1 Cross-sectional diagram of the "Incapacitator," from a DHS newsletter.

Source: S&T Snapshots. Department of Homeland Security: Science and Technology Directorate, DHS.

DOI: 10.4324/9781003091103-60

are just as advantageous to police officers as causing fear in the mind of a suspect potentially carrying a weapon. Other advantages of the strobe lights that have become extremely popular on police flashlights include:

- Diminishing an assailant's night adaptation
- Causing a disruption to the subject's vision, and affecting their ability to use force
- Providing a visual hurdle to aggression
- Decreasing the suspect's direct and peripheral vision.

2 LASER BEAMS

Directed energy weapons, better known as laser beams, can be used to kill a person by directing electromagnetic energy into a highly focused stream that can travel at the speed of light. A powerful laser beam could cause retinal damage and incapacitate large numbers of people; its blinding effect can be used to compromise video surveillance units when attempting to enter a restricted area. It is a federal crime to point a laser beam at an aircraft cockpit flying overhead, but the number of pilots reporting laser strikes has grown, making laser beams a menace endangering pilots.

Powerful laser beams pointed at the sky can strike an aircraft and potentially blind some of the 15,000 commercial pilots in the cockpits. This unexpected threat began in 2005 with only a few reported incidents and has, according to the Federal Aviation Administration (FAA), grown to several thousand reported incidents per year, including over 13,000 incidents in 2023. This increase occurred despite increased penalties for anyone caught doing it. The increasing number of strikes and afflicted airports prompted the FAA to create a website for pilots and anyone else to report when lasers are pointed at planes.

Low-powered laser pointers, the only type approved by the Food and Drug Administration (FDA), cannot burn the eye, although they can conceivably cause enough glare to temporarily obscure the runway on final approach, according to FAA tests. More powerful laser beam devices are increasingly becoming available and can cause far more problems, as reported by the Laser Institute of America, which promotes laser safety. Warnings had been issued that the FDA was not effectively regulating laser devices and that illegal, high-powered lasers are openly sold through online retailers. There are 100-mW pointers available 20 times more powerful than what the law allows and are delivered without the required safety devices. Laser pointers are supposed to bear labels advising users not to aim them at eyes or skin and are limited to 5 mW, which is equal in effect on the eye to several times the illumination of the noonday sun, painful, but not likely to be harmful in the long term.

3 RECONNAISSANCE ROBOTS

Increasingly, and for well over a decade now, robots are being used by private sector response agencies, and the military to perform visual reconnaissance, as well as – in some cases – perform duties related to recovery efforts in building collapses and bomb diffusion and disposal. This is something that has been encouraged by the United States Department of Defense, developed by the private sector, university research, and national labs. Figure 42.2 shows a UK version of one such robot.

4 FOG MACHINE

The fog machine was first introduced in the European security market by a Danish firm in 2006. It is essentially an expanded version of similar machines used in theatrical productions to create a fog-like atmosphere. Applying the same concept for the security industry, the fog would fill the entire room following the activation of a motion detector and at the same time send a signal to the alarm central station service provider. This atmosphere would supposedly disorient the thieves and

FIGURE 42.2 The Dragon Runner Bomb Disposal Robot operating during an exercise. (Used per The Open Government License version 1.0 (OGL v1.0); Attribution: Steve Dock/MOD)

prevent them from stealing merchandise and running out the closest exit door. Of course, the exit doors were no longer visible and by the time thieves could find the exit, police would have arrived and the thieves would be arrested.

Burglars are very aware that there is only a small window of opportunity once a burglary attempt begins before police arrive following the triggering of an alarm. Denied visibility, thieves become instantly disoriented and are forced to flee after finding the hidden exit doors. The dense fog is made up of water and food-grade glycol, is harmless to the skin and, after the area is ventilated, leaves no residue. Each fog "cannon" can be activated seven times, flooding a classroom-sized area before recharging.

Like many security systems that reappear and disappear every few years repeating failed concepts is common in the security industry. This fog deterrent has never become a marketable device because it is not in code compliance, a fact that some manufacturers never seem to understand. One version of fog machine even added a red dye, similar to that used in packs of bills given to robbers by a bank teller. The fog machine sprayed a red mist that not only floods the room but also leaves a residue on the burglars sending them fleeing. The manufacturer of the red dye fog machine stopped selling to the US market in 2008 after only one installation. Then in 2015, another foreign manufacturer tried again to gain a foothold in the American market but also failed for the same reason that the previous manufacturers faced.

A local alarm installation company in Miami, completely unaware of the fire codes (which is a typical shortcoming shared by almost every alarm installation firm), installed the latest system in 2015 and when the fog machine was accidentally triggered by a false alarm (another problem shared by alarm firms), the warehouse became flooded with fog within the promised 30 seconds. Fire trucks arrived five minutes later, and the alarm installer was fined $2,500 for activating the fire alarm instead of the security alarm.

It seems that the fog not only covers exit doors and merchandise, but it also covers the smoke detectors that immediately activate the fire alarm system. Neither municipalities nor the National

Fire Prevention Association allows fire department personnel to enter a building where a burglary may be taking place. Firefighters are trained to respond to fire situations, extinguish the fire, and save the building occupants. They are not trained to intervene in burglaries or capture robbers.

5 POWERFUL LOUD SIRENS

The fog machine fiasco is only one of many mistakes that have confronted the security industry since its early conception in 1970. Another example of technology gone wild is the burglary deterrent device developed by using a siren that generates unbearably loud sound waves, making it exceptionally painful for a burglar to remain in a room. It is almost impossible to locate the source of the sound as it bounces off the walls, even shattering fragile glassware. The sound level has to be carefully controlled since previous attempts to add loud sirens, or horns, to intrusion devices proved to be temporarily successful as a deterrent but posed a long-term danger to humans. The noise level can adversely affect hearing, damaging eardrums, which is what may have forced the manufacturers to remove the device from the marketplace.

The "loud noise" protection device describes a controversial protection system that had a short lifespan after being introduced in the mid-1980s. A very, very loud siren or horn was initially developed to overcome a serious shortcoming in all existing detection systems, whether it be simple door contacts or complex microwave or passive infrared motion detectors. When a loud noise was used in a retail store environment, particularly small-, one-, or two-room stores, the security systems were programmed to activate when anyone entered the premises after hours. The application of the noise in the range of 150 db would act as a deterrent and scare the intruder off the premises. The need for this device was especially suitable in suburban and rural areas since police response time was so long that the burglar (and stolen property) would be long gone by the time they arrived.

The sound deterrent was based on proven physical principles that pain from the noise would be so uncomfortable that most individuals would immediately leave the premises or experience possible permanent damage to their eardrums. Many security systems, like the loud siren, are theoretically good in concept, but often prove to be totally impractical in the real world. Prior to introducing the loud siren system to the public, a pilot program was started to test the concept and the results were far from successful. A store in the test group had closed for the day and a well-meaning passer-by thought he or she saw some suspicious activity in the empty store and reported it to the police. When the responding police arrived at the location, they entered the darkened store and unknowingly set off the loud noise alarm system. One officer was standing near the speaker when the alarm system activated the sound system, and his eardrum burst. Lawsuits followed, along with apologies and a settlement – early retirement for the officer, early closure of the manufacturer's factory.

6 POLICE SIRENS PENETRATE CAR DOORS

Another attempt to use "noise-based" systems to improve public safety was developed in 2006 by modifying the familiar loud police siren, a noise associated with police cars for generations living in urban settings. Pedestrians can not only hear but also feel the sound, which is emitted by a device called the "Rumbler." It was specifically designed to grab the attention of even the most distracted cell phone talking, text messaging or headphone-wearing pedestrian, or motorist with car windows rolled up. The idea was to get pedestrians and drivers out of the way of police cars in route to an emergency, especially at busy intersections. The Rumbler emits vibrations and low-frequency tones to penetrate enclosures and shake up nearby motorists so that they feel the sound waves rather than just hear the siren.

This police siren technology was borrowed from the popular souped-up car stereos that are loud enough to wake the dead. The Rumbler emits a low-frequency tone in the range of 180–360 hertz (between the 33rd and 46th keys on a standard piano keyboard), which penetrates hard surfaces like car doors and windows better than a high-frequency tone does. When this low-pitched sound is paired with the wail of a standard siren, the effect is hard to ignore. Some civilians have reacted by claiming the noise is so loud that it could be injurious to their health, but all tests have proven that complaint to be unfounded. The Rumbler is not louder than a standard siren, but it is, in fact, 10 db lower. Because the low-frequency sound waves penetrate cars better than those at a higher pitch, drivers experience the effect as being much louder than a standard siren. To spare police officers in the cars from any physical damage, the device cuts off after eight seconds. Regardless of the rumors, the siren does not cause eyeglasses to shatter.

7 DISPERSING CROWDS

An offshoot of the blasting sound device formerly used in-store alarm systems is the "sonic cannon," a military-grade device used by police departments to disperse crowds disrupting traffic. The sonic cannon is a long-range acoustic device that can shriek with repetitive blasts of noise at a volume of up to 152 db. Used by the military and law enforcement to disorient and disperse crowds, it was created partly as a response to a terrorist attack on the Navy destroyer Cole off the coast of Yemen in 2000.

Used at close range and at maximum volume for an excessive length of time, the sonic cannon can pose health risks ranging from loss of balance to eardrum rupture to permanent hearing damage. In 2009, it was used at a demonstration in Pittsburgh and caused severe hearing damage in a protester who sued the city and received a $72,000 settlement. That's the only recorded lawsuit resulting from sonic cannon use.

8 LONG-RANGE ACOUSTIC DEVICE

Loudspeaker units operate under the same principle as heavy-duty speakers used to make announcements at high school football games but can be built as powerful sound cannons known as Long-range Acoustic Devices, or LRAD. LRADs issue a series of sharp, piercing beeps that when directed at people less than ten feet away can be physically damaging and are considered by victims as excessive force, a violation of the First and Fourth Amendments, since anyone near the device developed migraines, sinus pain, dizziness, facial pressure, and ringing in their ears. The resultant lawsuits have caused some police departments to cease using the device. In 2004, the New York City Police Department purchased LRAD units to control demonstrators at the Republican National Convention, and in 2009, the Pittsburg police used it as a control device in conjunction with an international conference. It has not had many sales since police departments are reluctant to purchase this controversial device.

9 BREATHALYZER – ALCOHOL SENSOR

Breathalyzer. Drunk driving convictions vary significantly by jurisdiction even if the locations are within the same state. Different courts use different means of punishment, including fines, probation, educational classes, and even jail time, and drunk driving penalties are inconsistent and can vary greatly across states.

According to FBI statistics, almost 13,000 deaths result from accidents caused by legally drunk drivers each year, and over 1.4 million driving under the influence (DUI) offenders are arrested annually. Some convicted drunk drivers are sentenced to jail or required to use an ignition interlock Breathalyzer. The ignition interlock Breathalyzer requires a driver to blow into a mouthpiece on the device before starting or continuing to operate a vehicle. If the device detects a breath-alcohol amount above the preset level, the device will prevent the car from starting. One state requires drivers to rent one of these devices after the first DUI conviction. Alcohol-analyzing devices have allowed police officers to produce data routinely accepted in court, leading to a greater percentage of guilty pleas and fewer trials in drunk driving cases, reducing police and court costs.

There is also a growing number of individuals who are voluntarily using personal Breathalyzers to test themselves, family members, and friends for alcohol impairment. The cost is as little as $5 for a single-use device to $300 for a high-end digital device. Even if a testing unit isn't completely accurate, it still could have a very positive and sobering effect on people who believe that they are. Requiring personal Breathalyzer use after attending a party, where drinks were part of the festivities, has proven effective against drunk driving.

In 2015, a Breathalyzer was developed, which caused a sensation when featured on the television program *Shark Tank*. The device went on to make millions of dollars by being able to record breath-alcohol percentage by using a device called a Breathometer that was attached to a smartphone. The device was able to record percentages as low as.040 after the subject breathed into the device for five seconds within two inches.

There are several programs available that can be used with a smartphone to determine if someone has consumed enough alcohol to exceed the legal limit of .08% used in most states. Apps have also been developed in which users simply program in their age, weight, height, number, and type of alcoholic beverages consumed (beer, wine, vodka, etc.), and the period of time since the first drink was consumed. The data is then analyzed, and the alcohol level is displayed on the screen. The American Automobile Association considers the app and resulting data as providing less accurate readings than devices used by police because of less sophisticated technology. Also, a self-testing intoxicated person might not produce a good breath sample and thus the blood alcohol calculation might not be precise. It does not factor in drink variations, individual metabolism and/or what a person has eaten. However, it could be helpful for parents looking to enforce a zero-tolerance rule with their underage children.

Alcohol Monitoring. DUI convictions can lead to the suspension of drivers' licenses and, if they are repeat offenders, drivers may avoid jail but could be ordered to wear a monitoring device, technically known as a Secure Continuous Remote Alcohol Monitoring, or SCRAM bracelet. The bracelet's use is growing as more and more judges sentence offenders to either wear the bracelet or serve jail time. The small alcohol detection device operates like the radio frequency transmitter units attached to the ankles of anyone under house arrest and are activated when the person wanders outside a court-designated area.

The SCRAM bracelet skin sensor doesn't restrict movement, but it measures alcohol consumption by sensing alcohol levels through detecting changes in the skin's moisture. The battery-operated ankle device is activated if it is removed, or the wearer's sweat level reaches a preset point (around .02% alcohol in the sweat gland). Once activated, it continues to measure the wearer's blood alcohol every half hour. The information is sent, via the Internet, to a central recording location, which forwards the daily results to a security office or a local law enforcement agency.

Some jurisdictions that use SCRAM monitoring have recorded a 95% success rate among defendants who have worn the bracelet for the period stipulated by the judge.

Blood Samples (involuntarily taken). A traffic officer pulled over a car for speeding and quickly noticed the driver's bloodshot eyes, slurred speech, stumbling performance, and suspected that the driver was impaired. The driver refused to breathe into the Breathalyzer and did not agree to have his blood drawn for testing. The police officer was proceeding under an incorrect assumption that there was no requirement in the state to seek a warrant to take involuntary blood samples. The officer brought the driver to a nearby hospital where a blood sample was taken, indicating an alcohol level of 0.154, and the driver was arrested and subsequently convicted of a DUI. The case eventually went before the U.S. Supreme Court in 2013 because the blood sample was obtained without a warrant.

- The Supreme Court in the state where the encounter took place (Missouri) had thrown out the conviction because the arresting officer had been mistaken and, in a 1966 case, the U.S. Supreme Court had approved warrantless blood test only when "exigent circumstances" made it too difficult to reach a magistrate in time.

- No Warrant: The State defended the officer's action and claimed that a law requiring an officer to wait for a magistrate's approval was unrealistic since the alcohol level diminishes after a person stops drinking.
- Requiring Warrant: The defendant's lawyer in the case believed that the blood sample taken against his client's wishes was in violation of the Fourth Amendment against "unreasonable searches" and also pointed out that 25 states explicitly require warrants for involuntary blood draws.

The nine Supreme Court Justices tried to find a middle ground and acknowledged that the police could obtain a search warrant through a telephone call to an on-duty magistrate when a second judgment becomes necessary. On the other hand, it is not easy to get hold of someone (a magistrate) in a rural community at 3 in the morning. Requiring a warrant during off-hours would mean that the entire country had to operate like New York City where a court official is on duty at all times.

The Supreme Court voted 6 to 3 to require law enforcement officers to obtain a court-issued a warrant before taking blood involuntarily from a DUI suspect. A rule to suspend the Fourth Amendment would have allowed prosecutors to collect evidence to make their cases stronger by violating the Constitution. This ruling meant that the 25 states that allowed warrantless blood samples to be taken had done so illegally and previous convictions based on warrantless blood samples were thrown out.

10 DRUG SCANNER

Drug arrests have almost tripled since 2000, and police departments throughout the nation have been seeking a device, similar to the Breathalyzer, that operates quickly and accurately to test and identify confiscated drugs in the field. Traditional drug testing in the field, used by police for years to justify initial arrests and further investigations, relies on a $20 one-time test kit, which involves mixing a suspected drug with a small amount of chemical; courts require laboratory testing if the case goes to trial. Those standard, inexpensive kits can only test for a single drug, destroying the sample in the process. This means that an officer can eliminate the evidence while trying to identify a suspicious substance. A new method, using a laser light beam, can identify almost any controlled substance considered to be illegal to distribute or consume, using an amount as small as half the size of a grain of rice without destroying it.

These handheld lasers can immediately identify illegal drugs and could very well revolutionize how narcotics cases are investigated and prosecuted. The device should help officers quickly discern illicit substances at a time when police departments across the nation are experiencing a surge in new, difficult-to-identify designer drugs, such as psychoactive powders known as "bath salts." The laser drug tester relies on a technology called Ramen spectroscopy that is used regularly in many drug labs. The device throws a small laser light at a substance and each compound scatters that light in a slightly different pattern. The device then compares the pattern with those in its library to identify the substance. The cost is too high for some police departments and the evidence obtained using the device has yet to be accepted in court. There are some criminal justice experts who also caution against the early acceptance of this technology until law enforcement obtains some background in the science involving laser spectroscopy.

11 DNA MIST SPRAY

A police officer and a chemist (they were brothers) developed a variation of the DNA footprint that has already been used throughout Europe. It consists of a small spray device mounted over the entrance door in retail stores that is activated remotely by an employee when a robbery is attempted. The device, once activated, sprays a fine, barely visible mist laced with synthetic DNA on anyone walking under the device as they enter/exit the store. The basic goal of the spray device is prevention and scaring away criminals, not recovering stolen goods or capturing criminals. The system includes an important warning sign informing customers (and robbers) that they will be covered with traceable DNA if they attempt a robbery.

It has proven far more effective than the red dye residue wrapped inside stacks of currency used by some banks. The DNA mist is visible only under ultraviolet light and carries a unique DNA profile specific to the location that enables the police to match the thief with the store that was burglarized. The mist can be triggered by an employee by removing a 10-euro bill from a special clip kept behind the counter, eliminating the all too clumsy and dangerous act of pushing a holdup alarm button in full view of the robber.

Another very practical use for the DNA traceable signature is in the form of a soft crayon which is used to mark valuable items like computers or cameras to facilitate their identification when stolen goods are discovered by the police.

12 DUMMY TV AND HIDDEN CAMERAS

In the days when black and white closed-circuit TV (CCTV) surveillance systems were first introduced to the security industry (around 1978), there were dozens of manufacturers that supplied popular and inexpensive "dummy" cameras as an alternative to live video and CCTV surveillance devices. Many facilities, including bank lobbies, were reluctant to spend the necessary funds for a real camera system that required personnel to monitor them but recognized the deterrent value of the TV camera, fake or otherwise. The inexpensive dummy camera was simply an empty housing with a blinking red light, but it served as a psychological deterrent that required no maintenance or personnel to monitor the images. It was assumed that criminals could never be sure which cameras were "live" and which were dummies. They served their purpose but became obsolete as word spread among employees that the cameras weren't real, which usually took a few months.

There was little that could be done to keep the "secret" of dummy cameras from reaching the criminal element. Eventually, the empty housing units were abandoned or replaced with real CCTV systems once it was discovered that the high-level security expected was not actually being provided. Employees felt that the dummy units were "creating a false sense of security" when they were used as a deterrent and initiated expensive lawsuits. There are many recorded incidents where someone was physically attacked under the supposedly watchful eye of a CCTV camera – but the camera was a "dummy."

LAWSUITS. Many victims sued building owners for installing devices that gave a false sense of protection; attacks went unnoticed by authorities and eliminated any possibility of identifying the attacker. Some of these lawsuits resulted in awards exceeding $1 million and, by 1990, most fake cameras had disappeared from the marketplace.

An advanced camera system that is a cross between a dummy camera and a streaming video camera is a standalone digital motion camera with a memory chip. This camera system is hidden within a common-looking wall fixture to disguise the fact that an actual camera is recording the area, even though there are no cables required to transfer the images to the monitor. The device can be inserted in a wall clock, smoke, or motion detector, or even in a computer or wall-mounted thermostat. It is mainly mounted in areas where a hard-wired CCTV system would be impractical or too expensive (and too obvious). Mounting the camera in a housing that resembles a motion sensor – making it appear that the unit is used solely for nighttime intrusion detection – is a very popular implementation. These units require a battery and are programmed to take high-resolution digital still or motion pictures each time motion is detected. At an opportune time, or following a theft, the memory chip is removed and the stored images are conveniently displayed on a computer, exactly like every other digital camera. The combination of a detector and camera can also be mounted inside a hardcover book and placed on a shelf so as to be wholly inconspicuous.

13 DUMMY SAFES

Safe Deposit Boxes are probably the best place to keep important personal items and valuables such as cash and jewelry, since they are well protected in a bank vault. The problem with storing items in a bank vault is that it is inconvenient and impractical for frequently used items. The next best and most secure place for valuables is a metal safe built into the wall or locked to the floor in a closet. Prices for residential safes start at around $25 for a basic fire-resistant lockbox about the size of a loaf of bread. Luxury safes can be as big as an armoire, cost as much as $100,000, have built-in biometric locks that read fingerprints or irises, and may also contain a GPS antitheft system similar to LoJack, in the event that someone manages to carry it away. A third hiding place is called a "diversion safe" that might be a hollowed-out book or a less expected item like a can of soda, with space to hide a small family heirloom, gold necklace, diamond bracelet, etc.

Diversion Safes may not be fire-resistant and don't even have locks, so their value in protecting against burglary is highly questionable. They cost anywhere from $5 to $100 and are designed to look like various household objects. Familiar-looking cans, jars, and aerosol containers found in pantries and bathroom cabinets are typical "stealth" or diversion safes. Criminologists place little faith in these containers because burglars are also looking at these types of safes, so they know what to look for. Even with their limitations, they may be better than hiding valuables in coat pockets, shoeboxes, freezers, or buried under potted plants. The same experts who criticize diversion safes also advise against putting things inside of toilet tanks, cereal boxes, and inside medicine cabinets where thieves look for prescription drugs with resale value. The last place to hide a diamond necklace or a roll of bills is inside an empty bottle of OxyContin or Adderall.

One of the more popular hiding places is in a safe book, which costs about $30 for a ready-made version, though sometimes the book titles stand out too much. The best safe book is one that has been on the bookshelf for years and that nobody will ever read again. It is also one of the easiest diversion safes to make since there is no need for special tools and a real book tends to look more authentic than a fake, ready-made book. The selected book should be a hardcover with a title that someone is unlikely to pull from the bookshelf. The thicker the book, the more space there will be for valuables in the hollowed-out midsection. Place a piece of wax paper over the front and back of the book like it was a dust jacket in order to keep the covers clean while digging out the pages. Using a sponge paintbrush, liberally apply glue around the edges of the book to make the pages stick, but leave about ten pages in the front unglued to conceal the cavity to be carved inside. Place a heavy object on top of the book to compress the pages while the glue dries. After a few hours, or overnight, mark the area to be hollowed while leaving a border of about a half-inch around the inner hole in the book. Using a box cutter, cut out a few pages at a time because trying to cut through too many pages will result in messy, ragged edges. The last task is to brush glue on the interior edges of the cut pages and let the book dry for three or four hours.

14 DUMMY BOMBS

With every new terrorist threat, entrepreneurs are ready to create some new system or contraption to fulfill the need for a security solution. Airlines began purchasing "Dummy Bombs" in 2006 and used these mock bombs to test screening procedures against possible terrorist intrusion and as a training tool for the security staff at the X-ray machines which passengers pass through. The simulated bomb components are often installed in a computer toner cartridge or any other piece of equipment that terrorists may have used to circumvent airport screening machines. The fake bombs are sold to government agencies and corporate security companies in and outside the United States, though mainly to agencies like the US Transportation Security Administration and airports authorities throughout the world. Mock bombs vary in size and disguise, with the more popular devices being a foot long and placed in a container that will fit inside a carry-on suitcase.

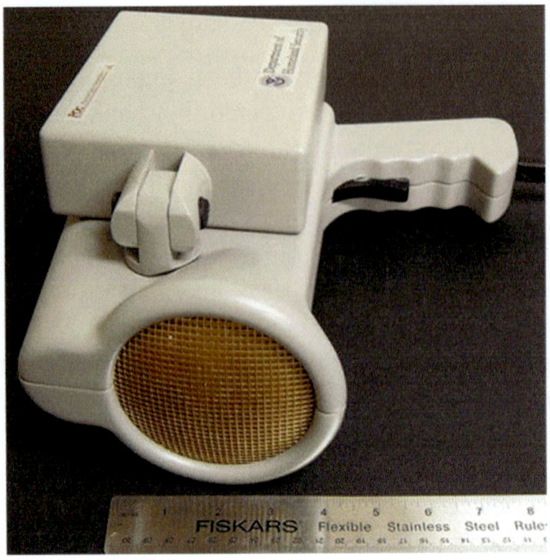

FIGURE 42.3 The LEXID X-ray imaging device.

15 LOBSTER EYE DETECTOR

Another brainchild of the Department of Homeland Security involves the appropriation of $1 million for a research project to develop what may well become the next weapon in the war on terror. The device is a handheld unit that can see through wood, concrete, and steel and is called LEXID (Lobster Eye X-ray Imaging Device) (Figure 42.3). The technology is based on the crustacean's uncanny ability to see through the dark, cloudy, deep-sea water. When fully developed, this device could be used by airport screeners, the Coast Guard, and border agents to identify dangerous items and/or potential threats. The LEXID radiates objects with tiny amounts of X-ray energy, modeled exactly like lobster eyes that look like small antennae, allowing the eyes to focus by reflection rather than by refraction (bending of light) as human eyes do. This unique optical geometric design allows lobsters to see in the dimmest light, and the "lobster-eye lens" technology focuses the X-ray images so that the device can actually see through a wall and project an image of what's on the other side. The device may never be perfected but it illustrates the inventiveness of DHS in seeking alternatives to existing technology.

16 SEE AROUND-A-CORNER CAMERA

An advancement over the lobster eye is a camera system that could capture three-dimensional images of objects even when they are hidden around the corner. The system offers a new approach to visualizing obscured objects or people by knitting together reflections of laser light pulses that are bounced off of walls and onto hidden objects to create the pictures. The device was developed by a technology group from Massachusetts Institute of Technology (MIT) that hopes to eventually miniaturize the concept to operate from an iPhone.

The camera system operates by emitting laser light in 50-femtosecond (quadrillionth of a second) pulses to various spots on the side-angled backing wall. The light pulses reflect off the obscured object and reflect back onto the wall. The camera collects these scattered reflections and feeds them into a computer, which knits them together like a jigsaw puzzle creating a three-dimensional picture of the hidden object within 15 millionths of a second. The resultant imagery is a slightly blobby-looking but recognizable reconstructed object; its edges are about one-tenth of an inch off the true measurements. The application of ultrafast computing has made the unimaginable possible that could not be conceived only a few years ago.

Similar technology is being applied in other fields with significant success:

Medical imaging called optical coherence tomography is used to peer tiny distances inside tumors or blood vessels also relies on reconstructing light bounces to create images.

Law enforcement researchers have explored efforts to use radar guns to see through walls for hostage situations, and the results have been encouraging.

17 SEE-THROUGH ROLLING STORE GATES

The first sign of a deteriorating neighborhood is when sheets of metal gates start to appear on storefronts and are pulled down at closing time to cover the windows and entrance door (usually the store is at street level – beneath residential apartments). Some city blocks have miles of these gates protecting businesses from after-hours vandalism – creating an urban symbol that the neighborhood is a dangerous place. The gates become even more unsightly when the predictable graffiti artists decorate the gates with obscenities. Once the gates appear, urban flight from inner cities to the suburbs by the residents follows. The menacing-looking metal gates start to define the street boundaries, and there doesn't seem to be any movement to remove the gates once installed.

One alternative that has proven to be successful is to install see-through vinyl slot sliding gates. The residents in the area can see into the lit store, there is less of a magnet for graffiti, and firefighters and police officers can take notice of unusual movement inside the store after hours. Some critics of the metal rolling gates have helped pass local ordinances mandating that storefronts in their community use see-through rather than solid gates. Some have claimed that the law infringes on decisions better left to the merchant, many of whom prefer the more secure, solid gates despite their unappealing appearance. These disputes are generally settled by grandfathering existing gates but requiring see-through plastic-type slat gates for new installations or replacements.

18 PROTECTIVE BULLETPROOF GLASS

A new type of protective glass was developed in 2015 in response to the mass shooting of 20 elementary school children and six staff members at a school in Newtown, Connecticut, in 2012. The glass, manufactured under the name School Guard Glass (https://schoolguardglass.com/), functioned more like the bulletproof glass found in banks and other high-security buildings than the tempered glass commonly used in car windows. The standard bulletproof glass is typically thick, heavy, and expensive, ranging from three-quarters of an inch to 3¾ inches in thickness; installing the glass in a door requires a heavy enough frame to support the weight of the glass. That traditional glass could cost $50,000 or, more apiece, a price that most schools cannot afford. The new School Guard Glass has an extremely strong center, is relatively thin and lightweight at a quarter-inch thick, and the price per door, including installation, is about $1,000 to $1,200.

Index

Note: **Bold** page numbers refer to tables and *italic* page numbers refer to figures.

C

E

F

K

L

M

For Product Safety Concerns and Information please contact our EU representative GPSR@taylorandfrancis.com Taylor & Francis Verlag GmbH, Kaufingerstraße 24, 80331 München, Germany